Kathryn E. Fishbaugh · Philippe Lognonné ·
François Raulin · David J. Des Marais · Oleg Korablev

Editors

Geology and Habitability of Terrestrial Planets

Introduction by Kathryn E. Fishbaugh, David J. Des Marais,
Oleg Korablev, Philippe Lognonné and François Raulin

Reprinted from *Space Science Reviews* Volume 129, Issues 1–3, 2007

 Springer

Kathryn E. Fishbaugh
International Space Science Institute (ISSI),
Hallerstrasse 6,
Bern CH-3012,
Switzerland

François Raulin
Laboratoire Interuniversitaire
des Systèmes Atmosphériques
(LISA, UMR CNRS),
Université Paris VII et Paris XII,
Val de Marne, 61 Av. du Général de Gaulle,
94010 Créteil Cedex,
France

Oleg Korablev
Russian Space Research Institute (IKI),
Profsoyuznaya 84/32,
117997 Moscow,
Russia

Philippe Lognonné
Equipe Etudies Spatiales et
Planétologie (IPGP),
Université de Paris VII,
4 Avenue Neptune,
94107 Saint Maur des Fossés,
France

David J. Des Marais
NASA Ames Research Center,
Mail Stop 239-4,
Moffett Field, CA 94035,
USA

Cover illustration: Bubbling hot spring in Yellowstone National Park, USA. Photo taken by Gordon Southam

Library of Congress Control Number: 2007934537

ISBN-978-0-387-74287-8   e-ISBN-978-0-387-74288-5

Printed on acid-free paper.

9 8 7 6 5 4 3 2 1

springer.com

Contents

ISSI Geology and Habitability of Terrestrial Planets Workshop

1. Mikhail Gerasimov
2. Agustin Chicarro
3. Helmut Lammer
4. Mike Carr
5. Rickard Lundin
6. Patrick Russell
7. Dave Des Marais
8. Geoff Collins
9. Sylvaine Farrachat
10. Nick Arndt
11. Gerda Horneck
12. Ralf Jaumann
13. François Forget
14. Alexandra Lefort
15. Charles Cockell
16. Stephan Labrosse
17. Tilman Spohn
18. Karim Benzerara
19. Kate Fishbaugh
20. Alex Halliday

21. Jean-Loup Bertaux
22. Kevin Zahnle
23. Frances Westall
24. Todd Stevens
25. Eric Gaidos
26. Philippe Lognonné
27. Cedric Gillman
28. Lucia Marinangeli
29. Gordon Southam
30. Sasha Basilevsky
31. Oleg Korablev
32. Jacques Laskar
33. Jörn Helbert
34. Lynn Rothschild
35. Peter van Theinen
36. Doris Breuer
37. Euan Nisbet
38. Vittorio Manno
39. Veronique Dehant

Participants not pictured: Willy Benz, Roger-Maurice Bonnet, Oliver Botta, Johannes Geiss, Iain Gilmour, Beda Hofmann, Reinald Kallenbach, James Moore, François Raulin, Christophe Sotin, Brigitte Fasler (ISSI secretary), Saliba Saliba (ISSI computer engineer), Silvia Wenger (assistant to ISSI Executive Director)

Space Sci Rev (2007) 129: 1–5
DOI 10.1007/s11214-007-9208-0

Introduction: A Multidisciplinary Approach to Habitability

Kathryn E. Fishbaugh · David J. Des Marais ·
Oleg Korablev · Philippe Lognonne · François Raulin

Received: 26 January 2007 / Accepted: 10 May 2007 /
Published online: 6 June 2007
© Springer Science+Business Media B.V. 2007

Developments during the early years of space exploration provided particular impetus toward articulating the concept of habitable planets beyond Earth. One of these developments was the collective results of the life-detection experiments conducted by the *Viking* missions, which demonstrated that an improved understanding of environmental context was crucial to the search for evidence of life on Mars. Another development was the expansion of the search for extraterrestrial intelligence and Frank Drake's articulation in 1961 of his famous Drake Equation (Drake and Sobel 1992), which identifies the parameters that collectively estimate the probability of detecting intelligent life elsewhere. A meeting entitled "Life in the Universe" was held in 1979 at NASA Ames Research Center (Billingham 1981). In addition to articulating topics in astronomy and biology, the participants disseminated key factors that sustain habitable planetary environments. They addressed, among other factors, stellar and orbital influences, planetary accretion and early development, atmospheres, climate, and the origins and evolution of continents and oceans. An additional strong impetus

K.E. Fishbaugh (✉)
International Space Science Institute (ISSI), Hallerstrasse 6, 3012 Bern, Switzerland
e-mail: kathryn.fishbaugh@issibern.ch

D.J. Des Marais
NASA Ames Research Center, Mail Stop 239-4, Moffett Field, CA 94035, USA

O. Korablev
Russian Space Research Institute (IKI), Profsoyuznaya 84/32, 117997 Moscow, Russia

P. Lognonne
Equipe Etudies Spatiales et Planétologie (IPGP), Université de Paris VII, 4 Avenue Neptune, 94107
Saint Maur des Fossés, France

F. Raulin
Laboratoire de Interuniversitaire des Systèmes Atmosphériques (LISA, UMR CNRS), Université
Paris VII et Paris XII, Val de Marne, 61, Avenue du Général de Gaulle, 94010, Créteil Cedex,
France

 Springer

to develop models of habitable planets orbiting other stars has been provided by ESA's *Darwin* and NASA's *Terrestrial Planet Finder* missions (e.g., Kaltenegger and Fridlund 2005; Des Marais et al. 2002). Several additional meetings and technical papers have now established that the search for life in the universe necessarily includes efforts to locate and characterize potentially habitable planetary environments.

This book has arisen from a unique workshop, entitled "Geology and Habitability of Terrestrial Planets", held at the International Space Science Institute (ISSI) in Bern, Switzerland, on 5–9 September 2005. The goal of this workshop, as articulated by the conveners, was "to define the influence of planetary geologic evolution on habitability and to assess the conditions necessary for life, using the Earth, Mars, and Venus as examples." The intent was to focus on the relationship between a planet's origin and evolution and its capacity to sustain habitable environments. Many previous books and conferences have dealt extensively with specific topics in astrobiology, such as the search for life on Mars (e.g., Tokano 2005) and for intelligent life in the universe (e.g., Ulmschneider 2006). A planet's geologic evolution and its relationship with its star are the main drivers for maintaining long-term habitability and for carrying life from apparition to eventual evolution. By focusing on the co-evolution of geology and habitability, this workshop has added a new dimension to the current focus of astrobiological studies and has combined insights from multiple relevant research disciplines into a broad-scale treatment of habitability.

The "Geology and Habitability of Terrestrial Planets" workshop clearly maintained the interdisciplinary character of ISSI workshops; 53 scientists from 12 countries collaborated to weave together geologic, atmospheric, geophysical, magnetospheric, and biological planetary studies. What makes this book truly unique is the cooperation of scientists from multiple disciplines on each chapter. Thus, these chapters tackle the topic of habitability from the viewpoints of multiple branches of planetary science in an interdisciplinary way, rather than representing a compendium of individual research papers presented at the conference. In some sense, however, this book can be considered a "conference proceedings" in that the authors have taken care to include many aspects of what was presented in formal talks and discussed by the participants as a group.

Defining the term "habitability" is not trivial. In this book, you will find "habitability" defined and addressed in myriad ways; no one definition can cover all aspects of the term. One could propose that a habitable planet lies within the so-called Habitable Zone around a star, a zone defined by temperature wherein water at the planet's surface would be expected to be liquid (Kasting et al. 1993). Yet such a definition would completely exclude the actual geologic and atmospheric conditions of the planet. An alternative definition could be that a habitable planet is one on (or within) which life can exist. Even this definition is lacking, as it does not account for the fact that life may only exist on that planet for a very short period of time. If life does not thrive for long enough to spread wide enough or to multiply sufficiently for later detection of even fossils or biomarkers, can the planet still be considered to have been habitable?

Our concepts of the attributes of life-sustaining planetary environments must be based, at least initially, upon the requirements of our own biosphere (Marais et al. 2003). Microorganisms tolerate greater environmental extremes than do plants and animals; therefore, biospheres having only microorganisms are expected to be more widespread than biospheres similar to our own. Microbial life requires chemical building blocks, utilizable sources of energy, and relatively stable conditions that can sustain liquid water at least intermittently. To search for evidence of these requirements on other planets and to understand our own origins, we must gain a fundamental understanding of planetary processes and history. The attributes of life arise not only from physical and chemical principles but also from the nature and evolution of the planetary environments in which life developed. Differences in the interplay between cosmic events and planetary processes might have shaped other life forms in ways that are far more diverse than occurred in our own biosphere. This book sets forth the necessary concepts and approaches to begin to comprehend that potentially enormous diversity.

Factors that determine the habitability of a planet are multifaceted and include but are certainly not limited to the presence or absence of plate tectonics, the presence or absence of a magnetosphere, the characteristics and evolution of the planet's atmosphere, the presence of liquid water, the presence of energy sources, variability in insolation, and even the means to transport nutrients to and wastes away from organisms. Some researchers are beginning to reassess the definition of the Habitable Zone around a star by taking some of these factors into account along with the liquid-water temperature requirement (Franck et al. 2000). It is evident that even the few factors listed here cover the entire spectrum of a planet's properties and are interrelated.

Without plate tectonics on Earth, mid-ocean ridges (a possible location for the origin of life (e.g., Russell and Hall 1997) and continental habitats would not exist. More importantly, CO_2 trapped in carbonate would not be recycled, but instead it could generate global warming incompatible with life and even with liquid water (as was the case for Venus). The Earth's magnetosphere shields our atmosphere from erosion by the solar wind, and the dynamo that generates this shield is maintained by plate tectonics; mantle viscosity is reduced by injection of water at subduction zones and by the efficiency of plate tectonics at cooling the mantle and hence the core (e.g., van Thienen et al. 2005). Both plate tectonics and the generation of a magnetosphere depend upon the nature of the internal dynamics of the planet.

Clearly, particular forms of life will utilize particular atmospheric chemistries. For example, the earliest forms of terrestrial life did not benefit from any oxygen (e.g., Canfield et al. 2000). The characteristics of a planet's atmosphere will depend on its position relative to its star, the number of volatile-bearing impacts it receives, the amount of volatile outgassing from volcanism (which is in turn, again, linked to internal dynamics), and other factors. Of course, many organisms (endoliths, chemotrophs, etc.) do not directly require an atmosphere at all and so may not even be directly affected by the atmosphere or lack thereof.

Liquid water and a source of energy are two important criteria for habitability according to life as we know it on Earth. Whether water exists on the planet depends upon many of the same factors that determine the atmospheric characteristics, such as temperature and distance from its star. Sources of energy can be varied: geothermal heat, insolation, volcanic activity, or even tidal heating (as in the case of Europa (e.g., Ross and Schubert 1987)). Timing is also crucially important. Even if a planet is likely to have liquid water on its surface (like Mars at various times in its past (e.g., Carr 1996)) and a sufficient energy source, this water and energy are of little consequence to habitability unless they existed long enough for any potential life to use them and thrive off of them.

Interestingly, it was suggested during workshop discussion that a flux of nutrients to and wastes away from a potential habitat is also crucially important to life. For example, if water pools in an area where it cannot leak into the groundwater system or escape via a channel and somehow be replenished, this closed system may not provide the necessary flux.

External factors also play a significant role in a planet's habitability. For example, if the obliquity of a planet changes with time, then the resultant changes in insolation will affect changes in climate. This is apparent in the imprint of the Earth's Milankovitch cycles on climate records gleaned from such sources as ice and deep-sea cores and in the modeled

effects of obliquity changes on Mars' climate (e.g., Forget et al. 2006; Mischna et al. 2003). These effects are more pronounced on Mars than on Earth since the Moon stabilizes Earth's rotation axis (Laskar et al. 1993).

This discussion illustrates that a planet's habitability depends upon a wide range of factors that cover the entire scope of a planet's characteristics. For this reason, a multidisciplinary book on the subject is a crucial step in our general understanding of habitability. The book chapters follow the themes of the workshop.

Chapter 1 (Southam et al. 2007) introduces habitability from a fundamental, biological perspective. This chapter delves into the questions of what requirements are necessary for life to exist and how geology creates those environments. The authors also tackle the difficult question of what is needed for life to originate in the first place, including speculation about the potential for a second genesis on Mars or Europa.

Chapter 2 (Zahnle et al. 2007) focuses on the earliest stages of Earth history and the development of habitable conditions. The authors pay particular attention to the effects of high impact rates. Chapter 3 (Nisbet et al. 2007) then covers a broad scope of topics, describing how the geologic evolution of a planet, both on the surface and in the near subsurface, creates habitable environments that evolve along with the geology.

Chapter 4 (Bertaux et al. 2007) narrows in on the importance to habitability of volatiles in the atmosphere, on the planet's surface (e.g., liquid water), and in rocks. This chapter borrows from the writing style of Galileo, framing the multidisciplinary parts as a discussion among famous founding fathers of science.

Chapter 5 (van Thienen et al. 2007) reviews the exchanges among a planet's interior dynamics, the atmosphere, and the magnetosphere and the ways in which these exchanges modify habitability. Chapter 6 explores these topics in a different light by addressing the stability of planetary atmospheres and their protection by the magnetosphere. This chapter has been divided into three parts, with an introduction by Lammer et al. (2007). Part 1 focuses on long-term solar variability and solar forcing of planets with magnetospheres (Lundin 2007). Part 2 (Kulikov et al. 2007) provides a comparative study of the early atmospheres of Earth, Mars, and Venus and the reasons for their consequent dissimilar evolutions. Part 3 (Dehant et al. 2007) then details the generation of magnetic dynamos and the protection of the atmosphere by the magnetosphere.

In the epilogue, Lognonne et al. (2007) discuss two main aspects of investigations of life in the universe—genesis and maintenance of habitability—providing directions for future studies and missions.

References

J.-L. Bertaux, M. Carr, D. Des Marais, E. Gaidos (2007), this volume

J. Billingham (Ed.), *Life in the Universe*. NASA Conference Publication Series, vol. 2156 (1981)

D. Canfield, K. Habicht, B. Thamdrup, Science **288**(5466), 658–661 (2000)

M. Carr, *Water on Mars* (Oxford University Press, New York, 1996)

V. Dehant, H. Lammer, T. Kulikov, J.-M. Griessmeier, D. Breuer, O. Verhoeven, Ö. Karatekin, T. van Hoolst, O. Korablev, P. Lognonne (2007), this volume

D. Des Marais, M. Harwitt, K. Jucks, K. Kasting, L. Lunine, D. Lin, S. Seager, J. Schneider, W. Tarub, N. Woolf, Astrobiology **2**(2), 153–181 (2002)

F. Drake, D. Sobel, *Is Anyone Out There? The Scientific Search for Extraterrestrial Intelligence* (Delacorte Press, New York, 1992)

F. Forget, R. Haberle, F. Montmessin, B. Levrard, J. Head, Science **311**(5759), 368–371 (2006)

S. Franck, A. Block, W.V. Bloh C. Bounama, H. Schellnhuber, Y. Svirezhev, Planet. Space Sci. **48**(11), 1099–1105 (2000)

L. Kaltenegger, M. Fridlund, Adv. Space Res. **36**, 1114–1122 (2005)

J.F. Kasting, D. Whitmire, R. Reynolds, Icarus **101**(1), 108–128 (1993)

Y. Kulikov, H. Lammer, H. Lichtenegger, T. Penz, D. Breuer, T. Spohn, R. Lundin, H. Biernat (2007), this volume

H. Lammer, V. Dehant, O. Korablev, R. Lundin (2007), this volume

J. Laskar, J. Joutel, P. Robutel, Nature **361**, 615–617 (1993)

P. Lognonne, D.D. Marais, F. Raulin, K. Fishbaugh (2007), this volume

R. Lundin (2007), this volume

D.D. Marais, L. Allamandola, S. Brenner, A. Boss, D. Deamer, P. Falkowski, J. Farmer, B. Hedges, B. Jakosky, A. Knoll, D. Liskowsky, V. Meadows, M. Meyer, C. Pilcher, K. Nealson, A. Spormann, J. Trent, W. Turner, N. Woolf, H. Yorke, Astrobiology **3**(2), 219–235 (2003)

M. Mischna, M. Richardson, R. Wilson, D. McCleese, J. Geophys. Res. 108E(06) (2003). doi: 10.1029/2003JE002051

E. Nisbet, K. Zahnle, J. Helbert, R. Jaumann, B. Hoffmann, K. Benzerara, F. Westall (2007). this volume

M. Ross, G. Schubert, Nature **325**, 133–134 (1987)

M. Russell, A. Hall, J. Geol. Soc. London **154**(3), 377–402 (1997)

G. Southam, L. Rothschild, F. Westall (2007), this volume

T. Tokano (Ed.), *Water on Mars and Life*, vol. XVI. Advances in Astrobiology and Biogeophysics Series, A. Brack, G. Horneck, M. Mayor, C. McKay (Eds.), (Springer, Dordrecht, 2005), 332 pp

P. Ulmschneider, *Intelligent Life in the Universe* (Springer, Dordrecht, 2006)

P. van Thienen, K. Benzerara, D. Breuer, C. Gillman, S. Labrosse, P. Lognonne, T. Spohn (2007), this volume

P. van Thienen, N. Vlaar, A.v.d. Berg Phys. Earth Planet. Interiors **150**, 287–315 (2005)

K. Zahnle, N. Arndt, C. Cockell, A. Halliday, E. Nisbet, F. Selsis, N. Sleep (2007), this volume

Space Sci Rev (2007) 129: 7–34
DOI 10.1007/s11214-007-9148-8

The Geology and Habitability of Terrestrial Planets: Fundamental Requirements for Life

G. Southam · L.J. Rothschild · F. Westall

Received: 23 August 2006 / Accepted: 18 January 2007 /
Published online: 10 May 2007
© Springer Science+Business Media, Inc. 2007

Abstract The current approach to the study of the origin of life and to the search for life elsewhere is based on two assumptions. First, life is a purely physical phenomenon closely linked to specific environmental conditions. From this, we hypothesise that when these environmental conditions are met, life will arise and evolve. If these assumptions are valid, the search for life elsewhere should be a matter of mapping what we know about the range of environments in which life can exist, and then simply trying to find these environments elsewhere. Second, life can be clearly distinguished from the non-living world. While a single feature of a living organism left in the rock record is not always sufficient to determine unequivocally whether life was present, life often leaves multiple structural, mineralogical and chemical biomarkers that, in sum, support a conclusion that life was present. Our understanding of the habitats that can sustain or have sustained life has grown tremendously with the characterisation of extremophiles. In this chapter, we highlight the range of environments that are known to harbour life on Earth, describe the environments that existed during the period of time when life originated on Earth, and compare these habitats to the suitable environments that are found elsewhere in our solar system, where life could have arisen and evolved.

Keywords Origin of life · Extremophiles · Earth · Life requirements

G. Southam (✉)
Department of Earth Sciences, University of Western Ontario, London, ON, Canada N6A5B7
e-mail: gsoutham@uwo.ca

L.J. Rothschild
National Aeronautics and Space Administration, Ames Research Center, Mail Stop 239-20,
Moffett Field, CA 94035-1000, USA

F. Westall
Centre de Biophysique Moléculaire, CNRS, Rue Charles Sadron, 45071 Orléans cedex 2, France

1 Introduction

We are concerned about whether or not we are alone in the universe, which may explain
our fascination with the possibility of life elsewhere (Des Marais and Walter 1999). While
the origin of life is one of the grand questions facing humanity, we do not have the absolute
answer for how it happened. However, since there was an origin in our solar system, there is
no reason to believe that there wouldn't be other origins elsewhere in the universe, perhaps
even in our solar system (Cady et al. 2003). In today's examination of our solar system, we
tend to separate the 'origins question', from that of habitability. Without worrying about the
origin of life, habitability can be more easily discussed in terms of life in extreme environments (can anything live there; see Table 1) and though beyond the scope of this chapter,
from a technological perspective—terraforming. Today, many view habitability from this

Table 1 Classification and examples of extremophiles (from Rothschild and Mancinelli 2001)

Environmental parameter	Type	Definition	Examples
temperature	hyperthermophile	growth >80°C	*Pyrolobus fumarii*, 113°C
	thermophile	growth 60–80°C	*Synechococcus lividis*
	mesophile	15–60°C	*Homo sapiens*
	psychrophile	<15°C	*Psychrobacter*, some insects
radiation			*Deinococcus radiodurans*
pressure	barophile	weight loving	unknown
	piezophile	pressure loving	for microbe, 130 MPa
gravity	hypergravity	>1 g	none known
	hypogravity	<1 g	none known
vacuum		tolerates vacuum (space devoid of matter)	tardigrades, insects, microbes, seeds
desiccation	xerophiles	anhydrobiotic	*Artemia salina*; nematodes, microbes, fungi, lichens
salinity	halophile	Salt loving (2–5 M NaCl)	*Halobacteriacea, Dunaliella salina*
pH	alkaliphile	pH >9	*Natronobacterium, Bacillus firmus* OF4, *Spirulina* spp. (all pH 10.5)
	acidophile	low pH loving	*Cyanidium caldarium*, *Ferroplasma* sp. (both pH 0)
oxygen tension	anaerobe	cannot tolerate O_2	*Methanococcus jannaschii*
	microaerophile	tolerates some O_2	*Clostridium*
	aerobe	requires O_2	*Homo sapiens*
chemical	gases		*Cyanidium caldarium* (pure CO_2)
extremes	metals	Can tolerate high concentrations of metal (metalotolerant)	*Ferroplasma acidarmanus* (Cu, As, Cd, Zn); *Ralstonia* sp. CH34 (Zn, Co, Cd, Hg, Pb)

 Springer

anthropocentric view; can we colonise another planet—Mars. However, from an extrasolar planetary perspective where terraforming is well beyond our current technology, our deliberations return to the requirements to produce an origin of life and, to the physical and chemical conditions that are capable of supporting the growth and reproduction of life forms once they have formed. For this, we only have Earth as an example. Initially, on any planet, the physical and chemical conditions would have controlled the origin of a biosphere. However, as soon as life began, it would have exerted its influence, reducing the entropy of the environments it inhabited. This chapter will focus on life's lessons from contemporary systems (Sect. 2), on the fundamental interactions between geology (and geochemistry; Sect. 3) and life (Sect. 4), and will introduce the possible relationship between life and the terrestrial planets (Sect. 5).

2 Lessons from Earth—Contemporary Systems

When addressing the fundamental aspects of life on Earth and its relationship to geology, we have focused on bacteria, which are considered to be a ubiquitous form of life, and exhibit the greatest molecular and metabolic diversity of any life form on Earth (Barns and Nierzwicki-Bauer 1997; Pace 1997; Reysenbach and Shock 2002). Bacteria are active in all surface and subsurface environments (Amy et al. 1992; Bechtel et al. 1996, 1998; Colwell et al. 1997; Lovley and Chapelle 1995; Parkes et al. 1994; Pedersen and Ekendahl 1990; Stevens et al. 1993), where they are only limited by the availability of liquid water (Navarro-Gonzalez et al. 2003) and temperatures between approx. $-20°C$ (permafrost; Rivkina et al. 2000) and $121°C$ (Kashefi and Lovley 2003). In natural systems, bacteria typically grow as biofilms on minerals (Konhauser et al. 1994), which are often facilitated by secondary, organic exopolymers that help bacteria bind to mineral surfaces (Fig. 1; see Costerton et al. 1995; Moser et al. 2003; Urrutia and Beveridge 1993; Wanger et al. 2006; Westall et al. 2000). This allows individual cells to persist in a chemical environment that supports their growth via a continuous 'stream' of nutrients that flows past them. The term nutrient is used broadly here to include a source (or several sources) of carbon, nitrogen, phosphorous, and energy, and an electron acceptor necessary to support growth. While bacteria do not have direct control over the chemistry of the fluid phase that they encounter in this situation, they can alter the chemistry of their surroundings (Emerson and Revsbech 1994; Purcell 1977; Revsbech and Jorgensen 1986), often to their benefit. For example, the formation and cell surface association of metal sulphides by dissimilatory sulphate reducing bacteria (Donald and Southam 1999) provides an oxygen 'sink' for these obligate anaerobes (an organism that is killed in the presence of oxygen).

Before we can consider the origin of life on Earth or the possibility of life elsewhere in the solar system, we need to understand the range of physical and chemical conditions that can harbour life, the chemical constituents that life can use to create cells and the energy generating reactions available to life to accomplish its activities.

2.1 Physical and Chemical Conditions

Bacteria can be classified according to the effect that various physical and chemical conditions have on their growth and survival. Understanding these environmental influences allows us to understand the distribution of microorganisms in natural systems.

In all environments, bacteria concentrate solutes within their cell envelopes and strategically obtain liquid water through osmosis resulting in the formation of cell turgor pressure

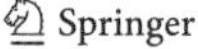 Springer

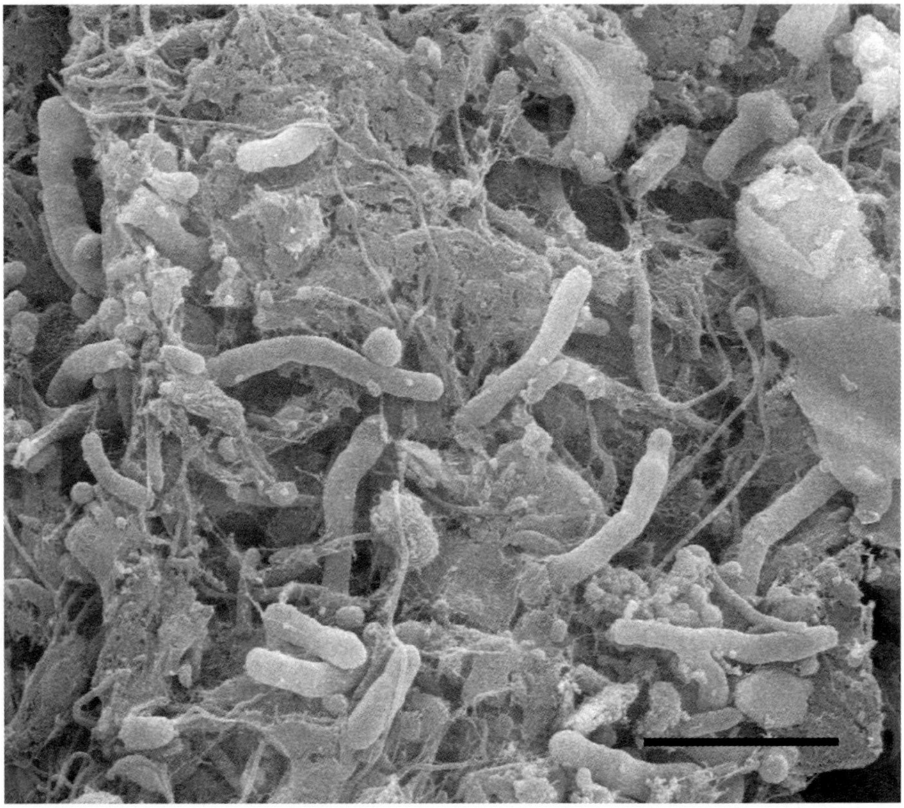

Fig. 1 A scanning electron micrograph of a biofilm recovered from a borehole, 3.1 km below land surface in the Dreifontein gold mine, Witwatersrand Basin, RSA. Note the intimate association between the bacteria, their exopolymer materials and their lithosphere habitat. Micrograph provided by M. Lengke. *Bar* equals 5 μm

(Kunte 2006). The bacterial cell envelope is designed to withstand the turgor pressure (the osmotic pressure resulting from the high concentration of molecules inside a cell), which can reach 15 atm (see Beveridge 1981 for an in depth discussion of cell envelope structure and chemistry). While most aquatic and sediment (including groundwater) environments possess an abundance of available liquid water, terrestrial environments are frequently water stressed. Hypersaline environments, like the Great Salt Lake, Utah, are continually under stress due to low water activity and are therefore considered to be examples of extreme environments, which will only support halophilic (or halotolerant) microorganisms.

In liquid water-rich environments, microorganisms can grow as extensive biofilms or as microbial mats (Nealson and Stahl 1997; Stahl 1994). However, under arid Earth surface conditions associated with both hot and cold desert environments (Chan et al. 2005; Friedmann and Ocampo 1976; Friedman 1980), bacteria often seek refuge in 'soils' or within rock to avoid desiccation (Fig. 2). These endolithic microbial communities display an intimate relationship with the geologic substrate they inhabit; even silicate minerals are susceptible to weathering (Hiebert and Bennet 1992). Silicate weathering is typically promoted via the formation of organic and inorganic acids, which are by products of metabolism (de los Ríos et al. 2003; Welsh et al. 2002). An added benefit to this growth strategy is the concentration

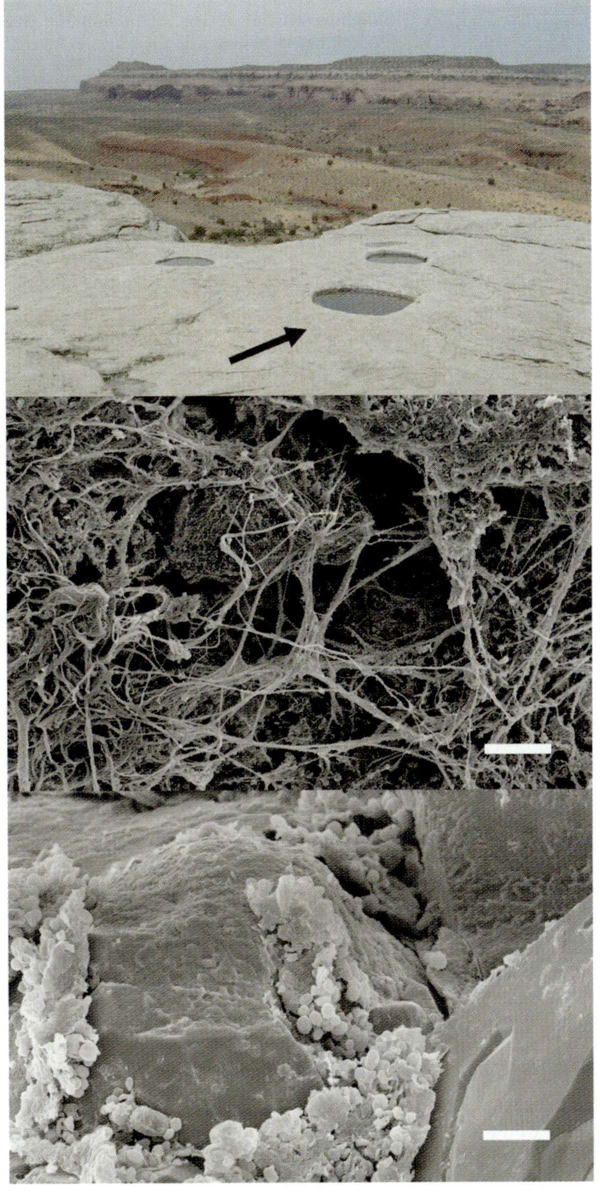

Fig. 2 A photograph of a pothole at Barlett Wash, near Moab, NM. Upon hydration, the previously desiccated biofilm in bottom of the pothole (the black lining seen in *top panel*) begins to grow and can be observed using scanning electron microscopy (*middle*). Growth of these surface exposed biofilms (*middle*) is presumably more ephemeral than the endolithic organisms found within the sandstone (*bottom*). Higher, eukaryotic organisms, such as snails are supported by this bacterially-based food chain (see Chan et al. 2005). Micrographs provided by A. Welty. The pothole indicated by the *arrow* is approximately 1 m in diameter; bars equal 50 μm and 10 μm, respectively

of essential elements such as iron and phosphate, which are limited in their abundance and are needed for growth to occur (Banerjee et al. 2000; Brantley et al. 2001; Rogers and Bennet 2004; Rogers et al. 1998; Taunton et al. 2000). Bennet et al. (1996, 2001) also proposed that microbial weathering of quartz grains, which are not viewed as a good substrate for bacterial growth, could be due to their need for inorganic micronutrients (Fe, Co, Zn, Mo, Cu, and Ni) that are often found in trace amounts in this silicate.

Most natural environments occur as circumneutral pH systems containing hydrogen ion concentrations of 100 μM, plus or minus an order of magnitude (log$_{10}$; pH 6 to 8). Therefore, most bacteria exhibit optimal activity in this pH range. However, even extreme environments possessing acidic (pH 1 to 4) or alkaline (pH 9 to 11) pH conditions will possess indigenous bacteria whose growth parameters reflect these extremes in environmental pH (Goodwin and Zeikus 1987). These bacteria are referred to as acidophilic (acidotolerant) or alkalophilic (alkalotolerant), respectively.

Temperature is generally an easily measurable physical condition that differentiates bacteria into broad groups (Stetter et al. 1990)). Temperature is also the most reliable measurement in microbial ecology because it is not affected by biogeochemical cycling on the scale of bacteria. Psychrophiles grow optimally at or below 15°C and have even demonstrated activity down to −20°C (Rivkina et al. 2000). Viable bacteria have also been recovered from million-year-old permafrost (Rivkina et al. 2000). Psychrotolerant bacteria grow optimally between 20 and 40°C but will grow at temperatures as low as 0°C. Mesophiles grow optimally between 15–45°C. Thermophiles have optimum growth temperatures above 45°C and hyperthermophiles grow optimally above 80°C. Thus, from a temperature perspective, bacteria can be found within ice or permafrost regions (note, bacteria are routinely stored in liquid N_2, −196°C) all the way up to hydrothermal fluids temperatures of 121°C.

Bacteria are subdivided into aerobic and anaerobic groups based on their oxygen requirements or tolerance (Chapelle 1993; Ehrlich 2002). However, there are also facultative anaerobic bacteria that utilise oxygen if it is available, then grow as anaerobes when oxygen is depleted resulting in less overall growth. The utilization of oxygen as a terminal electron acceptor enables bacteria to fully oxidise their respective source(s) of energy and thereby maximise their energy gain from the reaction. Most bacteria can utilise O_2 in respiration (reaction (1)), including obligate anaerobes:

$$C_6H_{12}O_6 + 6O_2 \rightarrow 6CO_2 + 6H_2O + \text{lots of energy} \qquad (1)$$

However, all organisms that grow successfully in the presence of oxygen, must have the capacity to detoxify the chemically reactive oxygen compounds (hydrogen peroxide, superoxide radical and the hydroxy radical) produced through the stepwise reduction of oxygen to water. Detoxification of these oxygen radicals requires a series of enzymes including superoxide dismutase, catalase and peroxidase (Atlas and Bartha 1997). The ability to tolerate oxygen relates to the presence or absence of each of these enzymes; anaerobes are simply unable to detoxify these reactive oxygen compounds. Measurements of dissolved oxygen and redox potential are important in determining which microbial processes are functioning within natural systems.

Constraints on bacterial metabolism focus on a single entities such as oxygen (aerobe vs. anaerobe), pH (acidophile vs. neutrophile) or temperature (psychrophile vs. thermophile). However, there are many examples of bacteria that can withstand multiple stressors. Thermoacidophiles are capable of growing at 85°C in pH 2 sulphuric acid solutions, and haloalkalophiles, can grow in a saturated sodium sulphate solution at pH 11.5 (see Fig. 3). It is in these environments that the diversity of bacteria and the limits of life are best appreciated. Within these environments, bacteria need to assimilate (incorporate) organic and inorganic nutrients to produce new cell biomass and to reproduce.

2.2 Assimilation (Biomass Forming) Reactions

A wide range of organic and inorganic nutrients can be assimilated by bacteria. Organic nutrients include nearly every organic monomer or polymer found in the living or diagenetically altered biosphere (carbohydrates, amino acids, nucleic acids and hydrocarbons). Inorganic nutrients include dissolved gases (carbon monoxide, carbon dioxide, hydrogen, nitrogen and dihydrogen sulphide), soluble cations (sodium, calcium, magnesium, ammonium,

Fig. 3 A photograph of a sulphur-containing hydrothermal spring (85°C, pH 2) in Monument Geyser Basin, Yellowstone National Park (*top*) and of Last Chance Lake, BC, an alkaline (pH 11.5), near-saturated sodium sulphate system. Extreme environments, such as these, are dominated by prokaryotic organisms representing both the Archaea and Bacteria domains

ferrous and ferric iron), base metals (chromium, nickel, copper, cobalt, zinc, lead, mercury), and soluble anions (chloride, nitrite, nitrate, hydrogen sulphide, sulphite, sulphate, phosphorous, selenate and arsenate). Depending on nutrient requirements, these compounds can be used in assimilation reactions to build new cells, or in dissimilatory processes to generate the energy needed to construct new biomass.

The small size (micrometer-scale) and resulting high surface-to-volume ratio of bacteria helps facilitate their growth. Access to nutrients is based on diffusion, therefore, the greater the surface-to-volume ratio, the greater the diffusion of nutrients into cells (Koch 1996; Pirie 1973; Purcell 1977; Beveridle 1988). Also, since individual bacteria only possess a mass of $\sim 1 \times 10^{-12}$ g (wet weight; Neidhardt et al. 1990), extremely low concentrations of dissolved nutrients can support bacterial growth. Bacteria are important components of the food chain because they utilise extremely low concentrations of inorganic and organic nutrients that are too dilute or essentially unavailable, to support higher, eukaryotic organisms (Fenchel and Jorgensen 1977).

Carbon utilization is one of the most important criteria by which bacterial populations are characterised. Heterotrophic bacteria require organic carbon for their metabolism and typically couple the oxidation of organic carbon to CO_2 with the reduction of inorganic dissolved species or minerals. The sources of carbon and energy for heterotrophs can range from simple organics such as asparagines, an amino acid (Goldman and Wilson 1977) or glucose, which can be used by most heterotrophs, to macromolecular materials such as cellulose (Ljungdahl and Erikson 1985). The ability of bacteria to metabolise organic com-

pounds in natural systems is also dependent on the chemical's nature, the composition and mineralogy of the soil (sediment) and the mechanism by which the chemical enters the environment (Knaebel et al. 1994). For example, Scow and Hutson (1992) found significant decreases in degradation of organic carbon in the presence of kaolinite, owing to the effects of adsorption and reduced diffusion from the clay surface.

Autotrophs are able to obtain their cellular carbon for biomass from inorganic sources (dissolved CO_2/HCO_3^-). Autotrophic organisms direct most of their energy toward the fixation (reduction) of CO_2 into organic carbon. They are extremely important to heterotrophs, since 'we' could not thrive without some basis to our food chain. For example, autotrophic sulphur oxidizing bacteria that synthesise carbohydrates at sea-floor hot springs (reaction (2)):

$$CO_2 + 4H_2S + O_2 \rightarrow CH_2O_{(n;a\ carbohydrate)} + 4S^\circ + 3H_2O \qquad (2)$$

are the base of the food chain in these lightless, extreme environments (Cavanaugh et al. 1981).

Nitrogen and phosphorous are limiting nutrients in most natural systems. However, bacteria can easily assimilate all forms of water soluble inorganic nitrogen and when this nitrogen is limiting, a wide range of bacteria possess nitrogenase activity and can fix atmospheric nitrogen (Brill 1975). Regarding phosphorus limiting conditions, microorganisms can produce a wide array or organic and inorganic acids that can enhance the dissolution of phosphorites such as apatite, promoting growth through the assimilation of this limiting nutrient (Rogers et al. 1998; Taunton et al. 2000).

The assimilation of materials into cells, forming a wide range of functional macromolecules such as deoxyribonucleic acid, ribonucleic acids and proteins (see Sect. 4.3) requires energy. Fortunately the near surface environment of the Earth, possessing liquid water below 121°C, is not a closed system, but has a range of energy inputs from within Earth and beyond, the sun.

2.3 Dissimilatory (Energy Generating) Reactions

The requirement of energy for biosynthetic reactions may be described according to their electron donor or source of reducing power and by their election acceptor; a redox couple.

Phototrophs utilise sunlight to produce ATP (adenosine triphosphate, the principal energy carrier of the cell) from otherwise chemically stable reduced compounds. Oxygenic phototrophs use light in combination with H_2O as their source of energy (reaction (3)):

$$6CO_2 + 6H_2O + sunlight \rightarrow C_6H_{12}O_6 + 6O_2 \qquad (3)$$

Anoxygenic phototrophs such as anaerobic green and purple sulphur bacteria derive their energy for growth from light and H_2S, producing elemental sulphur or sulphate as end-products of their metabolism. Chemolithotrophs (ammonia oxidisers, hydrogen oxidisers, Fe–Mn oxidisers, methylotrophs, and sulphur oxidisers) are aerobic bacteria that utilise inorganic chemicals (NH_4^+, H_2, Fe^{2+}, Mn^{2+}, CH_4 and H_2S) as their energy sources (electron donors); anaerobic chemolithotrophs include methanogens, which utilise H_2 as their source of energy. Heterotrophs that utilise organic carbon as their source of reducing power (and carbon) include both aerobic and anaerobic bacteria.

In natural systems, bacterially catalyzed redox processes are established through competition for available nutrients (assimilation) and through the efficiency of their respective energy generation mechanisms. Today, because of the presence of oxygen in our atmosphere

 Springer

and dissolved oxygen in surface and near-surface ground water, aerobes dominate these surface environments with anaerobic bacteria becoming increasingly important with depth below the water table. Below the aerobic zone, a progression of different anaerobes exists, most of which utilise organic matter for their metabolism (Chapelle and Lovely 1992; Chapelle 1993; Lovley and Chapelle 1995; Lovley and Klug 1983).

Under moderately reducing conditions, Fe- and Mn-reducing bacteria couple the oxidation of organic matter to the reductive dissolution of high-surface-area Fe- and Mn-oxyhydroxides (Coleman et al. 1993; Lovley and Phillips 1988; Myers and Nealson 1988). Below the zone of metal reduction, sulphate-reducing bacteria (SRB), oxidise low molecular weight organic compounds, utilizing sulphate as the terminal electron acceptor, forming and releasing hydrogen sulphide as a by-product of metabolism (Anderson et al. 1998; Bubela and McDonald 1969; Donald and Southam 1999; Skyring 1988). Bacterial methane production occurs under even more reducing conditions than bacterial sulphate reduction. Methanogenesis (reaction (4)) is also the dominant anaerobic bacterial process that cycles inorganic carbon back into the organic carbon pool:

$$2CO_2 + 6H_2 \rightarrow CH_2O_{(n;\ a\ carbohydrate)} + CH_4 + 3H_2O \tag{4}$$

Biogenic methane can migrate out of the zone of methanogenesis and become an organic carbon source (electron donor) to heterotrophic bacteria elsewhere. This redox zonation going from aerobic $\rightarrow$ Fe–Mn-reduction $\rightarrow$ sulphate reduction $\rightarrow$ methanogenesis is a universal phenomenon that occurs in anthropogenic and natural settings as diverse as landfills (Lyngkilde and Christensen 1992) and polluted aquifers (Bottrell et al. 1995) to marine sediment pore waters (Drever 1997), respectively.

As geochemical conditions become increasingly reducing, the amount of energy (as expressed by ΔG_r) available from each of the predominant biogeochemical reactions (redox zones noted above) decreases (Lovley and Phillips 1988; Chapelle and Lovely 1992; Chapelle 1993; Lovley and Chapelle 1995). Intuitively, all sedimentary environments will contain reactive organic matter (low molecular weight organic monomers; Jakobsen and Postma 1994), produced through the growth and death of microorganisms, and will generate this geochemical succession. While the geochemical responses that bacteria produce begin on a micrometer scale, (Emerson and Revsbech 1994; Revsbech and Jorgensen 1986), bacteria influence local, regional and global biogeochemical cycles. Life serves as a catalyst for these reactions; life does it at geologically low temperatures; and life does it faster than would happen abiotically.

Most current research on the diversity of life is focusing on phenotypic and genotypic studies of extreme environments, using thermodynamic calculations to highlight possible biogeochemical reactions that could support life (anaerobic methane oxidation; Thomsen et al. 2001). However, bacteria have recently demonstrated the capacity to produce electrically conductive nanowires under electron-acceptor limiting conditions (Gorby et al. 2006; Reguera et al. 2005) demonstrating that growth of some bacteria, which are not directly evident from the fluid chemistry, may still be possible via interspecies electron transfer (Gorby et al. 2006).

With an understanding of the physical and chemical conditions that can support life, we can now take a look at the Hadean–Archaean environment, and then consider the fundamental interactions between geology (and geochemistry) and life during the time when life originated. These biogeochemical processes and by-products are representative of the types of biomarkers that need to be examined in the search for contemporary or extant microbial life in our solar system (and beyond).

3 The Hadean–Archaean Environment

The Hadean–Archaean Earth presented a very different habitat from that of the modern planet (see reviews by Nisbet and Sleep 2001; Nisbet and Fowler 1999, 2004; Westall 2003, 2004). In particular, the period of late heavy bombardment between 4.0 and 3.8 billion years ago (bya; Lowe and Byerly 1986; Lowe et al. 2003; Kyte et al. 2003) represents a dramatic challenge to the origin of life. During this time period, the orbits of countless asteroids were shifted to the inner solar system as the gas-giants' orbits stabilised (Gomes et al. 2005), impacting the terrestrial planets. Did life originate before (or during) the period of heavy bombardment and if it did, how did it survive? The following description of the Archean habitat is what is thought to have existed during the origin of life.

3.1 The Hadean–Archaean Lithosphere

The Archaean possessed a variety of geological settings (recorded in Archaean greenstone belts) including: shallow water volcanic platforms with sediments, shallow and deep water basaltic provinces, shallow water to emergent arc sequences with bimodal volcanics, and subaerial pull-apart basins (Thurston and Ayres 2004). While all of these geological settings occur on Earth today, the hotter mantle meant that there was a greater amount of volcanic and hydrothermal activity, as well as faster recycling of the crust (Arndt 1994). The sedimentary environments were typically of a volcanic or chemical origin as well (De Vries 2004). Quartzitic sediments, indicating eroded granitic continental crust, did not appear until after 3.2 bya (Eriksson et al. 1994; Heubeck and Lowe 1999).

3.2 The Hadean–Archaean Hydrosphere

Evidence for thorough flushing and circulation of seawater through the crust (Paris et al. 1985; Duchač and Hanor 1987; Hofmann 2004) comes from geochemical investigations of fluid inclusions (de Ronde et al. 1994; de Ronde and Ebbesen 1996) and geochemical tracing of element movement (Orberger et al. 2006). Knauth and Lowe (2003) calculated that the Early-Mid Archaean seawater was strongly influenced by hot, hydrothermal fluids and that the temperature of the global ocean was between 50–80°C, and more probably in the upper end of this range. The slightly acidic nature of the Early-Mid Archaean ocean (Grotzinger and Kasting 1993; Orberger et al. 2006) would have enhanced abiotic weathering processes. With extensive liquid water-rock interactions, circulation, this Early-Mid Archaean seawater may have been enriched in minerals and had higher salinities compared to modern seawater (Channer et al. 1997; de Ronde and Ebbesen 1996; de Ronde et al. 1997; Van Kranendonk and Pirajno 2004). However, isotopic studies of fluid inclusions in Archean rocks suggest that the composition of the oceans has remained somewhat constant over time (Grotzinger and Kasting 1993; Holland 1984; Kasting 1993). In either seawater model, early microorganisms would have been at least halotolerant, if not halophilic, similar to marine organisms today. A marine (rather than a freshwater) origin of life is suggested by the general similarity in the abundance of trace elements in living systems with those in seawater (Goldsmith and Owen 2001). The main differences in Archaean seawater chemistry, compared to today, were its acidic pH and correspondingly higher concentrations of dissolved CO_2 (Grotzinger and Kasting 1993; Orberger et al. 2006), and its reducing chemistry (Kasting 1993; Van Kranendonk et al. 2003), which would have supported a higher concentration of soluble, ferrous iron and a lower concentration of sulphate (Habicht et al. 2002; Holland 1984, 1994).

 Springer

3.3 The Hadean–Archaean Atmosphere

The composition of the Earth's atmosphere at or before about 3.5 bya is the subject of much discussion. The primary atmosphere, generally considered to have been mostly composed of nitrogen, with minor amounts of carbon dioxide, carbon monoxide and methane (Holland 1984, 1994), was blown off during the moon forming impact about 100 My after the consolidation of Earth (see Fig. 4; timeline). The subsequent reducing to neutral atmosphere (Catling and Claire 2005; Mel'nik 1982; Rye et al. 1995; Kasting 1993) was composed of mostly CO_2, together with CO, NH_3, N_2, probably a certain amount of CH_4 (possibly in part biological; Pavlov et al. 2001) and a small amount of oxygen (<0.2% of present atmospheric levels; Kasting 1993). The low levels of reactive oxygen produced by photolysis (Kasting 1993), relative to the abundance of reduced chemical species released by active volcanism and hydrothermal activity (Arndt 1994; de Wit and Hart 1993), would result in corresponding reducing chemistries of the hydrosphere and lithosphere as well. However, this overall reducing chemistry does not mean that there were no oxidised chemical species present, which, from a biosphere perspective, should have been in sufficiently adequate supply to support energy-generating biogeochemical reactions (Lin et al. 2006; Wanger et al. 2006).

The early atmosphere may have presented a challenge to the development of the biosphere via UV exposure resulting from the lack of atmospheric ozone. While the Archaean Earth was certainly protected from the deleterious effects of solar wind by the magnetosphere (Ribas et al. 2005; as it is today), calculated UV radiation fluxes to the Archaean-Earth's surface suggest that they could have been up to three orders of magnitude greater than today (Cockell 2000), especially at lower wavelengths (Canuto et al. 1982). For example, highly damaging XUV fluxes at 3.5 bya are estimated to have been six times those of present levels (Ribas et al. 2005). Without any attenuating factors, such a flux would have made the exposed surface environments of the Archaean Earth uninhabitable. On the other hand, modelling of the evolution of main sequence G stars (Newman and Rood 1977; Gough 1981) suggests that the luminosity of our Sun during the Hadean-Early-Mid Archaean period on Earth would have been 25–30% lower than today, which translates into ∼35% less UV radiation in the 200–300 nm range (Zahnle and Walker 1982). Many other factors also influence the flux of UV radiation through an atmosphere, including: planet obliquity, length of day, gaseous composition of the atmosphere, and the presence of UV absorbers, such as dust (resulting from enhanced volcanic activity and bolide impacts during the Hadean; Lowe and Byerly 1986; Lowe et al. 2003; Kyte et al. 2003), sulphur and organic hazes (CH_4 smog; Lovelock 1988).

Carbon dioxide is an effective absorber of UV radiation below 200 nm, thus, even under a higher flux of radiation to the Hadean–Archaean Earth, a high CO_2 atmosphere would have helped attenuate any UV damage. Water vapour, which absorbs UV radiation in wavelengths up to 350 nm (Lammer et al. 2005) would have likely been abundant in the atmosphere of a more hydrologically active, warm Earth, protecting life. While UVB and UVC radiation can damage the DNA of organisms (Rontó et al. 2003), the length of day on the Early-Mid Archaean Earth was shorter (14 h), which would have resulted in lower UV dose per day (Walker et al. 1983). These early organisms were presumably radiation resistant, similar to many bacteria today (Daly et al. 1994) and would have had time, overnight, to repair the damage occasioned by UVB and UVC during the day. From modern exposed environments (Fig. 5), the outer layers of a microbial mat may consist of dead organisms that also provide protection for those living in the deeper layers (Garciia-Pichel et al. 1994). Ultraviolet light

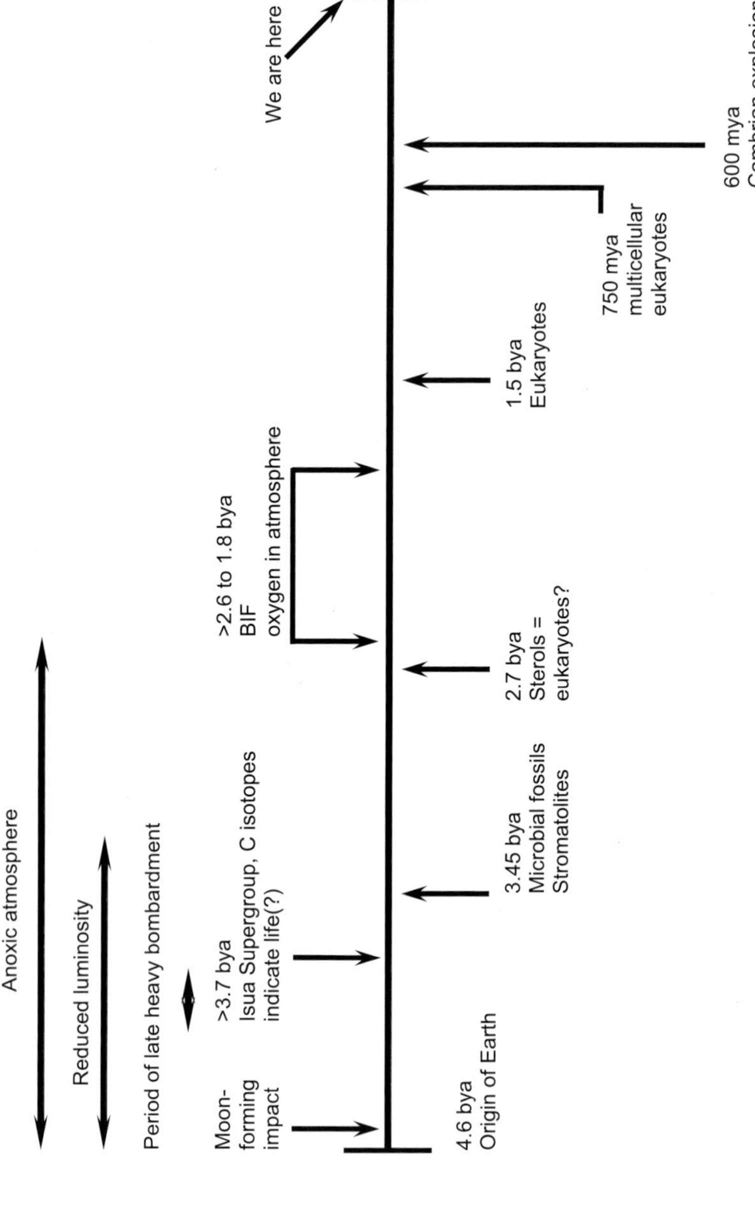

Fig. 4 A schematic representation of the Earth's timeline, demonstrating key geological and biological steps in Earth's evolution

Fig. 5 A photograph of a microbial mat from a wetland near Atlin, BC. The evaporites at the surface of the mat (*arrows*) and the surface exposed organisms protect the underlying biosphere from the harmful effects of UV radiation

is not considered to have been limiting factor in life's origin(s) or success (Westall et al. 2006).

Perhaps more important than its function as a UV block, the reaction water and UV radiation (a photolysis reaction) will produce hydrogen peroxide (Cockell 2000). The subsequent reaction between hydrogen peroxide and reduced inorganic compounds would have produced oxidised inorganic constituents such as sulphoxyanions. The presence of both reduced and oxidised chemical species may have actually been critical in establishing the redox conditions essential for life. Equally important to the early biosphere, the photolysis (breakdown) of water would have resulted in a flux of these redox reactive compounds.

In addition to their role in protecting life from UV radiation, atmospheric CO_2 (Mel'nik 1982; Rye et al. 1995; Walker et al. 1983) and other greenhouse gases (methane; Pavlov et al. 2001) would have been necessary to keep the Archaean Earth from freezing over (Kasting and Brown 1998) during the period of lower solar luminosity. Based on evidence from the rock record (Rye et al. 1995), as well as theoretical considerations of seafloor carbonation reactions (Sleep and Zahnle 2001), CO_2 would not have been abundant enough in the Archaean atmosphere to accomplish this; Kasting (1993) calculated that the partial pressure of CO_2 in the atmosphere was only about 1 bar for the 3.5 billion year old Earth. Therefore, there must have been a flux of methane, since it would have been lost via photochemical oxidation to carbon dioxide, for the Earth to have been habitable.

The environmental conditions within the Hadean–Archaean are generally within the habitable boundaries for contemporary life. Since life originated on Earth under these conditions, perhaps it can be assumed that the conditions that are habitable for life on a terrestrial planet are also those that are conducive to an origin of life.

4 Life on the (Hadean–?) Archaean Earth . . .

At least one of the planets in our solar system possessed conditions that were conducive to the formation of life. From a minimalist perspective, the physicochemical conditions include the presence of organic carbon, liquid water and redox conditions that are not only energetically favourable to support life but also in flux, providing a suitable chemostat for growth to occur once begun. Briefly, life will consist of organic carbon based cells occurring in

water, be composed of complex macromolecules, be able to generate energy, will possess a capacity to grow and reproduce, and will respond and adapt through successive generations to environmental conditions.

Geology is considered to have been important for the origin of life. Examples include the role of minerals in generating the energy necessary for life (Wächtershäuser 1988) in catalysing the formation of macromolecules such as nucleic acids (Cairns-Smith et al. 1992), and in forming non-living 'cell' structures (Fox 1969). The importance of a subaerial lithosphere, specifically, the alternate wetting and drying cycles associated with an emergent lithosphere for the origin of life (Cairns-Smith et al. 1992; Fox 1969) versus the theoretically possible (Knoll 1988) or impossible (Miller and Bada 1988) origin of life in hydrothermal vent environments on the ocean floor cannot be resolved in this chapter. Therefore, rather than discuss all of the known or hypothetical processes that would have to occur to produce the macromolecules required to construct a living cell, we take a simplified view, that the Earth fell within the spectrum of conditions in which life could originate and that any extraterrestrial body that is (or was a some point in its history) comparable to Earth when life originated, could also support (or have supported) an origin (Sect. 5).

4.1 Geology and Habitability

It is not possible to ascertain the exact time for the origin of life on Earth. Apart from the bolide impacts during the late heavy bombardment (Lowe and Byerly 1986; Lowe et al. 2003; Kyte et al. 2003), the Earth was generally habitable throughout the Hadean; the geological constraints described in Sect. 3 are only extreme if you take an anthropocentric view. Life at 80°C certainly appears to be extreme to us (see Fig. 3; Rothschild and Mancinelli 2001), however, it cannot be considered an extreme condition for life; it was simply the ambient environment at 3.5 bya (Konhauser et al. 2003).

If the period of late heavy bombardment (Gomes et al. 2005; Lowe and Byerly 1986; Lowe et al. 2003; Kyte et al. 2003) did sterilise Earth (boil away the oceans and melt or heat the crust to >121°C; Sleep et al. 1989; Nisbet and Sleep 2001; Ryder 2002) then life could have originated in as little as 100 million years (calculated by subtracting isotopic evidence of the presence of life at 3.7 bya (Rosing 1999) from the end of the planet sterilizing impacts at 3.8 bya). If the late heavy bombardment did not boil away all of the oceans, then life could have originated much earlier than 3.7 bya, and survived. Note, this period of late heavy bombardment is drastically different than an isolated impact, which can create a unique habitat, promoting life (Cockell et al. 2002).

From a microbiological point of view, the Hadean–Archaean environment would have possessed a wide range of nutrients and potential redox reactions that could have supported bacterial growth. The formation of organic carbon compounds (Miller 1953) and macromolecules on the Earth (Nussinov et al. 1997), similar to Titan today (Raulin et al. 2006), or the addition of extraterrestrial organic matter, similar to the Tagish Lake carbonaceous chondrite fall to Earth (Brown et al. 2000), would have provided abundant organic carbon (between 10^{24} to 10^{25} g C; Tingle 1998) for heterotrophic growth. With the reducing atmosphere, anaerobic heterotrophs would have been the dominant organisms during the Early-Mid Archaean (Westall and Southam 2006), until the available organic carbon ran out or until the evolution of oxygenic photosynthetic bacteria in the early Proterozoic (Kasting 1993).

Alternatively, lithoautotrophs, which make use of favourable redox contrasts to obtain their energy and carbon from inorganic sources, could have been the basis of the food chain (Nisbet and Sleep 2001). Their by-products and organic remains, which can be released after

cell death, can then be used by other microorganisms whose by-products, in turn, can be used by other microorganisms with yet another type of metabolism. Today, microenvironments are important for the growth of metabolically diverse bacteria in association with one another (Costerton et al. 1995); intuitively, they've always been important for the growth of these consortia. Major and trace elements, essential for life today would have been non-limiting due to active carbonic acid weathering of geologic material.

The Early-Mid Archaean environment would have presented many such possibilities for redox reactions in association with rocks and debris composed of basic to ultrabasic volcanics (with some rhyolites and dacites). These mineral suites have highly reactive surfaces, entrained gases; sulphate and nitrate compounds produced volcanically and through subsequent photochemical reactions; and include metal sulphide deposits, which have their own catalytic properties (Russell and Hall 1997). The coexistence of reducing and oxidised gases in the prebiotic atmosphere would have provided chemical energy. At some point, organisms like methanogens, which utilise the redox disequilibrium by combining H_2 and CO_2 to make methane (reaction (4)), must have evolved to exploit this (Catling and Claire 2005).

If the conditions on Hadean–Archaean Earth were replicated in the laboratory in a sealed culture vessel and inoculated with Earth materials collected from circumneutral, saline, anaerobic, hydrothermal environments (Aharon et al. 1992), the growth of several genera of bacteria would not be surprising. We cannot know for certain where these conditions fell in the spectrum between those ideal for life to originate versus those meeting the minimal requirements for life to appear. However, once the period of heavy bombardment ended, the Earth was a warm hospitable environment; it was habitable. The effect of life on geology and geochemistry would have begun at the micrometer scale of the microbes themselves. It would have taken time for life to colonise the entire planet and time for life to grow into a population large enough to exert its biogeochemical influence at the global scale.

4.2 The Effect of Life on Archaean Environmental Chemistry

Our understanding of the Archaean is limited by the number of samples remaining, by the preservation (fossilization) capacity of biological matter (Westall 1999; Westall et al. 1995), and by life as we know it today. Bacteria are thought to have inhabited the Earth for the last 3.5 billion years (Hofmann et al. 1999; Van Kranendonk et al. 2003; Walsh and Lowe 1985; Walsh and Westall 2003; Walter et al. 1980; Westall and Southam 2006; Westall et al. 2001). The vast majority of early life's traces, in the Pilbara, NW Australia and in Barberton, E. South Africa, occur in shallow water sediments (Lowe 1983; Westall et al. 2001). Although there is significant debate as to the biogenicity of a number of documented biosignatures, there is a significant body of bona fide evidence for the presence of bacterial life forms on the surface of the Earth in this period. The subtle, generally microscopic expression of life forms from a wide range of available habitats (Westall and Gerneke 1998; Westall and Southam 2006; Westall et al. 2001; Staudigel et al. 2006) suggest that they were widely distributed and that likely employed the less energy efficient anaerobic metabolism (Lin et al. 2006; Wanger et al. 2006). Even the identification of life's signatures vs. the abiogenic component in the small stromatolites occurring in the Early-Mid Archaean terrains is challenging (Allwood et al. 2006; Buick 1990; Garcia-Ruiz et al. 2003; Krumbein 1983; Lowe 1980; Schopf 1993; Schopf and Packer 1987; Schopf et al. 2002; Schoenberg et al. 2002; Westall and Southam 2006); indeed, they have been considered by some authors to be of purely abiogenic origin (Brasier et al. 2002; Buick et al. 1995; Grotzinger and Rothman 1996; Lowe 1994; van Zuilen et al. 2002).

 Springer

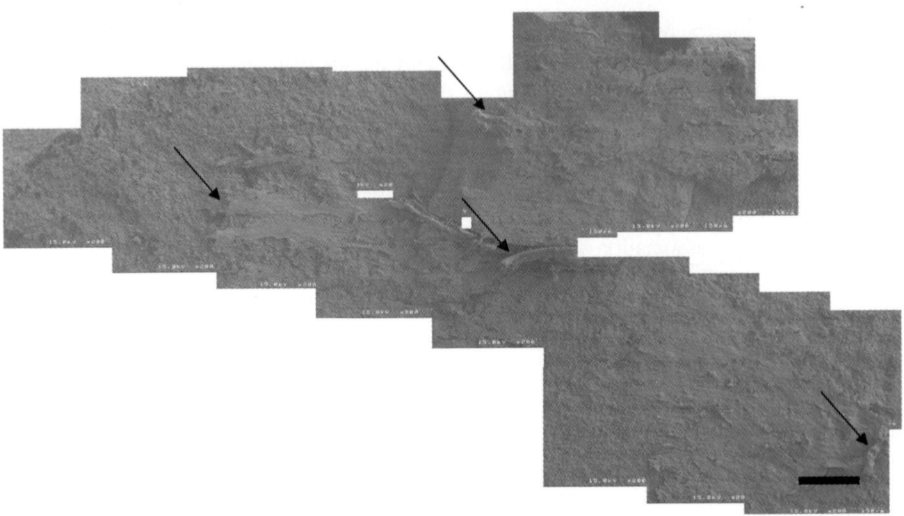

Fig. 6 A collage of scanning electron micrographs of a plan view of a microbial mat that was exposed across the fracture face of a sample of Josefsdal Chert from the Barberton greenstone belt (Westall et al. 2006). *Arrows* indicate the fibrous nature of the mat and regions where it is peeling off of the mineral substrate. *Bar* equals 250 μm

Although life seems to have been widely distributed in the Early-Mid Archaean epoch, based on the fossil record, it does not appear to have been volumetrically very important by today's standards. Westall et al. (2006; see Fig. 6) noted that, although common, the bacterial colonies in these sediments were very small, occurring as microcolonies composed of only a few to tens of cells, and the biofilms relatively thin, with the consequence that total organic carbon contents are very low (<0.01–0.05%, sometimes ranging up to 0.1–0.5% cf. Walsh and Lowe 1999). The stromatolitic constructions also possessed only microscopic biogenic components, orders of magnitude smaller than those of the Late Archaean–Proterozoic epochs. Thick accumulations of tabular microbial biofilms/mats are relatively rare (Walsh 2004; Westall et al. 2006). Only the organic-rich shales deposited in deeper waters, testify to a relatively abundant input of carbonaceous material from shallow water environments (Walsh and Lowe 1999; Tice and Lowe 2004) or from the planktonic photic zone (Hofmann 2004; Rosing 1999; Rosing and Frei 2004; Walsh and Lowe 1999). The main limitation to biomass development may have been the fact that oxygenic photosynthesis and the corresponding aerobic heterotrophs and lithotrophs had not yet developed (Des Marais 2000).

As soon as life began, it would have reduced the entropy of the environments it inhabited (Lyons et al. 2003). Structural and chemical fossils found in Archaean habitats demonstrate that the biosphere was already in an advanced evolutionary state by ~3.5 bya. Much of the strata preserved from this period appears to have been colonised by morphologically, and biochemically diverse bacteria that often produced unique mineralogical signatures (see review by Westall and Southam 2006; Staudigel et al. 2006). Bacteria possess a single, DNA chromosome (an archive of the cells' biochemical potential; some bacteria also possess extra-chromosomal elements, plasmids) that is replicated during cell division, with each daughter cell receiving one copy. Based on contemporary Bacteria and Archaea that produce about one mutation per 300 chromosome replications (Foster 2000), the progenote, even growing at a conservative 24 hour doubling time for a thermophile (Stetter et al. 1990),

 Springer

would have taken a molecularly divergent step within only 9 days. However, we do not know how long it would take to produce an autotroph from a heterotroph or vice versa. These molecularly diverse bacteria would have initially exerted a chemical influence at the nanometer (enzymatic and isotopic; Newman and Banfield 2002; Pinti and Hashizume 2001; Shen et al. 2001; Ueno et al. 2001a, 2001b, 2002, 2004) to micrometer (cellular or microcolony) scale (Marshall 1988; Roden et al. 2004; Sobolev and Roden 2002). Microbial colonies inhabiting pore spaces within liquid, water-logged sediments (Westall 2004, 2005; Westall et al. 2006) or within hydrothermal systems (Kakegawa 2001; Rasmussen 2000), and in corrosion pits in the vitreous outer layers of pillow lavas (endolithic habitats; Furnes et al. 2004, 2005; Staudigel et al. 2006) would have changed the chemistry of their surroundings and presumably promoted mineral weathering, creating both habitat and providing additional trace nutrients for growth. The eventual growth of stromatolites or microbial mats (large biomass containing features), formed at the surface of shallow water sediments in the photic zone (Allwood et al. 2006; Awramik and Sprinkle 1999; Byerly et al. 1986; Tice and Lowe 2004; Walsh 1992, 2004; Walter 1983; Westall et al. 2006) would have had a much more profound effect on their chemical surroundings than this cell-, to micro- (or mm-scale; Southam et al. 2001) colony stage, especially in evaporative environments (Chan et al. 2005), which would have been isolated from the bulk seawater chemical conditions and therefore, concentrated metabolic by-products and the (bio)geochemical effects of metabolism. Paleosol (Ohmoto 1996) and isotopic evidence (Eigenbrode and Freeman 2006) suggest that oxygenic photosynthesis may have originated as early as 2.9 billion years before present. The extension of these biogeochemical processes to geologically significant physical scales (hundreds of km), would have been needed to produce globally significant biogeochemically cycling.

Between 2.5 and 1.8 Gya, the global production of oxygen by oxygenic, photosynthetic cyanobacteria was consumed by the following biotic (Corstjens et al. 1992; Kappler and Newman 2004; reaction (5)) and abiotic (Langmuir 1997; reactions (5) and (6)) processes:

$$Fe^{2+} + \tfrac{1}{4}O_2 + H^+ \rightarrow Fe^{3+} + \tfrac{1}{2}H_2O \tag{5}$$

$$Fe^{3+} + 3H_2O \rightarrow Fe(OH)_{3(s)} + 3H^+ \tag{6}$$

This overall interaction, between iron and oxygen, is considered to be a detoxification mechanism that would have protected the early obligately anaerobic life on Earth. The reactivity of iron oxides with bacterial surfaces (Beveridge 1989) during this period in Earth's history would have presumably fossilised examples of most of the bacteria present in these ecosystems. With the consumption of available iron, these reactions could no longer bind all of the oxygen released by the biosphere, resulting in the release of large amounts of O_2 into the atmosphere. This buildup of oxygen would have led to a severe crisis for life on Earth, as oxygen radicals tend to react strongly with organic molecules, damaging biological, enzymatic activity. However, the formation of oxygen would have also created a unique opportunity for those organisms, which could utilise oxygen as a terminal electron acceptor. These aerobic heterotrophs would have been able to completely oxidise a wide range of organic compounds and generated more energy, been more competitive, than any of the anaerobes. During this time, anaerobic bacteria living in sheltered environments (isolated anoxic basins in the deep ocean or the terrestrial deep subsurface) were unaffected.

While microbial life has dominated Earth's history (some would argue that they are still the dominant life form today) and represents the most probable extraterrestrial life form that

we should expect to find, the development of macroscopic life forms deserves some consideration as an example of the timing and environmental conditions necessary to produce macroscopic organisms.

4.3 Evolution

The development of eukaryotic organisms (see Fig. 4) was reliant on both the geosphere for our origin and on the prokaryotic biosphere (influenced by the geosphere) for our evolution. Once the Earth became habitable/possessed liquid water (4.4 bya; Wilde et al. 2001), life, resembling contemporary prokaryotes, developed.

From 3.7 bya (or 3.5 bya from cellular fossils), prokaryotes ruled Earth for more than 2 billion years (see Fig. 4; Javaux et al. 2001), signifying that the origin of eukaryotes was the rate-limiting step in biological evolution on earth. The presence of sterols as a eukaryotic molecular marker (Brocks et al. 1999) is problematic because sterols are present in some bacteria (Hai et al. 1996). When combined with the endosymbiotic origin for eukaryotes (prokaryotic organisms were taken inside and protected/used by another cell; Sagan 1967), it is difficult to define exactly when eukaryotes evolved. Prokaryotes have left their biochemical signature in every living organism on earth.

From our study of biochemistry, we know that all life on Earth follows the 'Central Dogma' of biology where DNA → RNA → protein. Briefly, Deoxyribonucleic acid (DNA), the genetic archive of a cell is used to synthesise ribonucleic acids (the process is transcription). The RNA molecules provide a temporary template and the machinery to make proteins (translation), the catalysts of life. While multiple origins were presumably possible, based on our common biochemistry, only one life form ultimately colonised Earth. We refer to this organism as the progenote, representing the most competitive origin of life. Competition through energy generation and growth is more likely than the ability to produce toxic, biological compounds as the means for success, since antibiotics are directed towards a known or common metabolism (Cundliffe 2006). From this point forward, life proceeded to colonise Earth. The paleontological record of life shows an overall increase in complexity of structure through time. Beginning with single-celled prokaryotes growing as microcolonies and perhaps stromatolites from 3.45 bya (or perhaps, >3.7 bya; Rosing 1999) to present, to microbial mats consisting of large communities of prokaryotes (preserved in iron formations from 2.5–1.8 bya (Konhauser et al. 2002; Mel'nik 1982) and occurring in contemporary systems (Duhig et al. 1992; Edwards et al. 2004; Little et al. 2004), to eukaryotes occurring 1.5 bya to present (Javaux et al. 2001) and finally to multicellular organisms (plants and animals), 0.75 bya to present (Knoll et al. 2006).

Fossil evidence for multi-cellular eukaryotes at ~0.75 bya means that the step from unicellular to multi-cellular eukaryotes only took ~0.75 billion years. Thus, the prokaryotic-to-eukaryotic jump at 1.5 billion years before present, was the rate-limiting step in the evolution of life on Earth, presumably requiring the development of oxygenic photosynthesis and the oxygenation of Earth's surface to occur before eukaryotes came into existence. Long before plants and animals came along, our single-celled ancestors (prokaryotes and eukaryotes) had perfected all of the biochemical (enzymatic) processes needed for life. Multicellular eukaryotic life forms excelled via the establishment of genetic recombination (a shuffling of genes preventing populations of asexual organisms from accumulating deleterious mutations), sexual reproduction (Bernstein et al. 1981; Cleveland 1947; Margulis and Sagan 1986), and by the development of regulatory genes, allowing different kinds of cells to be produced from the same genetic information (tissue differentiation, especially hard tissues; Benton 2004).

 Springer

During the Vendian period, immediately before the Cambrian, a few organised, multicellular sea fauna were preserved in the fossil record; however, there was no hint of the abundance of forms and functions found in Cambrian fossil sediments (Knoll et al. 2006). In the early Paleozoic era, the Cambrian period, something remarkable happened, never again repeated in the history of life. In only 10 million years (~0.2% of Earth's history), all but one of the 30 animal phyla existing today appeared in the fossil record (Conway Morris 1989). This was the Cambrian Explosion. The succession, from the Vendian period to the Cambrian, represents a major transition in Earth's biological history from a microbially populated planet where competition was based on nutrient availability (Nealson and Stahl 1997), biochemistry, antibiotics (Cundliffe 2006) and simple grazing by protozoa (Eccleston-Parry and Leadbeater 1995) to a biosphere that was dominated by organisms that possessed teeth at the top of the food chain (Benton 2004). Such a macroscopic biosphere would be relatively easy to recognise compared to a contemporary or extant microscopic biosphere elsewhere in the solar system.

5 Life and the Terrestrial Planets

The widespread distribution of life preserved on Earth from 3.5 bya implies that life must have been present much earlier, perhaps as early as or well before 3.7 bya, when the rocks of Isua Supergroup and Akilia were formed. These rocks could potentially hold a record of this life (Schidlowski 1988, 2001; Rosing 1999; Rosing and Frei 2004) if the metamorphic overprinting (Fedo and Whitehouse 2002; McGregor and Mason 1977; Pflug and Jaeschke-Boyer 1979; van Zuilen et al. 2003, 2005) can be disentangled from the potential biogenic signal(s) (Lepland et al. 2002; Myers 2001; Papineau et al. 2005; Pinti et al. 2001; Rosing 1999). Regardless, the advanced, cellular level of evolution at 3.5 bya has important implications, not only for the timing of the origin of life on Earth but also, for the possible origin of life elsewhere in the solar system. Did life evolve soon after liquid water became available on Earth after the moon forming event (Canup and Asphaug 2001) and survive the period of heavy bombardment from 4.0–3.8 bya (Maher and Stevenson 1988; Mojzsis et al. 1996; Nisbet and Sleep 2001; Ryder 2002; Schoenberg et al. 2002; Sleep et al. 1989), or is the timing of life's origin immediately after? The predictions of mass annihilation in planet-sterilising impacts (Nisbet and Sleep 2001; Ryder 2002; Sleep et al. 1989) suggest that life arose after the heavy bombardment stopped. In this scenario, once Earth became habitable, life was able to reach a structural or cellular state of evolution comparable to contemporary organisms, in only a few hundred million years. Unless we're unique (McSween 1997), the ability of life to originate so quickly on Earth raises the possibility that life forms could have formed elsewhere in the solar system if conditions were habitable.

5.1 Mercury

Even though Mercury may possess water ice (Slade et al. 1992), it lacks an atmosphere and has extremely large temperature extremes (-150 to $400°C$), conceivably too drastic for a biosphere to cope with; therefore, it is not considered to be a good candidate from a habitable or an origins perspective.

5.2 Venus

Venus may have had $80–100°C$ liquid water on its' surface, very early in its history (Kasting et al. 1984), which is comparable to thermophilic environments that support life on Earth

today. Also, since it is a rocky planet, it is expected to have possessed many of the geochemical, redox conditions and processes that were favourable to an origin of life on Earth during the Hadean–Archaean. Therefore, we would have to consider it to have been habitable at this time and perhaps, even conducive to an origin of life. A potential problem with an origin on Venus could have been the acidification of it's hydrosphere. Organic carbon macromolecules, the building blocks of cells, are less stable under acidic conditions. Also, the circumneutral pH conditions inside all living cells, even acidophiles, is circumneutral (Matin et al. 1982) suggesting that life originated under circumneutral pH conditions. Today, the prospects for finding extant life on Venus are unlikely. The loss of water early in Venus' history (Chassefière 1997) and the runaway Greenhouse Effect (Kasting 1988), combined with the detrimental effect of oxidation of the planets' surface (Kasting et al. 1984) forming concentrated sulphuric acid clouds (Sill 1983), would have wiped-out any remnants of extant life preserved in its rock record through acid weathering.

5.3 Mars

Mars is a good candidate from both an origins and a habitability perspective. The geologic history for Mars's highlights three Epochs: the Noachian (~3.8 to 3.5 bya), which possesses terrain scarred by a lot of large impact craters and extensive flooding by liquid water (Carr 1987); the Hesperian (3.5 to 1.8 bya), possessing extensive lava plains; and the Amazonian (1.8 bya to present), which is comparable to Mars today, possessing minimal impact craters and some lava flows (Hartman and Neukum 2001). From a geochemical–mineralogical perspective, Bibring et al. (2006) proposed separating Mars into the Phyllosian, Theiikian and Siderikian Eras based on water-rock interactions. The presence of phyllosilicates (which are formed by aqueous alteration) in the Noachian terrains on Mars suggest that it must have had at least some standing water either at the surface or in the near subsurface early in its history (Phyllosian; Bibring et al. 2006). The environment on Mars during this time period, when volcanic, fluvial, and glacial resurfacing rates were all relatively high (Hartman and Neukum 2001) is comparable to the conditions when life originated on Earth. Mars was a rocky planet that had liquid water (Carr 1987; Hynek and Phillips 2001); it would also have had inputs of extraterrestrial organic carbon and prebiotic synthesis reactions similar those that are thought to have occurred on Earth, providing material to construct cells and suitable redox gradients to generate energy. Geochemical energy calculations (Jakosky and Shock 1998) indicate that Mars (over its entire history) could have generated a comparable amount of biomass to Earth's first 100 million years. The bulk of this geochemical energy would have been available during its early history when water was most abundant. Therefore, from our habitability on an Earth-like planet equalling the capacity to support an origin assumption, Mars would have met all of the fundamental requirements necessary for life. If life (based on Earth) can originate in as little as one- to several-hundred million years), it is reasonable to infer that Mars may have also supported an origin.

While its smaller size would imply that Mars cooled quicker than Earth, condensing steam to liquid water earlier than Earth did; could life have evolved on Mars before it did on Earth? Mojzsis et al. (2001) found oxygen-isotope evidence in ancient zircons for liquid water on Earth at 4.3 bya suggesting that Mars did not have much of a head start on the Earth. If life did originate on Mars first, could the transfer of planetary material (Horneck et al. 2002) from Mars during the period of late heavy bombardment have inoculated Earth? Whether or not this transfer occurred, the Noachian (Phyllosian) age rocks could harbour a record of fossilised 'bacteria' or life forms similar to bacteria, and mineralogical biomarkers similar to those produced by life on Earth (Banfield et al. 2001; Bazylinski 1996; Beveridge

1989; Cady et al. 2003; Ferris et al. 1986; Lowenstam 1981; Southam and Donald 1999; Toporski et al. 2002).

The subsequent climate change on Mars, from an alkaline, aqueous environment possessing a relatively thick atmosphere, to a more saline, acid-sulphate, aqueous environment (Theiikian Era; Bibring et al. 2006) possessing a less dense atmosphere due to the loss of water (Chassefière and Leblanc 2004) occurred after the period of heavy bombardment until approx. 3.5 bya (Bibring et al. 2006). This Martian, acid sulphate system is similar to acid mine drainage (Nordstrom and Southam 1997) or supergene enrichment (Enders et al. 2006) environments on Earth, and therefore is still habitable.

The Hesperian and Amazonian Epochs (Siderikian Era, from approx. 3.5 bya to present) is interpreted as being an essentially dry, weathering environment, the same as it is today (Bibring et al. 2006), which has produced the bulk of the anhydrous ferric oxides on Mars' surface. While liquid water is considered to have played a minor role during this Era, evidence of melting of the polar caps and the formation of gullies within the last few hundred million years (the Late Amazonian; Benito et al. 1997; Clifford 1987; Fishbaugh and Head 2002; Malin and Edgett 2000), perhaps even the last decade (Malin et al. 2006) suggests that Mars still possesses an active subsurface hydrosphere Thus, if life did become established on Mars during its first billion, relatively water rich years, it could still be present today in Mars' deep subsurface where the geologic flux of reduced or energetic redox species in groundwater could provide life support, similar to that provided to bacteria in Earth's deep subsurface (see Fig. 1; Wanger et al. 2006). Note, if origins are equated with habitability, life could have originated at any time in Mars' history. Even today, the water–rock–atmosphere dynamic that resulted in an origin on the Archaean Earth could be met by a subsurface vadose zone on Mars.

The detection of methane in Mars' contemporary atmosphere raises the possibility of a deep Martian biosphere, or at the very least, subsurface volcanic activity (Formisano et al. 2004; Krasnopolsky et al. 2004). Both are important to the life question, either as a direct biomarker or as a source of redox reactive chemical species. Certainly, Mars' cold, dry highly oxidising surface is detrimental to life, but the subsurface of could be acting as a refuge for an active biosphere. If life is established in the Mars' subsurface, it will likely persist, like life on Earth, until our Sun becomes a red giant.

We cannot expect Martian life, if it exists, to be exactly like that on Earth; however, from a structural and biogeochemical perspective, we should be able to differentiate materials that are of biological origin from those that are abiotic, just like we've done on Earth.

5.4 Europa

While not a terrestrial planet, Europa, the smallest of Jupiter's four planet-sized moons has drawn a lot of attention due to its ≤150 km thick water–ice shell (Carr et al. 1998). Based on Galileo imaging, near infrared reflectance spectroscopy, and gravity and magnetometer data, Kargel et al. (2000) described Europa as a carbonaceous chondrite that underwent a global primordial differentiation including degassing and hydrothermal activity, ultimately producing hydrated silicates and a magnesium- or sodium sulfate ocean at its surface. Infrared water ice absorption bands, and the near absence of impact craters indicate that the surface is ice rich and very young, perhaps only 30 million years old (Greeley et al. 2000). The young surface is an indication of liquid water, leading to speculation that an origin could have occurred in this dark, watery world (Carr et al. 1998). The presence of liquid water is thought to occur due to hydrothermal activity (Chyba 2000; Zolotov and Shock

2003), which is conducive to the formation of unique habitable zones on Earth, or to gravitational pumping (Greeley et al. 2000), which would produce a lower geochemical energy flux from the lithosphere but would still generate liquid water. In either case, throughout Europa's history, there would have been few oxidised chemical species (Hall et al. 1995), and therefore, less geochemical energy, via the formation of suitable redox gradients, available for life (Jakosky and Shock 1998). While Europa fulfills the habitability requirement for life, the difficulty with forming cells under sub-aqueous conditions on Earth (Miller and Bada 1988) would also apply to Europa, presumably preventing an origin.

6 Conclusions

While we do not know what the optimal conditions are to create life, it seems entirely reasonable that basic requirements include a rocky planet and liquid water, specifically occurring as a subaerial lithosphere-hydrosphere system. The required, liquid hydrosphere and its' concomitant hydrological cycle will constrain the temperature/pressure boundaries of the planet. These conditions were met on Venus soon after the planet formed, on Earth for approx. 1 billion years until life originated creating competition for nutrients, and on Mars, potentially since it cooled enough to possess liquid water. Regarding the time required to reach an origin, life would require a stable star possessing a main sequence of at least several hundred million years, once the above planetary criteria have been met. While Venus, Earth and Mars receive differing solar inputs, thus far, all have benefited from a main sequence star of 4.6 billion years. Conversely, habitable locations in our Solar System (places where life can be sustained) are not as restrictive as those where life could originate, and include Mars' subsurface today (Boston et al. 1992; if for some reason it did not support an origin), as well as Europa (Chyba and Phillips 2001; if a subaerial lithosphere is actually required for an origin).

7 Key Questions and Needed Observations

It is too late to ask many of the questions surrounding an origin of life on Earth. How exactly, did life originate on Earth? How long did it take for life to originate? Did life originate before or during the period of late heavy bombardment, and if it originated before the late heavy bombardment, how did it survive? However, the question concerning the possible origin(s) of life elsewhere in our solar system can be addressed. Are there fossilised remains of extant life on Mars? Is there contemporary life in Mars' deep-subsurface? Is there life on Europa—can life originate without a subaerial lithosphere? The examination and characterisation of Archaean Earth materials and Earth analogue systems by direct observation and remote sensing are necessary to ground truth the use of these methods beyond Earth. Remote sensing, robotic explorers and sample return, hopefully within the lifetime of the authors, are needed to address these questions.

Acknowledgements The authors wish to thank ISSI for organizing and hosting the workshop on "The Geology and Habitability of the Terrestrial Planets" and the other participants for a very enjoyable week talking science. We thank Francois Raulin and Kate Fishbaugh for providing constructive, critical reviews, which have contributed to a greatly improved manuscript.

References

P. Aharon, H.H. Roberts, R. Snelling, Geology **20**, 483–486 (1992)

A.C. Allwood, M.R. Walter, B.S. Kamber, C.P. Marshall, I.W. Burch, Nature **441**, 714–718 (2006)

P.S. Amy, D.L. Haldeman, D. Ringelberg, D.H. Hall, C. Russell, Appl. Environ. Microbiol. **58**, 3367–3373 (1992)

I.K. Anderson, J.H. Ashton, A.J. Boyce, A.E. Fallick, M.J. Russell, Econ. Geol. **93**, 535–563 (1998)

N.T. Arndt, in: *Archean Crustal Evolution*, ed. by K.C. Condie (Elsevier, Amsterdam, 1994), pp. 11–44

R.M. Atlas, R. Bartha, *Microbial Ecology: Fundamentals and Applications*, 4th edn. (Benjamin Cummings, 1997) 306 p.

S.M. Awramik, J. Sprinkle, Hist. Biol. **13**, 241–253 (1999)

M. Banerjee, B.A. Whitton, D.D. Wynn-Williams, Microb. Ecol. **39**, 80–91 (2000)

J.F. Banfield, J.W. Moreau, C.S. Chan, S.A. Welch, B. Little, Astrobiology **1**, 447–465 (2001)

S.M. Barns, S.A. Nierzwicki-Bauer, in: J.F. Banfield, K.H. Nealson (eds.), Geomicrobiology: Interactions between microbes and minerals. Rev. Mineral. **35**, 35–79 (1997)

D.A. Bazylinski, Chem. Ecol. **132**, 191–198 (1996)

A. Bechtel, Y.-N. Sieh, M. Pervaz, W. Puttmann, Geochim. Cosmochim. Acta **60**, 2833–2855 (1996)

A. Bechtel, M. Pervaz, W. Puttmann, Chem. Geol. **144**, 1–22 (1998)

G. Benito, F. Mediavilla, M. Fernandez, A. Marquez, J. Martinez, F. Anguita, Icarus **129**, 528–538 (1997)

P.C. Bennett, F.K. Hiebert, W.J. Choi, Chem. Geol. **132**, 45–53 (1996)

P.C. Bennett, J.A. Rogers, F.K. Hiebert, W.J. Choi, Geomicrobiol. J. **18**, 3–19 (2001)

M.J. Benton, *Vertebrate Palaeontology* (Blackwell, 2004), 455 p.

H. Bernstein, G.S. Byers, R.E. Michod, Am. Nat. **117**, 537–539 (1981)

T.J. Beveridge, Int. Rev. Cytol. **72**, 229–317 (1981)

T.J. Beveridge, Can. J. Microbiol. **34**, 363–372 (1988)

T.J. Beveridge, Annu. Rev. Microbiol. **43**, 147–171 (1989)

J.-P. Bibring, Y. Langevin, J.F. Mustard, F. Poulet, R. Arvidson, A. Gendrin, B. Gondet, N. Mangold, P. Pinet, F. Forget, the OMEGA team, Science **312**, 400–404 (2006)

P.J. Boston, M.V. Ivanov, C.P. McKay, Icarus **95**, 300–308 (1992)

S.H. Bottrell, P.J. Hayes, M. Nannon, G.M. Williams, Geomicrobiol. J. **13**, 75–90 (1995)

S.L. Brantley, L. Lierman, M. Bau, S. Wu, Geomicrobiol. J. **18**, 37–61 (2001)

M.D. Brasier, O.R. Green, A.P. Jephcoat, A.K. Kleppe, M. van Kranendonk, J.F. Lindsay, A. Steele, N. Grassineau, Nature **416**, 76–81 (2002)

W.J. Brill, Annu. Rev. Microbiol. **29**, 109–129 (1975)

J.J. Brocks, G.A. Logan, R. Buick, R.E. Summons, Science **285**, 1033–1036 (1999)

P.G. Brown, A.R. Hildebrand, M.E. Zolensky, M. Grady, R.N. Clayton, T.K. Mayeda, E. Tagliaferri, R. Spalding, N.D. MacRae, E.L. Hoffman, D.W. Mittlefehldt, J.F. Wacker, J.A. Bird, M.D. Campbell, R. Carpenter, H. Gingerich, M. Glatiotis, E. Greiner, M.J. Mazur, P.J.A. McCausland, H. Plotkin, T.R. Mazur, Science **290**, 320–325 (2000)

B. Bubela, J.A. McDonald, Nature **221**, 465–466 (1969)

R. Buick, Palaios **5**, 441–459 (1990)

R. Buick, D.I. Groves, J.S.R. Dunlop, D.R. Lowe, Geology **23**, 191–192 (1995)

G.R. Byerly, M.M. Walsh, D.L. Lowe, Nature **319**, 489–491 (1986)

S.L. Cady, J.D. Farmer, J.P. Grotzinger, W.J. Schopf, A. Steele, Astrobiology **3**, 351–368 (2003)

A.G. Cairns-Smith, A.J. Hall, M.J. Russell, Orig. Life Evol. Biospheres **22**, 161–180 (1992)

R.M. Canup, E. Asphaug, Nature **412**, 708–712 (2001)

V.M. Canuto, J.S. Levine, T.R. Augustsson, C.L. Imhoff, Nature **296**, 816–820 (1982)

M.H. Carr, Nature **326**, 30–35 (1987)

M.H. Carr, M.J.S. Belton, C.R. Chapman, M.E. Davies, P. Geisler, R. Greenberg, A.S. McEwen, B.R. Tufts, R. Greeley, R. Sullivan, J.W. Head, R.T. Pappalardo, K.P. Klaasen, T.V. Johnson, J. Kaufman, D. Senske, J. Moore, G. Neukum, G. Schubert, J.A. Burns, P. Thomas, J. Veverka, Nature **391**, 363–365 (1998)

D.C. Catling, M.W. Claire, Earth Planet. Sci. Lett. **237**, 1–20 (2005)

C.M. Cavanaugh, S.L. Gardiner, M.L. Jones, H.W. Jannasch, J.B. Waterbury, Science **213**, 340–342 (1981)

M. Chan, K. Moser, J.M. Davis, G. Southam, K. Hughes, T. Graham, Aquatic Geochem. **11**, 279–302 (2005)

D.M. Channer, C.E.J. de Ronde, E.T.C. Spooner, Earth Planet. Sci. Lett. **150**, 325–335 (1997)

F.H. Chapelle, *Ground-Water Microbiology and Geochemistry* (Wiley, New York, 1993), 448 p.

F.H. Chapelle, D.R. Lovely, Ground Water **30**, 29–36 (1992)

E. Chassefière, Icarus **126**, 229–232 (1997)

E. Chassefière, F. Leblanc, Planet. Space Sci. **52**, 1039–1058 (2004)

C.F. Chyba, Nature **403**, 381–382 (2000)

C.F. Chyba, C.B. Phillips, Proc. Nat. Acad. Sci. **98**, 801–804 (2001)

L.R. Cleveland, Science **105**, 287–289 (1947)

S.M. Clifford, J. Geophys. Res. **92**, 9135–9152 (1987)

C.S. Cockell, Planet. Space Sci. **4**, 203–214 (2000)

C.S. Cockell, P. Lee, G.R. Osinski, G. Horneck, P. Broady, Meteor. Planet. Sci. **37**, 1287–1298 (2002)

M.L. Coleman, D.B. Hedrick, D.R. Lovley, D.C. White, K. Pye, Nature **361**, 436–438 (1993)

F.S. Colwell, T.C. Onstott, M.E. Dilwiche, D. Chandler, J.K. Fredrickson, Q.-J. Yao, J.P. McKinley, D.R. Boone, R. Griffiths, T.J. Phelps, D. Ringelberg, D.C. White, L. LaFreniere, D. Balkwill, R.M. Lehman, J. Konisky, P.E. Long, FEMS Microbiol. Rev. **20**, 425–435 (1997)

S. Conway Morris, Science **246**, 339–346 (1989)

P.L.A.M. Corstjens, J.P.M. de Vrind, P. Westbroek, E.W. Vrind-de Jong, Appl. Environ. Microbiol. **58**, 450–454 (1992)

J.W. Costerton, Z. Lewandowski, D.E. Caldwell, D.R. Korber, H.M. Lappin-Scott, Annu. Rev. Microbiol. **49**, 711–745 (1995)

E. Cundliffe, J. Ind. Microbiol. Biotechnol. **33**, 500–506 (2006)

M.J. Daly, L. Ouyang, K.W. Minton, J. Bacteriol. **176**, 3508–3517 (1994)

A. de los Ríos, J. Wierzchos, L.G. Sancho, C. Ascaso, Environ. Microbiol. **5**, 231–237 (2003)

D.J. Des Marais, Science **289**, 1703–1705 (2000)

D.J. Des Marais, M.R. Walter, Annu. Rev. Ecol. Systemat. **30**, 397–420 (1999)

C.E.J. de Ronde, M.T. de Wit, E.T.C. Spooner, Geol. Soc. Am. Bull. **196**, 86–104 (1994)

C.E.J. de Ronde, D.M. Channer, K. Faure, C.J. Bray, E.T.C. Spooner, Geochim. Cosmochim. Acta **61**, 4025–4042 (1997)

C.E.J. de Ronde, T.W. Ebbesen, Geology **24**, 791–794 (1996)

S.T. De Vries, Early Archean sedimentary basins: depositional environment and hydrothermal systems. Examples from the Barberton and Coppin Gap greenstone belts. Geologica Ultraiectina, University of Utrecht, 2004, 159 p.

M.J. de Wit, R.A. Hart, Lithos **30**, 309–336 (1993)

R. Donald, G. Southam, Geochim. Cosmochim. Acta **63**, 2019–2023 (1999)

J.I. Drever, *The Geochemistry of Natural Waters*, 3rd edn. (Prentice-Hall, Upper Saddle River, 1997), 436 p.

K.C. Duchač, J.S. Hanor, Precambr. Res. **37**, 125–146 (1987)

N.C. Duhig, G.J. Davidson, J. Stolz, Aus. Geol. **20**, 511–514 (1992)

J.D. Eccleston-Parry, B.S.C. Leadbeater, Appl. Environ. Microbiol. **61**, 1033–1038 (1995)

K.J. Edwards, W. Bach, T.M. McCollom, D.R. Rogers, Geomicrobiol. J. **21**, 393–404 (2004)

H.L. Ehrlich, *Geomicrobiology*, 4th edn. (Marcel Dekker, New York, 2002), 800 p.

J.L. Eigenbrode, K.H. Freeman, Proc. Nat. Acad. Sci. **103**, 15759–15764 (2006)

D. Emerson, N.P. Revsbech, Appl. Environ. Microbiol. **60**, 4022–4031 (1994)

M.S. Enders, G. Southam, C. Knickerbocker, S.R. Titley, Econ. Geol. **101**, 59–70 (2006)

K.A. Eriksson, B. Krapez, P.W. Fralick, Earth Sci. Rev. **37**, 1–88 (1994)

C.M. Fedo, M.J. Whitehouse, Science **296**, 1448–1452 (2002)

T.M. Fenchel, B.B. Jorgensen, Adv. Microbiol. Ecol. **1**, 1–58 (1977)

F.G. Ferris, T.J. Beveridge, W.S. Fyfe, Nature **320**, 609–611 (1986)

K. Fishbaugh, J. Head, J. Geophys. Res. **107**(E3), (2002). doi:10.1029/2000JE001351

V. Formisano, S. Atreya, T. Encrenaz, N. Ignatiev, M. Giuranna, Science **306**, 1758–1761 (2004)

P. Foster, Proc. Nat. Acad. Sci. **96**, 7617–7618 (2000)

S.W. Fox, Naturwissenschaften **1**, 1–9 (1969)

E.I. Friedmann, Orig. Life **10**, 223–235 (1980)

E.I. Friedmann, R. Ocampo, Science **193**, 1247–1249 (1976)

H. Furnes, N.R. Banerjee, K. Muehlenbachs, H. Staudigel, M. de Wit, Science **304**, 578–581 (2004)

H. Furnes, N.R. Banerjee, K. Muehlenbachs, A. Kotninen, Precamb. Res. **136**, 125–137 (2005)

F. Garciia-Pichel, M. Mechling, R.W. Castenholz, Appl. Environ. Microbiol. **60**, 1500–1511 (1994)

J.M. Garcia-Ruiz, S.T. Hyde, A.M. Carnerup, A.G. Christy, M.J. van Krankendonk, N.J. Welham, Science **302**, 1194–1197 (2003)

M. Goldman, D.A. Wilson, FEMS Microbiol. Lett. **2**, 113–115 (1977)

D. Goldsmith, T. Owen, *The Search for Life in the Universe*, 3rd edn. (University Science Books, New York, 2001) 580 p.

R. Gomes, H.F. Levison, K. Tsiganis, A. Morbidelli, Nature **435**, 466–469 (2005)

S. Goodwin, J.G. Zeikus, J. Bacteriol. **169**, 2150–2157 (1987)

Y.A. Gorby, S. Yanina, J.S. McLean, K.M. Rosso, D. Moyles, A. Dohnalkova, T.J. Beveridge, I.S. Chang, B.H. Kim, K.S. Kim, D.E. Culley, S.B. Reed, M.F. Romine, D.A. Saffarini, E.A. Hill, L. Shi, D.A. Elias, D.W. Kennedy, G. Pinchuk, K. Watanabe, S. Ishii, B. Logan, K.H. Nealson, J.K. Fredrickson, Proc. Nat. Acad. Sci. **103**, 11358–11363 (2006)

D.O. Gough, Sol. Phys. **74**, 21–34 (1981)

R. Greeley, P. Figueredo, D. Williams, F. Chuang, J. Klemaszewski, S. Kadel, L. Prockter, R. Pappalardo, J. Head III, G. Collins, N. Spaun, R. Sullivan, J. Moore, D. Senske, B.R. Tufts, T. Johnson, M. Belton, K. Tanaka, J. Geophys. Res. **105**, 22559–22578 (2000)

J.P. Grotzinger, J.F. Kasting, J. Geol. **101**, 235–243 (1993)

J.P. Grotzinger, D.H. Rothman, Nature **383**, 423–425 (1996)

K.S. Habicht, M. Gade, B. Thamdrup, P. Berg, D.E. Canfield, Science **298**, 2372–2374 (2002)

T. Hai, B. Schneider, J. Schmidt, G. Adam, Phytochemistry **41**, 1083–1084 (1996)

D.T. Hall, D.F. Strobel, P.D. Feldman, M.A. McGrath, H.A. Weaver, Nature **373**, 677–679 (1995)

W.K. Hartmann, G. Neukum, Space Sci. Rev. **96**, 165–194 (2001)

C. Heubeck, D.R. Lowe, in: D.R. Lowe, G.R. Byerly (eds.), Geologic evolution of the Barberton greenstone belt, South Africa. Geol. Soc. Am. Special Publ. **329**, 259–286 (1999)

F. Hiebert, P.C. Bennett, Science **258**, 278–281 (1992)

A. Hofmann, in: *Abstract Volume, Field Forum on Processes on the Early Earth, Kaapvall Craton*, ed. by W.U. Reimold, A. Hofmann (2004), pp. 35–37

H.J. Hofmann, K. Grey, A.H. Hickman, R.I. Thorpe, Geol. Soc. Am. Bull. **111**, 1256–1262 (1999)

H.D. Holland, *The Chemical Evolution of the Atmosphere and Oceans* (Princeton Univ. Press, Princeton, 1984) 598 p.

H.D. Holland, in: *Early Life on Earth, Nobel Symposium No. 84*, ed. by S. Bengston (Columbia Univ. Press, New York, 1994), pp. 237–244

G. Horneck, C. Mileikowsky, J. Melosh, J.W. Wilson, F.A. Cucinotta, B. Gladman, in: *Astrobiology*, ed. by G. Horneck, C. Baumstark-Kahn (Springer, Berlin, 2002), pp. 57–78

B.M. Hynek, R.J. Phillips, Geology **29**, 407–410 (2001)

R. Jakobsen, D. Postma, Geology **22**, 1102–1106 (1994)

B.M. Jakosky, E.L. Shock, J. Geophys. Res. **103**, 19,359–19,364 (1998)

E.J. Javaux, A.H. Knoll, M.R. Walter, Nature **412**, 66–69 (2001)

T. Kakegawa, in: *Geochemistry and the Origin of Life*, ed. by S. Nakashsima, S. Maruyama, A. Brack, B.F. Windley (Academic, Tokyo, 2001), pp. 237–249

A. Kappler, D.K. Newman, Geochim. Cosmochim. Acta **68**, 1217–1226 (2004)

J.S. Kargel, J.Z. Kaye, J.W. Head, G.M. Marion, R. Sassen, J.K. Crowley, O.P. Ballesteros, S.A. Grant, D.L. Hogenboom, Icarus **148**, 226–265 (2000)

K. Kashefi, D.R. Lovley, Science **301**, 934 (2003)

J.F. Kasting, Icarus **74**, 472–494 (1988)

J.F. Kasting, Science **259**, 920–926 (1993)

J.F. Kasting, L.L. Brown, in: *The Molecular Origins of Life*, ed. by A. Brack (Cambridge University Press, 1998), pp. 35–56

J.F. Kasting, J.B. Pollack, T.P. Ackerman, Icarus **57**, 335–355 (1984)

D.B. Knaebel, T.W. Federle, D.C. McAvoy, J.R. Vestal, Appl. Environ. Microbiol. **60**, 4500–4508 (1994)

L.P. Knauth, D.R. Lowe, Geol. Soc. Am. Bull. **115**, 566–580 (2003)

A.H. Knoll, Science **239**, 199–200 (1988)

A.H. Knoll, M.R. Walter, G.M. Narbonne, N. Christie-Blick, Lethaia **39**, 13–30 (2006)

A.L. Koch, Annu. Rev. Microbiol. **50**, 317–348 (1996)

K.O. Konhauser, T. Hamade, R.C. Morris, F.G. Ferris, G. Southam, R. Raiswell, D. Canfield, Geology **30**, 1079–1082 (2002)

K.O. Konhauser, B. Jones, A.-L. Reysenbach, R.W. Renaut, Can. J. Earth Sci. **40**, 1713–1724 (2003)

K.O. Konhauser, S. Schultze-Lam, F.G. Ferris, W.S. Fyfe, F.J. Longstaffe, T.J. Beveridge, Appl. Environ. Microbiol. **60**, 549–553 (1994)

V.A. Krasnopolsky, J.P. Maillard, T.C. Owen, Icarus **172**, 537–547 (2004)

W.E. Krumbein, Precambr. Res. **20**, 493–531 (1983)

H.J. Kunte, Environ. Chem. **3**, 94–99 (2006)

F.T. Kyte, A. Shukolyukov, G.W. Lugmaor, D.R. Lowe, G.R. Byerly, Geology **31**, 283–286 (2003)

H. Lammer, N. Yu, T. Kulikov, M. Penz, H.K. Leitner, N. Biernat, V. Erkaev, Celest. Mech. Dyn. Astron. **92**, 273–285 (2005)

D. Langmuir, *Aqueous Environmental Geochemistry* (Prentice-Hall, Upper Saddle River, 1997), 600 p.

A. Lepland, G. Arrhenius, D. Cornell, Precambr. Res. **118**, 221–241 (2002)

L.-H. Lin, P.-L. Wang, D. Rumble, J. Lippmann-Pipke, E. Boice, L.M. Pratt, B. Sherwood Lollar, E.L. Brodie, T.C. Hazen, G.L. Andersen, T.Z. DeSantis, D.P. Moser, D. Kershaw, T.C. Onstott, Science **314**, 479–482 (2006)

C.T.S. Little, S.E.J. Glynn, R.A. Mills, Geomicrobiol. J. **21**, 415–429 (2004)

L.G. Ljungdahl, K.-E. Eriksson, Adv. Microbiol. Ecol. **8**, 237–299 (1985)

D.R. Lovley, F.H. Chapelle, Rev. Geophys. **33**, 365–381 (1995)

D.R. Lovley, M.J. Klug, Appl. Environ. Microbiol. **45**, 187–192 (1983)
D.R. Lovley, E.J.P. Phillips, Appl. Environ. Microbiol. **54**, 1472–1480 (1988)
J.E. Lovelock, *The Ages of Gaia* (Norton, New York, 1988), 305 p.
D.R. Lowe, Nature **284**, 441–443 (1980)
D.R. Lowe, Precambr. Res. **19**, 239–283 (1983)
D.R. Lowe, Geology **22**, 287–390 (1994)
D.R. Lowe, G.R. Byerly, Geology **14**, 83–86 (1986)
D.R. Lowe, G.R. Byerly, F.T. Kyte, A. Shukulyukov, F. Asaro, A. Krull, Astrobiology **3**, 7–48 (2003)
H.A. Lowenstam, Nature **211**, 1126–1131 (1981)
J. Lyngkilde, T.H. Christensen, J. Cont. Hydrol. **10**, 273–289 (1992)
T.W. Lyons, C.L. Zhang, C.S. Romanek, Chem. Geol. **195**, 1–4 (2003)
K.A. Maher, D.J. Stevenson, Nature **331**, 612–614 (1988)
M.C. Malin, K.S. Edgett, Science **288**, 2330–2335 (2000)
M.C. Malin, K.S. Edgett, L.V. Posiolova, S.M. McColley, E.Z. Noe Dobrea, Science **314**, 1573–1577 (2006)
L. Margulis, D. Sagan, *Origins of Sex: Three Billions Years of Recombination* (Yale Univ. Press, New Haven, 1986), 259 p.
K.C. Marshall, Can. J. Microbiol. **34**, 503–506 (1988)
A. Matin, B. Wilson, E. Zychlinsky, M. Matin, J. Bacteriol. **150**, 582–591 (1982)
V.R. McGregor, B. Mason, Am. Mineral. **62**, 887–904 (1977)
H.Y. McSween, Jr., *Fanfare for Earth: The Origin of Our Planet and Life* (St. Martin's Press, 1997), pp. 252
Y.P. Mel'nik, *Precambrian Banded Iron Formations* (Elsevier, New York, 1982), 310 p.
S.L. Miller, Science **117**, 528–529 (1953)
S.L. Miller, J.L. Bada, Nature **334**, 609–611 (1988)
S.J. Mojzsis, G. Arrhenius, K.D. McKeegan, T.M. Harrison, A.P. Nutman, C.R.L. Friend, Nature **384**, 55–59 (1996)
S.J. Mojzis, T.M. Harrison, R.T. Pidgeon, Nature, **409**, 178–181 (2001)
D.P. Moser, T.C. Onstott, J.K. Fredrickson, F.J. Brockman, K. Takai, D.L. Balkwill, G. Drake, S. Pfiffner, D.C. White, B.J. Baker, L.M. Pratt, J. Fong, B. Sherwood Lollar, G. Slater, T.J. Phelps, N. Spoelstra, M. DeFlaun, G. Southam, A.T. Welty, J. Hoek, Geomicrobiol. J. **20**, 517–548 (2003)
C.P. Myers, K.N. Nealson, Science **240**, 1319–1321 (1988)
J.S. Myers, Precambr. Res. **105**, 129–141 (2001)
R. Navarro-Gonzalez, F.A. Rainey, P. Molina, D.R. Bagaley, B.J. Hollen, J. de la Rosa, A.M. Small, R.C. Quinn, F.J. Grunthaner, L. Caceres, B. Gomez-Silva, C.P. McKay, Science **302**, 1018–1021 (2003)
K.H. Nealson, D.A. Stahl, in: J.F. Banfield, K.H. Nealson (eds.), Geomicrobiology: Interactions between microbes and minerals. Rev. Mineral. **35**, 5–34 (1997)
F.C. Neidhardt, J.L. Ingraham, M. Schaechter, *Physiology of the Bacterial Cell—A Molecular Approach* (Sinauer, Sunderland, 1990), 506 p.
D.K. Newman, J.F. Banfield, Science **296**, 1071–1077 (2002)
M.J. Newman, R.T. Rood, Science **198**, 1035–1037 (1977)
E.G. Nisbet, C.M.R. Fowler, Proc. Roy. Soc. Lond. B **266**, 2375–2382 (1999)
E.G. Nisbet, C.M.R. Fowler, in *Treatise on Geochemistry*, ed. by W.H. Schelsinger. Biogeochemistry, vol. 8 (Elsevier–Pergamon, Oxford, 2004), pp. 1–39
E.G. Nisbet, N.H. Sleep, Nature **409**, 1083–1091 (2001)
D.K. Nordstrom, G. Southam, Rev. Mineral. **35**, 361–390 (1997)
M.D. Nussinov, V.A. Otroshchenko, S. Santoli, BioSystems **42**, 111–118 (1997)
H. Ohmoto, Geology **24**, 1135–1138 (1996)
B. Orberger, V. Rouchon, F. Westall, S.T. de Vries, C. Wagner, D.L. Pinti, in: W.U. Reimold, R. Gibson (eds.), Processes on the Early Earth. Geol. Soc. Am. Bull. **405**, 132–151 (2006)
N.R. Pace, Science **276**, 734–740 (1997)
D. Papineau, S.J. Mojzsis, J.A. Karhu, B. Marty, Chem. Geol. **216**, 37–58 (2005)
I. Paris, I.G. Stanistreet, M.J. Hughes, J. Geol. **93**, 111–129 (1985)
R.J. Parkes, B.A. Cragg, S.J. Bale, J.M. Getliff, K. Goodman, P.A. Rochelle, J.C. Fry, A.J. Weightman, S.M. Harvey, Nature **371**, 410–411 (1994)
A.A. Pavlov, J.F. Kasting, L.L. Brown, K.A. Rages, R. Freedman, J. Geophys. Res. **105**, 11981–11990 (2001)
K. Pedersen, S. Ekendahl, Microb. Ecol. **20**, 37–52 (1990)
H.D. Pflug, H. Jaeschke-Boyer, Nature **280**, 483–486 (1979)
D.L. Pinti, K. Hashizume, Precambr. Res. **105**, 85–88 (2001)
D.L. Pinti, K. Hashizume, J. Matsuda, Geochim. Cosmochim. Acta **65**, 2301–2315 (2001)
N.W. Pirie, Annu. Rev. Microbiol. **27**, 119–131 (1973)
E. Purcell, Am. J. Phys. **45**, 3–11 (1977)
B. Rasmussen, Nature **405**, 676–679 (2000)

F. Raulin, Y. Bénilan, P. Coll, D. Coscia, M.-C. Gazeau, E. Hébrard, A. Jolly, C. Romanzin, R. Sternberg, Proc. Int. Soc. Opt. Eng. **6309**, 63090I (2006)

G. Reguera, K.D. McCarthy, T. Mehta, J.S. Nicoll, M.T. Tuominen, D.R. Lovley, Nature **435**, 1098–1101 (2005)

N.P. Revsbech, B.B. Jorgensen, Adv. Microb. Ecol. **9**, 293–353 (1986)

A.-L. Reysenbach, E. Shock, Science **296**, 1077–1082 (2002)

I. Ribas, E.F. Guinan, M. Güdel, M. Audard, Astrophys. J. **622**, 680–691 (2005)

E.M. Rivkina, E.I. Friedmann, C.P. McKay, D.A. Gilichinsky, Appl. Environ. Microbiol. **66**, 3230–3233 (2000)

E. Roden, D. Sobolev, B. Glazer, G. Luther, Geomicrobiol. J. **21**, 379–391 (2004)

J.R. Rogers, P.C. Bennett, Chem. Geol. **203**, 91–108 (2004)

J.R. Rogers, P.C. Bennett, W.J. Choi, Am. Mineral. **83**, 1532–1540 (1998)

G. Rontó, A. Bérces, H. Lammer, C.S. Cockell, G.J. Molina-Cuberos, M.R. Patel, F. Selsis, Photochem. Photobiol. **77**, 34–40 (2003)

M.T. Rosing, Science **283**, 674–676 (1999)

M.T. Rosing, R. Frei, Earth Planet. Sci. Lett. **217**, 237–244 (2004)

L.J. Rothschild, R.L. Mancinelli, Nature **409**, 1092–1101 (2001)

M.J. Russell, A.J. Hall, J. Geol. Soc. Lond. **154**, 377–402 (1997)

G. Ryder, J. Geophys. Res. **107**, E4, 5022 (2002). doi:10.1029/2001JE001583

R. Rye, P.H. Kuo, H.D. Holland, Nature **378**, 603–605 (1995)

L. Sagan (Margulis), J. Theor. Biol. **14**, 255–274 (1967)

M. Schidlowski, Nature **333**, 313–318 (1988)

M. Schidlowski, Precambr. Res. **106**, 117–134 (2001)

R. Schoenberg, B.S. Kamber, K.D. Collerson, M. Moorbath, Nature **418**, 403–405 (2002)

J.W. Schopf, Science **260**, 640–646 (1993)

J.W. Schopf, B.M. Packer, Science **237**, 70–73 (1987)

J.W. Schopf, A.B. Kudryavtsvev, D.G. Agresti, T.J. Wdowiak, A.D. Czaja, Nature **416**, 73–76 (2002)

K.M. Scow, J. Hutson, Soil Sci. Soc. Am. J. **56**, 119–127 (1992)

Y. Shen, R. Buick, D.E. Canfield, Nature **410**, 77–81 (2001)

G.T. Sill, Icarus **53**, 10–17 (1983)

G.W. Skyring, FEMS Microbiol. Ecol. **53**, 87–94 (1988)

M.A. Slade, B.J. Butler, D.O. Muhleman, Science **258**, 635–640 (1992)

N.H. Sleep, K. Zahnle, J. Geophys. Res. **106**, 1373–1399 (2001)

N.H. Sleep, K.J. Zahnle, J.F. Kasting, H.J. Morowitz, Nature **342**, 139–142 (1989)

D. Sobolev, E.E. Roden, Antonie van Leeuwenhoek **81**, 587–597 (2002)

G. Southam, R. Donald, Earth Sci. Rev. **48**, 251–264 (1999)

G. Southam, C. Brock, R. Donald, A. Röstad, Geology **29**, 47–50 (2001)

L.J. Stashl, in: *Biostabilization of Sediments*, ed. by W.E. Krumbein, D.M. Paterson, L.J. Stahl (Biblioteks und Informationssystem der Carl von Ossietzky Universität, Oldenburg, 1994), pp. 41–54

H. Staudigel, H. Furnes, N.R. Banerjee, Y. Dilek, K. Muehlenbachs, Geol. Soc. Am. Today **16**, 4–10 (2006)

K.O. Stetter, G. Fiala, G. Huber, R. Huber, A. Segerer, FEMS Microbiol. Rev. **75**, 117–124 (1990)

T.O. Stevens, J.P. McKinley, J.K. Fredrickson, Microb. Ecol. **25**, 35–50 (1993)

A.E. Taunton, S.A. Welch, J.F. Banfield, Chem. Geol. **169**, 371–382 (2000)

T.R. Thomsen, K. Finster, N.B. Ramsing, Appl. Environ. Microbiol. **67**, 1646–1656 (2001)

P.C. Thurston, L.D. Ayres, in: *The Precambrian Earth: Tempos and Events*, ed. by P.G. Eriksson, W. Altermann, D.R. Nelson, W.U. Mueller, O. Catuneanu (Elsevier, Amsterdam, 2004), pp. 311–334

M. Tice, D.R. Lowe, Nature **431**, 549–552 (2004)

T.N. Tingle, Chem. Geol. **147**, 3–10 (1998)

J.K.W. Toporski, F. Westall, K.A. Thomas-Keprta, A. Steele, D.S. McKay, Astrobiology **2**, 1–26 (2002)

Y. Ueno, S. Maruyama, Y. Isozaki, H. Yurimoto, in: *Geochemistry and the Origin of Life*, ed. by S. Nakashsima, S. Maruyama, A. Brack, B.F. Windley (Academic, Tokyo, 2001a), pp. 203–236

Y. Ueno, Y. Isozaki, H. Yurimoto, S. Maruyama, Int. Geol. Rev. **43**, 196–212 (2001b)

Y. Ueno, H. Yurimoto, H. Yoshioka, T. Komiya, S. Maruyama, Geochim. Cosmochim. Acta **66**, 1257–1268 (2002)

Y. Ueno, H. Yoshioka, S. Maruyama, Y. Isozaki, Geochim. Cosmochim. Acta **68**, 573–589 (2004)

M.M. Urrutia, T.J. Beveridge, Appl. Environ. Microbiol. **59**, 4323–4329 (1993)

M.J. Van Kranendonk, G.E. Webb, B.S. Kamber, Geobiology **1**, 91–108 (2003)

M.J. Van Kranendonk, F. Pirajno, Geochem. Explor., Environ., Analysis **4**, 253–278 (2004)

M. van Zuilen, A. Lepland, G. Arrhenius, Nature **418**, 627–630 (2002)

M. van Zuilen, A. Lepland, J. Teranes, J. Finarelli, M. Wahlen, G. Arrhenius, Precambr. Res. **126**, 331–348 (2003)

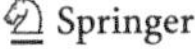 Springer

M. van Zuilen, K. Mathew, B. Wopenka, A. Lepland, K. Marti, A. Arrhenius, Geochim. Cosmochim. Acta **69**, 1241–1252 (2005)

G. Wächtershäuser, Syst. Appl. Microbiol. **10**, 207–210 (1988)

J.C. Walker, C. Klein, M. Schidlowski, J.W. Schopf, D.J. Stevenson, M.R. Walter, in: *Earth's Earliest Biosphere*, ed. by J.W. Shopf (Princeton Univ. Press, Princeton, 1983), pp. 260–290

M.M. Walsh, Precambr. Res. **54**, 271–293 (1992)

M.M. Walsh, Astrobiology **4**, 429–437 (2004)

M.M. Walsh, D.R. Lowe, Nature **314**, 530–532 (1985)

M.M. Walsh, D.R. Lowe, in: D.R. Lowe, G.R. Byerly (eds.), Geologic evolution of the Barberton greenstone belt, South Africa. Geol. Soc. Am. Spec. Pap. **329**, 115–132 (1999)

M.M. Walsh, F. Westall, in: *Fossil and Recent Biofilms*, ed. by W.E. Krumbein, T. Dornieden, M. Volkmann (Kluwer, Amsterdam, 2003), pp. 307–316

M.R. Walter, in: *Earth's Earliest Biosphere*, ed. by J.W. Schopf (Princeton Univ. Press, Princeton, 1983), pp. 187–213

M.R. Walter, R. Buick, J.S.R. Dunlop, Nature **284**, 443–445 (1980)

G. Wanger, T.C. Onstott, G. Southam, Geomicrobiol. J. **23**, 443–452 (2006)

S.A. Welch, A.E. Taunton, J.F. Banfield, Geomicrobiol. J. **19**, 343–367 (2002)

F. Westall, J. Geophys. Res., Planet. **104**, 16437–16451 (1999)

F. Westall, Comptes Rendus Palevol **2**, 485–501 (2003)

F. Westall, in: *Astrobiology: Future Perspectives*, ed. by P. Ehrenfreund, W.M. Irvine, T. Owen, L. Becker, J. Blank, J.R. Brucato, L. Colangeli, S. Derenne, A. Dutrey, D. Despois, A. Lazcano, F. Robert (Kluwer, Dordrecht, 2004), pp. 287–316

F. Westall, Science **434**, 366–367 (2005)

F. Westall, D. Gerneke, SPIE: Instrum., Methods Miss. Astrobiol. **3441**, 158–169 (1998)

F. Westall, G. Southam, AGU Geophys. Monogr. Ser. **164**, 283–304 (2006)

F. Westall, L. Boni, M.E. Guerzoni, Palaeontology **38**, 495–528 (1995)

F. Westall, A. Steele, J. Toporski, M. Walsh, C. Allen, S. Guidry, E. Gibson, D. McKay, H. Chafetz, J. Geophys. Res. Planets **105**, 24,511–24,527 (2000)

F. Westall, M.J. De Wit, J. Dann, S. Van Der Gaast, C. De Ronde, D. Gerneke, Precambr. Res. **106**, 94–112 (2001)

F. Westall, S.T. de Vries, W. Nijman, V. Rouchon, B. Orberger, V. Pearson, J. Watson, A. Verchovsky, I. Wright, J.-N. Rouzaud, D. Marchesini, S. Anne, in: W.U. Reimold, R. Gibson (eds.), Processes on the Early Earth. Geol. Soc. Am. Special Pap. **405**, 105–131 (2006a)

F. Westall, C.E.J. de Ronde, G. Southam, N. Grassineau, M. Colas, C. Cockell, H. Lammer, Philos. Trans. Roy. Soc. Lond. Ser. B **361**, 1857–1875 (2006b)

S.A. Wilde, J.W. Valley, W.H. Peck, C.M. Graham, Nature **409**, 175–178 (2001)

K.J. Zahnle, J.C.G. Walker, Rev. Geophys. Space Phys. **20**, 280–292 (1982)

M.Y. Zolotov, E.L. Shock, J. Geophys. Res. **108**, E4, 5022 (2003) doi:10.1029/2002JE001966

Space Sci Rev (2007) 129: 35–78
DOI 10.1007/s11214-007-9225-z

Emergence of a Habitable Planet

**Kevin Zahnle · Nick Arndt · Charles Cockell ·
Alex Halliday · Euan Nisbet · Franck Selsis ·
Norman H. Sleep**

Received: 16 March 2006 / Accepted: 17 January 2007 / Published online: 25 July 2007
© Springer Science+Business Media B.V. 2007

Abstract We address the first several hundred million years of Earth's history. The Moon-forming impact left Earth enveloped in a hot silicate atmosphere that cooled and condensed over $\sim$1,000 yrs. As it cooled the Earth degassed its volatiles into the atmosphere. It took another $\sim$2 Myrs for the magma ocean to freeze at the surface. The cooling rate was determined by atmospheric thermal blanketing. Tidal heating by the new Moon was a major energy source to the magma ocean. After the mantle solidified geothermal heat became climatologically insignificant, which allowed the steam atmosphere to condense, and left behind a $\sim$100 bar, $\sim$500 K CO_2 atmosphere. Thereafter cooling was governed by how quickly CO_2 was removed from the atmosphere. If subduction were efficient this could have

K. Zahnle (✉)
NASA Ames Research Center, MS 245-3, Moffett Field, CA 94035, USA
e-mail: Kevin.J.Zahnle@nasa.gov

N. Arndt
LGCA, University Joseph Fourier, 1381 rue de la Piscine, 38401 Grenoble, France

C. Cockell
Planetary and Space Sciences Research Institute, Open University, Milton Keynes, MK7 6AA, UK

A. Halliday
Department of Earth Sciences, University of Oxford, Oxford, OX1 3PR, UK

E. Nisbet
Department of Geology, Royal Holloway, University of London, Egham, Surrey, TW20 0EX, UK

F. Selsis
Ecole Normale Supérieure de Lyon, Centre de Recherche Astronomique de Lyon, 46 Allée d'Italie, 69364 Lyon, Cedex 07, France

F. Selsis
CNRS UMR 5574, Université de Lyon 1, Lyon, France

N.H. Sleep
Department of Geophysics, Stanford University, Stanford, CA 94305, USA

taken as little as 10 million years. In this case the faint young Sun suggests that a lifeless Earth should have been cold and its oceans white with ice. But if carbonate subduction were inefficient the CO_2 would have mostly stayed in the atmosphere, which would have kept the surface near $\sim$500 K for many tens of millions of years. Hydrous minerals are harder to subduct than carbonates and there is a good chance that the Hadean mantle was dry. Hadean heat flow was locally high enough to ensure that any ice cover would have been thin ($<$5 m) in places. Moreover hundreds or thousands of asteroid impacts would have been big enough to melt the ice triggering brief impact summers. We suggest that plate tectonics as it works now was inadequate to handle typical Hadean heat flows of 0.2–0.5 W/m^2. In its place we hypothesize a convecting mantle capped by a $\sim$100 km deep basaltic mush that was relatively permeable to heat flow. Recycling and distillation of hydrous basalts produced granitic rocks very early, which is consistent with preserved $>$4 Ga detrital zircons. If carbonates in oceanic crust subducted as quickly as they formed, Earth could have been habitable as early as 10–20 Myrs after the Moon-forming impact.

Keywords Hadean Earth · Moon-forming impact · Origin of Earth · Magma oceans · Planetary atmospheres · Late heavy bombardment

1 Introduction

Percival Lowell, the most influential popularizer of planetary science in America before Sagan, described in lively detail a planetology in which worlds formed hot and dried out as they aged (Lowell 1895, 1906, 1909). Large worlds cooled slowly, and were still evolutionarily young in 1895, "while in the moon we gaze upon the last sad age of decrepitude, a world almost sans air, sans sea, sans life, sans everything." One reason is that gases escape to space. "The maximum speed [a molecule] may attain Clerk–Maxwell deduced from the doctrine of chances to be seven-fold the average. What may happen to one, must eventually happen to all." Another reason presumes cooling. "As the [internal] heat dissipates, the body begins to solidify, starting with the crust. For cosmic purposes it undoubtedly still remains plastic, but cracks of relatively small size are both formed and persist. Into these the surface water seeps. With continued refrigeration the crust thickens, more cracks are opened, and more water given lodgement within, to the impoverishment of the seas." In many respects the modern story, if not the prose, broadly resembles Lowell's.

Lowell's speculations were rooted in Lord Kelvin's concepts of time. Kelvin derived the age of the Earth from the near surface thermal gradient (Kelvin 1895; Schubert et al. 2001; Wood and Halliday 2005). He made the explicit assumption that the Earth cooled by thermal conduction and the implicit assumption that the Earth harbored no unknown energy sources. He obtained an age for the Earth of 25 million years. Kelvin also computed the age of the Sun, in this case by presuming a convecting body for which gravitational contraction was the only energy source, and he obtained a similar age. These are the sort of coincidences that make for a robust theory, or at least a stubborn theorist, and Lowell was one among many to accept these arguments. In the context of Kelvin's history of brief time, monotonically cooling planets made sense: fate was ruled by the surface-to-volume ratio.

The discovery of radioactive heating triggered a relatively brief (and in retrospect ill-considered) counter-reaction in favor of a cold early Earth, in which the only primary source of heating was radioactive decay. In this story the slow internal build up of radiogenic heat eventually led to internal melting after hundreds of millions or even billions of years. A credible consequence of cold formation might be a hydrogen–methane–ammonia primary atmosphere (Urey 1951). Such an atmosphere would be conducive to the abiotic synthesis of complex organic molecules (Miller 1953). Cold formation got a foothold in textbooks, but the enormous gravitational energy released during accretion was never plausibly made to go away. Hot formation eventually returned to favor when it became more fully appreciated that accretion took the form of giant impacts (Safronov 1972; Wetherill 1985).

Of more moment to us here is that Lowell placed the origin of life in a Hadean realm of geothermal heat hidden from the Sun. Perhaps he saw no choice; 25 million years is not necessarily a lot of time. It is now known that the mantle cools by solid state convection, and that the Earth is more than 4.5 billion years old. This leaves plenty of time. Yet the suspicion remains widespread that life arose on Earth in a Hadean realm that is hidden from the rock record (Cloud 1988; Chyba 1993). The Hadean is important because it set the table for all that came later (*ibid*).

1.1 The Hadean Today

Today the Hadean is widely and enduringly pictured as a world of exuberant volcanism, exploding meteors, huge craters, infernal heat, and billowing sulfurous steams; that is, a world of fire and brimstone punctuated with blows to the head. In the background a red Moon looms gigantic in the sky. The popular image has given it a name that celebrates our mythic roots. As Kelvin and Lowell understood, a hot early Earth is an almost inevitable consequence of fast planetary growth. The apparent success of the Moon-forming impact hypothesis (Benz et al. 1986; Hartmann et al. 1986; Stevenson 1987; Canup and Righter 2000; Canup and Asphaug 2001; Canup 2004) has probably evaporated any lingering doubts. Earth as we know it emerged from a fog of silicate vapor.

1.2 Defining the Hadean

Discord confuses what "Hadean" means or should mean (Nisbet 1985, 1991, 1996). One choice has been to define the Hadean as the time before the first rock (currently the Acasta Gneiss, dated to 4.00–4.03 Ga, Bowring and Williams 1999). This puts the Hadean into the same category as the fastest mile or the tallest building. Another choice is to define it as the time before the first evidence of life. This definition was in use at one time. Before Cloud split it into the Hadean and the Archean Eons, there had been a lifeless "Azoic" Eon. "Archean" means "beginning" in the context of life (Nisbet 1982). This definition is consistent with geological convention but is open to endless debate over what constitutes evidence of life. Later, Cloud (1983, 1988) set the origin of life in the Hadean. A potentially useful definition is to synchronize the end of the Hadean with the end of the heavy bombardment of the inner solar system. This would encourage comparisons between planets. On the other hand, the end of the late bombardment is not (yet?) well defined as an instant in time, nor has it shown itself clearly in the terrestrial record. This leaves picking an arbitrary date. Cloud (1983) used 3.8 Ga, others have used 4.0 Ga. All of these definitions are in effect equal at present.

The Hadean record is not data rich. Any tale of the Hadean truly told would be so obscured with qualifications, caveats and prevarications that the reader would need a GPS system just to follow the narrative thread. We have opted instead to present a web of speculations in flat declarative sentences, constrained by basic physics when possible. This is the same point of view taken by Stevenson (1983) in an earlier essay on the topic. That our authoritative-seeming sentences often differ from Stevenson's authoritative-seeming sentences can be taken as a sign of progress.

 Springer

2 Astrophysical Context

2.1 The Interstellar Environment

Stars can form in dense clusters in which massive stars live short, brilliant lives, or they can form in quiet low-density suburbs where massive stars are rare. Massive stars dominate their environment. In general massive stars are very hot and extremely luminous and most of their light is emitted in the UV; such a star can emit 10^{10} times more UV than does our Sun. A nearby massive star can be a bigger source of ionizing UV radiation to the solar system than the Sun itself. Interstellar UV can drive photochemistry, and it can also photoevaporate the nebular disk from which the Sun and planets formed (Adams et al. 2004). Stellar UV can also drive off primary atmospheres of small planets. As massive stars hurry toward death they unleash enormous stellar winds that pollute the nebula with fresh products of stellar nucleosynthesis. The biggest stars end as supernovae. Supernovae provide the prime source of short-lived radionuclides such as ^{26}Al and ^{60}Fe. Astronomical observations of γ-rays from ^{26}Al decay imply that the current average ^{26}Al/^{27}Al ratio in the interstellar medium is 9×10^{-6} (Diehl et al. 2006). This is notably lower than the primordial solar system ratio of 5.25×10^{-5} (Bizzarro et al. 2004). The half-life of ^{26}Al is 7×10^{5} yrs. The implication is that the solar nebula was enriched with the products of a recent nearby supernova or supernovae. Evidently the Sun did not form in a quiet low-density suburb (Adams and Laughlin 2001). Nearby supernovae could have had other interesting effects on the Sun's environment. But massive stars destined for supernova last only a few million years (Arnett 1996). By the time the Sun reached the main sequence it was well entrenched in its suburban tract home. Any further speculation on these matters is beyond the scope of this essay.

2.2 The Faint Young Sun

According to the standard model, the Sun has steadily brightened since it arrived on the Main Sequence 4.52 billion years ago (the Zero-Age Main Sequence, or "ZAMS"). In the next billion years the Sun brightened from about 71% to 76% of its current luminosity. Standard solar evolution is shown in Fig. 1.

The faint young Sun imposes a stringent constraint on the climate of the young Earth (Ringwood 1961; Sagan and Mullen 1972). Without the addition of potent greenhouse gases the early Earth should have been at most times and places frozen over. This is important and will be discussed in more detail in the following.

The one way to make the young Sun brighter is to make it more massive than it is now. The Sun loses mass through the solar wind. At current rates the mass loss is tiny, amounting to only 0.01% of the Sun's mass over 4.5 Gyrs. To be as bright as it is now, the ZAMS Sun would have needed 6% more mass (Sackmann and Boothroyd 2003). This amount of mass loss far exceeds what is plausible. By studying stellar winds of a half-dozen Sun-like stars, Wood et al. (2002) found that a Sun-like star loses about 0.5% of its mass after it reaches the Main Sequence. This is too small to be important. Wood et al. (2005) have since characterized the winds of another half-dozen solar analogues. According to the newer study the total mass loss from the main sequence Sun was only $\sim$0.1% of its initial mass.

There is little evidence bounding mass loss from very young stars[1]. In 2002 Wood et al. argued that the empirical upper limit on X-ray flux implies a parallel upper limit on mass

---

[1] When stars are still accreting they generally have extremely large stellar winds, but these typically do not last more than a few million years at most, and given that the stars are accreting, the winds do not imply that the star is on net losing mass.

 Springer

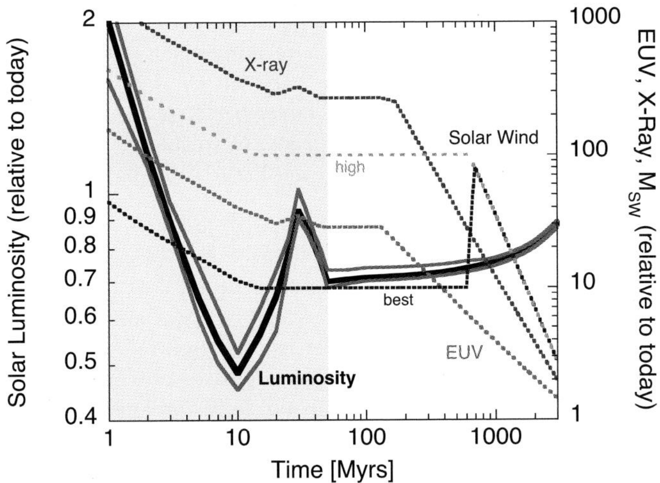

Fig. 1 The first 3 billion years of solar evolution. The solid curves show luminosity evolution. Main sequence luminosity evolution follows Sackmann et al. (1993). Pre-Main Sequence evolution (*shaded region*) is adapted from D'Antona and Mazzitelli (1994). The range of uncertainty is determined by mass loss. Preferred mass loss follows Wood et al. (2002, 2005). Sensitivity to mass loss is scaled from Sackmann and Boothroyd (2003). The upper bound on luminosity arbitrarily multiplies Wood et al.'s best estimate by a factor $4.56t^{-1}$, where t is the age of the Sun in Gyrs. Even with these relatively enormous solar winds the Sun's luminosity is barely affected. The solar wind, X-ray, and EUV evolutions (*broken curves*) follow Wood et al.'s recommendations and references therein. These latter are aspects of solar activity rather than solar luminosity—young stars are generally more active than the sedate modern Sun. The observed scatter in X-ray luminosities of young Sun-like stars (not shown) implies an order of magnitude uncertainty during the Hadean

loss rates; in 2005 they showed evidence that stellar winds may be *smaller* in stars younger than 0.7 Ga than they become later. This is rather surprising. Still, the data offer no support for a markedly more massive young Sun. The range of solar evolutions permitted by Wood et al.'s mass loss rates is shown in Fig. 1.

Often overlooked is that, irrespective of mass loss, the Sun's luminosity was far from constant in the 50 Myrs it took to contract to the main sequence[2]. Figure 1 includes a model of the Sun's pre-main-sequence evolution beginning at 1 Ma (D'Antona and Mazzitelli 1994). During the first few million years the Sun was brighter and redder than it is now. At 10 million years it was only half as bright as it is now, while at 30 million years it was almost precisely as bright as it is now. Thereafter the Sun faded to its ZAMS luminosity as the nuclear fires took over.

These time scales are comparable to the time scales currently seen as relevant to terrestrial planet accretion. Runaway growth of the first generation planets is thought to have taken no more than 1 Myr (Lissauer 1993; Chambers 2004). Planetary embryos, at first embedded in the nebula, would have emerged to see a bright red Sun. Earth and Venus were built by collisions between planetary embryos over some ~50 Myrs. As they grew the planets would have experienced major changes in solar luminosity. These changes were important because they determine the physical state of water in our part of the Solar System. As the Sun changed brightness the water condensation front would have

---

[2]If this time scale looks familiar, it is: it's Kelvin's time scale for gravitational contraction. This is the part of the Sun's evolution that predates the onset of significant nuclear fusion.

swept back and forth through the solar system. For a planet at Earth's distance from the Sun, at 2 Myr any water present would have been vapor, while at 10 Myr the water would have been ice, and ice would have stable at Venus. It is possible in Mars we are looking at a planet that is old enough to remember these times (Lunine et al. 2003; Halliday and Kleine 2006). In any event, the history of volatiles is sensitive to solar luminosity, and hence the eventual states of Venus and Earth would have been sensitive to the growth spurts of the young Sun.

2.3 The Active Young Sun

In contrast to luminosity in general, the active young Sun was a much stronger source of ultraviolet light, X-rays, and solar wind than it is today (see Fig. 1; see also the chapter by Kulikov et al. 2007, this issue). This inference is based on empirical observations of hundreds of young solar analogs. The theory is not fully developed, but in broad outline stellar activity (sunspots, flares, UV, X-rays, etc.) is directly related to the strength of the magnetic field, which in turn is generated from the star's rotation. As the star ages it loses angular momentum through the stellar wind. Solitary stars are like spinning tops. They all slow down.

3 The Age of the Earth and Solar System

There are no rocks surviving from the first 500 Myrs of Earth's history. The oldest zircon grain found thus far yields an age >150 Myrs after the start of the Solar System (Wilde et al. 2001). Therefore, deducing Earth's earliest history is strongly dependent on geochemistry, theory and comparison with other solar system objects using meteorites and returned samples. The Earth was formed through successive accretion events involving objects as large as other planets. As such the Earth has no simple "age" because it formed from combining earlier formed planetary objects which already had established their own differentiated reservoirs, including cores and atmospheres. We can determine the rate at which the Earth grew by making certain assumptions about the degree of mixing and equilibration between these planets as they coalesced. We can also define the start of the Solar System and this growth history very precisely. Chondrites are the most common form of meteorite landing on Earth. They are thought to represent early dust and debris from the circumstellar disk from which the planets grew. Most chondrites contain refractory Ca–Al-rich inclusions (CAIs) enriched in elements expected to condense at very high temperatures from a hot nebular gas. These are the oldest objects yet identified that formed in the Solar System. CAIs from the Efremovka chondrite have been dated by $^{235/238}$U–$^{207/206}$Pb at 4.5672 ± 0.0006 Ga (Amelin et al. 2002). This is the current best estimate of the start to the Solar System and hence defines a more precise slope to the meteorite isochron (called the "Geochron") first established by Patterson (1956) (Fig. 2). To sort out the growth history of planets it is necessary to use short-lived nuclides, dynamic simulations of planet formation and petrological constraints on likely core formation scenarios.

Short-lived nuclides provide a set of powerful tools for unraveling a precise chronology of the early solar system. The advantage of these is that the changes in daughter isotope can only take place over a restricted early time window; there is no correction for the effects of decay over the past 4.5 billion years. A disadvantage is that the parent isotope can no longer be measured. Hence its abundance at the start of the solar system must first be determined by comparing the isotopic composition of the daughter element in rocks and minerals of independently known age. Only then can it provide useful age constraints.

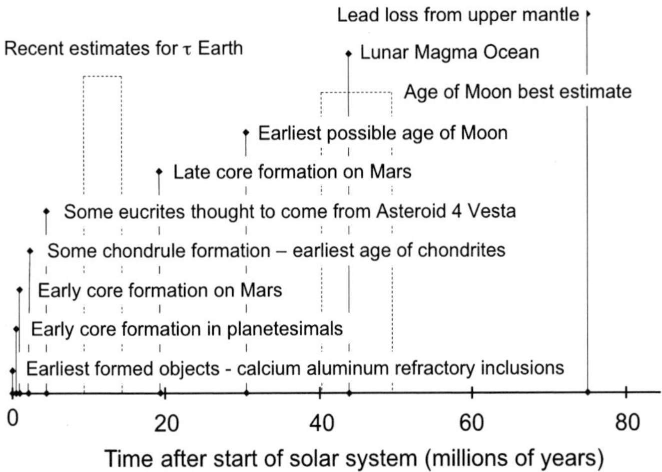

Fig. 2 The current best estimates for the time-scales over which very early inner solar system objects and the terrestrial planets formed. The approximated mean life of accretion (τ) is the time taken to achieve 63% growth at exponentially decreasing rates of growth. The *dashed lines* indicate the mean life for accretion deduced for the Earth based on W and Pb isotopes (Halliday 2003, 2004; Kleine et al. 2002; Yin et al. 2002). The earliest age of the Moon assumes separation from a reservoir with chondritic Hf/W (Kleine et al. 2002; Yin et al. 2002). The best estimates are based on the radiogenic ingrowth deduced for the interior of the Moon (Halliday 2003, 2004; Kleine et al. 2005b). See text for details of other sources. Based on a figure in Halliday and Kleine (2006)

The short-lived nuclides provide most of the information on the first 50 Myrs of the solar system. For example, as well as CAIs, most chondrites also contain chondrules, drop-shaped ultramafic objects with strange textures thought to reflect rapid heating, melting and quenching of pre-existing material in a dusty disk. Using ^{26}Al–^{26}Mg it has been shown that some of these chondrules formed as much as 1 to 3 million years after the start of the Solar System (Russell et al. 1996; Bizzarro et al. 2004) (Fig. 2). Therefore chondrites, the meteorites that contain chondrules, though primitive in composition, must have formed millions of years after the start of the solar system. This is interesting because simulations of planetary accretion indicate that dust should have accumulated into 1000 km-sized planetary embryos in just a few hundred thousand years—much less than the time indicated from chondrule formation. In fact we now have excellent isotopic evidence that a range of accretion styles were involved in the formation of the terrestrial planets. Before discussing this it is worth first explaining the theories behind the formation of Earth-like planets.

3.1 Planetary Accretion

A variety of theories have been advanced for how terrestrial planets form. For a recent review see Chambers (2004). In broad terms the rates of accretion of Earth-like planets will be affected by the amount of mass in the disk itself. If there is nebular gas present at the time of accretion the rates are faster. In fact the absence of nebular gas is also calculated to favor eccentric orbits, which gas would dampen (Agnor and Ward 2002). The presence of solar noble gases in the Earth and Mars is consistent with these requirements. In the simplest terms accretion of terrestrial planets is envisaged as taking place in four stages:

(1) Settling of circumstellar dust to the mid-plane of the disk.
(2) Growth of planetesimals up to ~1 km in size.
(3) Runaway growth of planetary embryos up to ~10^3 km in size.
(4) Oligarchic growth of larger objects through late-stage collisions.

Stage 1 takes place over time scales of thousands of years and provides a relatively dense plane of material from which the planets can grow. The second stage is the most poorly understood at present but is necessary in order to build objects that are of sufficient mass for gravity to play a major role. Planetesimals would need to be about a kilometer in size in order for the gravitationally driven stage 3 to start.

We do not know how stage 2 happens, although clearly it must. Scientists have succeeded in making fluffy aggregates from dust, but these are all less than a cm in size. How does one make something that is the size of a house or a stadium? One obvious suggestion is that some kind of glue was involved. Volatiles would not condense in the inner solar system. Not only were the pressures too low, but the temperatures were probably high because of heating as material was swept into the Sun (Boss 1990). An alternative is that, within a disk of dust and gas, collective effects can sort or gather particles into pockets of locally high density that might promote collisional coagulation or gravitational collapse (Weidenschilling and Cuzzi 1993; Cuzzi et al. 2005). Local separation and clumping of the material might also lead to larger scale gravitational instabilities, whereby an entire section of the disk has relatively high gravity and accumulates into a zone of concentrated mass (Ward 2000).

However they are formed, runaway growth builds these 1 km-sized objects into 1,000 km-sized objects. The bigger the object the larger it becomes until all of the material available within a given feeding zone or heliocentric distance is incorporated into planetary embryos. This is thought to take place within a few hundred thousand years (Kortenkamp et al. 2000). The ultimate size depends on the amount of material available. Using models for the density of the solar nebula it is possible that Mars-sized objects could originate in this fashion.

Building objects that are the size of the Earth is thought to require a more protracted history of collisions between such planetary embryos. Wetherill (1986) ran Monte Carlo simulations of terrestrial planetary growth and some runs with planets of the right size and distribution to be matches for Mercury, Venus, Earth and Mars. He monitored the time scales involved in these "successful" runs and found that most of the mass was accreted in the first 10 Myrs, but that significant accretion continued for much longer. Wetherill also tracked the provenance of material that built the terrestrial planets and showed that, in contrast to runaway growth, the feeding zone concept was flawed. The planetesimals and planetary embryos that built the Earth came from distances that extended over more than 2 AU. More recent calculations of solar system formation have yielded similar results (Canup and Agnor 2000; Raymond et al. 2004).

Such planetary collisions would have been catastrophic. The energy released is sufficient to raise the temperature of the Earth by thousands of degrees. The most widely held theory for the formation of the Moon is that there was such a catastrophic collision between a Mars-sized planet and the proto-Earth when it was approximately 90% of its current mass. The putative impactor planet, sometimes named "Theia" (the mother of Selene who was the goddess of the Moon), struck the proto-Earth with a glancing blow generating the angular momentum of the Earth–Moon system.

3.2 Tungsten Isotopic Tests for Earth Formation Models

The above models of planet formation differ with respect to timing and can therefore be evaluated using isotope geochemistry. The ^{182}Hf–^{182}W chronometer has been particularly useful

in determining the time-scales over which the planets formed. The principle of the technique is that Hf/W ratios are strongly fractionated by core formation because W normally is moderately siderophile whereas Hf is lithophile. Therefore, in the simplest of models, the W isotopic composition of a silicate reservoir such as the Earth's primitive mantle is a function of the timing of core formation. If, however, a planet grows over tens of millions of years, and if as it grows its core gets larger, as is nowadays assumed to be the case for the Earth, the W isotopic composition of the primitive mantle is a function of the rate of growth of the planet.

The silicate Earth has a ^{182}W abundance that is high relative to average solar system. This indicates that some portion of silicate Earth formed as a high Hf/W reservoir during the lifetime of ^{182}Hf. The growth of the Earth must have been protracted, otherwise the ^{182}W abundance would be much larger. How to interpret the results in terms of exact time scales depends on the models used. Because the ^{182}Hf–^{182}W chronometer has a half-life of just 8.9 Myrs, it is insensitive to changes that take place more than 60 Myrs after the start of the solar system. Therefore, the W isotopic composition of the silicate Earth has to be interpreted in terms of a relatively simple growth history of the Earth and its core that takes on board other scientific constraints. The mean life for the Earth assuming an exponentially decreasing rate of growth is 11 Myrs (Yin et al. 2002; Jakobsen 2005). This corresponds to the time taken to achieve 63% growth, which is in excellent agreement with the time scales inferred directly from the simulations of Wetherill (1986).

3.3 Comparisons with Other Objects

Although the Earth accreted over long time scales, the information from studying smaller objects is different. The most recent data for martian meteorites (Kleine et al. 2004; Foley et al. 2005) confirms earlier evidence (Lee and Halliday 1997) that accretion and core formation on Mars were fast. Some recent models (Halliday and Kleine 2006) place the time scale for formation of Mars at less than one million years (Fig. 2). If this is correct, Mars probably formed by a mechanism such as runaway growth, rather than by protracted collision-dominated oligarchic growth. In other words, Mars may represent a unique example of a large primitive planetary embryo with a totally different accretion history from that of the Earth. Lunine et al. (2003) drew a similar conclusion based on the low quantity and high D/H ratio of martian water.

A similar story is being recovered from iron meteorites. With extensive replication and better mass spectrometers very high precision can now be achieved on the W isotopic compositions of iron meteorites. The latest data for iron meteorites provide evidence that accretion and core formation were very short-lived (Kleine et al. 2005a; Markowski et al. 2006; Scherstén et al. 2006). In some cases planetesimal cores formed within 500,000 years of the start of the solar system (Fig. 2) (Markowski et al. 2006). Therefore, they too appear to represent examples of early planetary embryos, as predicted from dynamic theory.

Although there are variations, most giant impact simulations provide evidence that the Moon formed after the Earth had achieved approximately 90% of its current mass. From the W isotopic compositions of lunar samples it has been possible to determine that the Moon must have formed tens of million of years after the start of the solar system (Lee et al. 1997, 2002; Kleine et al. 2005b). All of these approaches yield similar time-scales of between about 30 and 55 Myrs after the start of the solar system providing strong support for the giant impact theory since such a late origin for an object of the size of the Moon is not readily explicable unless it formed from a previously formed planet.

 Springer

3.4 Core Formation, Accretion and the Early Earth

Mechanisms of core formation were originally based on the concept of metal segregation from a fully formed Earth via inter-granular percolation (Rushmer et al. 2000) and descending diapirs (Stevenson 1990). In a similar vein the models upon which we base our many ideas of partitioning of trace elements in a magma ocean assume a fully formed Earth undergoing metal segregation. Although these models form the backbone of thinking about the physical and chemical processes by which the Earth's core was formed, collisions between differentiated planetesimals and planets would result in core growth by core–core mixing (Yoshino et al. 2003; Halliday 2004).

These effects also will affect Hf/W chronometry. In calculating the time scales for the accretion of the Earth it is assumed that the entire W that is accreted is on average of chondritic composition. This bit is fine because planetary bodies are close to chondritic in Hf/W. However, it is also assumed that this composition mixes and isotopically equilibrates with the W in the silicate Earth. If instead a fraction of the incoming W is in metal from the impactor's core and this mixes with the metal in Earth's core, then the ^{182}Hf–^{182}W "age" of the Earth or its core will appear older than it really is (Halliday 2004).

The Pb isotopic composition of the silicate Earth, as estimated by various authors, plots to the right of the Geochron that defines the age of the solar system (Fig. 3). The standard explanation for this is that the Earth or its core formed late. Both U/Pb and Hf/W are fractionated by core formation because the parent is lithophile, whereas the daughter is siderophile or chalcophile. Therefore, if the Pb isotopic composition of the silicate Earth is modeled in the same manner as the W isotopic composition it should yield a similar result.

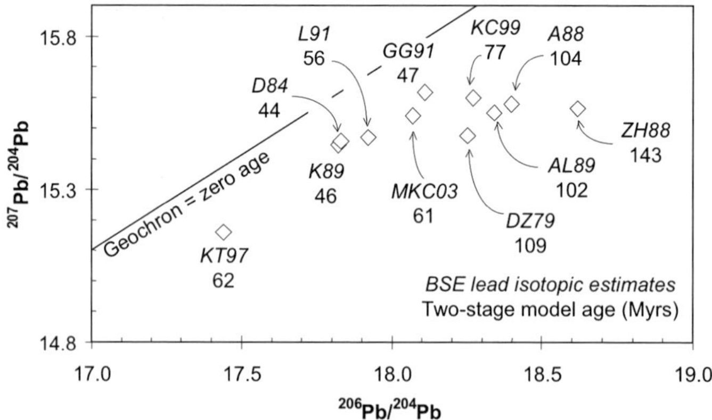

Fig. 3 Estimates of the lead isotopic composition of the bulk silicate Earth (BSE) plotted relative to the Geochron defined as the slope corresponding to the start of the solar system. All estimates plot to the right of this line. If any of these are nearly correct, they provide evidence of protracted accretion or core formation, or both. The times indicated in Myrs are the two-stage model ages of core formation assuming the same values for bulk earth parameters given by Halliday (2004) and Wood and Halliday (2005). Data from Doe and Zartman (1979), Davies (1984), Zartman and Haines (1988), Allègre et al. (1988), Allègre and Lewin (1989), Kwon et al. (1989), Liew et al. (1991), Galer and Goldstein (1991), Kramers and Tolstikhin (1997), Kamber and Collerson (1999), and Murphy et al. (2003). (Full references in Halliday 2004.) These Pb isotope estimates are all significantly longer than the ^{182}Hf–^{182}W estimate of 30 Myrs (Kleine et al. 2002; Yin et al. 2002) which may relate to the differing partitioning of W versus Pb during core segregation of metal versus sulphide (Wood and Halliday 2005)

When these two chronometers are compared a variable but uni-directional offset is found between the timing that is based on W and those based on any of the 11 estimates of the Pb isotopic composition of the silicate Earth (Fig. 3). For example, the two-stage ^{182}Hf–^{182}W model age of the Earth is 30 Myrs whereas the same model applied to $^{235/238}$U–$^{207/206}$Pb yields ages ranging between >40 and >100 Myrs (Fig. 3). It is of course possible that all the Pb isotope estimates are wrong. Given the difficulty with defining a meaningful average for the silicate Earth this is certainly a possibility. However, it is also a finding that is consistent with latest thinking on Earth's oxidation during growth. The transfer of W and Pb to the core may have changed, but not together, during the accretion history of the Earth (Wood and Halliday 2005). Tungsten is moderately siderophile but not chalcophile. The opposite is true for Pb. Sulfides would have formed following the cooling of the Earth after the Moon-forming impact. Removal of lead sulfide to the core may have been responsible for a late-stage increase in U/Pb that defines the Pb isotopic compositions observed. This being the case, the Pb isotopic composition of the bulk silicate Earth provides information on the time scales over which the Earth cooled following the Moon-forming impact. Very roughly speaking the Earth's upper mantle appears to have cooled from temperatures of about 7000 K at the time of the giant impact to about 3,000 K when sulphide would have become stable. The time scales inferred depend on which Pb isotopic estimate is deployed. However, they are all of the order of tens of millions of years after the Moon-forming impact.

3.5 The Age of Heroes

The gravitational energy released assembling Earth is roughly equal to what Earth receives from 200 million years of sunlight. But unlike sunlight, accretional energy arrived in hot lumps. During the hypothetical runaway phase the impacts must have come so frequently that the accretional pulses would have overlapped, with the individual heating events merging into a single geothermal heat flow that for climatological purposes would be like sunlight welling up from below. To this we add radioactive heating. Runaway accretion is fast enough that the short-lived radionuclides ^{26}Al and ^{60}Fe (half lives of 0.7 and 1.5 Myr, respectively) were still abundant. At their initial solar system abundances, radioactive aluminum and iron provide more heating to a Mars-size planet than does accretion. Even if runaway accretion were delayed 2 Myr, radioactive heating would still be comparable to accretional heating for a Moon-size body. Radioactive heating is most potent in small bodies that cannot cool effectively by solid state convection (Woolum and Cassen 1999). If the world is too large, radioactive heating by itself will not melt it (Stevenson 1983).

After a few million years the Sun is faint, short-lived radionuclides are no longer important, and collisions between protoplanets, although bigger than during runaway growth, are well separated in time. This is a different thermal regime. There is no way to spread accretional energy evenly over tens of millions of years. What happens instead is that most of the energy of the giant impacts is quickly radiated to space. After each giant impact the surface gets briefly very hot, but after a million years or so it cools to a point where the energy budget is again dominated by sunlight. If water is abundant, the cooling protoplanet spends some time after each impact perched in a runaway greenhouse state (Abe et al. 2000). For a planet the size of Earth struck by a planet the size of Mars the hot times can last for more than a million years (to be discussed in the following), with the duration depending on the energy of the impact and on the input from the Sun. With the background energy budget set by the (usually) faint Sun, a planet at Earth's distance should have frozen to an ice world if there weren't also an atmospheric greenhouse effect enormously stronger than that enjoyed by Earth today.

Another aspect of the giant impact phase of accretion is that more than half of the collisions are bounces rather than mergers (Agnor and Asphaug 2004; Asphaug et al. 2006). Bounces and mergers both can cause the loss of volatile materials, including geochemical volatiles. Impacts are especially prone to expel exposed surface materials; for example, a differentiated crust of an airless and oceanless planet. Thus there might be an expectation that depletions should depend on geochemical incompatibility or density. Here expectation meets mixed success. There is little evidence that refractory incompatible lithophile elements are depleted in Earth—excess ^{142}Nd might qualify (Boyet and Carlson 2005). But sorting by density between silicates and iron is obvious. In the extreme case one expects a planet like Mercury, where even the silicates have been lost relative to iron (Benz et al. 1988).

The heroic age of accretion ends with the Moon-forming impact. Although it was probably not the last big impact, it was probably the last time that Earth was hit by another planet.

4 Origin of the Atmosphere

Earth's atmosphere is often described as secondary, a choice of words that implies a history. The vanished primary atmosphere is defined as an atmosphere captured from the gases of the solar nebula, presumably by gravity. A primary atmosphere is overwhelmingly hydrogen. Other volatiles are present as hydrides (Urey 1951). Jupiter provides an example. It has been shown that a cool reduced primary atmosphere provides a good substrate for prebiotic chemistry (Miller 1953). Smaller planets like Earth could have captured significant primary atmospheres, depending on how long the nebula lasted (Hayashi et al. 1979; Mizuno et al. 1980; Sekiya et al. 1980a, 1980b, 1981; Sasaki 1990; Ikoma and Genda 2006). A primary atmosphere can be hot enough to melt the surface. Hayashi et al. (1979) showed that the surface temperature of a primary atmosphere of an Earth-mass planet would have been $\sim$4,000 K, scaling as the two-thirds power of the planet's mass. More recently Ikoma and Genda (2006), using more realistic opacities, revised the theoretical surface temperature of Earth's primary atmosphere downward to $\sim$3,000 K.

By hypothesis the secondary atmosphere was degassed from the solid Earth after the primary atmosphere was lost. In extreme form, a secondary atmosphere presumes that all Earth's volatiles were accreted in solid bodies akin to meteorites and were later degassed into a primordial vacuum after Earth heated up. In this way the idea that the oceans slowly grew over billions of years first got lodged in textbooks.

Traditional arguments for and against a primary atmosphere are based on the abundances of noble gases. Proponents of the primary atmosphere cite the isotopic composition of the noble gases, the presence of ^{3}He and isotopically solar neon inside the Earth (e.g., Harper and Jakobsen 1996), and the large amount of ^{36}Ar in the atmosphere of Venus (e.g., Genda and Abe 2005). Proponents cite isotopic evidence that massive hydrogen escape took place (Sasaki and Nakazawa 1988; Pepin 1991). If nebular gases circulated through the primary atmosphere, the nebula provided a vast reservoir of volatiles to react with the protoplanets embedded within (Lewis and Prinn 1984). A reduced protoplanet can acquire N, C, and S in refractory minerals (e.g., TiN) and so build up as something akin to the enstatite meteorites. Later, after the nebula is gone, hydrogen escape can oxidize the protoplanet.

Arguments against the retention of a significant primary atmosphere are apparent in the overall elemental pattern of the noble gases in planetary atmospheres, which resembles the

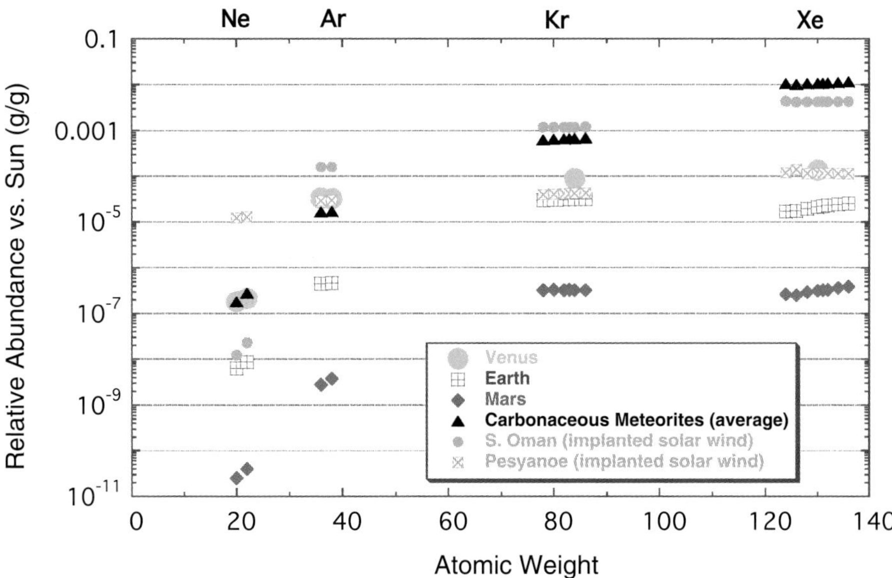

Fig. 4 The abundances of the isotopes of the noble gases (He not shown) relative to their abundances in the Sun. A purely solar abundance pattern would be a horizontal line on this plot. Apart from Xe, the noble gases on Earth and Mars resemble those in carbonaceous meteorites (Pepin 1991), although the planets have less of them. Xenon is discrepant both in quantity and isotopic pattern (which is obviously sloped even on this scale). Venus more closely resembles the implanted solar wind noble gases seen in the meteorites Pesyanoe and South Oman, although the data for Venus are poor (isotope ratios are effectively unconstrained, Kr is very uncertain, and Xe is an upper limit)

elemental pattern seen in meteorites better than it resembles the elemental pattern seen in the Sun, and the extremely low abundance of neon in the atmosphere compared to its abundance in the solar nebula (Fig. 4). First, compare Ne to nonradiogenic Ar. The ^{20}Ne/^{36}Ar ratio is >30 in the Sun but is typically ∼0.3 in planetary atmospheres. Hence, if both Ne and Ar were primary, Ne must have escaped 100 times more efficiently than Ar, and done so from Venus, Earth, Mars, and (apparently) even from the carbonaceous chondritic asteroid parent bodies (Mizuno and Wetherill 1984). Second, compare neon to nitrogen, which is the most volatile element apart from H and the noble gases. The solar N/Ne ratio is unity. In Earth's atmosphere that ratio is 86,000. Either Ne escaped 86,000 times faster than N, or the major source of N was in a condensate of some kind.

The failure of a primary atmosphere to account for neon does not mean that a secondary atmosphere degassed from the mantle into a vacuum. Most volatiles accreted to Earth in solids would have entered the atmosphere directly on impact (Jakosky and Ahrens 1979; Lange and Ahrens 1982; Ahrens et al. 1989). This would generally be the case for asteroids and meteors once collision velocities became high enough, and it would probably be the case for comets at pretty much any collision velocity. Moreover escape to space has been pervasive, and selective enough to affect every isotope differently (next section).

Cold comets provide a variant on the impact-degassed atmosphere. Cold comets are, by construction, low-temperature condensates from the solar nebula. Temperatures are presumed low enough that only H_2, He, and Ne fail to condense in significant quantities. There is evidence that such comets exist now and once existed in large numbers. The most impressive argument is Jupiter's across-the-board volatile enrichment (Owen et al. 1999). Dynam-

ical arguments suggest that such comets are unlikely to have contributed significantly to the C, N, or water inventories of the terrestrial planets, unless they did so in the presence of the solar nebula before Jupiter was fully formed, but they could have supplied the noble gases (Zahnle 1998). Presumably these frosty comets would have degassed their volatiles directly to the atmosphere on impact.

Impact degassing has a velocity threshold that is high enough that it occurs only after planet-sized bodies have formed. Impact degassing is unlikely to have been efficient during the runaway phase of protoplanet growth. On the other hand, the first-generation protoplanets grew so quickly that many of them melted. Thus we might expect general degassing to occur from the first generation of protoplanets simply because they were hot. The net effect is that even the planetary embryos probably had atmospheres and, if far enough from the Sun, hydrospheres. Because these protoplanets were small, once the nebula cleared and they became exposed to the Sun they would have suffered terrible volatile losses to space. It is possible that some of the signatures of atmospheric escape that we perceive in the atmospheres of the planets were established very early in the life of the solar system in the atmospheres of long-vanished little earths.

4.1 Escape and the Noble Gases

The active young Sun was a powerful source of ultraviolet radiation (Fig. 1). Far UV wavelengths between 100 and 200 nm are absorbed by water and CO_2 and cause these molecules to break up into atoms or into simpler, more transparent molecules such as H_2 and CO. The survivors are in general poor infrared coolants. The more energetic Extreme UV ($\lambda < 100$ nm) is strongly absorbed by everything. Absorption takes place at very high altitudes where, if the FUV and EUV fluxes are big, only poor infrared coolants remain. Some of these matters are discussed in the chapter by Kulikov et al. (2007, this issue). Without effective coolants the EUV makes the thin gas very hot, and if hydrogen is relatively abundant the gas can satisfy conservation of energy by driving a wind of hydrogen into space. This is called hydrodynamic hydrogen escape and it appears to have been an important process in sculpting the atmospheres of the terrestrial planets (Sekiya et al. 1980ba, 1981; Watson et al. 1981).

Evidence that Earth experienced vigorous hydrodynamic hydrogen escape is preserved in the mass fractionated isotopes of neon and xenon. Mantle neon is isotopically lighter than atmospheric neon; this can be readily explained by escape (Ozima and Podosek 2002). Xenon is more interesting because it is the heaviest gas found in the atmosphere. Yet atmospheric xenon is strongly mass fractionated compared to any of its plausible solar system sources (Fig. 4). In principle vigorous hydrodynamic hydrogen escape can produce the observed isotopic fractionation (Sekiya et al. 1980b; Zahnle and Kasting 1986; Hunten et al. 1987; Sasaki and Nakazawa 1988; Pepin 1991). The required hydrogen flux is high but within the range permitted by EUV emission from the active young Sun. However, the model predicts that gases lighter than Xe (i.e., all of them) should also escape. But krypton is not mass fractionated, and it is relatively more abundant than xenon. How might xenon escape leaving krypton behind? Sasaki and Nakazawa (1988) and Pepin (1991) suggested that fractionated xenon is a remnant of the lost primary atmosphere. Argon and krypton are later replenished by degassing from the planet's interior (i.e., they are secondary), but xenon in the Earth is presumed to enter the core or into high-pressure silicates. Another possibility is that xenon escaped as an ion. Xenon is the only noble gas more easily ionized than hydrogen. In a hydrogen wind Xe would be ionized but Kr would not. If hydrogen ions escaped, as would be possible along open magnetic field lines, strong Coulomb interactions would drag any other ions along. In this way Xe can be the only noble gas to escape.

Radiogenic xenon supplies two additional arguments for large-scale escape. Some 7% of atmospheric ^{129}Xe is from the decay of radioactive ^{129}I (half-life 15.7 Myr). The atmospheric inventory of radiogenic ^{129}Xe is about 0.8% of what it should be given the primordial abundance of ^{129}I in meteorites. This means that, over the course of accretion, Earth and the protoplanets, planetesimals, km-size bodies, loose boulders, grains and dust motes that built it lost 99.2% of their ^{129}Xe (Porcelli and Pepin 2000).

Another radiogenic xenon is the product of spontaneous fission of the very nearly extinct ^{244}Pu (half-life 81 Myr). Plutonium was never abundant but its abundance in primary solar system materials is known. Fission yields a spectrum of neutron-rich daughter nuclei. Five xenon isotopes—129, 131, 132, 134, 136—can be produced this way. It is difficult to separate ^{244}Pu fission products in air from nonradiogenic xenon in air, or from confusion with the similar products from spontaneous fission of ^{238}U, but fissogenic Xe is seen unambiguously in mantle samples. The interesting point is that most of the fission xenon is missing from the atmosphere (Ozima and Podosek 2002, pp. 235–241). For the atmosphere we have two model-dependent estimates that a disinterested student can regard as upper limits. Expressed in terms of ^{136}Xe (the most fissogenic xenon isotope), Pepin (1991) concluded that 4.6% of the ^{136}Xe in air is fissogenic, while Igarashi (1995) found that 2.8% is fissogenic. Both models have questionable features. Pepin (1991) made use of a hypothetical primordial xenon ("U–Xe") that is distinctly depleted in fissogenic isotopes when compared to solar wind samples, while Igarashi (1995) treated the composition of fissogenic isotopes as unconstrained by the known fission yields from ^{244}Pu or ^{238}U. This problem is currently unresolved. But even Pepin's fission xenon is only 30% of what we would expect of a thoroughly degassed mantle with chondritic Pu. It is possible that Xe was still being lost (either to space, the core, or the deep mantle) more than 200 million years after the meteorites were made.

4.2 Water

The most probable original source of Earth's water was ice, either condensed locally in the planetesimals from which the bulk of Earth was made (Abe et al. 2000), or in more distant planetesimals scattered from what is now the asteroid belt (Morbidelli et al. 2000; Raymond et al. 2004), or in comets. This is not to say that this water was in the form of ice when it reached the Earth. Rather, much of the water was in the form of hydrous silicate minerals. Chemical reactions between silicates and meltwater inside the planetesimals were the source of the hydrated silicate minerals that are abundant in many meteorites (Bunch and Chang 1980). Other possible sources of water involve oxidation of organic molecules or of primary H_2 (Abe et al. 2000). If the water came in planetesimals, either of local origin or from the asteroid belt, the water came with the building blocks of the Earth, and therefore likely predated the Moon-forming impact. Dynamical simulations (Levison and Duncan 1997; Levison et al. 2001) show that after Jupiter formed the chance that an object from the outer solar system hits Earth is $\sim 3 \times 10^{-7}$. This means that an outer solar system source of water either predated the formation of Jupiter (and therefore predated the Moon-forming impact), or that by happenstance Earth was struck by a Pluto-sized body. If there were once 10^4 Plutos (the summed masses of Uranus and Neptune), the chance that a Pluto hit Earth after Jupiter formed is only $\sim 0.3\%$. The odds favor water's co-accretion.

5 After the Moon-Forming Impact

The Moon-forming impact is presently thought to have occurred at around 30–50 Ma (Fig. 2). By coincidence the Sun reached the main sequence $\sim$50 Ma (Fig. 1). This is a good place to take up Earths' story.

Most of the mantle was melted by the Moon-forming impact, and $\sim$20% was vaporized (Stevenson 1987; Canup 2004). Strong heating occurred extensively: throughout the hemisphere that was hit, everywhere in the upper mantle where impact ejecta fell back into it, and in the deep mantle because of the energy released by merging the two planets' cores. If appreciable solid mantle survived the impact it probably sank into contact with the hot core and was melted. Canup's (2004) simulations show large parts of the mantle heated by less than 1,000 K, large parts of the mantle heated by more than 4,000 K, and substantial mantle heated to all temperatures in between. She also found that the surface is hottest ($\sim$8,000 K), with the silicates captured from Theia being especially hot. Even higher temperatures are found at the top of the core, mostly in materials that came from Theia's core. That Theian materials would be especially hot on Earth is unsurprising given their high pre-impact kinetic energies (as measured against the center of mass frame).

The Moon-forming impact may or may not have expelled a significant fraction of Earth's pre-existing volatiles, and the Earth may or may not have had abundant volatiles to lose. It is generally agreed that the volatiles on the side of Earth that got hit were lost, but it is an open question how volatiles on the other side could be lost. Theory suggests that the answer depends on whether there had been a deep liquid water ocean on the surface. It is possible that a thin atmosphere over a thick water ocean could be expelled. Otherwise the atmosphere is retained (Genda and Abe 2003, 2005). One notes that water is retained in either event. The view taken here is that the proto-Earth did have water oceans (cf. Abe et al. 2000) and that the atmosphere was incompletely expelled.

Canup's numerical simulations give no indication that significant escape from Earth occurs in the Moon-forming impact. At 8,000 K there would be some thermal escape, but radiative cooling is extremely competitive: it takes less than a day to cool 100 bars of silicate vapor at 8,000 K. On such prompt time scales escape is possible only if the gas carries within it all the energy it needs to escape. At 8,000 K this implies that a quantitatively escaping gas would have to have had a mean molecular weight of less than 3, which is plausible for a hydrogen atmosphere but unachievable otherwise. Thus a giant impact provides thermal energy sufficient to dissipate a primary atmosphere but not enough to dissipate a secondary atmosphere. Such events are too abrupt to produce isotopic fractionation. Presumably Earth's primary atmosphere was removed by an earlier giant impact long before the Moon was made.

5.1 A Fog of Rock Vapor

Once the atmosphere settled down it was mostly rock vapor topped by silicate clouds at $\sim$2,500 K. For a thousand years the silicate clouds defined the visible face of the planet. The Earth might have looked something like a small star or a fiery Jupiter wrapped in incandescent clouds. Silicates condensed and rained out at a rate of about a meter a day. Mixed into the atmosphere, at first as relatively minor constituents but becoming increasingly prominent as the silicates dropped out, were any volatiles initially in the mantle plus the air that survived the impact. Because convective cooling requires that every parcel be brought to the cloudtops to cool, it seems likely that the mantle would have degassed. This phase is labeled "rock vapor" in Fig. 5.

 Springer

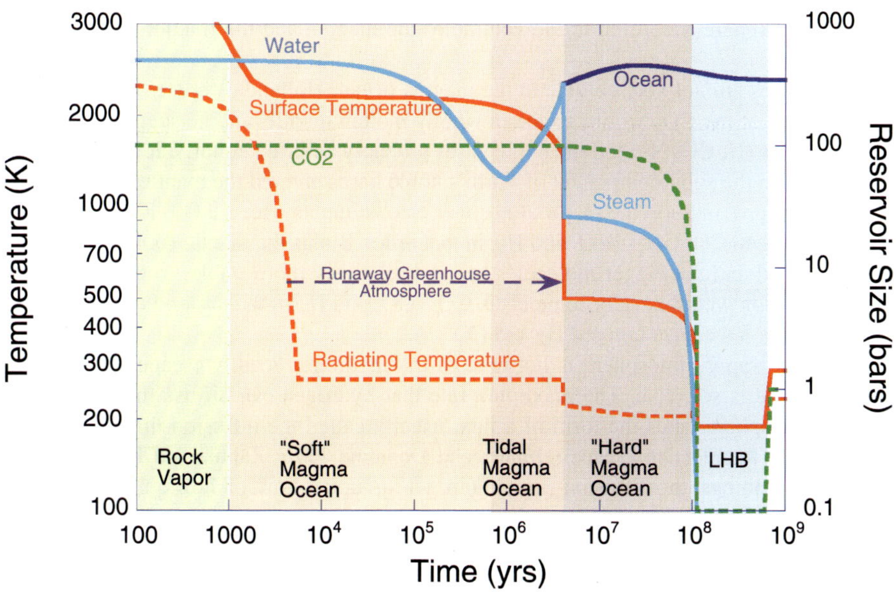

Fig. 5 A cartoon history of temperature, water, and CO_2 during the Hadean. The Hadean begins with the Moon-forming impact. For 1,000 years Earth is enveloped in rock vapor. CO_2 and other gases are presumed to degas from the convecting silicate gas, while water mostly partitions into the interior. A substantial greenhouse effect and tidal heating maintain the magma ocean for some 2 million years. Most of Earth's water and the rest of the CO_2 degassed as the mantle solidified. After the mantle solidified the steam atmosphere condensed to form a warm ($\sim$500 K) water ocean under $\sim$100 bars of CO_2. This warm, wet early Earth would have lasted while Earth's CO_2 stayed in the atmosphere. In this illustration CO_2 is assumed to subduct into the mantle on either a 10 Myr (*solid curves*) or a 100 Myr (*dotted curves*) time scale. The asymptotic CO_2 partial pressure is assumed to be controlled at low levels by chemical weathering of oceanic crust and abundant ultramafic impact ejecta. Prior to the origin of life, in the absence of an abundant potent greenhouse gas, the surface should have been ice covered and very cold, although occasional impacts brought brief thaws. Finally, after the late bombardment, the CO_2 is allowed to return to $\sim$1 bar levels in order that the surface be clement; this too is arbitrary

After the silicates condensed what remained was a hot atmosphere over a deep magma ocean. The composition of the atmosphere in detail depends on the oxygen fugacity of the silicates, the temperature, the solubilities, and the chemical inventories of the different volatile elements (Holland 1984; Abe et al. 2000). Popular buffers are QFM (quartz-fayalite-magnetite) and IW (iron-wustite). The IW buffer is about 100 times more reducing at 1,500–2,000 K (Lodders and Fegley 1998). Elementary calculations (see Holland 1984) indicate that, for an oxygen fugacity bounded by these buffers, the CO_2/CO ratio is between 0.5 and 5 at 1,500 K and between 0.2 and 2 at 2,000 K. Similarly, the H_2O/H_2 ratio would be between 30 and 300 at 1,500 K and between 1 and 10 at 2,000 K. On the other hand, Sasaki (1990) and Abe et al. (2000) both suggested that the ratio of H_2O to H_2 was between 0.1 and 0.3 for IW at 1,500 K, which leaves a hundred-fold disagreement that is hard to understand.

Water is relatively soluble in silicate melt (Holland 1984, pp. 81–82; Matsui and Abe 1986. Holland referred to Rubey 1951; evidently this is not a new idea). As the magma freezes the water is expelled (Zahnle et al. 1988). In preparing Fig. 5 we used a solubility of $x_s = 6 \times 10^{-4} \sqrt{pH_2O}$ (bars) (Zahnle et al. 1988). The square root dependence indicates that H_2O dissolves in silicate as OH^-. For Fig. 5 we partition the water between the mantle

and the atmosphere according to the volume of the magma and the fraction of the surface that is liquid.

Other gases are not very soluble in the magma ocean. Holland estimates that somewhat less than half of the CO_2 would go into a wholly molten mantle, and he suggests that other gases, such as H_2, CO, N_2, and the noble gases probably behave similarly. It is possible that this was when the solar component of Earth's noble gases entered the mantle.

To first approximation the major atmospheric constituents over a 2,000 K magma ocean would have been CO, CO_2, H_2O, and H_2, in that order. But as the mantle cooled most of the H_2O degassed and the bigger molecules became relatively more stable, so that at 1,500 K the composition would have become H_2O, CO_2, CO, and H_2 in that order. For simplicity in preparing Fig. 5 we have ignored H_2 and CO.

A silicate vapor atmosphere is hot enough that hydrogen readily escapes to space, although escape is selective. The maximum rate that hydrogen can diffuse through a background atmosphere takes the form of a flux, usually called the diffusion-limited flux, that sets the limit to selective escape (Hunten and Donahue 1976; Zahnle and Kasting 1986). Among other things, the diffusion-limited flux sets an upper limit on how quickly hydrogen can be separated from oxygen. For example, for Earth or Venus, it takes at least 20 million years to separate the hydrogen from the oxygen in an ocean of water. Over 1,000 years this is obviously a negligible amount of oxidation. Over 2 million years Earth might lose as much as 10% of its hydrogen. For our purposes this is probably negligible; in any event we still have enough. But one hesitates to say the same for Venus, which because of proximity to the Sun spends more time in the runaway greenhouse state. The lead author (for one) suspects that hydrogen escape may have doomed Venus very early. For hydrogen escape to play a significant part in oxidizing the Earth it needs to have taken place over many tens or hundreds of millions of years while Earth was more-or-less normal.

The post-silicate atmosphere may also have contained moderately volatile elements such as cadmium. The most abundant of these are S, Na, Zn, Cl, and K. These may not fully condense until after the magma ocean freezes. We might therefore expect the first crust to be enriched in these elements. Mass balance would imply an early chalcophilic crust a few km thick. Presumably the hot new ocean would interact with the crust preferentially. A salty sea seems slated from the start.

5.2 Steam Atmospheres and Magma Oceans

While the magma ocean was everywhere hotter than the liquidus convective cooling was extremely fast (Abe 1993; Solomatov 2000). A crude estimate of the thermal energy available to the magma ocean is to assume that the whole mantle was on average 800 K hotter than the liquidus (roughly the difference between the condensation temperature at the cloudtops and the melting temperature), so that it contained $\sim 4 \times 10^{30}$ J of readily accessible heat. To this can be added another $\sim 2 \times 10^{30}$ J of excess heat in the core. Together these correspond to $\sim 20\%$ of the impact energy, which may be a little high.

Tidal dissipation complicates the budget by providing an energy source that is of the same order of magnitude as the thermal energy. If the Moon formed just beyond the Roche limit (Kokubo et al. 2000), it would have formed at ~ 3 Earth radii and Earth's day would have been ~ 5 hours long. The energy dissipated inside the Earth while raising the Moon's orbit provides another 3×10^{30} J. We will return to tidal heating in Sect. 5.2.1.

The rate that Earth cooled after the silicate clouds condensed is determined by thermal blanketing by the atmosphere and by the surface temperature of the magma ocean. Figure 6, adapted from Abe et al. (2000), illustrates the effect. If the surface is cool enough and the

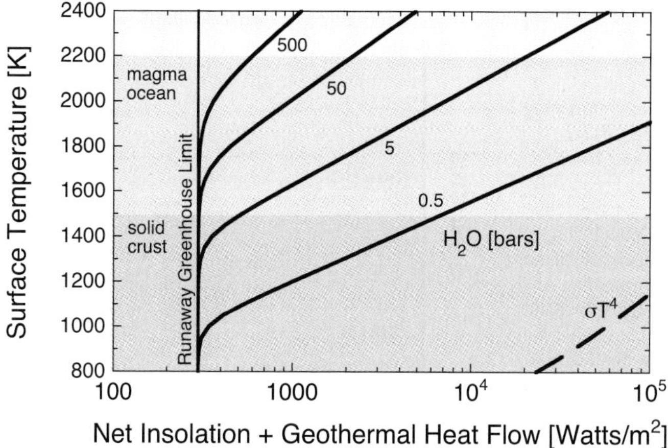

Fig. 6 Radiative cooling rates from a steam atmosphere over a magma ocean. The radiated heat is equal to the sum of absorbed sunlight (net insolation) and geothermal heat flow. The plot shows the surface temperature as a function of radiated heat for different amounts of atmospheric H_2O (adapted from Abe et al. 2000). The radiated heat is the sum of absorbed sunlight (net insolation) and geothermal heat flow. The different *curves* are labeled by the amount of H_2O in the atmosphere (in bars). The runaway greenhouse threshold is indicated. This is the maximum rate that a steam atmosphere can radiate if condensed water is present. If at least 30 bars of water are present (a tenth of an ocean), the runaway greenhouse threshold applies even over a magma ocean. Note that the radiative cooling rate is always much smaller than the σT^4 of a planet without an atmosphere

atmosphere thick enough, the thermal blanketing can be extremely effective. But as the temperature increases the atmosphere becomes more transparent to thermal radiation from the surface, and it becomes harder for water to condense. This was shown by Kasting (1988) and by Abe and Matsui (1988) and exploited in some 1980s models of terrestrial planet accretion (Abe and Matsui 1988; Zahnle et al. 1988). If water does condense, then the familiar runaway greenhouse limit applies and the rate of radiative cooling is very slow. A dry atmosphere is not subject to the runaway greenhouse limit because it has already run away.

The atmosphere stays in a runaway greenhouse state while the magma ocean is fluid and the geothermal heat flow is high enough. Given at least 20 bars of water, the surface can be held at the melting point of rock by a stabilizing negative feedback between water vapor's control over the surface temperature and water's solubility in the liquid magma (Abe and Matsui 1988; Zahnle et al. 1988). To illustrate the feedback, add water to the atmosphere. This raises the surface temperature. The fraction of the surface covered with liquid magma increases. Hence more water dissolves in the magma, and the extra water is removed from the atmosphere.

Once water condensed the atmosphere entered the runaway greenhouse state[3], in which thermal radiation is emitted to space at a fixed rate of $\sim$310 W/m^2 (Fig. 6) set by the physical and optical properties of water (Kasting 1988; Abe 1988). For 500 bars of water (which, including the mantle, approximates Earth's total inventory today), water begins to condense

---

[3]The usual runaway greenhouse refers to the response of a wet planet to too much sunlight, in which case the oceans evaporate into steam. It can be told as a cautionary tale: "Don't go too close to the Sun, or you'll end up like Venus." In detail the runaway greenhouse is best understood in terms of the thermal radiation a planet emits to space. The "runaway greenhouse limit" refers to the particular rate of thermal emission where the phase change between oceans and steam takes place.

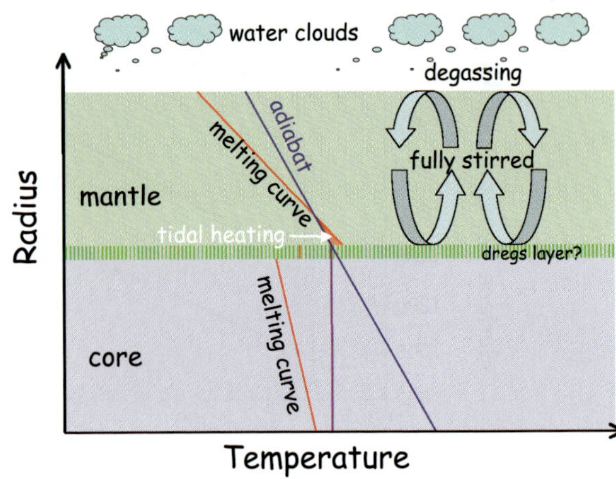

Fig. 7 A cartoon illustrating conditions inside the Earth ca. 1 million years after the Moon-forming impact. Schematic adiabats and melting curves are indicated. Tidal heating is concentrated at the base of the mantle. The mostly liquid mantle is probably fully stirred and equilibrated with the atmosphere. Temperature in the mantle follows the adiabat. At this early time the core is liquid but probably not convective. It is more likely that is heated from above by conduction. Therefore there is no reason to expect a substantial magnetic field. Slag layers may form at the top and bottom of the mantle

when the surface temperature drops below 1,800 K (Fig. 6). Absorbed sunlight provides 120–170 W/m^2, depending on the albedo. The difference, 140–190 W/m^2, is made up by the geothermal heat flow associated with secular cooling[4]. If water is not the most abundant gas, the runaway greenhouse threshold can rise to $\sim$400 W/m^2 (Nakajima et al. 1992), for which the geothermal heat flow would be 230–280 W/m^2. If there is less than 20 bars of water, water does not condense until after the magma ocean has frozen, but this is not a plausible state for Earth. For specificity we will use 140 W/m^2 (equivalent to a 30% albedo) as the amount of geothermal heat flow required to maintain the runaway state.

5.2.1 Tidal Heating

The mantle freezes from the bottom up because the melting curve is steeper than the adiabat. This is illustrated by Fig. 7. Consequently tidal heating is concentrated at the bottom of the mantle. Fast-growth of Rayleigh–Taylor instabilities guarantees that the mantle's temperature profile obeys the adiabat (Solomatov 2000).

Viscous damping of tidal motions generates heat. Therefore tidal heating occurs most strongly in materials that are solid but close to melting. This introduces the possibility of a governing feedback that works through the dependence of viscosity on temperature. If tidal dissipation exceeds what the atmosphere can radiate, the excess heat raises the temperature, which lowers the viscosity, which in turn lowers the rate of tidal dissipation. This looks like a stable feedback.

It follows that, while tidal dissipation was important, the base of the mantle was solid but the rest of it was fluid, and tidal heating generated almost all of the thermal energy radiated to space. In the limit of an asymptotically thick steam atmosphere, tidal dissipation would have been regulated to generating heat at the runaway greenhouse limit of $\sim$140 W/m^2.

The runaway greenhouse limit is a good approximation as the magma ocean cooled, but it is not as good an approximation when the magma ocean was much hotter than the liquidus and most of the water was dissolved in the melt. In preparing the mantle cooling history

---

[4]If the geothermal heat flow is too low to support the runaway greenhouse, rain falls, oceans accumulate, and the surface of the mantle would be cold. A cold surface contradicts the assumption of a liquid magma ocean.

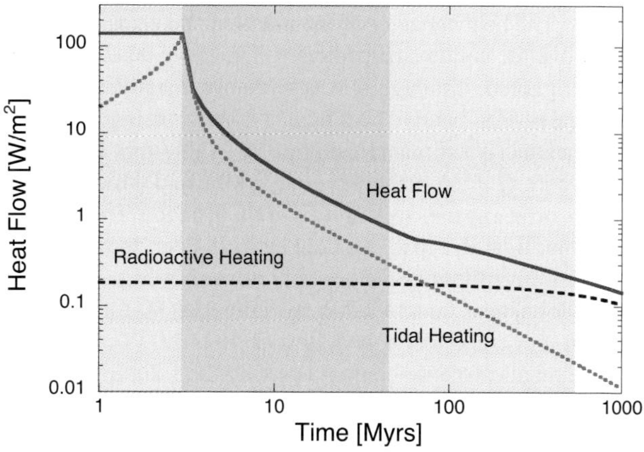

Fig. 8 A cartoon history of mean heat flow during the first billion years after the Moon-forming impact. Epochs correspond to those in Fig. 5. Tidal heating plays an important role in prolonging the magma ocean. Tidal forcing wanes as the Moon evolves away from the Earth. Thereafter heat flow is controlled by convection of the solid mantle. By 4.4 Ga the global average heat flow would have been ~0.5 W/m^2. Later in the Hadean typical heat flows would have been 0.2–0.3 W/m^2, not enormously larger than what they are now. For comparison heat flow today is 0.065 W/m^2 through the continents and 0.1 W/m^2 through the ocean crust. Computational assumptions are given in the text

shown in Fig. 5 we have self-consistently used (1) the cooling rates from Fig. 6; (2) water's solubility in liquid basalt; (3) assumed that water is confined to the molten fraction of the mantle; (4) assumed a total Earth inventory of 500 bars of water; and (5) assumed a heat capacity of 1.2×10^7 ergs/g/K and a heat of fusion of 4×10^9 ergs/g for the melt. Tidal heating was arbitrarily concentrated toward the beginning of the magma ocean.

In this model, after ~1.4 million years the Moon will have evolved far enough away from the Earth that tidal dissipation[5] drops below the ~140 W/m^2 threshold. Thereafter secular cooling takes over, and a freezing front rises through the mantle until it reaches the surface. This transition marks the end of the liquid magma ocean (Fig. 5). As the mantle freezes the solubility feedback tries to keep the surface molten by degassing water. In some models most of Earth's water is degassed in a terminal event like this (Zahnle et al. 1988). But once the mantle solidified heat flow fell under rheological control (Solomatov 2000), dropping to (unknown) levels well below those required to sustain a runaway greenhouse atmosphere (Fig. 8). Thereafter water condensed and rained out at ~1 meter per year until the oceans returned[6].

In preparing Fig. 5 we assume that heat flow decays inversely as the one-third power of viscosity after the collapse of the runaway greenhouse. This is like conventional parameterized convection in the limit that the only temperature dependence to matter is viscosity's. Viscosity varies by at least 15 orders of magnitude over a small temperature range when near the solidus (Liebske et al. 2005). We arbitrarily assume that viscosity is an exponential function of temperature, with e-folding scales of 43 K below the solidus and 3.3 K above the

---

[5]For conservative values of the quality factor Q. Q is the ratio of power in the tides to dissipation. Low Q implies lossy tides.

[6]The net rainfall rate of ~1 m/yr refers to the global mean difference between rainfall and evaporation. It is not obvious what a rain gauge would report. Ishiwatari et al. (2002) computed average rainfall rates exceeding 10 m/yr in a GCM study of a runaway greenhouse atmosphere, but their simulated climates are far out of balance, and feature cold poles and a lot of hot, dry air. A passively cooling steam atmosphere may be too bland to demand heavy rain.

solidus. The heat capacity of the mantle is taken as 6×10^{27} J/K and 1.4×10^{28} J/K below and above the solidus, respectively. The latter takes into account latent heat of fusion. The core is ignored, although it was probably a significant heat source to the mantle on these time scales. Radioactive heating is 0.2 W/m^2 using conventional K, Th, and U abundances. Tidal heating is estimated using a "Q" value that is at first 10-fold more dissipative than Io's (where $Q = 36$, Schubert et al. 2001), and which linearly increases to Io's value as the Moon's orbit expands to half its current distance. These assumptions take their cue from our sense that tidal dissipation ought to have been most efficient before the mantle hardened. An illustrative thermal history is presented in Fig. 8. The mantle reaches the solidus at 20 Myr. Global heat flow is 0.5 W/m^2 after 100 Myr and drops to 0.2 W/m^2 by the end of the Hadean. For comparison, heat flow on Io is 2.0 W/m^2 (Spencer et al. 2000).

The mantle volatile content is set by the solubility (in the melt) or the stability (of minerals in a solid) at the surface, at least while the mantle is strongly convective. The surface acts like a cold trap, and while the mantle remains strongly convective, every parcel visits the surface. What this means for water is that, if hydrous minerals are unstable at the surface, even the deep mantle dries out, irrespective of the stability of water-bearing minerals at high pressure. What this means for gases that are sparingly soluble in the melt is that they degas as the mantle freezes. Hence the gases that were left behind in the mantle represent a small sample of the atmosphere at the time of the last major magma ocean.

The argument that the mantle should rid itself of water and other volatiles as it froze implicitly assumes whole mantle convection. If instead the young mantle convected in separate layers such that the deep mantle was isolated, the greater stability of hydrous phases at high pressure may be relevant. This opens the possibility that substantial amounts of water, initially incorporated as solute, could be stored indefinitely in the lower mantle. Later, when layered convection broke down, the stored water would emerge and be degassed. In this way it is possible for the oceanic volume to grow over time. Such a model must make a host of other predictions. Conventional layered convection, in which the lower mantle remains convectively isolated over the whole history of the planet, renders the composition of the lower mantle irrelevant to the history of volatiles at the surface.

5.2.2 History of the Lunar Orbit

The history of the lunar orbit has been lucidly discussed many times (Goldreich 1966; Touma and Wisdom 1994, 1998). Figure 9 is not a substitute for these studies. Rather it merely shows the distance to the Moon and the length of the day during the Hadean for the tidal heating history shown in Fig. 8. Figure 9 assumes a circular orbit and conservation of angular momentum of the Earth–Moon system.

The lunar orbit is inclined by $\sim 5°$ to the ecliptic. Integrations of the lunar orbit backward in time indicate that the 5° inclination to the ecliptic today maps directly into a $\sim 10°$ inclination to Earth's equator when the Moon was near Earth (Goldreich 1966; Touma and Wisdom 1994). An inclined birth-orbit is deeply puzzling, because the giant impact origin of the Moon generates the Moon from a debris disk that revolves in the Earth's equator, and the disorderly precession of inclined orbits causes collisions that ultimately drive all the debris into orderly equatorial orbits.

Touma and Wisdom (1998) suggested that the Moon acquired its inclination via two resonances that occur early in the evolution of the lunar orbit. The first of these occurs when lunar perigee precesses with a period of one year (the resonance is between perigee and perihelion). This resonance pumps up eccentricity. The second resonance is between the year and the combined precessions of perigee and the Moon's inclination with respect to Earth's equator. This resonance converts eccentricity into inclination.

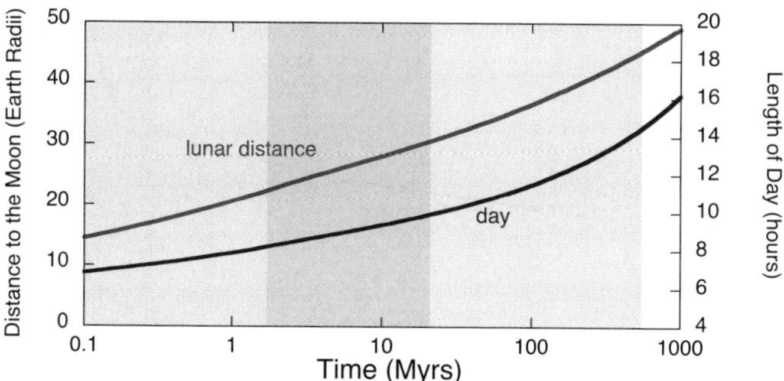

Fig. 9 The distance to the Moon and the length of Earth's day in the Hadean after the Moon-forming impact. Here the rate of evolution is at first controlled by the runaway greenhouse effect. This is 100–1,000 times slower than tidal evolution would be in the absence of an atmosphere (cf. Touma and Wisdom 1998)

To capture the lunar orbit in these resonances requires that the lunar orbit evolve at least 2 orders of magnitude slower than conventional models predict. Thermal blanketing by a steam atmosphere can do this. At the runaway greenhouse limit, tidal evolution is slowed by 3 orders of magnitude compared to conventional models. Thermal blanketing also unbalances the rates of tidal dissipation in the Earth and Moon, which meets another of Touma and Wisdom's (1994) requirements. Thus the history of the lunar orbit both sets a lower limit on the duration of Earth's magma ocean and suggests the presence of a significant amount of water on Earth at the time of the Moon-forming impact.

5.3 Hot Water

The state of the atmosphere after the deluge depends on how much CO_2 was available. If most of Earth's modern CO_2 inventory were in the atmosphere as CO_2, the surface temperature would have been $\sim$500 K (Fig. 10). Presumably carbonate rock formed quickly, but the capacity of the oceans and ocean crust to store carbonate is limited, and the bulk of the CO_2 remained in the atmosphere until the carbonates were subducted into the mantle or unless or until there were stable continental platforms on which to put it (Sleep et al. 2001).

Today most subducted carbonate enters the mantle rather than erupting through arc volcanoes. Was subduction more efficient in a hot mantle? Higher temperatures made carbonates less stable, but the lower viscosity let foundering crustal blocks sink more quickly. If they sank quickly enough they would have taken their carbonates to the bottom of the mantle, thereby scavenging the atmosphere of its CO_2.

We follow the discussion of Sleep et al. (2001). It is unlikely that the seafloor itself at any one time could have held more than $\sim$10 bars of CO_2 as carbonates. This estimate comes from assuming that the seafloor is ultramafic and hydrothermally altered to a depth of 500 m, and that most of the available cations were used to make carbonates. This is a small fraction of the $>$100 bar planetary inventory. At some stage—after 20 million years in the cartoon—global heat flow waned to 1 W/m^2. This crust would have resembled 1 Ma ocean crust on Earth today. It was a significant CO_2 sink. Carbonate now forms within the uppermost hundreds of meters of young oceanic crust. If there was no CO_2 degassing from foundered crust, the global resurfacing time of 1 Ma implies that it could have taken as little as 10 Ma to remove 100 bars of CO_2 from the atmosphere and inject it into the mantle. How long it actually took to remove the CO_2 depended on how efficiently carbonate was subducted. For specificity in preparing Fig. 5 we assume this takes either 10 Myr or 100 Myr, but we cannot

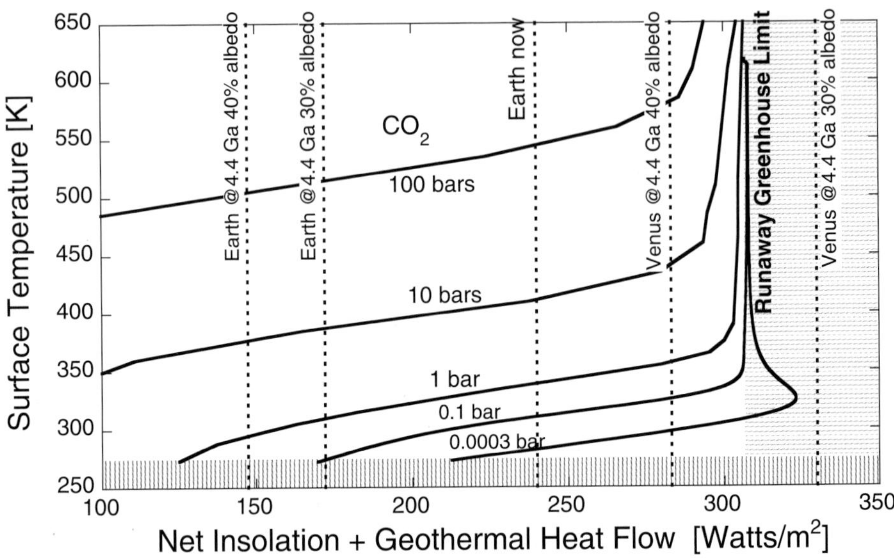

Fig. 10 The H_2O–CO_2 greenhouse. The plot shows the surface temperature as a function of radiated heat for different amounts of atmospheric CO_2 (after Abe 1993). The albedo is the fraction of sunlight that is not absorbed (the appropriate albedo to use is the Bond albedo, which refers to all sunlight visible and invisible). Modern Earth has an albedo of 30%. Net insolations for Earth and Venus ca. 4.5 Ga (after the Sun reached the main sequence) are shown at 30% and 40% albedo. Earth entered the runaway greenhouse state only ephemerally after big impacts that generated big pulses of geothermal heat. For example, after the Moon-forming impact the atmosphere would have been in a runaway greenhouse state for $\sim$2 million years, during which the heat flow would have made up the difference between net insolation and the runaway greenhouse limit. A plausible trajectory takes Earth from $\sim$100 bars of CO_2 and 40% albedo down to 0.1–1 bar and 30% albedo, at which point the oceans ice over and albedo jumps. Note that CO_2 does not by itself cause a runaway. Also note that Venus would enter the runaway state when its albedo dropped below 35%

guarantee that it even happens at all. However, we note that there is no obvious buffer on CO_2 levels other than those producing hot ($\sim$500 K) and cool or cold ($\sim$270 K) climates (Sleep et al. 2001).

By contrast hydrated minerals are not very stable at high temperatures and low pressures. If they survive a fast passage to the mantle they may not have stayed there. They ought to have formed water-rich melts at the base of the magma ocean that ascended as proto-granitic plumes. We therefore expect that water was mostly partitioned into surface reservoirs during the magma ocean, and that the early oceans were if anything deeper than the oceans later became.

6 Hadean Geography and Geodynamics

Earth today has a global mean heat flow of 0.086 W/m², which much exceeds heating by radioactive decay (0.040 W/m²). The mismatch is even bigger if the continents aren't included in the accounting. The mismatch shows us that mantle convection is not well described by textbook boundary layer theory, which predicts that heating and cooling are nearly equal. To fix this with plate tectonics requires taking the strength of the plates into account (Sleep 2007). When this is done, the model of plate tectonics that gives the observed behavior predicts that heat flow is nearly independent of the mantle's temperature. This has several

interesting consequences, including possible non-monotonic thermal histories[7]. Plate tectonics of this kind cannot handle a heat flow much greater than 0.1–0.2 W/m^2 (Sleep 2007). Something else is needed between the end of the magma ocean and the advent of plate tectonics.

Jupiter's volcanically active moon Io provides an interesting albeit imperfect analogy. Io is dry, its surface gravity is only 1.8 m/s^2, and its heat is generated by tides. But the heat flow is big. Io's global mean heat flow is observed to be 2.0 W/m^2 (Spencer et al. 2000). This is higher than what we expect for the Hadean Earth after the first 10 Ma. Io cools itself by lava flows, layer upon layer, each thin and each well cooled. Cooling by flood volcanism is very efficient (Sleep and Langan 1981). The old cold flows sink slowly back into the crust. This leaves a thick cold crust atop a hot mushy ocean (Keszthelyi et al. 1999). Thermal emission indicates that Io's lava flows are very hot when erupted, $>1,600$ K. The high temperature suggests an ultramafic composition (McEwen et al. 1998), which implies that Io's mantle is not highly differentiated (Keszthelyi et al. 1999). In particular Io does not seem to have generated a compositionally distinctive crust of low-melting-point magmas. Cooling by lava flows explains this as well, because differentiation occurs only on the scale of the individual lava flows, which are thin (Sleep and Langan 1981).

We might expect similar behavior on Earth for a few tens of millions of years after the Moon-forming impact when heat flows were still high and tidal heating was still important. A heat flow of 1 W/m^2 equates to a resurfacing rate of $\sim$1 cm/yr (i.e., 100 km in 10 Myr). On Earth the ultramafic lavas would react with abundant water and abundant CO_2 to make hydrous minerals and quite a lot of H_2, and possibly CH_4. When recycled the hydrous ultramafics would make bonanites, and these upon dehydration become dense enough to sink into the mantle, thereby erasing much of the evidence. In any event we do not expect that Earth spent a long time in such a state, assuming that it did enter such a state, because heat flows >1 W/m^2 cannot be long sustained given that radioactive heating was only $\sim$0.2 W/m^2 and the Moon was receding.

We speculate that the missing link between Io-like hyperactive volcanism and modern plate tectonics was a basaltic mush ocean about 100 km thick (Fig. 11). Modern fast ridge axes provide an analogy. Magma entering the axis freezes quickly (Sinton and Detrick 1992). The bulk of the "magma" chamber is mostly crystalline mush at the basalt solidus; it is almost melt-free and when fresh flows like a glacier. Only a thin (tens of meters) lens of fully molten rock exists at the axis. The heat flow at the ridge axis can be estimated from the dimensions of the magma chamber, the latent heat of fusion, and the spreading rate (Sleep et al. 2001). The fastest spreading ridge on Earth today corresponds to a heat flow of 40 W/m^2. Such rapid cooling was unsustainable globally for more than a few million years. When the magma ocean froze to mush, it was only a few 100 K hotter than the modern mantle.

The 100 km thickness of the basalt layer is self-regulated by the stability field of garnet, resulting in a phase change that makes pressurized basalt more dense than mantle. The bottom of a thicker pile would spontaneously sink into the mantle. The hot basaltic mush is topped by a colder frozen basaltic crust. In a thick mush ocean the material is solid by the time it sinks to the base of the ocean. Some of it gets heated by the hot mantle underneath and melts a little. It is buoyant and rises. This is solid state convection with a "solid-fluid" that is somewhat less viscous than the mantle but a lot more viscous than melt. Blocks of the base of the solid crust founder as at ridge axes. In this regime the basalt is fluid enough

---

[7]If heat flow in plate tectonics is constant, the mantle reached a peak temperature at 2.7 Ga.

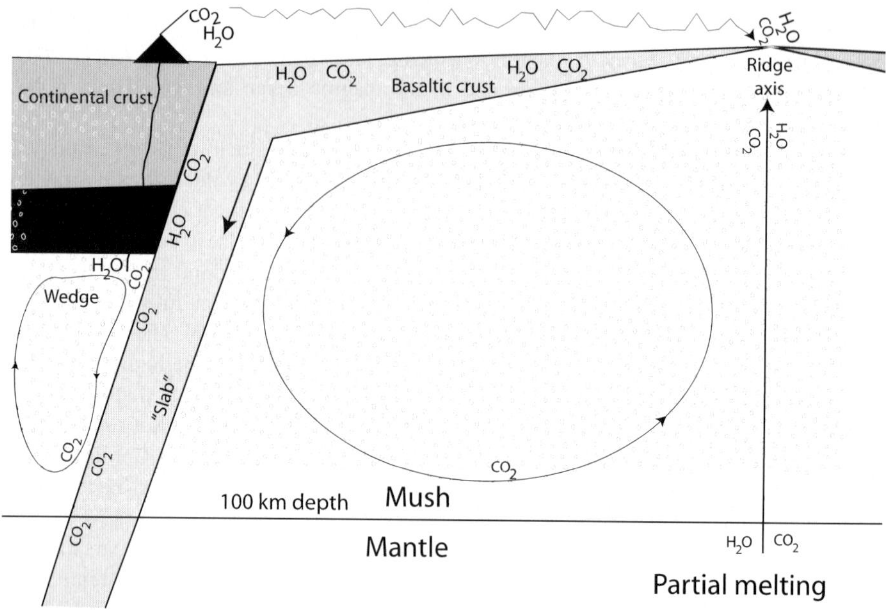

Fig. 11 Cartoon of crustal cycling in the Hadean. A basalt mush ocean sits atop the solid mantle. Heat flow is determined by convection in the solid mantle. The less viscous basalt does not substantially impede heat flow. The basalt ocean spreads and subducts much like the modern oceanic lithosphere. The basalt ocean is fed by partial melting of the underlying mantle and drained by subsidence into the mantle, the latter enabled by the phase change to garnet that occurs at a depth of $\sim$100 km (after Sleep 2007)

that the bottleneck on heat flow is set by solid state convection of the underlying mantle. The viscosity dependence of the upper mantle provides the thermostat. The basalt is fed by partial melting of the mantle and is drained by subduction of solid sinking slabs. The overall behavior of the basalt—upwelling at ridges and sinking of slabs—resembles plate tectonics, especially at the surface, but global heat flow does not depend upon subduction.

Impact churning of the seafloor suggests that Hadean basalts would be more deeply and thoroughly hydrated than their modern analogs. Upon sinking the basalts would be heated and the resulting hydrous melts would erupt as granitic rocks. These would be much more voluminous than now both because the extent of hydration was greater and because the speed of the recycling was greater. Kamber et al. (2005) emphasized that early "granites" would carry such a high abundance of radioactive elements that they would melt themselves if thick or deeply buried, which implies that early granites would tend to segregate themselves to the surface.

Neither heat flow nor the interior temperature need have decreased monotonically with time (Sleep 2000; Stevenson 2003; van Thienen et al. 2007, this issue). The surface heat flow as a function of mantle temperature may be multivalued, with a high heat flow from a mush ocean and relatively low heat flow for plate tectonics. It is possible that the lowest heat flow achievable with a mush ocean exceeds the highest heat flow achievable in plate tectonics, in which case the system either overcooled or alternated between regimes. Relatively brief episodes of mush ocean might have been key periods of rapid continent generation.

 Springer

6.1 Zircons

The chief source of terrestrial data for the Hadean are ancient detrital zircons found in Archean and Proterozoic quartzites in the Jack Hills of Western Australia (Amelin et al. 2000; Wilde et al. 2001; Mojzsis et al. 2001; Valley et al. 2002; Cavosie et al. 2004; Cavosie et al. 2004; Harrison et al. 2005; Valley et al. 2005). Zircons are $ZrSiO_4$ crystals that are renowned for their durability. Zircon crystals readily incorporate uranium and hafnium. U–Pb dating gives accurate ages that can be as old as 4.4 Ga. The old zircons resemble those from modern granites (sensu latu) (Cavosie et al. 2004 and references therein). In particular, some of the granites apparently formed at the expense of sediments derived from meteoric weathering. Oxygen isotopes in ancient zircons provide compelling evidence that rocks on Earth were being chemically altered by liquid water before 4.2 Ga and probably before 4.3 Ga (Wilde et al. 2001; Mojzsis et al. 2001; Valley et al. 2002; Cavosie et al. 2005). The zircons are silent on whether the water was 273 K or 500 K, but they suggest that Earth's oceans were in place by 4.2 Ga.

Radiogenic Hf in 4.01 to 4.37 Ga zircons suggests that Lu, a more incompatible element than Hf and therefore quicker to segregate into a granitic crust, was already separating from Hf at 4.5 Ga (Amelin et al. 2000; Harrison et al. 2005). This result is somewhat controversial because interpreting the hafnium depends on the age of the zircon, but old zircons typically have many ages, and the hafnium is harvested from volumes that are large enough to sample several different ages. Nevertheless the least radiogenic hafnium demands separation before 4.4 Ga. Overall, the Lu–Hf data suggest that continents of a sort were already a significant presence on Earth's surface within a hundred million years of the Moon-forming impact. Segregation may even imply subaerial weathering in order to separate Lu from Hf in the sediments from which zircons were forged.

The existence of old zircons implies that there were places near the surface where zircons could be protected from subduction for hundreds of millions of years. Cavosie et al. (2004) presented a constraint on the vigor of crustal recycling processes after 4.4 Ga. Age gaps and clusters exist within their sample suite, just like with a modern orogenic belt. If further sampling confirms this finding, terrane-scale regions experienced tens of million year periods of quiescence. The age gaps between 4.4 and 3.8 Ga are 50–100 m.y., compared with 500–1000 m.y. for the subsequent history, including modern zircon suites. Taken at face value, this implies that crustal recycling rates were $\sim$10 times what they are now. The corresponding heat flow was $\sim$3 times the present, or 0.2 W/m^2. This cooling rate is consistent with those shown in Fig. 8. The lithosphere would have been 40 km thick, enough like the modern Earth to show plate-like characteristics. In any case, tectonics after 4.4 Ga were sluggish enough that some Hadean continental crust survived to at least the end of the Archean.

As a counterbalance to the natural tendency to over-interpret the zircons, we should not forget that until now ancient detrital zircons have been found in only one place on Earth. There is no guarantee that this one occurrence is representative, or that ancient continents were more than a local anomaly.

It is sometimes suggested that the impacts of the late bombardment are the reason why there are so few Hadean rocks, but if so the mechanism was probably indirect. First, both the Moon and Mars retain ancient surfaces despite impacts. Second, the expected (i.e., most probable) total energy released by late Hadean impacts on Earth would have been on the order of 10^{29} J, an order of magnitude less than the total geothermal energy over the same interval. Basic thermodynamics suggest that the geothermal forces did more work (Stevenson 1983). Cavosie et al. (2004) found no evidence of impacts in their sample of detrital zircons going back to $\sim$4.4 Ga. This includes zircons with metamorphic rims, which might

show annealed shock features. On the other hand, impact statistics are described by power laws in which much of the energy is concentrated in the single largest event. If this event was itself extreme (we are deep in the realm of small number statistics here), then we could imagine the whole rock record erased by a single impact, with an estimated likelihood on the order of 10% (such an impact would vaporize the oceans and melt much of the exposed crust). The greater likelihood is that impacts played a major role in preparing early continental material for subduction, but it was Earth itself that erased its history.

6.2 Continental Crust

The debate about the timing of crustal growth rumbles quietly on. In the 1970s and 1980s Armstrong (1981) challenged the widely held belief that continental crust has grown continuously and that little crust existed before about 3.8 Ga. He suggested that the crust had reached its present volume by the end of the Hadean and that recycling has counterbalanced growth ever since. Since then some geochemists continue to argue for crustal growth (e.g. Coltice et al. 2000), but several factors have tilted the balance Armstrong's way. Widespread discoveries of ultrahigh pressure metamorphism (Coleman and Wang 2005) and seismic-tectonic evidence of subduction of lower continental crust in the Himalayas have provided a credible mechanism of crustal recycling (von Huene and Scholl 1991; Jahn et al. 1999; Elburg et al. 2004). Relicts of very old continental crust continue to be found in all Archean cratons. The outcrops of 4 Ga Acasta gneiss total at most a few square kilometers and represent only a minute fraction of the original Acasta continent: the rest was lost by recycling or reworking. And the Mt. Narryer zircons record the existence of Hadean continents from which no rock remains; yet the survival of these zircons for a billion years at the surface of the unstable Hadean Earth requires the persistence of a stable platform which can only have been continental lithosphere.

6.3 Ocean Volume and Depth

Many Archean volcanic rocks appear to have erupted under water onto a continental substrate. At Kambalda in Australia, for example, pillow basalts (which indicates underwater eruption) contain old zircons and geochemical signatures that record assimilation of old continental crust; and the Belingwe belt in Zimbabwe is made up of a shallow- to deep-water sequence of volcanic and sedimentary rocks that unconformably overlies an older granitic basement. It seems that at the time these rocks formed during the late Archean, oceans flooded much of the continental crust.

The simplest explanation is that the volume of the oceans was greater. Earlier we described how hydrogen is lost to space decreasing the Earth's water content. The total amount of water was greater in the Hadean than at present. But the volume of the oceans depends also on the partitioning of the water between the surface and the mantle and this depends on mantle temperatures. At present water and other volatiles migrate to the surface in partial melts and are returned to the mantle in subducting plates. Much of the altered oceanic crust that contains the water dehydrates at shallow depth. Most of the water escapes to the surface in subduction-related magmas and only a small fraction penetrates deeper. The dehydration reactions are temperature dependant and the plate is stripped of water shallower and more efficiently when the mantle is hotter. If the mantle cooled progressively though Earth history, the ocean volume should have progressively declined; while if the mantle hit a maximum temperature in the mid to late Archean (Sleep 2000), the proportion of water at the surface should also have been greatest during the mid to late Archean.

7 Hadean Atmosphere and Climate

Hades can also suggest icy wastes trapped in perpetual winter. It is not as certain that Earth was at times bitterly cold as it is certain that Earth was once infernally hot, but the argument that a lifeless young Earth should have been very cold when not very hot is good. The key point is that the young Sun was much fainter than it is now (Fig. 1). If a snowball Earth seems plausible in the Neoproterozoic, when the Sun was 96% as bright as it is now, it should seem more plausible when the Sun was just 71% as bright as it is now. A warm Hadean Earth needs either enormous geothermal heat flow or abundant greenhouse gases. As discussed earlier, geothermal heat was comparable to insolation during accretion, and at times much bigger, but its role was confined to aftermaths of big collisions. Geothermal heat was probably climatologically insignificant after 4.5 Ga. Of greenhouse gases the only good candidates are CO_2 and CH_4[8]. Methane can help provided that there are reducing agents and catalysts to generate it from CO_2 and H_2O. On Earth today methane is mostly made by biology. Methane is a good candidate for keeping Earth warm once it teemed with life, but it is not clear what the catalysts for making it would be when Earth was lifeless (Shock et al. 2000).

That leaves CO_2. It takes about a bar of CO_2 in the atmosphere to provide enough greenhouse warming to stabilize liquid water at the surface (Fig. 10). Although this is only about 0.5% of Earth's carbon inventory, it is 3,000 times more than is there today. CO_2 would have been scoured from the Hadean oceans by chemical reactions with abundant ultramafic volcanics and impact ejecta to make carbonate rocks (Koster van Groos 1988; Zahnle and Sleep 2002). The relatively fast time scales governing the early Hadean CO_2 cycle suggest that oceanic CO_2 was controlled by a fast crust–mantle cycle. The first question is whether carbonates were subducted into the mantle. If they were, the CO_2 atmosphere would have been thin and the surface very cold. If not, and if there were no continents on which to store carbonate rocks, the CO_2 would have remained in the atmosphere (Sleep et al. 2001) and the surface would have approached 500 K (Fig. 10). Sleep et al. (2001) did not identify a mechanism to sustain CO_2 at the intermediate $\sim$1 bar level needed to maintain a clement climate, which is not to claim that such a mechanism does not exist.

7.1 Ice

Ice thickness on a snowball Earth is a matter of active debate (McKay 2000; Warren et al. 2002; Pierrehumbert 2004; McKay 2004; Pollard and Kasting 2005). Thick ice solutions occur when the ice is white and opaque, full of bubbles and flaws. In a thick ice solution, ice thickness is determined by thermal conduction of geothermal heat. At a Hadean heat flow of 0.3 W/m^2, the ice would be 300 m thick. Thin ice solutions occur near the equator if the ice is transparent (black ice). In a thin ice solution, ice thickness is determined by thermal conduction of sunlight transmitted through the ice. Thin ice is predicted to be 1–5 m thick and easily broken up by winds. Using standard algorithms for sea ice, Pollard and Kasting estimated that a thin ice solution would typically present 2% open water. This is more than enough to permit efficient gas exchange between the atmosphere and ocean (Sleep and Zahnle 2001). Available evidence suggests that sea ice freezes as black ice if it freezes slowly (McKay 2004; Pollard and Kasting 2005).

---

[8]Water vapor is in fact the most important greenhouse gas, but it is a dependent variable, mobilized from oceans and ice sheets as needed.

The chief data-based argument favoring thick ice stems from a particular theory for how the cap carbonates formed. The NeoProterozoic snowball Earth events end with several-meter-thick carbonate beds of anomalous isotopic composition. The cap carbonates can be explained as volcanic CO_2 that had accumulated in the atmosphere to levels great enough to melt the ice (Pierrehumbert 2004). The chief data-based argument that the ice was thin is that the evolutionary record of life on Earth seems unaffected by the Neoproterozoic snowball events (Knoll 2003), which is surprising given that an ice shell thick enough to segregate oceanic and atmospheric CO_2 reservoirs would also be thick enough to extinguish most photosynthetic organisms[9].

In the Hadean, the fainter Sun favors thick ice, while Earth's faster rotation and generally higher heat flows favor thin ice. For an early Hadean heat flow of 1 W/m^2, the thick ice solution would be only 100 m thick. Even when the average heat flow was 0.3 W/m^2, heat flow would not have been the same everywhere, and we might reasonably expect substantial areas with heat flows in the range of 1 to 10 W/m^2, thinning the thick ice to 10–100 m. Io provides some guidance here. Much of Io's heat flow is concentrated in a few hot spots. The biggest volcano generates $\sim 1.2 \times 10^{13}$ W (Spencer et al. 2000). This one volcano accounts for 15% of the global heat flow. The heat flow around it is equivalent to 10 W/m^2 over an area 1,000 km across. A heat flow of 10 W/m^2 implies that even the thick ice solution would be only 10 m thick. Thin ice solutions, which depend on diffusion of sunlight through the ice, are independent of heat flow once established, but obviously could be triggered by high local heat flow.

As the global heat flow declined the prospects for thick ice improved. One can imagine a thermostatic negative feedback, roughly analogous to the stabilizing negative feedback we posited for water over an ocean of magma, in this case operating between atmospheric greenhouse gases and the surface areas covered by thin and thick ice (Pollard and Kasting 2005). In this feedback volcanic CO_2 would build up to the point where there are enough regions of thin ice near the equator to permit adequate ocean-air gas exchange. The ocean crust would provide the sink on CO_2.

7.2 Doubts

Our cold early Earth is a product of pure reason. We start from the best established datum—the faint Sun—and conclude that the Hadean ought to have been cold provided that the seafloor consumed CO_2. But in fact the temperature of the Archean, and by extrapolation of the Hadean, is a topic of vigorous debate. The major sources of data are inferred weathering rates and oxygen isotopes. Proponents of a cool Archean cite evidence that Archean weathering rates were not markedly different from today's (Holland 1984; Condie et al. 2001; Sleep and Hessler 2006), while proponents of a hot Archean make similar arguments for their side (Schwartzman 2002). Weathering rates depend on many things including temperature; there does not yet appear to be anything approaching a consensus.

The oxygen isotopes are more directly interpreted as a thermometer. Presumptively ocean-deposited carbonates and silicates show a progressive increase in oxygen isotopic fractionation over the history of the Earth. This can be interpreted as a change of seawater composition (traced to changing temperatures of rock-water interactions), as a product of diagenesis (the older samples are on average more cooked), or as a direct measure of

---

[9]Plausible refugia in hot springs or in endolithic communities on islands or continents make complete extinction unlikely (Kirschvink 1992).

decreasing seawater temperatures (so that the precipitates become increasingly more fractionated with time). If the latter, the mid-Archean oceans would have been $\sim$340 K (Knauth and Lowe 2003). Such high temperatures require a potent greenhouse effect. Three bars of CO_2 would be needed (less if supplemented by a lot of CH_4).

This sort of atmosphere is most likely if (1) the seafloor sink was insignificant (so that continental weathering was the main CO_2 sink); (2) emergent continents were few (to shrink the sink); and (3) biology has always been an important catalyst of chemical reactions between CO_2 and continental rocks (Schwartzman 2002). By presumption the catalytic weathering powers of biology have improved as life evolved. In this picture the history of atmospheric CO_2 presents an inverted image of the history of biological evolution (*ibid*). These arguments imply that the Hadean would also have been hot if the Hadean seafloor was not a significant sink of CO_2.

8 The Late Bombardment

A major scientific result of the Apollo program is that the Moon was hit by several 100-km-size asteroids and by hundreds of 10-km-size asteroids ca. 3.9 Ga (Wilhelms 1987). Earth was hit at the same time, and because Earth's effective cross section is 20 times bigger than the Moon's, Earth was hit 20 times as often. Not only was Earth hit by a hundred 100-km asteroids (or comets), it was also hit by a dozen bodies bigger than any to hit the Moon. Here we speak of probabilities in lieu of direct evidence, but the biggest asteroid likely to hit the Earth ca. 3.9 Ga would have been comparable to Vesta or Pallas; that is, as big as any asteroid now in the asteroid belt. Whether these impacts were the tail end of a sustained bombardment dating back to the accretion of the planets or whether they record a catastrophe associated with a rearrangement of the architecture of the solar system (e.g. Gomes et al. 2005) is contentious but obviously of some importance to the Hadean environment.

8.1 The Lunar Record

Debate over the lunar cratering record has tended to emphasize two extreme views. Hartmann et al. (2000) and Ryder et al. (2000) provided recent reviews written from the perspectives of active participants in this debate, while Chyba (1991) and Bogard (1995) provided careful reviews written from more centrist perspectives. Different conceptions of the lunar cratering record are illustrated in Fig. 12.

At one end, Hartmann (1975) and Wilhelms (1987) presumed that the observed basins and craters mark the end of a monotonically declining impact flux that extrapolates smoothly back in time to the origin of the Moon. Such enormous hidden impact fluxes present serious challenges. There isn't enough contamination of the lunar crust by extreme siderophiles (elements such as iridium that partition strongly into the core) that would have been abundant in the impacting bodies (Sleep et al. 1989; Chyba 1991; Ryder 2003), and the ancient anorthositic crust has not been obliterated, as would be the case if the Moon were saturated with Imbrium-sized basins (Baldwin 1987a, 1987b). Mars too shows little evidence of a hidden history of catastrophic cryptic craters. The preservation of a 4.5 Ga martian rock—the famous martian meteorite ALH84001, which was found in Antarctica—is inconsistent with the ancient Martian surface being pulverized into oblivion. And how did the differentiated asteroid Vesta survive with its basaltic crust intact? Moreover, it is difficult to point to any otherwise inexplicable data that might be explained by extremely high impact fluxes predating the observable crater record.

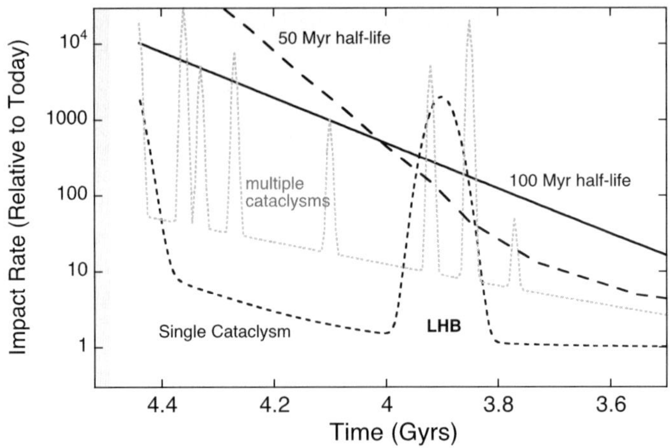

Fig. 12 Four concepts of the late lunar bombardment. The "100 Myr half-life" is Neukum's standard lunar crater curve for times after 3.86 Ga. It is calibrated to crater counts and surface ages from Apollo landing sites and the Imbrium impact basin (Neukum and Ivanov 1994; Neukum et al. 2001). The steeper "50 Myr half-life" extrapolation uses the same data but it also uses the young 3.92 Ga age for the Nectaris impact basin and crater densities on still older but undated surfaces (after Wilhelms 1987). The "single cataclysm" is a schematic but quantitatively representative late cataclysm as advocated by Graham Ryder (2002, 2003). "Multiple cataclysms" scatters several cataclysms over the Hadean (Tera et al. 1974). Available data do not favor the more aggressive 50 Myr half-life before 4.0 Ga. The terrestrial and Vestan impact records favor the higher standard impact rates ca. 3.2–3.5 Ga

The other extreme view is that the late lunar bombardment was an actual event confined to a relatively short period of time. This hypothetical event was named the "late lunar cataclysm" by Tera et al. (1974). The late lunar cataclysm remains an important and testable hypothesis. In its most extreme form, Ryder (2002, 2003) limited the total lunar impact record to just those craters and basins that survive today, and he stuffed them all into a 50–100 Myr window ca. 3.9 Ga. In Ryder's view, impacts may have been as infrequent before 3.95 Ga as they are now.

Both extreme points of view are represented on Fig. 12. Although a cartoon, the figure is quantitatively faithful. The cataclysm is shown either as a single event (Ryder) or as several events (Tera et al. 1974). Also shown on the Fig. 12 is Neukum's "standard" lunar cratering history, which is in wide use in discussions of inner solar system chronometry (Neukum and Ivanov 1994).

There is some evidence that sides with the cataclysm. Tera et al. (1974) gave several arguments, the most durable of which is lead-based. Taylor (1993, p. 172) gave an updated version with shortened error bars. The argument is that a ^{207}Pb–^{206}Pb plot for lunar highland breccias gives a straight line—a mixing line between end-members at ∼4.46 Ga and ∼3.86 Ga—that is distinct from the curved concordia that one would see if all intermediate ages were represented. Tera et al. interpreted this as evidence for a lead mobilization event ca. 3.9 Ga, which they attributed to impact shock metamorphism.

Lunar basins Imbrium (3.85), Serenitatis (3.89), Crisium (3.91), and Nectaris (3.92) have been independently dated (see Ryder et al. 2000) and, if these dates are correct, clearly cluster around 3.9 Ga. Crater densities superposed on the ejecta blankets from these basins imply that ca. 3.85 Ga the impact flux decayed with a half-life of ∼50 Myr. The four big nearside basins also provide an obvious source of ∼3.9 Ga impact metamorphosed nearside highlands rocks. Whether farside lunar breccias share this common age is unknown.

 Springer

Histograms of Apollo data suggest an impact spike but don't demand it (Wetherill 1975; Bogard 1995). An alternative to a cataclysm is to argue that the Imbrium impact reset all the radiometric clocks (Haskins 1998). Another alternative chronology dates Imbrium at 3.77 Ga (Stadermann et al. 1991), a blasphemy that Ryder rejected without comment.

Cohen et al. (2000) avoided Apollo sampling biases by using lunar meteorites recovered in Antarctica. They found a scattering of dates, none older than about 3.9 Ga. They argued that the distribution of ages supports a cataclysm at 3.9 Ga. However, using Antarctic meteorites introduces a different bias, one that favors strong competent target surfaces (Melosh 1989; Warren et al. 1989; Gladman 1997). All but one of the meteorites from Mars were ejected from young basalts—clearly a biased sample. Lunar meteorites are much smaller and no more numerous than martian meteorites. The straightforward story is that it is difficult to eject rocks from the Moon into space by impact because strong rocks are rare near the surface, and it is more difficult to launch old rocks because old strong rocks are even more rare.

8.2 Other Sources of Data

The asteroid Vesta is a $\sim$500 km diameter fully differentiated world (core, mantle, basaltic crust) that resembles a miniature terrestrial planet. It is located near the inner edge of the asteroid belt and is a significant source of meteorites. These meteorites carry a record of strong shock events that correspond to the late bombardment of Vesta. These shock ages are spread between 3.4 and 4.1 Ga (Bogard 1995). The record at Vesta seems inconsistent with an inner-solar-system-wide late cataclysm.

The Archean Earth also retains evidence that big impacts were still relatively frequent as late as 3.2 Ga (Lowe and Byerly 1986; Lowe et al. 1989; Byerly et al. 2002; Kyte et al. 2003). The evidence takes the form of spherule beds, which from their siderophile element abundances were almost certainly impact generated. There were at least four of these events between 3.2 and 3.5 Ga. The extant spherule beds are all much thicker than those left behind by the K/T impact. To the extent that one can extrapolate worldwide catastrophes from a small number of samples, all four appear to have been bigger than the K/T event. Extrapolation suggests that the thickest spherule bed corresponds to an impact on Earth 50–300 times bigger than the K/T (Kyte et al. 2003), making it as big as the Imbrium impact on the Moon. Such an impact is big enough to boil off 40 meters of ocean water. To our knowledge there is no evidence of such a catastrophe, although the geologic record of the time is scanty and controversy surrounds even the most basic issues.

8.3 Theorists Prefer Cataclysms

Öpik–Arnold simulations of orbital evolution suggested that the characteristic time scale for sweep up of stray debris in the inner solar system was some $\sim$100–200 Ma years (Wetherill 1975). This result implied that a monotonic decline in the impact rate was easy to explain, but cataclysm required something special. Wetherill (1975) showed that collisional disruption of a Vesta-sized asteroid is a 10^{11} year event in the current solar system, which makes a collisional source of the late bombardment a $\sim$1% chance event. Wetherill suggested that tidal disruption of a Vesta-sized body passing Earth is more likely.

Öpik–Arnold simulations use Monte Carlo techniques to perform otherwise impractical computations. When more powerful computers made direct numerical integrations possible, it was found that Öpik–Arnold methods greatly underestimate the chance that a stray body hits the Sun. The longest-lived inner solar system reservoir has only a $\sim$40 Myr half-life

(Morbidelli et al. 2002). This is too short to explain a monotonic impact history. A suitable longer-lived source of more distant asteroids in unstable orbits has yet to be identified, although there may be reasonable candidates in the outer asteroid belt or beyond that haven't been fully studied.

There are a lot of dynamical theories that can explain a spike in the impact rate. In general these posit a rearrangement of the architecture of the solar system taking place ca. 3.95 Ga. A good recent example is a series of papers by Gomes et al. (2005) that posit Saturn and Jupiter evolving through the 2 : 1 resonance[10]. Another story posits Uranus and Neptune forming between Saturn and Jupiter, with both being scattered out to their present locations (Thommes et al. 2002). A third possibility is that a small planet in the asteroid belt was ejected by Jupiter after a suitable interval, producing a rain of asteroids (Chambers and Lissauer 2002).

In the theories the rearrangement itself is a brief event, and precisely when it occurs is up to the discretion of the modelers, although a time scale of hundreds of millions of years is physically plausible. The key consequence of rearranging the planets for the late bombardment is that Jupiter moves, and when Jupiter moves, the resonances that disturb the asteroid belt move with it. The net effect of moving major resonances into a previously stable part of the asteroid belt is to unleash an asteroid shower onto the inner solar system (Levison et al. 2001). The natural time scale of the asteroid shower is some 20–100 million years (Levison et al. 2001). The magnitude of the asteroid shower depends on the mass of the primordial asteroid belt near important resonances, but is plausibly that of the lunar late bombardment.

It is also probable that moving the planets triggers a comet shower into the inner solar system (Levison et al. 2001). The comet shower is brief, lasting some 10–20 Myr (Levison et al. 2001). In these theories Uranus and Neptune come to rest by landing in a thick primordial belt of planetesimals. The magnitude of the comet shower is estimated by multiplying the presumed mass of the belt (30 Earth masses) by the probability that a stray body in the vicinity of Uranus or Neptune will hit the Moon ($\sim 10^{-8}$, Levison et al. 2001). The predicted comet flux of 2×10^{18} kg is about a tenth of the inferred mass of the lunar LHB (Chyba 1991; Ryder 2002; Zahnle and Sleep 2006), which is close enough to be interesting, but two orders of magnitude short of what it is needed to give Earth its oceans.

On balance, we prefer cataclysms over monotonic decay. To our minds the most telling argument against a huge unseen Hadean impact flux is that it doesn't explain anything else in the solar system that needs explaining; it's a kind of dead end. By contrast, a cataclysm (or cataclysms) fits in well with current concepts of how a solar system might evolve. Mechanisms devised to generate a cataclysm have suggested explanations for other unexplained solar system properties. Background impact rates before a cataclysm would have been much higher than they were afterward, because there were vastly more stray bodies in the solar system before the cataclysm cleared them away. An issue is that the impact rate after 3.9 Ga decays more slowly than theory predicts. In our opinion this failure is better attributed to incompleteness in the theory than to a fundamental flaw.

---

[10]The idea is that Saturn was formed closer to the Sun and evolved outward. When Saturn's year was twice Jupiter's year (2 : 1), the resonance between their orbits caused havoc that launched showers of asteroids and comets into the inner solar system while quickly driving Saturn further from the Sun. The theory explains several traits of the solar system.

8.4 Transient Environmental Effects of Impacts

Today, the total number of prokaryotes on Earth is estimated to be 4–6 $\times 10^{30}$ cells (Whitman et al. 1998). The nature of the first micro-organisms on Earth and how they survived the early conditions on Earth is necessarily a matter of some conjecture, but asteroid and comet impacts must have played an important role in the emergence of life and its prebiotic precursors.

The evidence of heavy bombardment in the Hadean suggests that impact events played a role in defining the physical characteristics of the early Earth, and thus the physiological traits of organisms that would have been required to colonize ecological niches periodically subjected to these events. The emergence of life, or at least the isotopic evidence for its presence, occurs soon after the period of late bombardment (e.g., Schidlowski 1988), suggesting that prebiotic or biological processes were ongoing during bombardment and radiated rapidly as the impact flux declined.

Impactors larger than 500 km could have boiled the entire ocean. Based on the statistical properties of lunar basin-forming impactors, we expect some zero to four impacts big enough to evaporate the oceans and heat the surface to the melting point (Fig. 13) between the time of the formation of the Earth and $\sim$3.8 Ga ago. In such events life evolving in the oceans may have been extirpated (Sleep et al. 1989; Zahnle and Sleep 1997). These events may have favored life in the deep regions of the Earth below the oceans. Russell and Arndt suggested that prebiotic evolution in hydrothermal vents would have created acetogenic precursors to life, which may have migrated into the ocean floor. The presence of these precursors in the ocean floor would have allowed for the evolution of life in environments protected from impacts (Russell and Arndt 2005).

The occasional boiling of the oceans provides a compelling explanation for the hyperthermophilic root of the phylogenetic "tree of life" (Pace et al. 1986; Lake 1988; Maher and Stevenson 1988; Sleep et al. 1989). Of course, the hyperthermophilic root of life does not suggest that life originated in hot conditions, nor does the first organism need to have been a hyperthermophile; the tree of life merely suggests a bottleneck resulted in the survival of hyperthermophiles that led to the diversity of life on Earth today. Impact events, by periodically boiling the oceans and providing a globally distributed source of heat, may have caused this bottleneck. But if life did originate and evolve to something like its current complexity in hydrothermal systems at the bottoms of oceans, life should be widespread in our solar system, independently evolving wherever one finds hydrothermal systems charged with simple C- and N-containing molecules. Several icy moons meet or have met these criteria; many of the larger asteroids, comets, and Kuiper Belt Objects have met them too. Hydrothermal origin of life is therefore a testable hypothesis.

It is not clear that, even if early impactors had extirpated the entire biosphere by boiling away the oceans and heating the early crust, life would have been completely reset. Impacts can lift surface rocks into orbit essentially unshocked and unheated (Melosh 1989). Going into orbit might be viable strategy (Sleep and Zahnle 1998; Mileikowsky et al. 2000; Wells et al. 2003). Modeling results suggests that life could have been launched in rocks into space, to return to the Earth several thousand years later and reseed the planet (Wells et al. 2003). Wells's data suggest that with an initial cell population of 10^3–10^5 cells/kg, at least one cell in this material would return after 3,000–5,000 years following a sterilizing impact. Qualitatively similar conclusions have recently been obtained by Gladman et al. (2005) who showed that 1% of the impact ejected material might eventually return to Earth.

Recent experiments on the shock survival of the microorganisms, *Bacillus subtilis* and *Rhodococcus erythropolis* suggest that they can survive high shock pressures (up to 78 GPa)

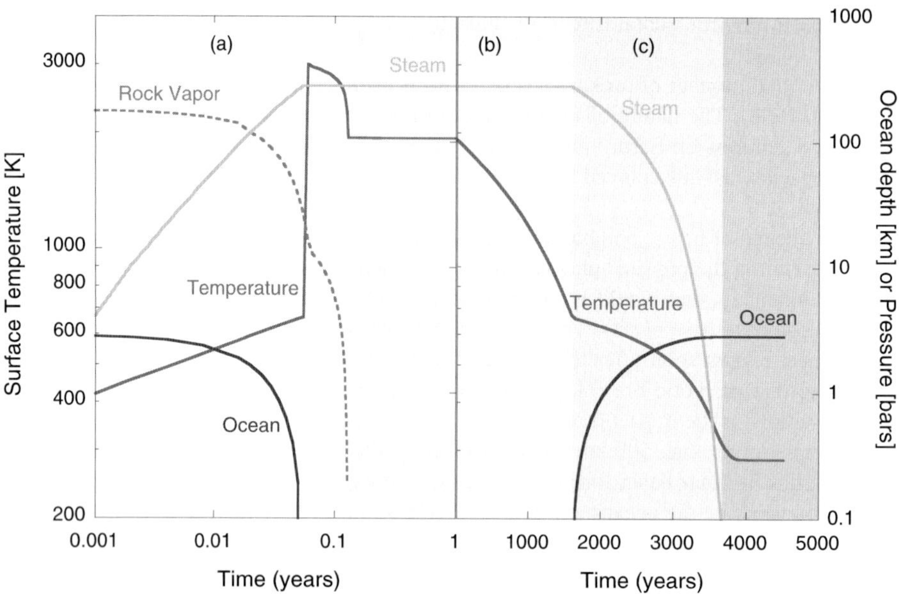

Fig. 13 History of an ocean vaporizing impact. (*a*) The impact produces 100 bars of rock vapor. Somewhat more than half the energy initially present in the rock vapor is spent boiling water off the surface of the ocean, the rest is radiated to space at an effective temperature of $\sim$2300 K. (*b*) Once the rock vapor has condensed the steam cools and forms clouds. Thereafter cool cloudtops ensure that Earth cools no faster than the runaway greenhouse threshold, with an effective radiating temperature of 270 K. (*c*) The steam atmosphere becomes cool enough for rain to reach the surface. Some examples of what happens after smaller impacts on Earth are shown in the chapter by Nisbet et al. (2007, this issue).

associated with launch from a planetary surface (Burchell et al. 2004). The low temperatures inferred from the mineralogy of the interior of meteorites from Mars (Weiss et al. 2000) also suggest that meteorites can re-enter the Earth's atmosphere while maintaining internal temperatures low enough to preserve viable organisms. Depending on how long they remain in space and the degree to which their DNA is damaged by radiation exposure, it seems that escape from Hadean impacts and later return to Earth is plausible.

After about 3.8 Ga, impact events would have had effects confined to boiling of the surface layers of the oceans. The boiling of the surface layers has been suggested to have delayed the evolution of photosynthesis since a sufficient flux of light for this mode of metabolism requires living in the top $\sim$200 m of the oceans. Maher and Stevenson (1988) also pointed out that episodic darkness caused by the injection of dust and rock into the atmosphere would have been inhibitory to photosynthesis by blocking a source of light. These events need not have prevented the establishment of photosynthesis, however.

Smaller impacts are less potent but there would have been hundreds capable of melting oceanic ice sheets. Each big impact triggers a brief impact summer. Impacts big enough to melt the ice occurred on a $\sim$1 Myr time scale (Fig. 14).

The other potentially life-changing aspect of late impacts is their power to drive chemistry (Kasting 1989). An obvious agent of change would have been the big iron asteroid. These exist; indeed they were probably rather plentiful. Chemical reactions of H_2O and CO_2 with iron can generate considerable amounts of reduced gases, H_2 and CH_4 in particular. Methane is both a greenhouse gas and a desirable starting point for prebiotic chemistry.

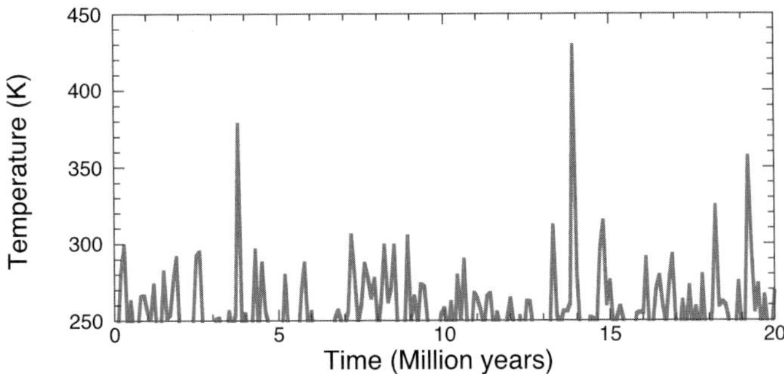

Fig. 14 An exemplary Monte Carlo slice of the Hadean during the LHB. Big impacts occasionally melt the ice and create temporarily warm or even hot conditions. The figure shows the warmest events occurring over a random 20 million year interval; each point plotted is the warmest event in a 100,000 year interval

The aftermath of a big iron impact seems an especially auspicious time to try to generate life.

Hydrogen and methane are also generated by serpentinization of ultramafic impact ejecta. We estimated that ca. 3.85 Ga impact ejecta were mobilizing ferrous iron at an inconstant rate on the order of 10^{13} moles/yr (Zahnle and Sleep 2002). The generic serpentinization reaction is $3FeO + H_2O \rightarrow Fe_3O_4 + H_2$, so that at 3.85 Ga the H_2 source might have been on the order of 3×10^{12} moles/yr (about an order of magnitude larger than the current source). In the aftermath of big impacts the H_2 source would have been at least an order of magnitude larger still and, especially when coupled with a similarly enhanced CH_4 source, would have had profound influence on atmospheric chemistry.

9 Conclusion

In this essay we first set the stage for the Earth, and then we track Earth through the Hadean. The general point of view is to bracket the Hadean Earth between boundary conditions. The Moon-forming impact provides the initial condition, while a smooth transition from the Hadean into the early and mid-Archean almost a billion years later provides the end point.

The embryos of the terrestrial planets formed quickly, perhaps less than 1 million years after the origin of the solar system. Fast formation implies that they captured primary nebular atmospheres. Evidence of primary atmospheres is preserved in the noble gases. The colder embryos also held indigenous stores of water ice or hydrous minerals made from melted ice. Collisions and mergers between the embryos built up a smaller number of larger ones, which by this point can be called planets. Much of this growth took place after the solar nebula was dispersed. Collisions between the planets and a generally corrosive environment outside the nebular cocoon caused the planets to lose volatiles as they evolved. Losses were most severe for the smaller planets and the planets nearest the Sun. These losses were offset for planets that accreted planets or embryos or stray bodies from the colder, more-distant reaches of the solar system, where condensed volatiles were more abundant. This picture provides plausible context for the collision between a Mars-sized planet that had lost its volatiles and a larger, Earth-sized planet that had held on to its own.

The Moon-forming impact took place some 40–50 Ma after the solar system formed. The aftermath left Earth enveloped in a hot silicate atmosphere. At this time Earth's effective radiating temperature—∼2,500 K—was set by the optical depth of silicate condensates. Fast

cooling at high temperature ensured that the silicate vapor atmosphere did not last long. The silicates would have condensed and receded into the depths of the planet within $\sim$1,000 yrs. Volatiles would have partitioned according to their solubility in silicate melt, with much of the H_2O partitioning into the mantle, while CO_2 and most other gases were left behind as a deep thick atmosphere.

Thereafter thermal blanketing by the atmosphere reduced the effective radiating temperature dramatically. The latter gradually approached the asymptotic $\sim$300 K radiating temperature of a water vapor atmosphere. The resulting cooling rate was orders of magnitude slower than for an airless planet with a magma surface. We estimate that, because of thermal blanketing, it took would have taken $\sim$2 Myrs for the surface of a magma ocean to freeze. During this time tidal heating by the new Moon was a major energy source in the mantle. But because the atmosphere controlled the rate of mantle cooling, thermal blanketing would have controlled the pace of lunar orbital evolution. The slow lunar orbital evolution that results can help explain how the Moon acquired its inclination.

Water was expelled from the mantle as the magma ocean froze. This generated a thick steam atmosphere. The steam atmosphere kept the surface near the melting point until the mantle dried out. Thereafter the mantle solidified and geothermal heat was no longer climatologically significant. Without the support of geothermal heat the steam condensed and rained out over a relatively short time, forming oceans of hot water. At this point the atmosphere was dominated by $\sim$100 bars of CO_2, and the surface temperature was $\sim$500 K. Further cooling was governed by how quickly CO_2 was removed from the atmosphere. Vigorous hydrothermal circulation through the oceanic crust and rapid mantle turnover could have removed 100 bars of CO_2 from the atmosphere in as little as 10 million years. However, the balance of the CO_2 cycle remains obscure, which makes it difficult to predict how quickly this would have happened and what the asymptotic atmospheric CO_2 level would be. If carbonate subduction were efficient, and its scavenging from the atmosphere and ocean promoted by abundant ultramafic crust and abundant impact ejecta, a lifeless Earth should have been cold and its oceans white with ice. But if carbonate subduction were inefficient, most of the CO_2 could have stayed in the atmosphere and kept surface temperatures near $\sim$500 K for many tens of millions of years. Intermediate states exist but require a finely tuned CO_2 cycle. Hydrous minerals entered the hot mantle less easily than carbonates, which suggests that the mantle was dry and Earth's water was mostly partitioned into oceans.

The transition between vigorous magma ocean convection and modern plate tectonics is unlikely to have been simple or direct. Plate tectonics as it works now has difficulties with heat flows much greater than 0.1 W/m^2 from a mantle distinctly warmer than it is now; it was probably inadequate to handle typical Hadean heat flows of 0.2–0.5 W/m^2. In place of plate tectonics we suggest that the mantle was topped by a $\sim$100 km deep basaltic mush that, unlike modern plates, was relatively permeable to heat flow. This picture resembles the idealized physics of parameterized convection, more so than plate tectonics does. Recycling and distillation of hydrous basalts produced granitic rocks very early, which is consistent with preserved >4 Ga detrital zircons.

Earth could have been habitable as early as 10–20 Myrs after the Moon-forming impact if CO_2 entered the mantle efficiently. But in the absence of potent greenhouse gases, the faint Sun suggests that the Hadean would not have enjoyed a stable, pleasant climate. Simple models suggest that the modest 1 bar CO_2 atmosphere needed to maintain a clement climate represents an unlikely balancing point for a lifeless planet: the simple models favor either too little CO_2 (efficient subduction) or too much (inefficient subduction). Of course these are just models; moreover the real world would have presented a range of subduction environments that might have averaged out just right. Nevertheless we take the view that a lifeless Hadean

Earth was likely ice-covered at most times and places, that the ice was thin near the equator or over regions of locally high heat flow, with the thin ice broken by lanes and patches of open water, and that major impacts induced hundreds or thousands of transient impact summers.

As we emphasized in the Introduction, the Hadean is not data rich. The only points of surety are that the impact that made the Moon melted and superheated the Earth; that some 700 million years later recognizably terran rocks began to leave a permanent record of their existence; and that some zircons forged in the knowledge of liquid water found refuge there, so that today they can tell their tale. The rest of it is variously informed speculation intended as a framework (or foil) for further work. A few topics stand out. One is the importance of tidal heating to the thermal and perhaps the compositional evolution of the mantle as the magma ocean froze. Another topic is the mechanism of mantle convection between the freezing of the magma ocean and the onset of plate tectonics. A possibly related topic is how continents formed (or did not form) under these conditions. What was the clock that inserted a $\sim$30 Myr time delay between the Hf–W and U–Pb dating systems? Was it related to tectonic style or continent formation? Did the cooling mantle generate persistent slag layers and dregs layers at the top and bottom of the mantle? Did the Earth long maintain an ultramafic crust? Another general topic is how fast carbon dioxide was removed from the atmosphere into the mantle. This depends on the details of carbonate mineral stability as crustal blocks foundered into the mantle. It controls the evolution of the climate. Was there a feedback mechanism to provided the right amount of greenhouse warming to counter the faint Sun, or was the Hadean Earth really as cold and icy as we have surmised? These are just a few of the issues raised here. But the main goal for future research is simply to describe the full diversity of the Hadean surface and near surface environments. Until then we won't be close to an answer to the big question: just what was it about the Hadean that made it the Garden of Eden?

Acknowledgements   We thank Yutaka Abe, Tom Ahrens, Luke Dones, Mikhail Gerasimkhov, Jim Kasting, Joe Kirschvink, Paul Knauth, Hal Levison, Jack Lissauer, Don Lowe, Chris McKay, Mark Marley, Steve Mojzcis, David Morrison, Bill Nelson, Francis Nimmo, Ian Parsons, Bob Pepin, David Schwartzman, John Spencer, and John Valley for discussions, insights, and explanations of thorny problems that for the most part they didn't realize would come to this. KZ thanks the NASA Exobiology Program, the NASA Planetary Geology and Geophysics Program, and the National Astrobiological Institute for support. NTA acknowledges the help of the French CNRS and the ArchEnviron program of the European Science Foundation. NS acknowledges support by NSF grant EAR-0406658.

References

Y. Abe, Proc. 21st ISAS Lun. Planet Symp. (1988), pp. 225–231

Y. Abe, Lithos **30**, 223–235 (1993)

Y. Abe, T. Matsui, J. Atm. Sci. **45**, 3081–3101 (1988)

Y. Abe, E. Ohtani, T. Okuchi, K. Righter, M. Drake, in *Origin of the Earth and Moon*, ed. by R.M. Canup, K. Righter (University of Arizona Press, Tucson, 2000), pp. 413–433

F.C. Adams, G. Laughlin, Icarus **150**, 151–162 (2001)

F.C. Adams, D. Hollenbach, G. Laughlin, U. Gorti, Astrophys. J. **611**, 360–379 (2004)

C.B. Agnor, W.R. Ward, Astrophys. J. **567**, 579–586 (2002)

C. Agnor, E. Asphaug, Astrophys. J. **613**, L157–L160 (2004)

T.J. Ahrens, J.D. O'Keefe, M.A. Lange, in *Origin and Evolution of Planetary and Satellite Atmospheres*, ed. by S. Atreya, J.B. Pollack, M.S. Matthews (The University of Arizona Press, Tucson, 1989), pp. 328–385

C.J. Allégre, E. Lewin, Earth Planet. Sci. Lett. **96**, 61–88 (1989)

C.J. Allégre, E. Lewin, B. Dupré, Chem. Geol. **70**, 211–234 (1988)

Y. Amelin, D.-C. Lee, A.N. Halliday, Geochim. Cosmochim. Acta **64**, 4205–4225 (2000)

Y. Amelin, A.N. Krot, I.D. Hutcheon, A.A. Ulyanov, Science **297**, 1678–1683 (2002)

R.L. Armstrong, Phil. Trans. Roy. Soc., Lond. Ser. A **301**, 443–472 (1981)

D. Arnett, *Supernovae and Nucleosynthesis. An Investigation of the History of Matter, from the Big Bang to the Present* (Princeton University Press, Princeton, 1996)

E. Asphaug, C. Agnor, Q. Williams, Nature **439**, 155–160 (2006)

R.B. Baldwin, Icarus **71**, 1–18 (1987a)

R.B. Baldwin, Icarus **71**, 19–29 (1987b)

W. Benz, W.L. Slattery, A.G.W. Cameron, Icarus **66**, 515–535 (1986)

W. Benz, W.L. Slattery, A.G.W. Cameron, Icarus **74**, 516–528 (1988)

M. Bizzarro, J.A. Baker, H. Haack, Nature **431**, 275–278 (2004)

D.D. Bogard, Meteoritics **30**, 244–268 (1995)

A.P. Boss, in *Origin of the Earth*, ed. by H.E. Newsom, J.H. Jones (Oxford University Press, Oxford, 1990), pp. 3–15

S.A. Bowring, I.S. Williams, Contributions Mineral. Petrol. **134**, 3–16 (1999)

M. Boyet, R.W. Carlson, Science **309**, 576–581 (2005)

T.E. Bunch, S. Chang, Geochim. Cosmochim. Acta **44**, 1543–1577 (1980)

M.J. Burchell, J.R. Mann, A.W. Bunch, Mon. Not. Roy. Astron. Soc. **352**, 1273–1278 (2004)

G.R. Byerly, D.R. Lowe, J.L. Wooden, Science **297**, 1325–1327 (2002)

R.M. Canup, Icarus **168**, 433–456 (2004)

R.M. Canup, C.B. Agnor, in *Origin of the Earth and Moon*, ed. by R.M. Canup, K. Righter (Univ. of Arizona Press, Tucson, 2000)

R.M. Canup, E. Asphaug, Nature **421**, 708–712 (2001)

R.M. Canup, K. Righter, *Origin of Earth and Moon* (University of Arizona Press, Tucson, 2000)

A.J. Cavosie, S.A. Wilde, D. Liu, P.W. Weiblen, J.W. Valley, Precambr. Res. **135**, 251–279 (2004)

A.J. Cavosie, J.W. Valley, S.A. Wilde, E.I.M.F., Earth Planet Sci. Lett. **235**, 663–681 (2005)

J.E. Chambers, Earth Planet. Sci. Lett. **223**, 241–252 (2004)

J.E. Chambers, J.J. Lissauer, Lun. Planet. Sci. Conf. **33** (2002), abstract 1093

C.F. Chyba, Icarus **92**, 217–233 (1991)

C.F. Chyba, Geochim. Cosmochim. Acta **57**, 3351–3358 (1993)

P. Cloud, in *Earth's Earliest Biosphere*, ed. by J.W. Schopf (Princeton University Press, Princeton, 1983), pp. 14–31

P. Cloud, *Oasis in Space* (Norton, New York, 1988)

B.A. Cohen, T.D. Swindle, D.A. Kring, Science **290**, 1754–1756 (2000)

R.G. Coleman, X. Wang (eds.), *Ultrahigh Pressure Metamorphism* (Cambridge University Press, Cambridge, 2005)

N. Coltice, F. Albarède, P. Gillet, Science **288**, 845–847 (2000)

K.C. Condie, D.J. DesMarais, D. Abbott, Precambr. Res. **106**, 239–260 (2001)

J.N. Cuzzi, F.J. Ciesla, M.I. Petaev, A.N. Krot, E.R.D. Scott, S.J. Weidenschilling, in *Chondrites and the Protoplanetary Disk*, ed. by A.N. Krot, E.R.D. Scott, B. Reipurth. ASP Conference Series, vol. 341 (2005)

F. D'Antona, I. Mazzitelli, Astrophys. J. Suppl. **90**, 467–500 (1994)

G.F. Davies, J. Geophys. Res. **89**, 6017–6040 (1984)

R. Diehl et al., Nature **439**, 45–47 (2006)

B.R. Doe, R.E. Zartman, in *Geochemistry of Hydrothermal Ore Deposits*, ed. by H.L. Barnes (Wiley, New York, 1979), pp. 22–70.

M.A. Elburg, M.J. van Bergen, J.D. Foden, Geology **32**, 41–44 (2004)

C.N. Foley, M. Wadhwa, L.E. Borg, P.E. Janney, R. Hines, T.L. Grove, Geochim. Cosmochim. Acta **69**, 4557–4571 (2005)

S.J.G. Galer, S.L. Goldstein, Terra Abstr. **3**, 485–486 (1991)

H. Genda, Y. Abe, Icarus **164**, 149–162 (2003)

H. Genda, Y. Abe, Nature **433**, 842–844 (2005)

P. Goldreich, Rev. Geophys. **4**, 411–439 (1966)

R. Gomes, H.F. Levison, K. Tsiganis, A. Morbidelli, Nature **435**, 466–469 (2005)

B.J. Gladman, Icarus **130**, 228–246 (1997)

B.J. Gladman, L. Dones, H.F. Levison, J.A. Burns, Astrobiology **5**, 483–496 (2005)

A.N. Halliday, in *Meteorites, Comets and Planets*, ed. by A.M. Davis. Treatise on Geochemistry, vol. 1 (H.D. Holland and K.K. Turekian) (Elsevier-Pergamon, Oxford, 2003), pp. 509–557

A.N. Halliday, Nature **427**, 505–509 (2004)

A.N. Halliday, T. Kleine, in *Meteorites and the Early Solar System II*, ed. by D. Lauretta, L. Leshin, H. MacSween (Univ. of Arizona Press, 2006, in press)

C.L. Harper, S.B. Jakobsen, Science **273**, 1814–1818 (1996)

T.M. Harrison, J. Blichert-Toft, W. Müller, P. Holden, F. Albarede, S.J. Mojzsis, Science **310**, 1947–1950 (2005)

W.K. Hartmann, Icarus **24**, 181–187 (1975)

W.K. Hartmann, R.J. Phillips, G.J. Taylor (eds.), *Origin of the Moon* (Lunar and Planetary Institute, Houston, 1986)

W.K. Hartmann, L. Dones, G. Ryder, D. Grinspoon, in *Origin of the Earth and Moon*, ed. by R.M. Canup, K. Righter (University of Arizona Press, Tucson, 2000), pp. 493–512

L.A. Haskins, J. Geophys. Res. **103**, 1679 (1998)

C. Hayashi, K. Nakazawa, H. Mizuno, Earth Planet. Sci. Lett. **43**, 22–28 (1979)

H.D. Holland, *The Chemical Evolution of the Atmosphere and Oceans* (Princeton Univ. Press, Princeton, 1984)

D.M. Hunten, T.M. Donahue, Annu. Rev. Earth Planet. Sci. **4**, 265–292 (1976)

D.M. Hunten, R.O. Pepin, J.C.G. Walker, Icarus **69**, 532–549 (1987)

G. Igarashi, in *Volatiles in the Earth and Solar System*, ed. by K. Farley, AIP Conf. Proc. vol. 341. 1995, pp.70–80

M. Ikoma, H. Genda, Astrophys. J. **648**, 696–706 (2006)

M. Ikoma, H. Genda, Astrophys. J. **649** (2006, in press)

M. Ishiwatari, S.-I. Takehiro, K. Nakajima, Y.-Y. Hayashi, J. Atmos. Sci. **59**, 3223–3238 (2002)

B.M. Jahn, F.Y. Wu, C.H. Lo, C.H. Tsai, Chem. Geol. **157**, 119–146 (1999)

S.B. Jakobsen, Annu. Rev. Earth Planet. Sci. **33**, 531–570 (2005)

B.M. Jakosky, T.J. Ahrens, *Proc. Lun. Planet. Sci. Conf. 10* (Pergamon, New York, 1979), pp. 2729–2739

B.S. Kamber, K.D. Collerson, J. Geophys. Res. **104**, 25479–25491 (1999)

B.S. Kamber, M.T. Whitehouse, R. Bolhar, S. Moorbath, Earth Planet. Sci. Lett. **240**, 276–290 (2005)

J.F. Kasting, Icarus **74**, 472–494 (1988)

J.F. Kasting, Orig. Life Evol. Biosphere **20**, 199–231 (1989)

Kelvin, Nature **51**, 438–440 (1895)

L. Keszthelyi, A.S. McEwen, G.J. Taylor, Icarus **141**, 415–419 (1999)

J.L. Kirschvink, in *The Proterozoic Biosphere*, ed. by J.W. Schopf, C. Klein (Cambridge University Press, Cambridge, 1992), pp. 51–52.

T. Kleine, K. Mezger, C. Münker, H. Palme, Nature **418**, 952–955 (2002)

T. Kleine, K. Mezger, C. Munker, H. Palme, A. Bischoff, Geochim. Cosmochim. Acta **68**, 2935–2946 (2004)

T. Kleine, K. Mezger, H. Palme, E. Scherer, C. Münker, Geochim. Cosmochim. Acta **69**, 5805–5818 (2005a)

T. Kleine, H. Palme, K. Mezger, A.N. Halliday, Science **310**, 1671–1674 (2005b)

L.P. Knauth, D.R. Lowe, GSA Bull. **115**, 566–580 (2003)

A.H. Knoll, *Life on a Young Planet: The First Three Billion Years of Evolution on Earth* (Princeton Univ. Press, Princeton, 2003)

E. Kokubo, S. Ida, J. Makino, Icarus **148**, 419–436 (2000)

S.J. Kortenkamp, E. Kokubo, S.J. Weidenschilling, in *Origin of the Earth and Moon*, ed. by R.M. Canup, K. Righter (The Univ. of Arizona Press, Tucson, 2000), pp. 85–100

A.F. Koster van Groos, J. Geophys. Res. **93**, 8952–8958 (1988)

J.D. Kramers, I.N. Tolstikhin, Chem. Geol. **139**, 75–110 (1997)

Y.N. Kulikov, H. Lammer, I. Herbert, M. Lichtenegger, T. Penz, D. Breuer, T. Spohn, R. Lundin, H.K. Biernat (2007), this issue, doi:10.1007/s11214-007-9192-4

S.-T. Kwon, G.R. Tilton, M.H. Grnnenfelder, in *Carbonatites—Genesis and Evolution*, ed. by K. Bell (Unwin-Hyman, London, 1989), pp. 360–387

F.T. Kyte, A. Shukolyukov, G.W. Lugmair, D.R. Lowe, G. Byerly, Geology **31**, 283–286 (2003)

J.A. Lake, Nature **331**, 184–186 (1988)

M.A. Lange, T.J. Ahrens, Icarus **51**, 96–120 (1982)

D.-C. Lee, A.N. Halliday, Nature **388**, 854–857 (1997)

D.-C. Lee, A.N. Halliday, G.A. Snyder, L.A. Taylor, Science **278**, 1098–1103 (1997)

D.-C. Lee, A.N. Halliday, I. Leya, R. Wieler, U. Wiechert, Earth Planet. Sci. Lett. **198**, 267–274 (2002)

H.F. Levison, M. Duncan, Icarus **127**, 3–16 (1997)

H.F. Levison, L. Dones, C.R. Chapman, S.A. Stern, M.J. Duncan, K.J. Zahnle, Icarus **151**, 286–306 (2001)

J.S. Lewis, R.G. Prinn, *Planets and Their Atmospheres—Origin and Evolution* (Academic, Orlando, 1984)

C. Liebske, B. Schmickler, H. Terasaki, B.T. Poe, A. Suzuki, R. Ando, K. Funakoshi, D.C. Rubie, Earth Planet. Sci. Lett. **240**, 589–604 (2005)

T.C. Liew, C.C. Milisenda, A.W. Hofmann, Geol. Rundsch. **80**, 279–288 (1991)

J.J. Lissauer, Annu. Rev. Astron. Astrophys. **31**, 129–174 (1993)

K. Lodders, B. Fegley, *The Planetary Scientist's Companion* (Oxford Univ. Press, Oxford, 1998)

D.R. Lowe, G.R. Byerly, Geology **11**, 668–671 (1986)

D.R. Lowe, G.R. Byerly, F. Asaro, F. Kyte, Science **245**, 959–962 (1989)

P. Lowell, *Mars* (Houghton, Mifflin and Co., Boston, 1895)

P. Lowell, *Mars and Its Canals* (MacMillan, New York, 1906)

P. Lowell, *The Evolution of Worlds* (MacMillan, New York, 1909)

J.I. Lunine, A. Morbidelli, J. Chambers, L.A. Leshin, Icarus **165**, 1–8 (2003)

K.A. Maher, D.J. Stevenson, Nature **331**, 612–614 (1988)

A. Markowski, G. Quitté, A.N. Halliday, T. Kleine, Earth Planet. Sci. Lett. **242**, 1–15 (2006)

T. Matsui, Y. Abe, Nature **319**, 303–305 (1986)

C.P. McKay, Geophys. Res. Lett. **27**, 2153–2156 (2000)

C.P. McKay, in *The Extreme Proterozoic: Geology, Geochemistry, and Climate*, ed. by G.S. Jenkins, M.A.S. McMenamin, C.P. McKay, L. Sohl. Geophysical Monograph, vol. 146 (American Geophysical Union, Washington, 2004), pp. 193–198

A.S. McEwen et al., Science **281**, 87–90 (1998)

H.J. Melosh, *Impact Cratering: A Geological Process* (Oxford University Press, New York, 1989)

C. Mileikowsky, F.A. Cucinotta, J.W. Wilson, B. Gladman, G. Horneck, L. Lindgren, J. Melosh, H. Rickman, M. Valtonen, J.Q. Zheng, Planet. Space Sci. **48**, 1107–1115 (2000)

S.L. Miller, Science **117**, 528–529 (1953)

H. Mizuno, K. Nakazawa, C. Hayashi, Earth Planet. Sci. Lett. **50**, 202–210 (1980)

H. Mizuno, G.W. Wetherill, Icarus **59**, 74–86 (1984)

S.J. Mojzsis, T.M. Harrison, R.T. Pidgeon, Nature **409**, 178–181 (2001)

A. Morbidelli, J. Chambers, J.I. Lunine, J.-M. Petit, G.B. Valsecchi, F. Robert, K.E. Cyr, Meteor. Planet. Sci. **35**, 1309–1320 (2000)

A. Morbidelli, J.-M. Petit, B. Gladman, J.E. Chambers, Meteor. Planet. Sci. **36**, 371–380

D.T. Murphy, B.S. Kamber, K.D. Collerson, J. Petrol. **44**, 39–53 (2003)

S. Nakajima, Y.-Y. Hayashi, Y. Abe, J. Atmospheric Sci. **49**, 2256–2266 (1992)

G. Neukum, B.A. Ivanov, in *Hazards due to Comets and Asteroids*, ed. by T. Gehrels (University of Arizona Press, Tucson, 1994), pp. 359–416

G. Neukum, B.A. Ivanov, W.K. Hartmann, in *Chronology and Evolution of Mars*, ed. by R. Kallenbach, J. Gleiss (Kluwer, Dordrecht, 2001), pp. 55–86, reprinted from Space Sci. Rev. **96**(1–4)

E.G. Nisbet, Precambr. Res. **19**, 111–118 (1982)

E.G. Nisbet, Geol. Mag. **122**, 84–94 (1985)

E.G. Nisbet, Episodes **14**, 327–331 (1991)

E.G. Nisbet, Nature **380**, 291 (1996)

E.G. Nisbet et al., Space Sci. Rev. (2007), this issue, doi:10.1007/s11214-007-9175-5

T. Owen, P. Mahaffy, H.B. Niemann, S. Atreya, T. Donahue, A. Bar-Nun, I. de Pater, Nature **402**, 269–270 (1999)

M. Ozima, F.A. Podosek, *Noble Gas Geochemistry* (Cambridge University Press, Cambridge, 2002)

N. Pace, G.J. Olsen, C.R. Woese, Cell **45**, 325–326 (1986)

C.C. Patterson, Geochim. Cosmochim. Acta **10**, 230–237 (1956)

R.O. Pepin, Icarus **92**, 2–79 (1991)

R.T. Pierrehumbert, Nature **429**, 646–649 (2004)

D. Pollard, J.F. Kasting, J. Geophys. Res. **110**, C07010 (2005). doi:10.1029/2004JC002525

D. Porcelli, R.O. Pepin, in *Origin of the Earth and Moon*, ed. by R.M. Canup, K. Righter (The Univ. of Arizona Press, Tucson, 2000), pp. 435–458

S.N. Raymond, T.R. Quinn, J.I. Lunine, Icarus **168**, 1–17 (2004)

A.E. Ringwood, Geochim. Cosmochim. Acta **21**, 295–296 (1961)

W.W. Rubey, Bull. Geol. Soc. Am. **62**, 1111–1147 (1951)

T. Rushmer, W.G. Minarik, G.J. Taylor, in *Origin of the Earth and Moon*, ed. by R.M. Canup, K. Righter (The Univ. of Arizona Press, Tucson, 2000), pp. 227–243

M.J. Russell, N.T. Arndt, Biogeosciences **2**, 97–111 (2005)

S.S. Russell, G. Srinivasan, G.R. Huss, G.J. Wasserburg, G.J. MacPherson, Science **273**, 757–762 (1996)

G. Ryder, C. Koeberl, S. Mojzsis, in *Origin of the Earth and Moon*, ed. by R.M. Canup, K. Righter (The Univ. of Arizona Press, Tucson, 2000), pp. 475–492

G. Ryder, J. Geophys. Res. **107**, 10 (2002). doi:1029/2001JE001583

G. Ryder, Astrobiology **3**, 3–6 (2003)

I.-J. Sackmann, A.I. Boothroyd, Astrophys. J. **583**, 1024–1039 (2003)

I.-J. Sackmann, A.I. Boothroyd, K.E. Kraemer, Astrophys. J. **418**, 457–468 (1993)

V.S. Safronov, Evolution of the protoplanetary cloud and formation of the Earth and the Planets. NASA TT F-677, 1972

C. Sagan, G. Mullen, Science **177**, 52–56 (1972)

S. Sasaki, in *Origin of the Earth*, ed. by H.E. Newsom, J.H. Jones (Oxford University Press, Oxford, 1990), pp. 195–209

S. Sasaki, K. Nakazawa, Earth Planet. Sci. Lett. **89**, 323–334 (1988)

A. Scherstén, T. Elliott, C. Hawkesworth, S.S. Russell, O. Masarik, Earth Planet. Sci. Lett. **241**, 530–542 (2006)

M. Schidlowski, Nature **333**, 313–318 (1988)

G. Schubert, D. Turcotte, P. Olson, *Mantle Convection in the Earth and Planets* (Cambridge University Press, 2001)

D. Schwartzman, *Life, Temperature, and the Earth* (Columbia University Press, 2002), 272 pp.

M. Sekiya, K. Nakazawa, C. Hayashi, Earth Planet. Sci. Lett. **50**, 197–201 (1980a)

M. Sekiya, K. Nakazawa, C. Hayashi, Prog. Theor. Phys. **64**, 1968–1985 (1980b)

M. Sekiya, C. Hayashi, K. Kanazawa, Prog. Theor. Phys. **66**, 1301–1316 (1981)

E.L. Shock, J.P. Amend, M.Y. Zolotov, in *Origin of Earth and Moon*, ed. by R.M. Canup, K. Righter (University of Arizona Press, Tucson, 2000), pp. 527–543

J.M. Sinton, R.S. Detrick, J. Geophys. Res. **97**, 197–216 (1992)

N.H. Sleep, J. Geophys. Res. **105**(17), 563–17578 (2000)

N.H. Sleep, in *Treatise on Geophysics, vol. 9. Evolution of the Earth*, ed. by D.J. Stevenson (Elsevier, 2007)

N.H. Sleep, A.M. Hessler, Earth Planet. Sci. Lett. **241**, 594–602 (2006)

N.H. Sleep, R.T. Langan, Adv. Geophys. **23**, 1–23 (1981)

N.H. Sleep, K. Zahnle, J. Geophys. Res. **103**, 28529–28544 (1998)

N.H. Sleep, K. Zahnle, J. Geophys. Res. **106**, 1373–1400 (2001)

N.H. Sleep, K.J. Zahnle, J.F. Kasting, H.J. Morowitz, Nature **342**, 139–142 (1989)

N.H. Sleep, K.J. Zahnle, P.S. Neuhoff, Proc. Nat. Acad. Sci. **98**, 3666–3672 (2001)

V.S. Solomatov, in *Origin of the Earth and Moon*, ed. by R.M. Canup, K. Righter (University of Arizona Press, Tucson, 2000), pp. 413–433

J.R. Spencer, J.A. Rathbun, L.D. Travis, L.K. Tamppari, L. Barnard, T.Z. Martin, A.S. McEwen, Science **288**, 1198–1201 (2000)

F.J. Stadermann, E. Heusser, E.K. Jessberger, S. Lingner, D. Stöffler, Geochim. Cosmochim. Acta **55**, 2339–2349 (1991)

D.J. Stevenson, in *Earth's Earliest Biosphere*, ed. by J.W. Schopf (Princeton University Press, Princeton, 1983), pp. 32–40

D.J. Stevenson, Annu. Rev. Earth Planet. Sci. **15**, 271–315 (1987)

D.J. Stevenson, in *Origin of the Earth*, ed. by H.E. Newsom, J.H. Jones (Oxford University Press, London, 1990), pp. 231–249

D.J. Stevenson, Earth Planet. Sci. Lett. **208**, 1–11 (2003)

S.R. Taylor, *Solar System Evolution* (Cambridge University Press, Cambridge, 1993)

F. Tera, D.A. Papanastassiou, G.J. Wasserburg, Earth Planet. Sci. Lett. **22**, 1–21 (1974)

E.W. Thommes, M.J. Duncan, H.F. Levison, Astron. J. **123**, 2862–2883 (2002)

J. Touma, in *Origin of Earth and Moon*, ed. by R.M. Canup, K. Righter (University of Arizona Press, Tucson, 2000), pp. 165–178

J. Touma, J. Wisdom, Astron. J. **108**, 1943–1961 (1994)

J. Touma, J. Wisdom, Astron. J. **115**, 1653–1663 (1998)

H.C. Urey, Geochim. Cosmochim. Acta **1**, 209–277 (1951)

J.W. Valley, W.H. Peck, E.M. King, S.A. Wilde, Geology **30**, 351–354 (2002)

J.W. Valley, J.S. Lackey, A.J. Cavosie, C.C. Clechenko, M.J. Spicuzza, M.A.S. Basei, I.N. Bindeman, V.P. Ferreira, A.N. Sial, E.M. King, W.H. Peck, A.K. Sinha, C.S. Wei, Contributions Mineral. Petrol. **150**, 561–580 (2005)

P. van Thienen et al., Space Sci. Rev. (2007), this issue, doi:10.1007/s11214-007-9149-7

R. von Huene, D.W. Scholl, Rev. Geophys. **29**, 279–316 (1991)

W.R. Ward, in *Origin of the Earth and Moon*, ed. by R.M. Canup, K. Righter (The Univ. of Arizona Press, Tucson, 2000), pp. 65–289

P.H. Warren, E.A. Jerde, G.W. Kallemeyn, Earth Planet. Sci. Lett. **91**, 245–260 (1989)

S.G. Warren, R.E. Brandt, T.C. Grenfell, C.P. McKay, J. Geophys. Res. **107**(C10), 3167 (2002). doi:10.1029/2001JC001123

A.J. Watson, T.M. Donahue, J.C.G. Walker, Icarus **48**, 150–166 (1981)

S.J. Weidenschilling, J.N. Cuzzi, in *Protostars and Planets III*, ed. by E.H. Levy, J.I. Lunine (Univ. of Arizona Press, Tucson, 1993), pp. 1031–1060

B.P. Weiss, J.L. Kirshvink, F.J. Baudenbacher, H. Vali, N.T. Peters, F.A. Macdonald, J.P. Wikswo, Science **290**, 791–795 (2000)

L.E. Wells, J.C. Armstrong, G. Gonzalez, Icarus **162**, 38–46 (2003)

G.W. Wetherill, Proc. Lunar Sci. Conf. **6**, 1539–1561 (1975)

G.W. Wetherill, Science **228**, 877–879 (1985)

G.W. Wetherill, in *Origin of the Moon*, ed. by W.K. Hartmann et al. (Lunar Planetary Institute, Houston, 1986), pp. 519–550

W.B. Whitman, D.C. Coleman, W.J. Wiebe, Proc. Natl. Acad. Sci. **95**, 6578–6583 (1998)

S.A. Wilde, J.W. Valley, W.H. Peck, C.M. Graham, Nature **409**, 175–178 (2001)

D.E. Wilhelms, The Geologic History of the Moon, U.S.G.S. Professional Paper 1348, 1987

D.S. Woolum, P. Cassen, Meteor. Planet. Sci. **34**, 897–907 (1999)

B.E. Wood, H.-R. Muller, G.P. Zank, J.L. Linsky, Astrophys. J. **574**, 412–425 (2002)

B.E. Wood, H.-R. Muller, J.L. Linsky, G.P. Zank, S. Redfield, Astrophys. J. **628**, L143–L146 (2005)

B.J. Wood, A.N. Halliday, Nature **437**, 1345–1348 (2005)

Q.Z. Yin, S.B. Jacobsen, K. Yamashita, J. Blichert-Toft, P. Télouk, F. Albarède, Nature **418**, 949–952 (2002)

T. Yoshino, M.J. Walter, T. Katsura, Nature **422**, 154–157 (2003)

K.J. Zahnle, in *Origins*, ed. by C.E. Woodward, J.M. Shull, H.A. Thronson. ASP Conference Series, vol. 148, 1998, pp. 364–391

K. Zahnle, J. Kasting, Icarus **68**, 462–480 (1986)

K.J. Zahnle, J.F. Kasting, J.B. Pollack, Icarus **74**, 62–97 (1988)

K.J. Zahnle, N.H. Sleep, in *The Early Earth: Physical Chemical and Biological Development*, ed. by C.M. Fowler, C.J. Ebinger, C.J. Hawkesworth. Geol. Soc. London Spec. Pub., 199, 2002, pp. 231–257

K.J. Zahnle, N.H. Sleep, in *Comets and the Origin and Evolution of Life*, ed. by P. Thomas, C.F. Chyba, C.P. McKay (Springer, 1997), pp. 175–208

K. Zahnle, N.H. Sleep, in *Comets and the Origin and Evolution of Life*, 2nd edn, ed. by P.J. Thomas, R.D. Hicks, C.F. Chyba, C.P. McKay (Springer, Berlin, 2006), pp. 207–251

R.E. Zartman, S.M. Haines, Geochim. Cosmochim, Acta **52**, 1327–1339 (1988)

Space Sci Rev (2007) 129: 79–121
DOI 10.1007/s11214-007-9175-5

Creating Habitable Zones, at all Scales, from Planets to Mud Micro-Habitats, on Earth and on Mars

Euan Nisbet · Kevin Zahnle · M.V. Gerasimov · Jörn Helbert ·
Ralf Jaumann · Beda A. Hofmann · Karim Benzerara · Frances Westall

Received: 2 November 2006 / Accepted: 22 March 2007 /
Published online: 1 August 2007
© Springer Science+Business Media, Inc. 2007

Abstract The factors that create a habitable planet are considered at all scales, from planetary inventories to micro-habitats in soft sediments and intangibles such as habitat linkage. The possibility of habitability first comes about during accretion, as a product of the processes of impact and volatile inventory history. To create habitability water is essential, not only for life but to aid the continual tectonic reworking and erosion that supply key redox contrasts and biochemical substrates to sustain habitability. Mud or soft sediment may be a biochemical prerequisite, to provide accessible substrate and protection. Once life begins, the habitat is widened by the activity of life, both by its management of the greenhouse and

E. Nisbet (✉)
Department of Geology, Royal Holloway, University of London, Egham, Surrey, TW20 0EX, UK
e-mail: e.nisbet@gl.rhul.ac.uk

K. Zahnle
NASA Ames Research Center, MS 245-3, Moffett Field, CA 94035, USA
e-mail: kzahnle@mail.arc.nasa.gov

M.V. Gerasimov
Space Research Institute, Russian Academy of Science, Profsoyuznaya st., 84/32, Moscow 117997,
Russian Federation
e-mail: mgerasim@mx.iki.rssi.ru

J. Helbert · R. Jaumann
Deutsches Zentrum für Luft- und Raumfahrt (DLR), Institute of Planetary Research,
Rutherfordstr. 2, 12489 Berlin, Germany

B.A. Hofmann
Naturhistorisches Museum Bern, Bernastrasse 15, 3005 Bern, Switzerland

K. Benzerara
Institut de Minéralogie et de Physique des Milieux Condensés, UMR 7590 CNRS & IPGP,
140 rue de Lourmel, 75015 Paris, France

F. Westall
Centre de Biophysique Moléculaire, CNRS, Rue Charles Sadron, 45071 Orléans cedex 2, France

by partitioning reductants (e.g. dead organic matter) and oxidants (including waste products). Potential Martian habitats are discussed: by comparison with Earth there are many potential environmental settings on Mars in which life may once have occurred, or may even continue to exist. The long-term evolution of habitability in the Solar System is considered.

Keywords Mars · Water · Fluvial erosion · Habitable surfaces · Rock cycle

1 Introduction

This chapter follows the creation of habitability on all scales, on Earth and on Mars. It examines the creation of habitability on a planetary scale, then the formation of a suitable surface environment, then focuses down to the formation of local zones, such as muds, where habitation could occur. Habitability is not just macro-suitability, having a planet in the right place next to the right star. It is also micro-scale: the availability of cell-sized, cell-friendly habitats. Moreover, habitability must be sustained over billions of years—fortunately life has self-sustaining character.

The focus is on the one planet we know to be habitable: the Earth. Mars is also discussed towards the end of the chapter. The implications range more widely: to be habitable, extrasolar planets will need to be suitable both on the grand planetary scale, and on the micro scale.

A habitable planet must be able to sustain life over eons. The fundamental environmental requirements for habitability include providing favorable conditions for the assembly of organic molecules and energy sources to sustain metabolism. Less obvious is that far more than offering a single specific environment (which will obviously be short-lived), habitability needs diversity. To continue, life needs an ongoing suite of environments that communicate through the exchange of materials and which are sustained in time.

A habitable planet must "live" geologically. It must renew its surface. It has to be active volcanically and tectonically. Habitable environments must provide a highly flexible exchange medium for chemical and biochemical components. Although it is conceivable that other solvents could carry out this role (e.g. ammonia), the obvious (and possibly unique) medium for exchange is water (Brack 2002). If so, the planetary environment must sustain liquid water. Water's physical prerequisites constrain the abundance and distribution of habitable zones, managing every aspect from hydrothermal systems to UV control.

During the earliest Hadean, the primary construction of the planets was accomplished. The sites for the potential houses for life were marked out, each with its distinct power service from the Sun's light. The foundations accreted, planetesimal by planetesimal; the bodies were put up by much bashing and crashing. The plaster was thrown onto the walls. Then in the mid-Hadean finishing-work on the various rooms for life was begun. What was rough and uninhabitable slowly took on the appearance of a living room, where life could make itself comfortable. Life appeared and claimed its habitat—at least on Earth, perhaps on Mars; just possibly on Venus. And once it had arrived, life began changing the room immediately: not only did it fill the room with furniture, but it began to replaster the walls and eventually even to remodel the foundations themselves.

In this chapter we follow the story from the end of the mid- to late-Hadean, around 4.2 to 4 Ga ago, when the main work of accretion was completed but some bombardment was still continuing, to the beginnings of what we recognise as a 'modern' oxic habitat, ~2.3 Ga ago. The central focus of the account is water and its interactions with the planet and with life. The first part of the chapter discusses the degassing of the oceans and the interaction between water and rock. Then the water-rich habitable environments are explored, and the

 Springer

sources of the chemical disequilibrium that supported early life. The early record of life on Earth is explored: what signatures has early life left in the rock record and in our own genes in modern life? What impact did life have in remodeling its home? Mars is explored—are there such habitats there. Finally, the issue of sustained habitability is discussed.

The Zimbabwean geologist A.M. Macgregor (1927), pointed out that the modern atmosphere of the Earth is a biological construction. What came before presumably had far more carbon dioxide; life remade the air. Macgregor also pointed out that in changing the air, life also reshaped the controls on greenhouse temperature. It risked changing life-giving water into hard rock, setting off glaciation. In general, life shapes, changes and sustains its own conditions.

2 The Planetary Scale: Creating Habitability

For long-term habitability by carbon-based life, a planet needs both:

1) Liquid water to provide a medium of chemical exchange (Brack 2002), and
2) Sustained thermodynamic disequilibrium between chemical species in close spatial proximity. The latter is most effectively done by utilizing solar photons, which are grossly out of equilibrium with conditions on the Earth.

Water is the most abundant condensable material in the cosmos. It is arguably the likeliest chemical compound to be delivered to a new planetary surface. Here it is taken as granted that the early Earth and many other planetary bodies in the solar system had significant water inventories at the end of accretion.

The problem for the biologist is that at least some of this water must be on the surface of the planet, not dissolved in its interior. Indeed, for many years the lifespan of oceans remained an unsolved problem for geologists over time, why does the water not simply sink into the ground and vanish? That is exactly what it has mostly done on Mars. Only with the discovery of plate tectonics did the answer come for Earth—because the Earth's interior is hot, hydrated oceanic crust does indeed sink in, but the water is recycled by partial melting in subduction zones.

Water exists as a liquid at pressures above 6.1 mbar and temperatures above 273 K. On planetary surfaces temperatures are determined by the solar irradiance, the planet's albedo (reflectivity), and the atmosphere's greenhouse effect. Both the albedo and the greenhouse effect depend in turn on the response of water to surface temperature. The albedo is sensitive to ice, snow and clouds, while the greenhouse effect is dominated by the opacity of water vapor to thermal radiation. The chief independent variables are the irradiance and the abundance of other greenhouse gases. The latter can be something of a wild card, but the irradiance is a simple well-defined function of the stellar luminosity (L) and distance (D) from the star: L/D^2. Thus, phase conditions suitable for liquid water depend on the distance from the central star. Conservative estimates for the innermost and outermost distances where liquid water is stable in our solar system are 0.95 and 1.37 AU, respectively (Kasting et al. 1993). This range of distances has been called the "habitable zone".

It is noteworthy that this habitable zone moves outwards from the Sun over time. This is because the Sun becomes brighter as it ages. The Sun is now about 30% brighter than it was 4 Ga (see Fig. 1, Zahnle et al. chapter, this volume) ago. We will discuss the deep past and the deep future in more detail at the end of this chapter.

The inner edge of the habitable zone is hard to finesse short of painting the planet white, but the outer limit is negotiable, provided that there be greenhouse gases enough. Popular

 Springer

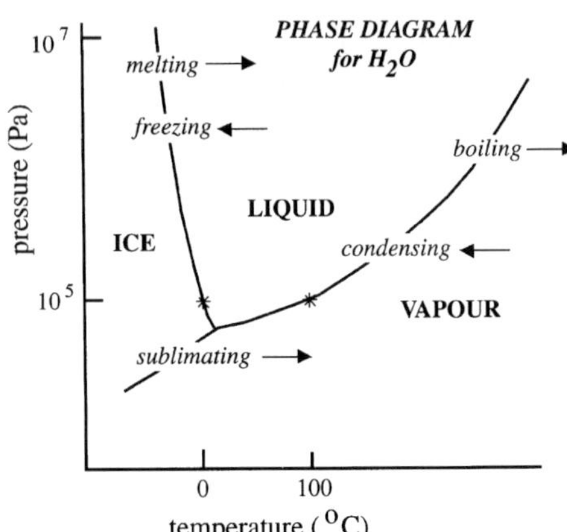

Fig. 1 Phases of H_2O (triple point is at 6.1 mbar and 273 K)

candidates include CO_2, NH_3, and CH_4, but there are many more a technological civilization can manufacture new ones (e.g. C_3F_8) by the dozens. Stevenson (1999) pointed that with enough hydrogen in the atmosphere there may be no effective outer limit at all.

Another option is provided by internal heat sources (such as radioactive or tidal heating). With heat welling up from below planetary crusts can provide protected insulated subsurface conditions that allow a liquid phase of H_2O in planetary bodies even outside the habitable zone of liquid water. Therefore subsurface habitable environments may be a major target in searching for extant and extinct extraterrestrial life, if we could access them, for example by targeting outflow regions. Otherwise, the search must be confined to surficial life and the superficial qualities of life, until deep drilling on Mars or Europa becomes possible. Outcrops with evidence of fossil subsurface life may be found on present planetary surfaces, however, because to some extent impact craters exhume underlying material. Shallow drilling (<5 m) in Martian gully regions might also be a possibility to look for organics.

3 Making a Habitable Zone: The Planetary Skin in the Hadean

Chapter 2 has shown how the planet was assembled. By 4.4 Ga ago it had a cool surface with liquid water: the first prerequisite for habitability.

A cool surface is not enough. There must be chemical differences that life can exploit. On the Hadean Earth and early Mars this was provided by the thermodynamic difference between atmosphere/hydrosphere chemical species created by solar photons, entering the top of the atmosphere, and those interacting with magma from the interior. The interior provided air, and below it water, and cycled that water in hydrothermal systems. The exterior setting in the solar system provided energy—photons—that reworked the components of the ocean/atmosphere system, creating chemical species out of equilibrium with the interior.

The temperature of the interior is crucial. On modern Mars, it is low. On early Earth it was high. The potential temperature of the mantle is the temperature of a piece of the mantle if it were suddenly brought unmelted to the surface. Immediately after the Moon-forming event, temperature would have been very high, above 2500 K. By the late Archean the "potential temperature" of the mantle ("potential temperature" is a thermodynamic term defined as the temperature at which magma would rise adiabatically to the surface) had fallen to about 1800 K (Nisbet et al. 1993). Today the potential temperature is about 1600 K.

To form a solid lid on the Earth's surface, over a molten interior, the surface temperature must fall below about 2000 K (Sleep et al. 2001). By the time a first proto-crust formed, the potential temperature was about 2500 K (Sleep et al. 2001). This was probably the potential temperature before 4.4 Ga ago, at a time a few millions of years after the Moon-forming event. Sleep et al. (2001) suggest that this could have been around 3 million years after the global melting event, assuming a massive CO_2 atmosphere did not exist. There would then follow an interval during which the surface heat flow from the interior, at around 70–100 W m^{-2}, would have been sufficient to maintain equable surface conditions around 30°C. For comparison, the modern Earth receives a very roughly comparable heat input from the Sun. However, such an environment would be barely habitable—the crustal rind would be some tens of meters thick, with molten rock below. This would not be mechanically stable, and any newly-born living cells would be soon tipped into the inferno by minor quakes.

3.1 Hadean Surface Temperature: The Cool Early Earth

The oldest material on the Earth is found in the interiors of single crystals of zircon, some of which formed as early as 4.4 Ga ago (Wilde et al. 2001). From the $\delta^{18}O$ values of these zircons, some broad generalizations about conditions on the planetary surface conditions can be inferred. Their titanium contents, used as a geothermometer, provide evidence of formation in water-rich granitic magmas (Watson and Harrison 2005). The only unusual aspect of the 4.4 Ga zircons is their age. In all other respects they are closely similar to later (Archean) zircons. From this similarity it is inferred that they incorporated oxygen from water that had been in contact with the planetary surface.

The chain of logic is as follows: 1) $\delta^{18}O$ values in zircons are nearly constant for the long time interval 4.4 to 2.6 Ga ago. 2) liquid water oceans must have been present throughout the 2.6 Ga to 3.8 Ga period, to deposit the clastic and chemical sediments that are ubiquitous in the Archean record. 3) Zircons made in the late Archean incorporate water from hydrothermal systems linked to liquid oceans. Ergo 4) the Hadean zircons, that are closely similar to the Archean examples, also incorporate oxygen from liquid oceans. Thus the conclusion is that liquid water existed as early as 4.4 Ga ago.

From this type of reasoning, Valley et al. (2002) inferred a surface temperature of lower than 200°C at 4.4 Ga. This is consistent with the finding of Sleep et al. (2001), that if the entire CO_2 inventory were placed in the air, the temperature would have been around 230°C.

Sleep et al. (2001) showed the transition from massive CO_2 greenhouse to cool surface temperature probably took place rapidly but not instantaneously. Widespread surface conditions hospitable to hyperthermophilic microbial organisms (say around ∼100°C) probably persisted for a few million years (range <1 to ∼20 million years). Thus for a brief period, probably around 1 million years but possibly as long as 20 million years, it might be possible to maintain surface temperatures around 100°C, suitable for hyperthermopile life. After the close of this interval, the system would cool and the surface temperature would begin to drop, eventually towards glacial conditions. In a late Hadean icehouse, hyperthermophile habitats with temperatures around ∼100°C would be tightly restricted to the hydrothermal systems in the vicinity of volcanic activity.

Nevertheless, the 'brief' <1 to ∼20 million years of widespread warm oceans is a very long time, especially when thought of in terms of microbial generations. Any line of living organisms born at this early time would be preadapted to take refuge in hydrothermal systems as the world ocean cooled. Such an early hyperthermophile biota would be likely to suffer one or more ocean-boiling meteorite impacts before the end of the Hadean at ∼4 Ga. Nevertheless, cells might survive a global heating event if they had refuge in the rock, in hydrothermal systems.

 Springer

3.2 The Lively Interior: The Contribution from Volcanism

Hadean volcanism would have occurred in an already formed ocean, as demonstrated by the zircon oxygen isotope evidence (Valley et al. 2002; Wilde et al. 2001). After each mid-Hadean impact (a 10 km-size body each 0.1 Ma at that time), massive quantities of basaltic and komatiitic ejecta would have been thrown out and then landed back into the seas. Massive komatiitic volcanism was probably occurring, locally intensified in the aftermath of major impacts. Consequently there would be large-scale interaction of ocean water with hot mafic rock. This hydrothermal interaction would have been much more rapid than today: perhaps such that every million years or less a volume of water equal to the volume of the oceans would pass through hydrothermal systems.

The consequence of this would be pH and geochemical control on the water body. Close to hot vents, hydrothermal fluids would be acidic, but komatiitic volcanism was very widespread and the dominant cooler exiting flow may have been more alkaline. As far as prebiotic conditions are concerned, it is likely there was an array of settings, from highly alkaline around cool hydrothermal systems in ageing ultramafic lavas, to highly acid in proximal settings to active basaltic volcano vents.

Elsewhere in this book (Chap. 1) we argue that, other things equal, the faint young Sun would render surface conditions cool to icy, except immediately after very large-scale impact events. Ice is a good insulator, and ambient geothermal heat would be trapped underneath it. Thus an ocean that was 5 km thick would not be frozen solid. Even today, Antarctica has huge lakes under the ice for this reason: the ambient geotherm. Volcanic vents would add to this ambient heat, and would have provided warm oases, where liquid water might reach the surface. Typical Hadean heat flows would have kept the ice $\sim$100 m thin, and over ridge axes or hotspots the ice would have thin enough ($<$10 m) that sunlight would have played a major role in the ice sheet's heat budget.

Even if the bulk of the planetary surface was ice, some pools would exist around the abundant active volcanoes (especially given the higher potential temperature of the mantle) or as leads between jostling ice masses. The zircon evidence for subduction implies that crustal blocks sank toward the mantle, which suggests that new crust was forming at the surface (as opposed to say a stagnant lid regime). We argue elsewhere in this book that Hadean tectonics would have superficially resembled plate tectonics in having sea-floor spreading and subduction. Hadean mid-ocean ridges may have been komatiitic, with huge far-running lava flows. Hence large pools of liquid water would be expected along the lengths of the mid-ocean ridge systems, perhaps under thick icy cover. Because mid-ocean ridges are linear and connected, there could have been major warm sub-ice lakes.

On very early Venus, D/H evidence implies a warm or hot ocean existed, liquid throughout, perhaps several km deep (Watson et al. 1984). On Mars, conditions would be more like the Hadean Earth, with short-lived hot phases after major impacts, but with a puddle ocean and active tectonics (Sleep 1994). The volcanism would be less vigorous and perhaps somewhat cooler—dominantly mafic rather than ultramafic. Heat flow would have declined faster. Impacts would create major topographic features such as depressions and lakes. Impacts would also cause porosity in subsurface rocks, which would be taken up by hydrothermal fluids. On Mars this porosity would extend several km deep. On Earth, with higher gravity and temperatures, porosity would anneal better.

Around hydrothermal system vents in all three planets, muds would have formed. On early Earth muds were probably dominantly palagonitic, rich in smectite clays (nontronite, saponite, etc.) characteristic of pillow debris. However the presence of old zircons implies that locally at least there were granitoid bodies, and thus it is likely that some muds were felsic-derived clays.

 Springer

4 Impact Modeling—Tilling the Surface: Processing the Habitability Zone

A young planet experiences major impacts. Each such major impact would have tilled the surface of the planet. Very large impacts destroy extant habitability, though may increase it later. More modest impacts can enhance habitability. Craters exhume material and thus investigation by drilling may not always be warranted: the experiment is already done.

The history of impacts on early Earth is discussed in the chapter by Zahnle et al. in this book. The Moon is the chief source of data. The face of the Moon has been sculpted by uncounted myriads of impact craters, ranging from vast basins thousands of kilometers across to the microscopic. Earth, because it is bigger, was hit by twenty projectiles for every one that hit the Moon, and with a high degree of confidence we can assert that Earth was hit by several projectiles that were much bigger than any that ever hit the Moon. The big ones posed a serious recurrent hazard to life on Earth (Zahnle and Sleep 1997, 2006).

Figure 2 shows a model realization of the aftermath of a very large impact on Earth. The model is based on conservation of energy and the radiative properties of steam atmospheres. It is akin to, but simpler than, the model of Earth's cooling history following the Moon-forming impact described in the chapter by Zahnle et al. (this volume). The key features of the model is that it presumes that a significant fraction of the impact energy of a very large impact is invested in making and spreading rock vapor around the planet. The rock vapor cools by radiating upward to space or downward onto the seas. In a big impact the latter is the bigger term because the rock vapor atmosphere is hotter at the bottom than at the top. As the rock vapor cools it condenses into rock raindrops that fall out and quench in the seas. The energy absorbed by the seas—the thermal radiation and the cooled molten rain—evaporates water. The water vapor builds up in the atmosphere until the energy of the rock vapor is exhausted. Thereafter the steam atmosphere cools radiatively until the excess water

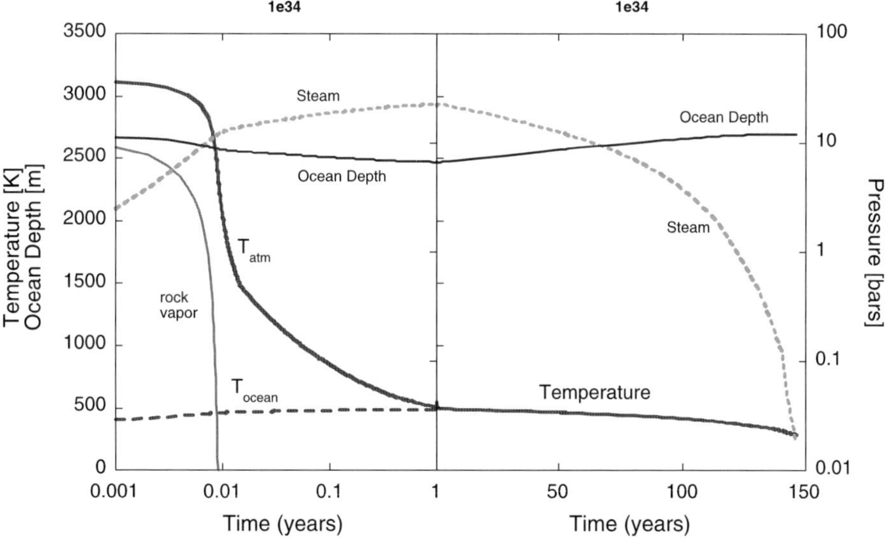

Fig. 2 Aftermath of a S. Pole-Aitken or Hellas scale impact on Earth. The energy released is 10^{27} joules. The figure plots ocean depth, sea surface temperature, atmospheric temperature, and the gas pressures of rock vapor and steam while in the atmosphere. The modern terrestrial atmosphere contains 1 bar of nitrogen and oxygen

vapor has fully rained out. How long this takes is governed by the thermal energy of the atmosphere, the latent heat of condensation of the water vapor, and the radiative physics of the runaway greenhouse effect (see the chapter by Zahnle et al., this volume). More details describing these sorts of models are given by Zahnle and Sleep (1997, 2006) and Sleep and Zahnle (1998).

The impact modelled in Fig. 2 is comparable to the largest impacts recorded in the extant crust of the solar system: those which made the South Pole-Aitken basin on the Moon (2500 km wide, 13 km deep) or the Hellas Planitia (2100 km wide, 8 km deep) basin on Mars. The energy released in these impacts was on the order of 10^{27} joules. The figure plots ocean depth, sea surface temperature, atmospheric temperature, and the amount of steam and rock vapor in the atmosphere as a function of time.

The response occurs on two basic time scales—a period of a few days during which rock vapor is present in the atmosphere, and a longer period of more than 100 years during which the atmosphere contains abundant hot steam. Energy in the rock vapor evaporates a few hundred meters of seawater. The surface waters of the ocean heat to nearly 500 K, but the deep waters can remain cool. Hot fresh rainwater pools at the top of the ocean, so that the ocean is stably stratified both by temperature and by composition.

An impact on this scale would have been disastrous for photosynthetic life: in short, the habitat zone would be eliminated. However, hydrothermally-hosted life in deep water might have survived.

Figure 3 shows the results of a slightly smaller impact, on the scale of that which made the ~1000 km wide lunar basins, such as Mare Imbrium or Orientale. These basins are prominent because they host lunar maria—dark basaltic lavas that flooded the low parts of the Moon and make up the pattern of light and dark known as the "Man in the Moon". The energy released is 2×10^{26} joules. As before, there are two basic time scales – a day or so during which rock vapor is present in the atmosphere, and a longer period of 30 years

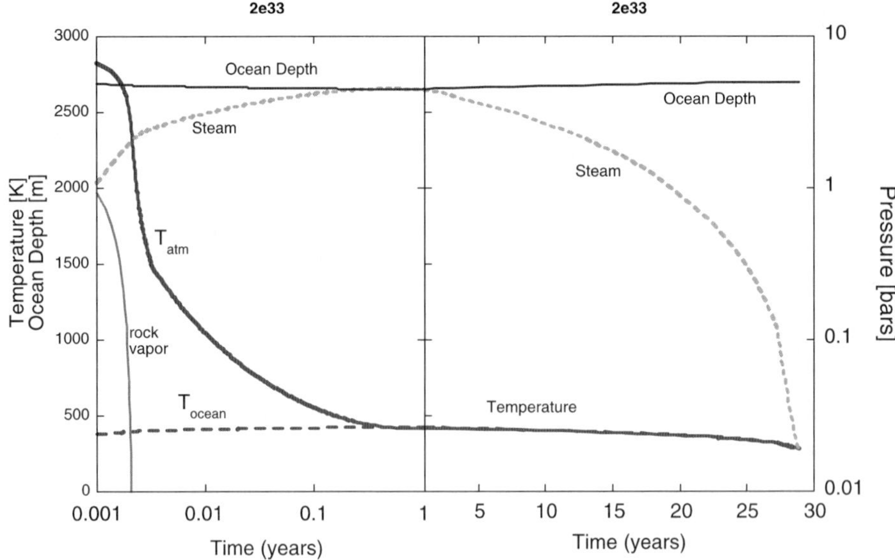

Fig. 3 Aftermath of an Imbrium or Orientale scale impact on Earth. The energy released is 2×10^{26} joules. The figure plots ocean depth, sea surface temperature, atmospheric temperature, and the amount of steam and rock vapor in the atmosphere

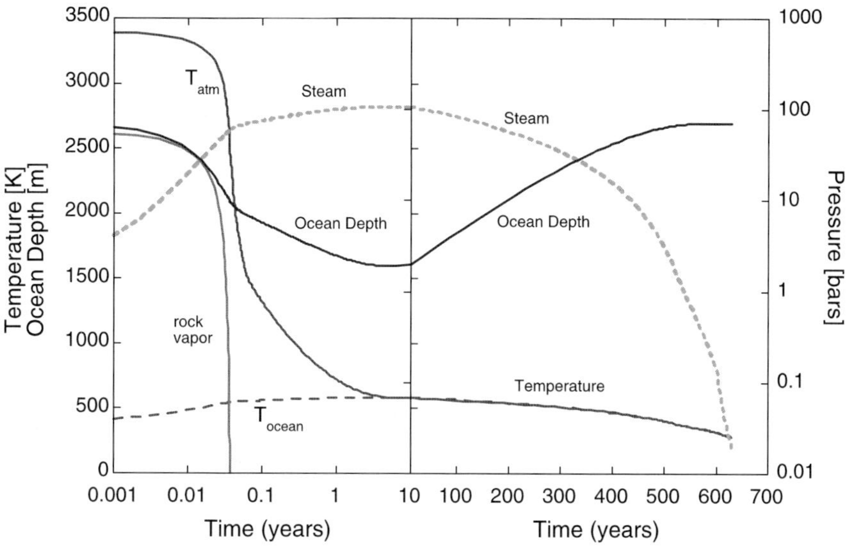

Fig. 4 Aftermath of a hypothetical 5×10^{27} joule impact of on Earth. A small number of impacts on this scale are likely on Earth at the time when the lunar Imbrium and Orientale impact basins formed. The figure plots ocean depth, sea surface temperature, atmospheric temperature, and the amount of steam and rock vapor in the atmosphere

during which the atmosphere contains abundant hot steam. In this case, energy in the rock vapor evaporates only 50 meters of seawater. The surface waters of the ocean heat far less, to nearly 150°C, but the deep waters remain cool. As before hot fresh rain waters pool at the top of the ocean, so that the ocean is stably stratified both by temperature and by composition, though less so than in the case of the larger impact. Even so, contemporary photosynthetic communities may have fared ill.

Figure 4 considers an impact even bigger than the Hellas and the S. Pole/Aitken events. The figure illustrates the aftermath of a hypothetical 5×10^{27} joule impact on Earth. A small number of impacts on this scale (perhaps a half-dozen) are likely on Earth at the time when the lunar Imbrium and Orientale impact basins formed. Aside from the bigger scale of the event, the figure is directly analogous to Figs. 2 and 3. The two basic time scales here are a few weeks for rock vapor and 600 years for an atmosphere of hot steam. There is enough energy in the rock vapor to evaporate more than a kilometer of seawater. The surface waters of the ocean heat to 550 K. Boiled brine sinks into the ocean and contains enough energy to heat the unevaporated sea waters by 50 K. Whether the deep oceans actually do get hot depends on whether the oceans mix. The deep waters are at first cold and dense and therefore stable against convection. However the oceans may mix by other means, for example through shear instabilities if currents are strong, as they might be expected to be. Later, hot fresh rainwater would tend to pool on top of the remnant salty ocean. Again, static stability would inhibit convection, and it is possible albeit not guaranteed that the hot fresh waters would long remain afloat atop the colder denser remnant waters. Ecological consequences would be severe and mesophilic ecosystems may not survive.

 Springer

5 Adding Components of Habitability to the Planetary Skin: The Impact Heritage on the Surface Inventory

Giant impacts totally rework the planetary body (Sleep et al. 2001). The largest (bodies of hundreds of km diameter) create a magma ocean and a huge steam or molten silicate atmosphere. Were one to have occurred in the past 3.5 Ga (billion years), life would have been eliminated. Large impacts (say 5–50 km diameter) do not terminate microbial life, but can cause mass extinction bottlenecks, even among prokaryotes. Impacts on a smaller scale do not produce any global changes but can provide changes in local environment, which can be favorable for habitability (warm ponds, nutrients, etc.) (Sleep et al. 2001).

The volume of a crater produced by a hypervelocity impact is hundreds of times larger than that of the projectile. Most of that volume is fractured rocks, which fall to surface near the crater. Solid state shocked rocks can contribute to habitability by creation of multiple rock cracks with fresh (i.e. chemically active) surfaces, moderate warming, and possibly by some effects of shock metamorphism (decomposition of volatile bearing minerals, mobility of rock-forming elements, etc.).

The volume of melted rock is from several to tens of times of the volume of the projectile. Impact-induced melting releases incorporated volatile components. Part of the melt is deposited in large craters as sufficiently well mixed and rapidly solidified bodies, which provide long term heating of subsurface crater material stimulating hydrothermal activity. For typical asteroidal impacts (with velocities of 10–25 km/s) silicates are not completely vaporized (Pierazzo and Melosh 2000). Partial vaporization creates a gas phase of volatile elements and leaves refractory elements in the residual melt. The vapor is about equal in mass to the projectile. This vapor forms fireball ejecta with starting temperatures up to 5000–6000 K and pressures up to several thousands of bars. As the vapor plume expands it cools, and condenses.

Gerasimov et al. (1999) have shown that the end products after expansion and quenching at high-temperature appear to be formed in strong disequilibrium, quite different from products expected in the normal planetary environment. Once formed, the fireball products are widely dispersed over the planetary surface. This must have provided a major input of chemically reactive components on the Hadean surface, capable of providing the thermodynamic contrasts to support metabolism.

5.1 Impact-Produced Gases

Low-velocity impacts decompose volatile-bearing minerals to release H_2O, CO_2 and SO_2. Gerasimov et al. (1984) showed that impact vaporization thermally releases oxygen as O_2 and O into the plume. Though its lifetime would have been extremely short some oxygen may have escaped into the wider atmosphere. Carbon gases are also released in this process, mainly in the form of CO and CO_2 in a ratio of about $1:1$, close to the thermodynamic equilibrium ratio (Gerasimov et al. 2002). SO_2, COS, H_2S, and CS_2 also occur in impact processes (Ivanov et al. 1996; Gerasimov et al. 1994a, 1994b, 1997). Nitrogen is released from impacts mainly as N_2 and some oxides NO_x. Synthesis of HCN and traces of CH_3CN occurs. The amount of nitrogen in the impact-released gases is limited by its trace concentration in meteorites and surface rocks. More nitrogen can be involved in impact chemistry by interaction of a projectile or expanding plume with a nitrogen-containing atmosphere. Passage of a large asteroid through the Earth's atmosphere may in theory have produced nitrous oxides and hydrogen cyanide in an air

 Springer

shock wave. Acidic rains from dissolution of nitrous oxides in the water could be sufficient to increase ocean acidity and to perturb biological activity (Fegley et al. 1986; Prinn and Fegley 1987).

5.1.1 Redox State of Impact-Induced Gases

Gas mixtures formed after impacts are in disequilibrium compared to normal conditions. In addition to oxidized components the gases contain significant quantities of reduced components, including the simultaneous presence of H_2 and O_2, SO_2 and H_2S, CO_2 and CH_4. The most abundant reduced gases in quenched mixtures are H_2, H_2S, CH_4, and light hydrocarbons up to C_6H_6. Coexistence of O_2 and H_2 in high temperature silicate vapors is supported by thermodynamic calculations (Gerasimov 2002).

Iron is present as Fe^0, Fe^{+2}, and Fe^{+3} reflecting complex redox processes in the vapor. Formation of metallic iron is accompanied by the increase of the portion of Fe^{+3} compared to the starting sample. The main rock-forming element, silicon, has complex redox behavior, detected in Si^0, Si^{+2}, and Si^{+4} states (Yakovlev et al. 1993).

5.1.2 Formation of Organics During an Impact

Organic molecules can be synthesized in impact-induced vapors. Reported impact-generated hydrocarbons include: CH_4, C_2H_2, C_2H_4, C_2H_6, C_3H_4, C_3H_6, C_3H_8, C_4H_{2-8}, C_4H_{10}, and C_6H_6 (Mukhin et al. 1989). Production of unsaturated hydrocarbons is noticeably higher than of saturated. The only experimentally proven oxygen-containing organic molecule is likely to be CH_3CHO. The output of organics in and after an impact is correlated with the total amount of C and H in the starting material of the impact. The mechanism of hydrocarbon synthesis is still unclear. Possibly Fischer–Tropsch synthesis occurs (Zolotov and Shock 2000; Sekine et al. 2003), involving reaction of carbon monoxide and molecular hydrogen (abundant in the vapor). This reaction uses surfaces of condensing particles as catalysts and goes different ways depending on the type of catalysis (Chichibabin 1953)

$$nCO + 2nH_2 \xrightarrow{Ni,Co} C_nH_{2n} + nH_2O,$$

$$2nCO + nH_2 \xrightarrow{Fe} C_nH_{2n} + nCO_2.$$

The role of surface catalysis is also supported by observed synthesis of sufficient amount of complex kerogen-like organics bound to condensates, which were insoluble in solvents but detected by C–C and C–H bonding (Gerasimov et al. 2002).

5.1.3 Impact-Generated Silicate Vapor

Most of an impact-generated plume is silicate vapor. This mostly condenses into nanoscale particles during expansion and cooling of the plume (see Fig. 5). The cumulative mass of these particles is comparable to the mass of the projectile. Impacts on a scale similar to the putative K/T event would provide global dispersion of these particles, generating a dusty atmosphere, affecting global albedo and hence climate.

Analysis of experimentally produced condensates shows that the solids made in the experiment have structures that are highly out of equilibrium. Such disequilibrium products of late Hadean impacts may have provided a useful supply of reductants and oxidants to the first living communities.

 Springer

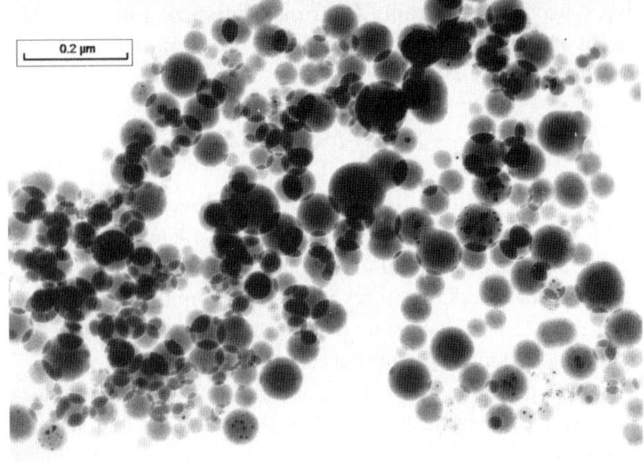

Fig. 5 Transmission electron microscope image of typical condensed particles. These are formed in an expanding vapor plume after high-temperature vaporization of a pyroxene (from Gerasimov et al. 1999)

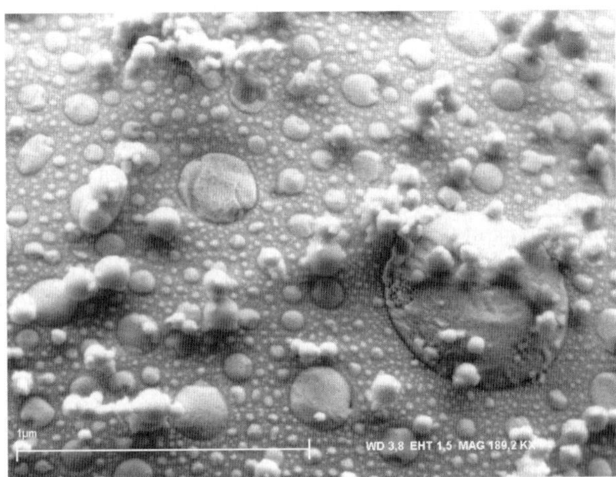

Fig. 6 Scanning electron microscope image of Fe-metal and FeS (*large*) liquates on the surface of $\sim$9 μm glass spherule. From experiment on high-temperature pulse heating of carbonaceous chondrites (Murchison). The *scale bar* is 1 μm (from Yakovlev et al. 2002)

X-ray photo-electron spectroscopy analysis indicates the simultaneous presence of structures with isolated, chain, and framework polymerization of silicon–oxygen tetrahedrons (Gerasimov et al. 2002). Multivalence elements usually exist in collected condensates in all possible states. Redox behavior is also experimentally observed in impact condensates containing the main rock-forming element, silicon, which has been detected in Si^0, Si^{+2}, and Si^{+4} states.

Iron in impact condensates as Fe^0, Fe^{+2}, and Fe^{+3} may have been important in habitability by providing redox contrast. Metallic iron particles absorb siderophile elements and other reduced elements (Si, Ni, S, Mn, Pt, etc.). Such compositions in turn provide catalysts for various properties. Impact reduction of iron into a metallic state proceeds with formation of metallic nanoparticles. These are immiscible and segregate into a separate phase from the silicate melts . They can be dispersed into the vapor or found on the surface of silicate droplets (see Fig. 6).

 Springer

6 Habitability Zones on the Surface: The Volume of the Oceans, the Emergence of Land

We now move from considering the inventory of materials to a discussion of the wider environmental constraints put on habitability by planetary tectonics.

Significant Freeboard, that is, the height which continents protrude from the ocean, is probably necessary for habitability. The oldest large body of rock, Greenland's Isua belt (up to ~3.8 Ga), includes clastic sedimentary material—i.e. liquid (presumably water) was present. Unlike Venus, which has a unimodal relief, the Earth's surface is divided into continents and ocean basins. Sleep (2005) has pointed out the lucky coincidence that Earth, as always the Goldilocks planet, has just enough water that tectonics and erosion/deposition tend together to maintain continents that marginally rise above sea level. Continents are not inundated, nor are they immense and dry, as on Mars. Hynes (2001) showed a relationship between continental crustal thickness, continental crustal volume, and mantle temperature. Buoyancy of the oceanic lithosphere is controlled by the amount of melting at the ridges and mean age of the ocean floor (themselves functions of mantle temperature).

Galer and Mezger (1998) investigated the antagonistic relationship between uplift and rifting processes driven by tectonics and the erosion and deposition of the sedimentary cycle. They inferred that the balance between the two was the main mechanism that kept continental freeboard constant, by regulating continental thickness. From an assessment of the geological settings of ten relatively undisturbed greenstone belts, they inferred that at 3 Ga ago the mean continental thickness was ~46 km at the time of crustal stabilization, and the mean thickness of oceanic crust was around 14 ± 2 km. This thickness of oceanic crust tallies with evidence for a somewhat hotter mid-Archean potential temperature of the mantle (Nisbet et al. 1993).

As far as habitability is concerned, it is clear that some continents have protruded from the Earth's oceans ever since the mid-Hadean (Nisbet 1987). In searching other planets, this may be a useful though not perhaps necessary criterion: that there is freeboard.

7 The Small Scale—From the Volcano to the Ion: Supply of Key Components for Biochemistry

Life is small, especially single-celled microbial life. Habitability is defined on the scale of a cell, not a planet. A planet is only habitable if it is habitable at the very smallest scale. There must be a supply of key components that is accessible to life that is restricted in movement on the micron scale.

At the chemical core of biochemistry are the metal centers of the metalloenzymes. These are 'house-keeping' proteins, many of which may have origins in the very greatest antiquity. Iron in particular is central to a wide array of enzymes that appear to be of very great antiquity. Typically Fe is in association with S. The iron enzymes include catalases, peroxidases, ferredoxins, oxidases, and all nitrogenases. Nickel examples include enzymes such as the hydrogenases (used in dealing with hydrogen and probably very ancient) and urease (essential to the nitrogen cycle). Carbon monoxide dehydrogenase, which is at the center of the acetyl-coA metabolic pathway, contains nickel, zinc, iron and molybdenum. Methanogens use coenzyme F_{430} and hydrogenase, both nickel enzymes: thus nickel is essential to methanogenesis. Nitrogenase (at the core of the nitrogen cycle), is an Fe-Mo enzyme. A particularly interesting class of enzymes includes metal–4N groups. The best known of these include chlorophyll and bacteriochlorophyll, which have Mg at the center,

and haem, with Fe. Metal-4N groups occur in alkaline conditions, such as those in hydrothermal systems around komatiites.

The presence of Mo in sediment has been seen as evidence for oxic waters in deep time (Siebert et al. 2005), as oxic conditions are needed to mobilize the Mo. There that there are other ways to mobilize Mo: thiosulfate, which is plausible in Archean oceans, will mobilize reduced Mo. Arsenic resistance may be ancient (Gihring et al. 2003) and may even pre-date the last common ancestor.

All these house-keeping proteins suggest a geological origin where the metals and S would be in abundant supply. The first life could not seek out metals in the way modern life does. The metals must have intruded into the organism: i.e. for the 'unskilled' organism to incorporate metal ions, the setting must have had abundant and accessible supplies, such that entry of metal into the cell was inevitable, not a rare accident. The most probable setting is in and around hydrothermal systems. Fe–S, for example, immediately suggests hydrothermal settings, as does the Fe–Mo association. Moreover, the commonness of metal–S groups immediately suggests sulphides to a geologist: hydrothermal settings.

The high potential temperature of the Hadean mantle is relevant here. If the temperature of the mantle in the late Hadean was, say, around 1800 K, then komatiitic volcanism would be extremely common. Komatiites are highly magnesian: they are very hot; they contain abundant Ni. During submarine eruptions, hydrothermal activity probably begins very early, as a komatiite lava flows, creating Ni–S rich layers in the cooling lava. The high MgO content means that the fluids can be very alkaline. We speculate that Ni-enzymes and also Mg–4N complexes first evolved in such settings. Had the mantle been cooler (i.e. no komatiites), then such settings would not have been possible. In other words, the high potential temperature and the consequent presence of komatiites may have made the Earth much more habitable than a similar but colder planet, more conducive to the origin of life as we know than the cooler Earth we live on now.

Saito et al. (2003) have studied the trace metal preferences of cyanobacteria. They showed that trace metal preferences in cyanobacteria are consistent with their evolution in a sulphidic ocean setting. Saito et al. (2003) suggested a variant on the hydrothermal hypothesis put forward above. They pointed out that in an early anoxic ocean, dominated by high concentrations of Fe, and with abundant Fe supply, and dissolved sulphidic species, the relative availability of trace metals would have been similar to the availability in a sulphidic system. In order, the availability would have been Fe > Mn, Ni, Co $\gg$ Cd, Zn, Cu. Saito et al. (2003) suggested it is possible that strong aqueous metal-sulphide complexes were as important as mineral precipitation in incorporating metal enzymes into biology. Possibly marine geochemistry and marine biology have co-evolved.

8 The Small Muddy Doorway of Habitability

There must have been a pre-biotic 'doorway' (not a window, but an entry point) of habitability, when life could begin. But this doorway would have been small. It would only have been open where there was local thermodynamic disequilibrium, created by geological and atmospheric/space contrasts.

Figures 7 and 8 illustrate this. The late Hadean surface environment would typically support only small and local redox contrasts. Volcanisms would have been prolific, given the mantle's high potential temperature. Volatiles such as SO_x that were degassed from volcanoes would have been oxidized in an atmosphere that was photolytically losing H to space: this would provide a somewhat more oxidized ocean. Lavas erupting would provide a

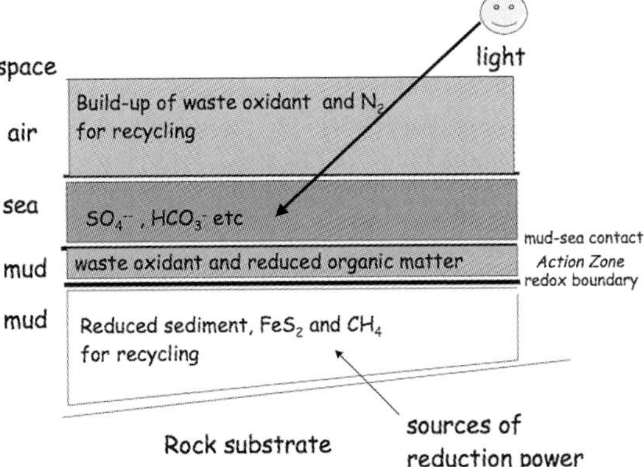

Fig. 7 Mud—the habitat

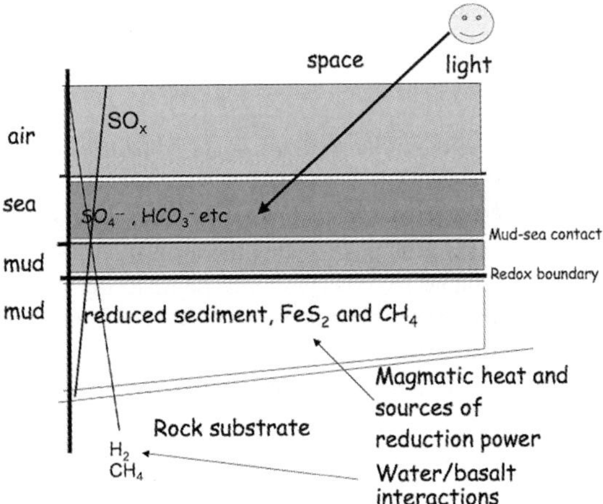

Fig. 8 Thermodynamic contrast before life. The atmosphere would have been marginally more oxidized (because of H escape to space and more generally from photolysis) while the interior would be more reduced

somewhat more reduced geological substrate. Methane haze and CO from photolysis would also be accessible.

A major early source of redox contrast may have been supplied by the hot lava via the intermediary of H_2. When lava and seawater interact, hydrolysis of ferrous oxide components in rocks produces hydrogen, e.g.:

$$3Fe_2SiO_4 + 2H_2O = 2Fe_3O_4 + 3SiO_2 + 2H_2.$$

The liberation of H_2 is always associated with formation of magnetite as this phase stabilizes ferric iron at low redox potentials. Epidote too does this. Hydrothermal systems similarly can produce inorganic methane.

The crucial site of contrast would be the ocean/mud interface, especially in and around hydrothermal systems. 'Mud' is the clay of life. In the Hadean, muds would have been sparse and magnesian–saponitic and nontronitic, not aluminous kaolins. Perhaps it was in the alkaline muds around hydrothermal vents from a recent seafloor komatiite flow, that key early housekeeping metal–4N proteins like the cytochrome family first formed. In such settings, brucite would scavenge borate and phosphate from seawater. Abiotic formation of pentoses, particularly ribose, would be promoted by the high pH (Holm et al. 2004) and boron (Ricardo et al. 2004).

 Springer

The presence of redox contrast between different parts of the habitat is essential. An ongoing contrast is only sustainable if geological processes continually introduce supplies of relatively oxidized and relatively reduced species. The supply of reduced species in a hydrothermal system, especially one that involved very hot ultramafic rocks, would be abundant. Both H_2 and CH_4 would be likely. However, the supply of relatively oxidized material must depend on photons. Oxidized species would originate in the air. They would have to exchange with ocean waters by rainfall and wave bubbles, and then the supply to the putative early cells would depend on the interaction between water currents and the water/mud interface.

Sulfate, thiosulfate, elemental sulfur, and perhaps carbonyl sulfide (OCS) would be the most likely carriers of redox contrasts. They would all rain out of the atmosphere into the ocean. Both thiosulfate and OCS can help to mobilize metals. On the modern Earth, sulfur in arc volcanoes is in part from subduction of sulfide originally precipitated in the ocean crust by hydrothermal circulation, which in turn derived the S from sulfate. In the modern ocean the sulfate is mostly in this oxidation state because of photosynthesis. In the pre-photosynthetic planet, sulfate would thus not necessarily be so readily available.

9 Exploiting the Muds: Widening the Habitability of the Earliest Archean Ecosphere

Life makes its own habitat. One of the many paradoxes of life is that the planet is habitable because it is inhabited. Today the atmosphere is an almost wholly biological construct: the nitrogen fluxed by nitrifying and denitrifying bacteria, the oxygen and carbon dioxide cycled by photosynthesis and respiration, the methane by methanogens and methanotrophs. Even the abiotic argon is arguably biologically controlled, as it is brought to the surface via granitoid magmas, that need subducted water, and water is stabilised on the planetary surface by the biological thermostat.

The arrival of life immediately makes the planet change: life generates its own habitability. This is because life in one niche of habitability immediately creates new niches of habitability. Imagine, for example, a tiny population of first living cells existing in a small vesicle in porous rock on the outer edges of a hydrothermal system, exploiting a supply of H_2 from water reacting with the rock, or using the redox contrast between a) incoming fluxes of slightly more reduced S species in the hydrothermal fluids and b) intermittent inputs of slightly more oxidized ambient seawaters. For the first few generations, the population will expand until the local habitat's resources are exploited. But there will be much cell death, either in 'drought' periods when one or the other inputs is absent for too long, or simply because of the Malthusian corrective that the population expands until some cells are driven to the edge.

Once there is cell death, there is reduced carbon lying on the floor—a new niche of habitability. There will be hydrogen coming up from the water-rock system below: thus there will be a new niche of habitability, using H_2 to turn the dead bodies into methane. In the air, methane haze and CO from photolysis provide substrate for life. The result of life's activities is to focus the redox boundaries more sharply: that in turn defines the zone of habitability and increases the productivity of the system, which goes from a disseminated one-niche to a set-apart two-niche system. Dead bodies foster life.

Sustained habitability makes sophisticated use of dead bodies. Like the ancient Greek hero Trygaeus, who saved Peace by riding to the Moon mounted backwards on a dung beetle, which he fuelled himself, life rises to the heavens on its own excreta. The use of hydrogen to recycle dead bodies made of organic carbon compounds will generate methane. Methane

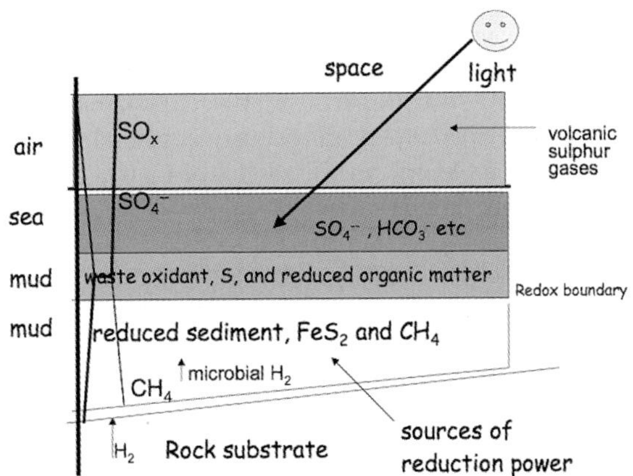

Fig. 9 Remaking the habitability of the mud, before photosynthesis. Even before photosynthesis began, life would begin to rearrange redox gradients in the mud. Dead organic matter would collect in the lower mud, making it more reducing (i.e. more CH$_4$ and H$_2$S, less SO$_4$) while upward release of more oxidized species such as CO$_2$ would make the upper environment more oxic. Microbial organisms typically focus redox contrast, and create a sharply defined zone where life flourishes

in turn is a fuel to consortia of cells that can use it to reduce sulfate: a new niche arises. By early Archean time, perhaps in the interval 4 Ga to 3.8 Ga ago, before the Isua belt was laid down, the pre-photosynthetic community could likely have created a very diverse set of niches, remaking the planet to be far more habitable than before life (Fig. 9). This biosphere would have been non-photosynthetic, depending purely on available redox contrast. In other words, the habitable part of the planet would have been limited to the close proximity of volcanic hydrothermal systems.

10 Habitability of Highly Mineralizing Environments

There is a potential paradox here: on one hand, for all the reasons above, volcanic hydrothermal systems were probably the first habitable zones on the early Earth; on the other hand, they are dangerous for life, because of the lethal effects of mineral precipitation on cells. Fossilization of microbial cells possibly occurs by precipitation of many different minerals: e.g. silica (e.g. Westall et al. 1995; Toporski et al. 2002; Benning et al. 2004), manganese oxide (Tebo et al. 2004), calcium phosphate (Benzerara et al. 2004a, 2004b). Cells can be totally entombed in the precipitates (Benning et al. 2004; Benzerara et al. 2004a, 2004b).

The ecological and evolutionary implications of fossilization are profound (see Caldwell and Caldwell 2004 for a conceptual view of these issues). Mineralization processes have a major influence on habitability. The traditional view considers microbes to be purely *"passive"* in the precipitation of the minerals: microbes modify indirectly the chemical conditions of the surrounding environment by their metabolic activity and hence foster mineral precipitation. This is substantiated by chemical/thermodynamical modeling (e.g. Frankel and Bazylinski 2003).

It is possible that the ability to biomineralize may only be a side-effect of metabolism. It does not provide a selective advantage to a microbe. It may actually be disadvantageous: the precipitation of few hundred nanometer thick layers of minerals around cells may limit diffusion of nutrients necessary for life; the formation of nanometer-sized crystals may disrupt cellular structures and hence be lethal. The fitnesses (i.e. an individual's ability to propagate its genes) of a microbe promoting biomineralization vs. that of a microbe inhibiting biomineralization have, however, never been measured.

Some strategies developed by microbes to inhibit precipitation of minerals on their membrane might be operating in highly mineralized environments. One example is the formation

 Springer

of sheaths, which are extracellular tubes surrounding microbes and that offer preferential nucleation sites for crystals. Microbial cells can get rid of these tubes and form new sheaths (Phoenix et al. 2000; Konhauser et al. 2001). Emerson and Ghiorse (1992) showed that sheathless variants arise spontaneously in laboratory cultures if predation and mineral precipitation are no longer present. Schultze-Lam et al. (1992) proposed that some cyanobacteria synthesize protein surface layers (S-layers), which provide nucleation sites for calcite precipitation and that they can shed when encrusted in mineral precipitates.

There is however no conclusive evidence that those strategies are more developed in microbes inhabiting highly mineralizing environments than anywhere else. Moreover, the comparison of the various studies on mineral precipitation on bacterial cells is confusing regarding whether this process is lethal or not (see Kappler et al. 2005). Some authors suggest that biomineralization on cell walls occurs during or after death of cells (e.g. Wierzchos et al. 2005), while others report the existence of viable cells encrusted by minerals (e.g. Phoenix et al. 2001; Tebo et al. 2004).

Seemingly passive biomineralization may actually provide an advantage to microorganisms. First of all, some studies contest that mineral precipitation on cells is always a passive mechanism (Castanier et al. 1999). They suggest that some species have a better ability than others to precipitate calcium phosphates or calcium carbonates (Castanier et al. 1999). Could this ability to precipitate minerals be advantageous in some cases?

Several potential advantages have been proposed for extracellular biomineralization: detoxification of toxic heavy metals, reactive oxygen species, UV light, predation or viruses; protection against immune system for pathogenic microbes; protection against grazing; storage of an electron acceptor for later use in anaerobic respiration; scavenging of micronutrient trace metals (e.g. Sommer et al. 2003; Mire et al. 2004; Tebo et al. 2004; Ghiorse 1984). Chan et al. (2004) suggested that extracellular iron oxide precipitation may provide energy to microbial cells. Whereas this has almost never been proposed for minerals like silica, calcium carbonates or calcium phosphates, it is usually considered that lead phosphate precipitation by bacteria is a detoxification process. Microbially driven calcium phosphate precipitation, though, shares many similarities with biomineralization of lead phosphate. It is thus reasonable to consider that microbial calcium phosphate precipitation may have a similar ecological 'status' as lead phosphate precipitation and such biomineralization processes may have played a role in the habitability of highly mineralizing seemingly "toxic" environments.

11 Sub-Surface Habitability

Modern marine microbial life does not only live in the water and water/mud interface. It is prolific down to several kilometers below the terrestrial surface and the seafloor. This is a large habitat, that preserved a wide anoxic environment even when the air became oxic.

The limit for life essentially is given by the increasing temperature and probably corresponds broadly to the 120°C isotherm (Kashevi and Lovley 2003) or, in the case of Earth, a depth of 2–12 km (typically 4 km), depending on the geotectonic setting. Both sediments (Aitken et al. 2004; D'Hondt et al. 2002, 2004; Parkes et al. 1994, 2005; Pedersen 2000) and igneous rocks (Moser et al. 2005; Pedersen 2000; Stevens 1997; Stevens and Mckinley 1995; Stevens et al. 1993) are inhabited by microbes. Colonized voids may range from microscopic pores to caves (Boston et al. 2001) and include active faults (Moser et al. 2005).

In muds and sediments undergoing diagenesis, very abundant microbial life occurs, living by exploiting the buried archive of more and less-reduced material: reacting dead organic matter with sulfate, for example, in processes such as anaerobic methane oxidation.

 Springer

In igneous rocks, microbial colonization has not just been documented as recent activity but there is also evidence of a fossil record of subsurface life (Furnes et al. 1999; Furnes and Staudigel 1999; Hofmann and Farmer 2000; Kretzschmar 1982; Mckinley et al. 2000; Trewin and Knoll 1999). There is free energy in basaltic glass relative to crystals. It has been proposed that endolithic microorganisms can form long channels in basaltic glass and use the oxidation of Fe(II) as an energy source (e.g. Furnes et al. 2005). Although the biogenicity of those structures is debated (Brasier et al. 2005), it is ascertained that microbes can promote dissolution and hence affect habitability (Rogers and Bennett 2004; Benzerara et al. 2005).

Compared to surface environments on Earth, the subsurface microbial biomass has very low productivity, but can be very long-lived and cell populations can be abundant (though many orders of magnitude less than on the surface). Most subsurface life is indirectly dependent on the surrounding inventory of surface derived oxidants or reduced organic debris. Even in the case of absence of surface-derived oxidants, microbial life based on methanogenesis using abiotically produced hydrogen (from water-rock interactions) and CO_2 is possible (Stevens and Mckinley 1995; Chapelle et al. 2002). It has been estimated (Pedersen 2000) that the subsurface biomass is similar to the combined continental and marine biomass. On a planet subject to surface catastrophe—for example a cometary or meteorite impact, or undergoing a snowball, the subsurface community provides habitability that protects the long-term continuity of life. Also, where surface conditions have degraded over time (e.g. Mars), the subsurface may still provide a water-rich (Burr et al. 2002; Clifford and Parker 2001) habitat for microbes long after the demise of all surface life. Moreover, peroxide reacting with iron or relict organic matter can provide a photolysis-based niche (e.g. on Mars).

The crucial factor for subsurface life is the presence of suitable redox couples to maintain a source of energy. Since the very action of microbial life will, in a closed system, eventually lead to a state close to chemical equilibrium, an input of oxidants from outside is required (Weiss et al. 2000). In sediments and fractured igneous rocks, diffusion and/or advective flow of water containing surface derived oxidants is required. This, in addition to temperature, is the limiting factor for subsurface life. However, non-terrestrial sources of redox contrast will in general be different from those on Earth. Here, even the mid-ocean ridge black smoker communities are dependent on surface oxygenic photosynthesis (which provides free oxygen and sulfate to seawater, the main oxidants reaching the hydrothermal community).

On a planet without oxygenic photosynthesis, or even without an ocean, redox contrast would be harder to find but may be related to photolysis reactions in the atmosphere as in the case of early Earth. The Martian subsurface is an obvious candidate to host such settings with accessible redox contrast, for example where large units of methane clathrate (produced abiotically by water-rock reactions with hot lava) are in proximity with oxic waters circulating to the surface.

Subsurface life on Earth is very ancient. Buried carbon and sulfur in the Belingwe belt (Grassineau et al. 2006) has probably gone through microbial processing during diagenesis. It is likely that subsurface sediments and igneous rocks have hosted microbial life, including anaerobic methane oxidizers, ever since the early-mid-Archean (Westall et al. 2006aa). On present Earth, microbial life is ubiquitous in sediments to depths of hundreds of meters and in igneous rocks to several km.

Subsurface habitats likely exist on other planets. Even where the surface is barely habitable, subsurface environments may be suitable for microbial life in a similar manner as on

Earth. The deep terrestrial subsurface is relatively good analogue for such subsurface habitats on other planetary bodies, such as Mars. Deep fluids in both cases are low in oxygen and largely controlled by water–rock interaction. The major difference is the presence of abundant oxidants at the Earth's surface, while the deep Martian subsurface is largely sealed from surface water circulation by permafrost. Besides the possibility of present subsurface life on other planetary bodies, the palaeosubsurface is a promising area to search for extinct life. Any former cavity in a rock must be considered a potential palaeohabitat. Non-sedimentary rock formations, especially where signs of aqueous alteration are present, must not be neglected in the search for former life. Suitable biosignatures allowing the recognition of such ancient subsurface life may be of morphological (well-preserved filaments) or geochemical nature (e.g. evidence of low-T sulfate reduction).

In this context, the discovery of methane in the Martian atmosphere (Formisano et al. 2004) is interesting. It could be a microbial waste product. However, an abiogenic source is also plausible—formed, possibly many millions of years ago, by hydrothermal systems around volcanism, and recently sweated out of clathrates by seasonal cycling. Given the ubiquitous oxidation of the Martian surface, sites of methane venting, whether the gas is of abiotic or biogenic origin, could support a marginal community of methane oxidizers.

12 The Intangible Air, Connector of Habitability: The Methane Greenhouse

In particular, methanogens would have been important in the pre-photosynthetic Earth. The methane would have had no obvious early biological sink (except perhaps anaerobic methane oxidation, but limited by sulfate supply), and methane would thus have collected in the atmosphere. Methane would have warmed the planet. Just possibly, the Earth is habitable because it is inhabited: possibly the methane produced by early life saved us from an eternal snowball (Pavlov et al. 2001).

Such a world would sequester large amounts of methane in clathrate hydrate in icy sediments. Large sequestrations can cause burst-out catastrophes: e.g. where a volcano erupted over a major clathrate-trapped pool of free gas. These very large methane bursts could then trigger global warming and release of the other clathrates (as may have taken place more recently, in the Paleocene/Eocene thermal maximum).

Over the eons, the carbon cycle has continuously released oxidizing power (Hayes and Waldbauer 2006). Most of this oxidizing power is now represented by Fe^{3+} that has accumulated in the crust or been returned to the mantle via subduction, but about 3% is the oxidized atmosphere (Hayes and Waldbauer 2006) and the rest sulfate. Nevertheless, in the sediments methanogens can still dominate.

Walker (1987) suggested that at times the Archean biosphere was 'inverted', with the atmosphere more reduced than the muds (Fig. 10). In the long run, the contrast between the more-reduced mantle and the more oxidized air, subject to solar photons and hydrogen escape to space (Catling et al. 2001), would restore the normal state-of-affairs, but potentially, Walker-world inversion events could have been sustained for long periods, when for some time the methane-rich atmosphere was more reduced than the mud. Such events would trigger major evolutionary changes. Mass extinction is unlikely in a microbial planet where, via winds and ocean currents, even one surviving cell can repopulate a niche very rapidly, but major changes such as a redox inversion could spur very rapid evolution into new niches.

 Springer

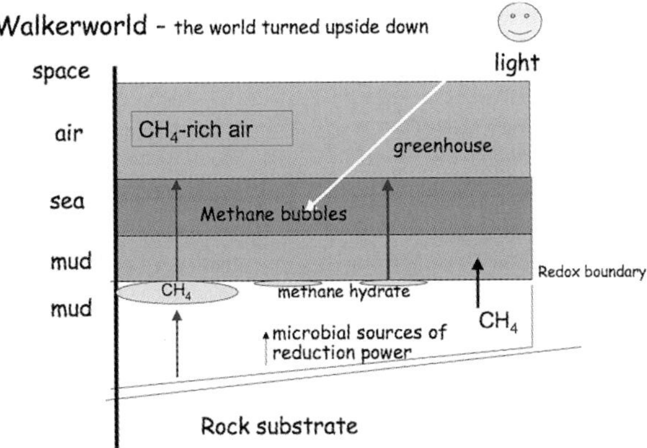

Fig. 10 Walkerworld—the upside-down biosphere. From time to time in the Archean, catastrophic release of methane from mud may have overwhelmed the slight oxidation reservoir of the atmosphere, creating an environment in which the atmosphere was more reducing than the magmagenic interior of the planet

13 Reworking the Habitable Mud: Anoxygenic and Oxygenic Photosynthesis, and Burrowing Metazoa

The history of the changes imposed by photosynthesis is beyond the scope of this chapter. Nevertheless, a brief comment is tempting. The evolution of *anoxygenic* photosynthesis, perhaps prior to 3.8 Ga ago, processing species such as H_2, H_2S, NH_3, CH_4, FeO, etc., would have vastly increased the habitability options. By 3.4 Ga ago, photosynthesis was well-established (Tice and Lowe 2004; Westall et al. 2006aa) presumably anoxygenic as the atmosphere was anoxic. Compared to today, the air was relatively CO_2-rich 3.2 billion years ago (Hessler et al. 2004).

Oxygenic photosynthesis, when it arrived, was a much more dangerous proposition than anoxygenic light-capturing. Waste oxygen made life banished anaerobes from the surface, and risked freezing the planet by destroying the methane greenhouse. This would have reduced habitability briefly. In the longer run, the vast increase in productivity (Kharecha et al. 2005) would create a much greater habitat, both on the seafloor and in the water.

The arrival of the nematode would have been the next dramatic change in habitability. By burrowing and fluxing fluid through the mud, the presence of worms very greatly expands the productivity and hence habitability of the sea floor (Fig. 11). Though the overall range of habitability would probably not have changed much (bacteria are deep in the sediment), the fluxing of nutrient and redox power between sea and mud would have increased markedly. Although microbial activity extends deep into the sediment column, on a microbial world photosynthesis and hence the associated major zone of high productivity is confined to a thin skin a few millimeters thick on an otherwise dead planet. In contrast, a biosphere with worms enlarges the high-productivity surface-linked community by pumping fluids around. The zone of prosperity becomes several meters thick, or more in good muds: the production of the zone of habitability inflates like the universe itself. Of course, the burial biosphere is much deeper, as much as several kilometers thick (and tens or hundreds of kilometers wide) in sedimentary basins filled with cool organic-rich, but here productivity is much lower, even if cell counts are high.

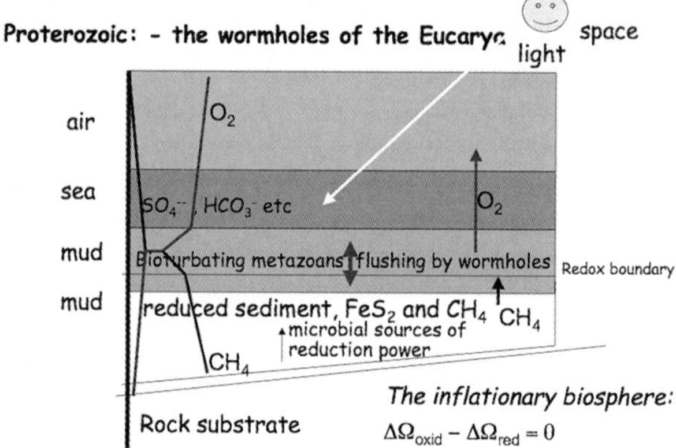

Fig. 11 Wormholes—the inflation of habitability. Where worms pump fluid up and down in the sediment, the access of deeper microbial consortia to oxidant is much improved. Cycling of chemical species is enhanced. This process creates a more productive deeper (reduced) zone and a matchingly larger more oxic zone. The interface is more focussed and there the redox gradient becomes sharper. The net redox sum remains the same, but the two opposite reservoirs increase

Bioturbation began in the Proterozoic - waste oxidant was flushed round into mud again, widening the biosphere

14 Habitability of Another World—Is There a Rock Cycle on Mars that Can Sustain Life?

Minerals and rocks are stable only under the conditions at which they form, though kinetics and activation energies permit sustained disequilibrium (if not, diamond rings would vaporise to CO_2). Changing these conditions will initiate metamorphism of the rock and its minerals. Therein is disequilibrium, and resources for life. On Earth plate tectonics regularly remodels the face of the planet and is central to the sustained supply of ions and disequilibrium for biogeochemistry. Is there also evidence for plate tectonics and related disequilibrium among surface rocks on Mars?

The discovery of magnetic lineation in the southern Martian highlands (Connerney et al. 1999) suggested that Mars experienced an early era of plate tectonics, forming surface material by crustal spreading in the presence of a reversing dynamo (Connerney et al. 2005). Sleep (1994) argues that tectonic features on Mars such as Gordii Dorsum resemble plate break-up margins and that the Tharsis Montes volcanoes, which are located at the south to north dichotomy boundary began as volcanic centers along a Martian subduction zone. This suggests plate tectonics produced the northern lowlands of Mars. Possibly on early Mars plates of cooled basaltic material might have subducted into a mafic-gabbro mush mantle. Water release would then cause voluminous tonalitic (TTG) melts.

The magnetic lineation is correlated with the oldest surface units, implying the dynamo's demise very early in the Martian evolution (Nimmo and Stevenson 2000). Yet there is evidence (Head et al. 2001; Frey et al. 2002) suggesting considerable crust formation after the inferred end of plate tectonics at about 4 Ga ago. However, early-ceszation plate tectonics models are difficult to reconcile with such substantial formation of post plate tectonics crust. The more efficient the cooling, and the more thermal power is available for the dynamo, the less likely is the production of crust after ceszation of the plate tectonics regime (Breuer and Spohn 2003).

Van Thienen et al. (2004), using buoyancy arguments, suggest plate tectonics was unlikely during early Martian history. There is only a relatively low operational temperature window for plates (1300–1400°C), outside which plates do not attain enough negative buoyancy to subduct on Mars, within reasonable time scales.

The lower gravity of Mars allows deeper hydrothermal circulation through cracks and hence more hydration of oceanic crust so that water is more easily subducted than on the

Earth (Sleep 1994). However, the chemistry of the SNC meteorites has been interpreted as implying Martian mantle has been dry since core formation (Dreibus and Wänke 1987, 1989; Carr and Wänke 1992). On the other hand, cooled basaltic material might subduct into a mafic-gabbro magma ocean and due to lower gradient with depth on Mars slabs could go down hundreds of kilometers (Warren 1993; Bridges and Warren 2006). The subducted water might be responsible for extensive volcanic eruptions.

Lenardice et al. (2004) postulated a model tectonic history of Mars. In this, Mars initially underwent active lid tectonics, driving a geodynamo and producing the southern highlands through some mixture of melt generation and crustal accretion. The growth of the crust increased the mantle temperature, lead to the ceszation of plate tectonics, with the tectonic style transforming into a stagnant lid regime. If the model is correct, plate tectonics was active, if at all, only for a very short period of less than 300 Million years at the beginning of the planet's history.

Evidence for rock diversity on Mars resulting from recycling induced by plate tectonics is hard to find—yet this is central to habitability. If Mars did not cycle its chemistry, it would soon have become uninhabitable, an unsuitable substrate, when life was starved of nutrient supply.

14.1 Martian Volcanism

The bulk composition of the SNCs as well as the bulk composition of the Martian soil is basaltic (McSween 1994; Rieder et al. 1997a, 2004; McSween et al. 2004; Mustard et al. 2005; Yen et al. 2005) with no evidence for alterations induced by subduction. Some rocks at the Mars Pathfinder landing site are characterized by higher SiO_2 contents $>55\%$ (Rieder et al. 1997b; McSween et al. 1999) indicating a more andesitic composition (although the measurement of higher SiO_2 composition could also be due to some form of alteration rind on the rocks, similar to desert varnish). On Earth, icelandites originate from fractionated crystallization of basaltic magma in rifting anorogenic environment. The composition of Martian andesitic basalts is broadly comparable to icelandites (McSween et al. 1999). Thus, the composition of Martian basalts shows no direct evidence of formation under aqueous conditions, but the high SiO_2 content may indicate a volatile rich basaltic component. However, Bridges and Warren (2006) report on the evidence of water in the evolution of shergottites give by their lithium chemistry and Dann et al. (2001) showed, that crystallization of shergottitic melts under water saturated conditions can lead to Fe and Al_2O_3 depletion and SiO_2 enrichment to produce andesitic-like melts. Re-melting of hydrated mafic rock in a mush magma ocean might have supported these processes.

Thus, although there is no clear evidence for subduction, there is abundant evidence of volcanism that could have supported hydrothermal systems that were habitable.

15 Water-Rich Environments on Mars

Volatiles are crucial to habitability (see Chap. 4 of this book). Not only are they essential for life, they also shape the enviroment. Although we do not have evidence for a rock cycle driven by plate tectonics, alteration of rocks due to weathering and transportation is common on Mars. The surface geology exhibits giant outflow channels which start fully developed at discrete sources and expand for hundreds of kilometers. (e.g. Sharp and Malin 1975; Baker et al. 1992; Carr 1996; Jaumann et al. 2002) branching down-stream and having various streamlined bedforms on their floors. Valley networks are smaller but nevertheless

show fluvial characteristics. Their origin is drainage dependent (e.g. Sharp and Malin 1975; Pieri 1980; Carr 1996; Jaumann et al. 2002), with various glacial and periglacial features (e.g. Squyres and Carr 1986; Kargel and Storm 1992; Neukum et al. 2004). There is evidence for continuing activity (Malin et al. 2006).

Erosion on Mars was most likely driven by wind and water, either in liquid form or as ice. Fluviatile phases are old and date back mostly to the Noachian and early Hesperian times (e.g. Baker et al. 1992; Carr 1996; Jaumann et al. 2005). Glacial activities are ongoing, even in younger Amazonian times. In parts they may be recent (Neukum et al. 2004; Head et al. 2005; Hauber et al. 2005b). Polar layered deposits occur. Mars had and has a hydrologic cycle that varied in intensity with time. However, compared to Earth the erosion on Mars is not well developed as indicated by the relatively rare distribution of fluvial and glacial features and low erosion rates (e.g. Arvidson et al. 1979; Carddock and Maxwell 1993; Golombek and Bridges 2000; Jaumann et al. 2005), which are a few orders of magnitude lower than on Earth. Nevertheless fluvial and glacial erosion induces a rock cycle driven by sediment production through weathering followed by transportation and re-deposition (Fig. 12).

Aeolian processes also contribute to erosion, by cycling the finest fraction of material. Mechanical weathering produces dust and sand-like material, which is redistributed by wind and deposited in dunes (Fig. 13).

Chemical alteration of rocks is found only in localized concentrations of hydrated phyllosilicates and sulfates such as gypsum and kieserite (Bibring et al. 2005; Langevin et al. 2005; Arvidson et al. 2005; Gendrin et al. 2005), mostly in northern circumpolar regions, layered terrain in Valles Marineris and in Sinus Meridiani, indicating aqueous environments but a minor role of water in modifying surface rocks. Due to the high cation to sulfur ratio in typical mafic rocks the sulfur needs to oxidize rapidly compared to clay formation so that CaO and NaO do not neutralize the acid. Therefore these environments are weak biomarkers, as is jarosite.

Sedimentary rocks are exposed in the Meridiani region at the Mars Exploration Rover Opportunity landing site. These sediments are flat lying, finely laminated with fine-scale cross lamination in some areas. They are sulfur rich and contain abundant sulfate salts (Squyres et al. 2004). These rocks are thought to be a mixture of chemical and siliciclastic sediments formed by episodic inundation of the shallow surface by water in combination with evaporation and desiccation probably under acid-sulfate conditions (Klingelhöfer et al. 2004; Squyres et al. 2004). Embedded hematite-rich concretions within these sediments also indicate the involvement of water. Sulfur, chlorine and bromine indicate chemical mobility and separation during a period of aqueous activity (Haskin et al. 2005; Gellert et al. 2004). Although there is strong evidence for chemical alteration of rocks at the Meridiani site, other places on Mars, particularly the Pathfinder and Gusev sites, indicate that physical weathering might have played a more significant role than chemical weathering in the rock alteration process on Mars (Morris et al. 2005).

The thin Martian atmosphere is not able to hold much water (less than 10 μm) vapor and thus is a very water-poor environment. Current Martian climatic conditions of about 6 mbar and seasonal average temperatures below freezing, prohibit water on the surface in most Martian surface regions and for the most time of the year. Small amounts of water may be feasible on the surface during summer daytime temperatures and in low laying areas, especially when perihelion occurs in Mars's northern hemisphere. That said, today's water occurs mostly in the atmosphere and at the poles. The water content in the atmosphere is about 1–2 km^3, which is equivalent to a 10 μm global water layer. One meter of equivalent global water layer translates to 1.44×10^5 km^3. In contrast, the water in the Earth's atmosphere would cover Mars with a global layer of 10 cm.

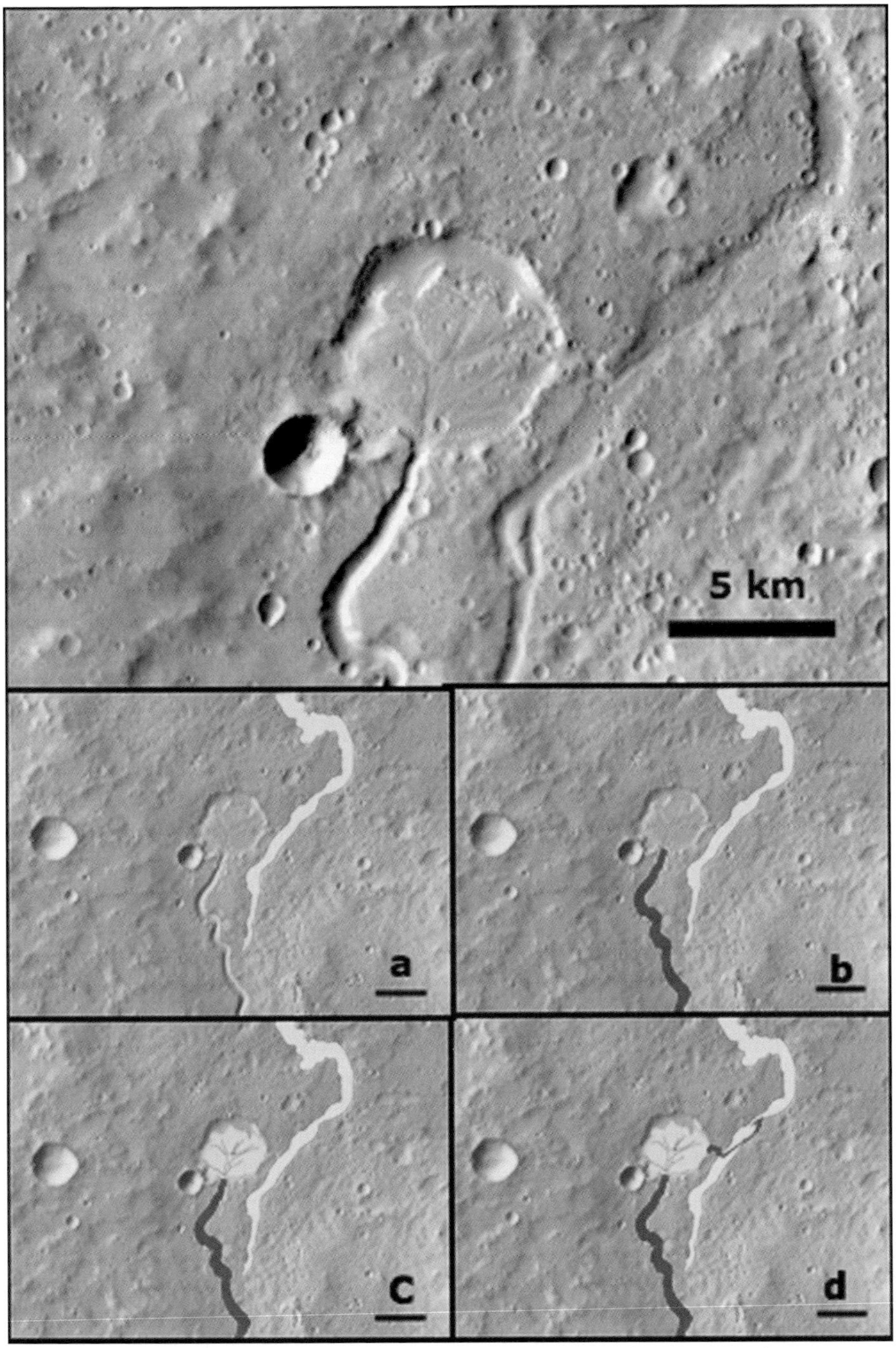

Fig. 12 Valley and delta in Xanthe Terra on Mars (Hauber et al. 2005a). Remnants of an eroded older channel (**a**) are superimposed by a younger channel (**b**) that ends with a delta feature (**c**) in a nearby crater, which shows a small spill of and outflow into the older channel (**d**) indicate different periods of fluvial erosion and deposition (Mars Express HRSC images, ESA/DLR/FUB HSRC orbit 905, 8.65°N 48°W *scale bar* 5 km)

Fig. 13 Barchan dunes of fine
grained dark material on Mars.
(MGS MOC Image, MSSS.
MOC FHA00451, 292.93°W
8.83°N, NASA/JPL/MSSS, MGS
release no. MOC2-88, 11 March
1999)

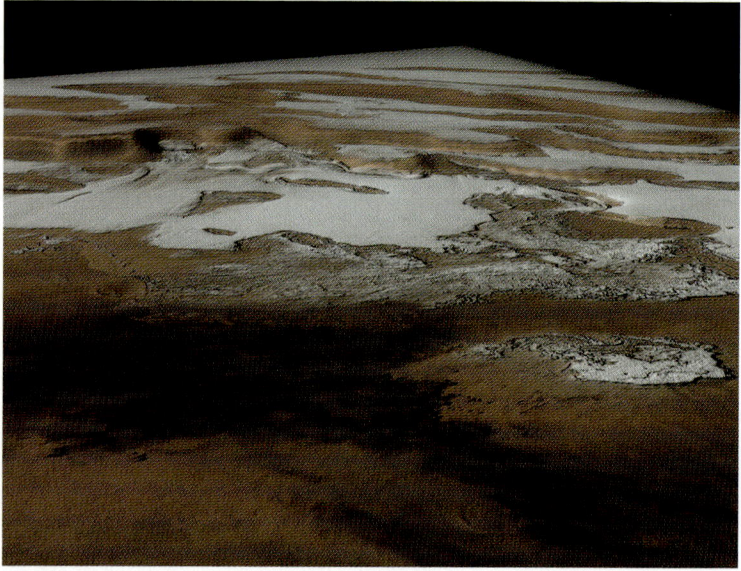

Fig. 14 Ice and layered terrain of the Martian North Pole. (Mars Express HRSC image ESA/DLR/FUB
(G. Neukum) HSRC orbit 1167, 78°N 117°E, width of image in the front ~100 km)

Apart from CO_2, the polar caps contain water ice about equivalent a global water layer
about 0.6 m thick. In addition the polar layered terrains (Fig. 14) are assumed to contain
the equivalent of a global water layer between 6–29 m deep (Kieffer and Zent 1992). The
uppermost few meters of the Martian regolith contain about 35 ± 15 wt% H_2O, at latitudes
>60°, as measured by the High Energy Neutron Detectors of the Mars Odyssey Mission
(Feldman et al. 2002) and about 1 wt% in regions of lower latitude (Bieman et al. 1977;
Boynton et al. 2002). The H_2O in the upper regolith is assumed to be not liquid but either
pore ice, adsorbed water (Möhlmann 2004) or mineral bound water (Boynton et al. 2002).
The total inventory of surface and near surface water is thus the equivalent of a global water
layer about 7–30 m deep.

 Springer

Clifford (1993) estimated the theoretical pore volume of the upper 8.5 km of the Martian crust at about 7.8×10^7 km^3 comparing to about 540 m equivalent global water layer. As the Martian geothermal heat flow decreases with time the thickenss of the permafrost (the cryosphere) increases, thus reducing the pore volume for liquid water. The thickness of the cryosphere is $z = \kappa\,[(T_m - T_a)/Q]$, with $\kappa =$ thermal conductivity, $T_m =$ melting temperature of ground ice, $T_a =$ mean annual temperature and $Q =$ geothermal heat flux. The geothermal heat flux was about 150 mW/m^2 at the beginning and has decreased to about 30 m W/m^2 today (Spohn et al. 2001) resulting in a mean increase of the cryosphere (the permafrost layer) from about 0.5 km to over 3 km. If water is present in the cryosphere it is frozen. However, according to the theoretical pore volume of 7.8×10^7 km^3, between 5.7×10^7 km^3 and 2.9×10^7 km^3 should have been available over time for ground water beneath the cryosphere, i.e. equivalent to a global water layer of about 400 m to 200 m. Thus, the Martian subsurface has a high potential for water rich environments, either as ice in the cryosphere, or as liquid groundwater beneath it.

However, was there enough water available on Mars to fill the pore volume? There is much evidence for this on the surface of Mars, which exhibits fluvial and glacial erosion. The amount of water involved in this surface erosion has been estimated by a number of authors (e.g. Carr 1987, 1996; Clifford 1993; Baker et al. 1991; Baker 2001; Jaumann et al. 2002, 2005) using eroded volumes of material, erosion rates and water to sediment ratios. Although the different estimates vary between equivalent global water layer of 80 m to 1200 m, they suggest there should have been water available to fill at least part of the available pore volume.

Valley networks (Fig. 15), chaotic terrain and outflow channels (Fig. 16) are the surface expression of subsurface water release or precipitation episodes. However, almost

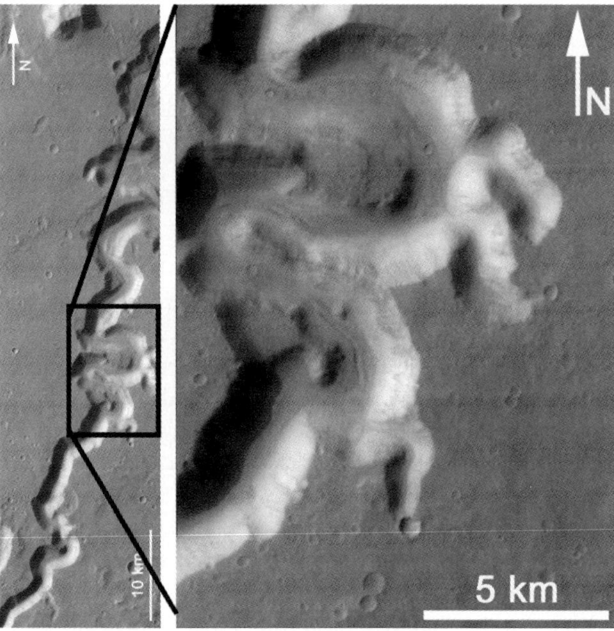

Fig. 15 Nanedi Vallis on Mars. Meander, terraces and interior channels indicate fluvial origin of the sinuous surface feature. (Mars Express HRSC image ESA/DLR/FUB (G. Neukum) HSRC orbit 905, 5.25°N 311.8°E)

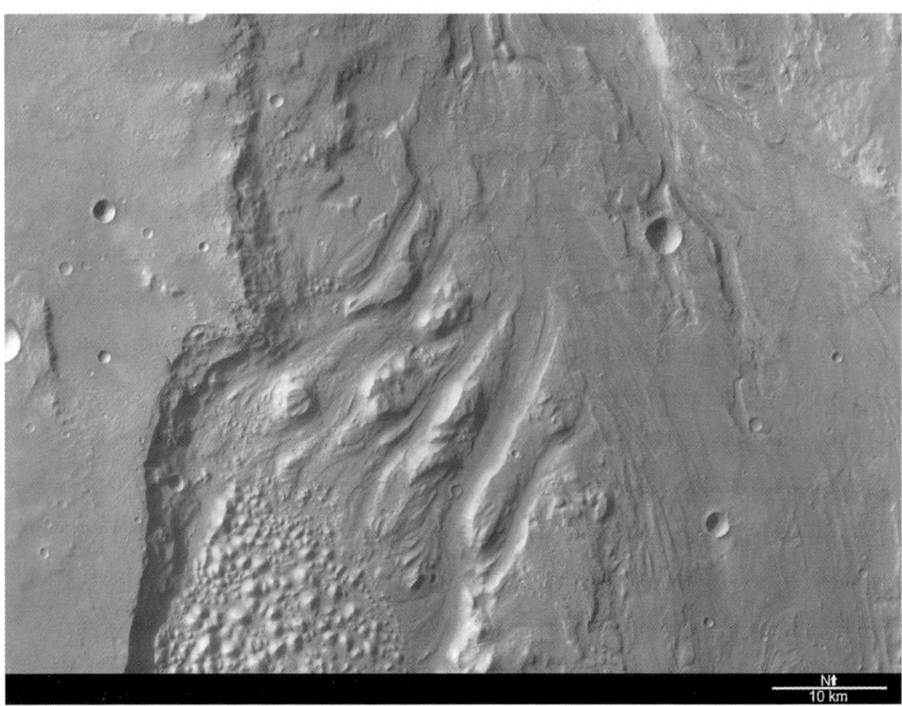

Fig. 16 Outflow channels in Iani Chaos on Mars. Large scale erosion and flow features indicate rapid water release. (Mars Express HRSC image ESA/DLR/FUB (G. Neukum) HSRC orbits 923 and 924, 1.5°N 342.2°E)

all of these features date back to 3.6 Ga or more ago (e.g. Scott and Dohm 1992; Carr 1995; Masson et al. 2001; Jaumann et al. 2002), indicating that significant surficial water environments outside the poles are very ancient and inactive today. Nevertheless, these fluvial features together with sedimentary features (e.g. Malin and Edgett 2000a; Squyres et al. 2004) (Fig. 17), show that water rich, and by implication, habitable environments on the surface existed in early Martian history.

However, all the fluvial and sedimentary features on Mars are relatively poorly developed, indicating inadequate drainage and erosion (Carr 1995; Carr and Malin 2000; Jaumann et al. 2005), and thus restricted time frames for water being present on the surface. This is consistent with the distribution of hydrated minerals (Bibring et al. 2005).

Relatively young erosion features (a few tens to hundreds of million years old) on Mars appear to be of glacial origin (Neukum et al. 2004; Hauber et al. 2005b; Head et al. 2005). Recent water related surface features are restricted to small scale gullies (Malin and Edgett 2000b) and wet debris flows (Reiss and Jaumann 2003). The gully morphologies are most consistent with a formation from near surface aquifers (Heldman and Mellon 2004). The discovery of very recently formed light toned gully deposits suggest, that liquid water flowed on the surface of Mars during the last decade (Malin et al. 2006).

In summary, although there are rare local recent water environments outside the polar regions, most of the water is expected in the porous subsurface and all traces of significant amounts of water on the surface are old (Fig. 18).

 Springer

Fig. 17 Layered deposits in the Valles Marineris on Mars indicate sedimentation. The layers are composed of the sulfates gypsum and kieserite (Bibring et al. 2005). (Mars Express HRSC image ESA/DLR/FUB (G. Neukum) HSRC orbit 243, 4.4°S 297.7°E length of deposit 60 km)

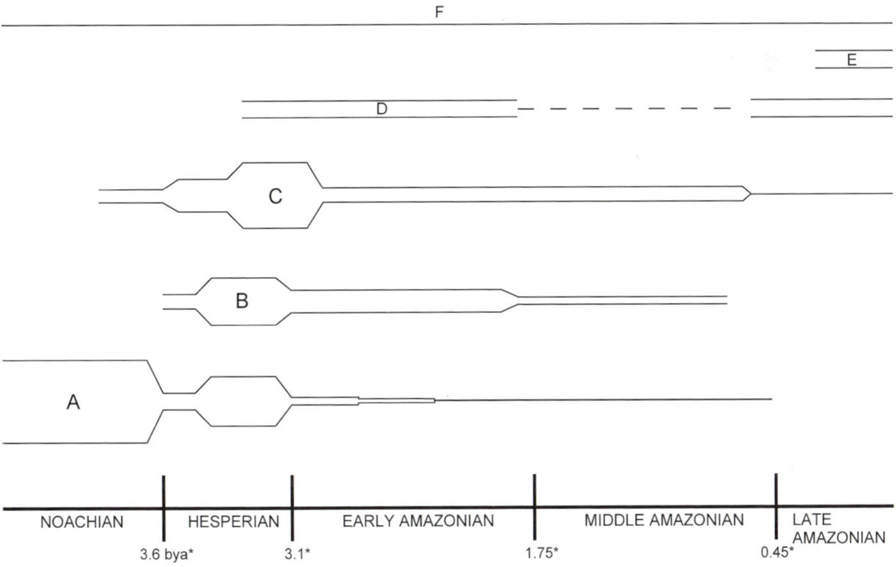

Fig. 18 Evolution of surface water and ice throughout the Martian history. *A* ... valley networks; *B* ... lacustrine deposits; *C* ... outflow channels; *D* ... glacial and periglacial deposits; *E* ... polar caps, low latitude glaciers gullies; *F* ... aeolian activity. (For references see e.g. Carr 1996; Malin and Edgett 2000a, 2000b; Masson et al. 2001; Jaumann et al. 2002; Head et al. 2003; Jaumann 2003; Head et al. 2005; Hauber et al. 2005b)

16 Crustal Geothermal Processes on Mars

Creating a habitable environment is a complex process involving a wide variety of interacting processes. But an energy source is the first prerequisite for any biological activity. The other prerequisite is temperature: life needs liquid. The seasonal average surface temperature is below freezing (though summer daytime temperatures can be above the freezing point

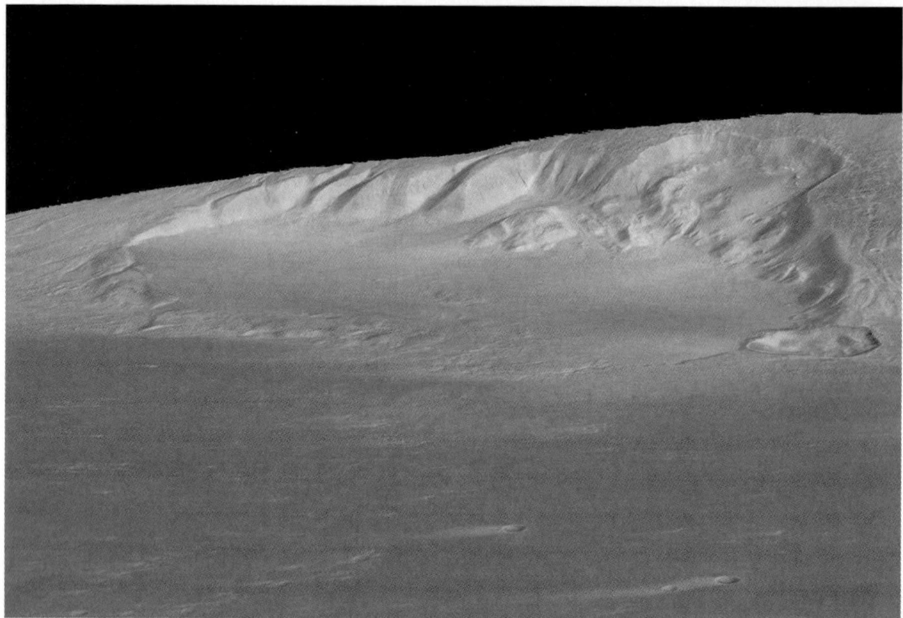

Fig. 19 Thermal modeling (Helbert et al. 2005) indicates that these glacial deposit on the northwestern flank of Hecates Tholus still contain ice (Hauber et al. 2005b). (Mars Express HRSC image ESA/DLR/FUB (G. Neukum) Location 148.8°E 32.8°N (*center of image*))

of liquid water, though not above the triple point). In the past and in the future, as orbital cycles produce periods of more intense local solar forcing, Mars has more clement episodes. Can life survive between periods of clement conditions by retreating underground?

Unfortunately our knowledge of geothermal processes or general thermal processes in the crust or even in the near surface of Mars is still very limited. Calculated geotherms (strictly termed Aresotherms, for Ares, Mars, as compared to Geos, Earth) from measurements of heat producing elements in Martian samples show that the interior is cold for many kilometers down. Cold, of course, is a subjective term: Hoffman (2001) estimated a crustal temperature increase with depth of 6.4–10.6 K/km.

Volcanoes provide activity and habitat. There is ample morphological evidence for continuous but episodic volcanic activity over the geological history of Mars—especially now with the image dataset obtained by Mars Express (for example Neukum et al. 2004). The youngest ages determined by crater size-frequency measurements are about 2 Ma ago, suggesting that the volcanoes are potentially still active today (Werner 2005). While we have not yet observed any modern volcanic activity it seems that the likelihood for localized hot spot activity or hydrothermal systems is high.

New infrared spectrometers (TES on Mars Global Surveyor, THEMIS on Mars Odyssey and OMEGA and PFS on Mars Express) have permitted a more detailed insight into the thermophysical properties of the surface of Mars (Christensen et al. 2004; Putzig et al. 2005; Formisano et al. 2005). Our knowledge of these properties is however still limited to the first few millimeters of the surface. Everything below this is accessible only by modeling. The stability of subsurface ice and the possibility of extant stagnant ice cores within glacial deposits remain controversial. It has been shown recently using thermophysical modeling (Helbert et al. 2005) that a morphologically identified glacial deposit on the northwestern

Springer

Fig. 20 A steaming fumarole on Mount Erebus with a 10 m ice-tower precipitated from H$_2$O-rich volcanic gas (from Hoffman and Kyle 2003)

Fig. 21 Inside a grotto below an ice-tower (from Hoffman and Kyle 2003)

flanks of Hecates Tholus (Fig. 19) (Hauber et al. 2005b) might still contain a stagnant ice core. Head et al. (2005) report several units on Mars which morphological evidence suggests may be glacial deposits. Many of these units are found on the flanks of volcanic edifices.

Hoffman and Kyle (2003) suggested the ice towers of Mt. Erebus (Figs. 20, 21) as analogues of biological refuges on Mars. They combined the idea of still existing near surface ice deposits with the assumption that there is still some localized volcanic activity on Mars today.

There are several examples from Mars that show a direct interaction between lava and ice in the geological history of Mars. The most obvious cases are the rootless cones seen in the northern lowlands. HRSC images show direct and violent interaction in the relatively recent geological history, for example at the scarps of Olympus Mons. Mars today is in relatively dormant phase, and any interactions which might be occurring today are presumably on a much less dynamic scale. Nevertheless, they may be driving local hydrothermal systems. Studying the geothermal processes in the first few tens to hundreds of meters below the surface of Mars today might thus uncover a wide variety of new habitats where biological activity may survive on this cold and dry planet.

17 The Evidence that Habitability Was Achieved: Early Life and Its Signatures

We know the Earth was habitable, but if we are to seek out the habitability of other planets, we must ask "what evidence do we have on Earth that habitability was actually achieved?". It is this type of evidence that we must use as a guide in seeking out life elsewhere.

Unfortunately the very dynamism of the Earth, essential for the upkeep of its habitability, is also the cause of the destruction of the earliest history of life. Plate tectonic recycling of the earliest continents and oceanic crust has effectively erased almost the first billion years of rock record. Although ancient zircons dating back to about 4.4. Ga exist (Wilde et al. 2001), and although ancient gneisses from about 4.0 Ga are exposed in Canada (Bowring and Housh 1995) and 3.7–3.8 Ga) metasediments occur at Isua, Greenland, the oldest low metamorphic grade sedimentary rocks are only ~3.5 Ga old. Sedimentary rocks are important repositories of life's history because of their formation in aqueous environments: as we have seen above, water is the "spring of life" (Brack 2002).

It is not the aim of this chapter to air the heated debates surrounding the oldest traces of life (see review in Westall and Southam 2006). Suffice it to say that volcanic and chemical sediments in the ancient terrains (~3.5–3.3 Ga) of the Pilbara in NW Australia and Barberton in South Africa contain a plethora of traces pertaining to life—carbon and nitrogen isotopic signatures, as well as macroscopic to microscopic traces in the form of stromatolites, microbial mats, biofilms and colonies of microorganisms. It is interesting to relate the occurrence of these traces to the habitability concept (Fig. 22).

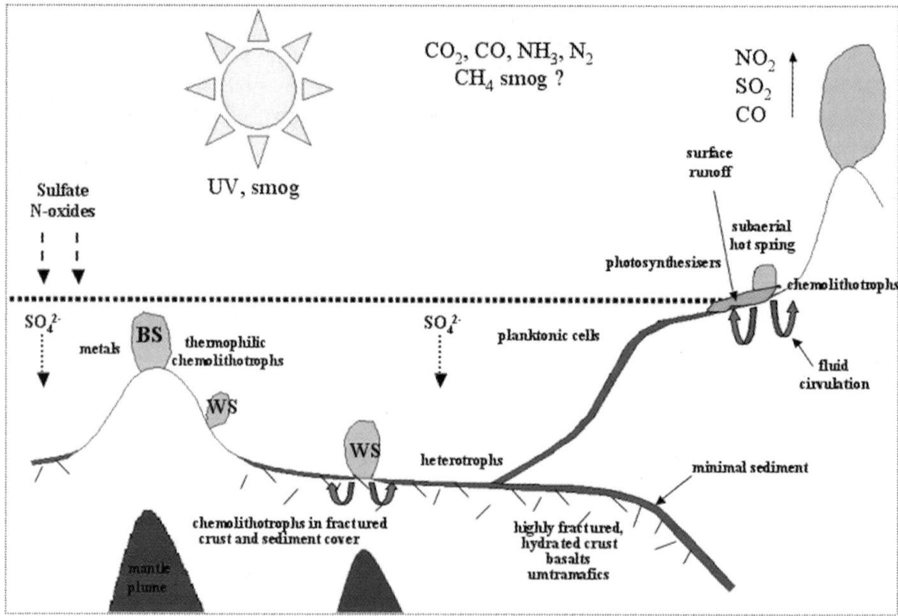

Fig. 22 Sketch showing the potential microbial habitats in the early-Mid Archean period. *BS*—black smoker; *WS*—white smoker. In the deep sea area, the potential habitats include the fractured, hydrated crust, the thin layer of sediments covering the crust, the black and white smokers, and the water column. In the shallow water regions around the emerged land masses (mainly volcano tops and emerged platforms), again sediment and fractured crust provide potential habitats, but the surfaces (subaqueous and subaerial sediment, lava, mineral) exposed to the sunlight represent habitats for potential photosynthesizers. Sketch based on Nisbet and Sleep (2001) and Westall and Southam (2006)

 Springer

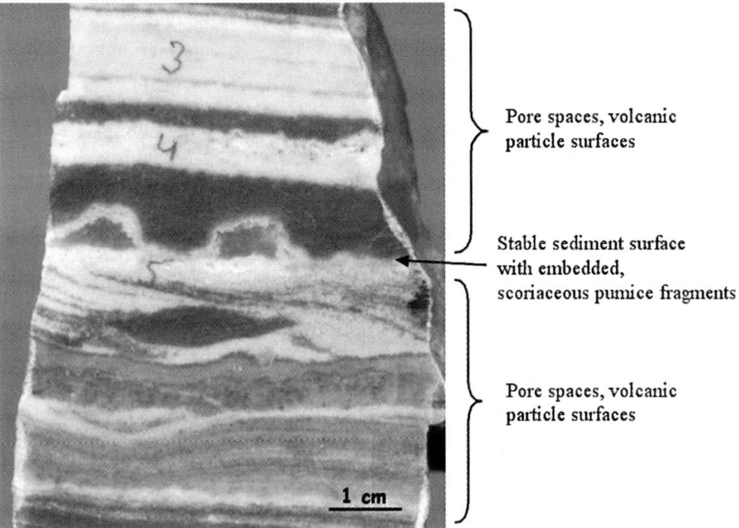

Fig. 23 Microhabitats in shallow water, water-logged coastal, volcanic sediments from the "Kitty's Gap Chert", Pilbara, W. Australia (3.446 Ga). Two main microhabitats are shown—the pore spaces and particle surfaces in the subsurface (ideal habitat for chemolithotrophic microganisms) and a sediment surface that was in the sunlight zone and stable for a short period of time (habitat for mat-building photosynthesisers)

We have seen above that the microhabitats available to life on early Earth were varied: the most protected being subsurface niches, sheltered from UV radiation and the occasional impact (cf. Lowe et al. 2003): the flux of UVb and UVc (DNA-damaging ultra-violet, the radiation wavelengths most damaging to the genetic machinery of organisms), is estimated to have been up to 1,000 times greater than today (Cockell 2000).

Clear evidence of life based apparently on a chemolithotrophic metabolism (whereby inorganic sources of carbon and energy are used) is to be found in the waterlogged, subsurface sediments of volcanic sands and silts deposited in a mudflat environment on the edges of a shallow water basin reported by De Vries (2004); Westall et al. (2006a); and Orberger et al. (2006) (Figs. 23, 24). These sediments host colonies of coccoidal microorganisms on the surfaces of the volcanic grains. Both shallow and deeper water marine sediments contain the resedimented remains of planktonic prokaryotes and shallow water microbial mats (Walsh 1992; Tice and Lowe 2004; Westall et al. 2006aa). Hydrothermal chert veins contain clots and wisps of carbonaceous material of enigmatic origin but with isotopic carbon and nitrogen signatures that are consistent with microbial fractionation (Ueno et al. 2004; Pinti and Hashizume 2001; Pinti et al. 2001; Orberger et al. 2006). It has been suggested that this type of carbon is a purely abiogenic Fischer–Tropf synthesis (van Zuilen et al. 2002, 2005; see Westall and Southam 2006 for a discussion). Other putative traces of life in a subsurface environment, namely corrosion pits in the vitreous surfaces of pillow lavas (Furnes et al. 2004), have been evoked above. The possibility of microbial involvement in the formation of "stalactites" in chert-filled vadose zone cracks in these Early Archean sediments is also under investigation (Hofmann et al. 2006).

Despite the supposedly hostile conditions reigning at the surface of the Earth in the Archean (high UVb and UVc radiation and impact extinction), there are numerous indications that life was not particularly perturbed and that biological activity not only survived but flourished. Apart from the few resedimented remains of shallow water microbial mats,

 Springer

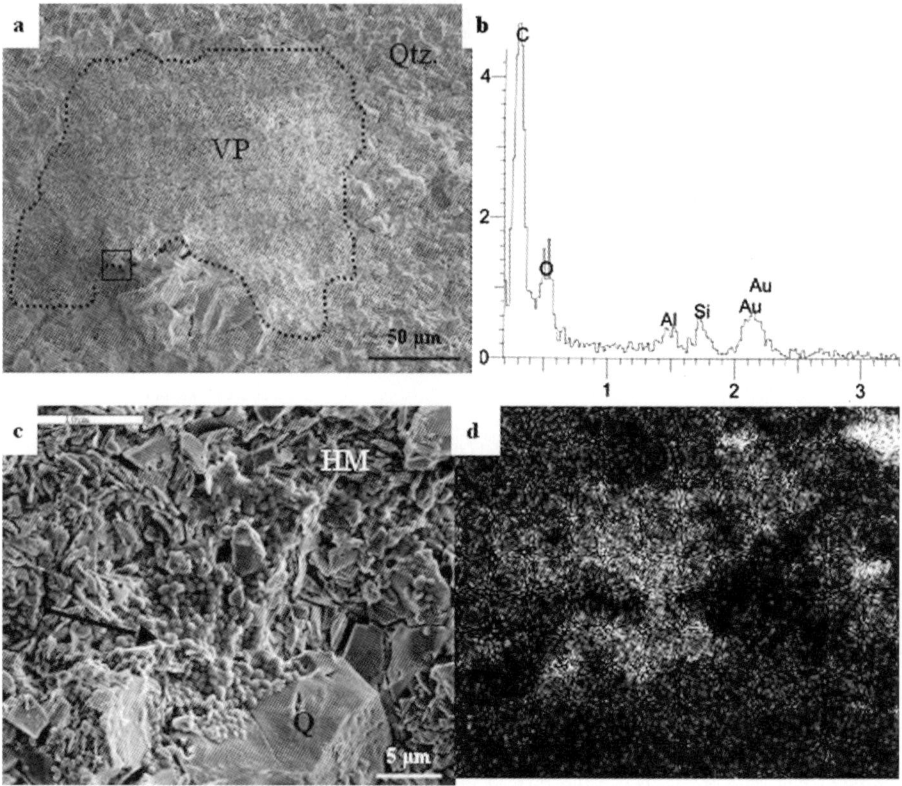

Fig. 24 Silicified microbial colonies in the "Kitty's Gap Chert" sediments. A volcanic particle (**a**) hosts colonies (**c**) of carbonaceous (**b**) coccoidal microorganisms (**d**) (Westall et al. 2006aa)

the silicified remains of mid- to late Archean in situ mats are relatively common (Walsh 1992, 2004; Westall et al. 2006b).

In general, all that remains of the mats are compacted layers of carbonaceous-rich horizons in which occasionally stray filaments or bundles of filaments are trapped in a silica matrix. Only where rapid fossilization caused by the deposition of a protective layer of silica, as well as silica impregnation, has occurred, are these mats preserved well enough to provide further information useful for environmental interpretations. Such a case is a filamentous mat that formed 3.33 Ga ago on a beach surface under evaporitic conditions (Westall et al. 2006b, 2006c) (Fig. 25). Given the apparently healthy appearance of the microorganisms prior to their silicification (from an outflow of hydrothermal silica that effectively killed the mat off), it is clear that they were well adapted for coping with adverse environmental conditions.

Another interesting characteristic of this and other microbial mats from this period is that they were most likely formed by anoxygenic photosynthesisers (Westall 2003; Walsh 2004; Tice and Lowe 2004; Westall et al. 2006b); indeed, mat-forming behaviour is, today, typical of photosynthetic microorganisms. Mats readily form even in ephemeral environments (Stahl 1994). The silicified Early Archean beach mat described above is just one such ephemeral environment. These mats are, however, poorly developed in the sense that, even though anoxygenic photosynthetic metabolisms provide far more energy and therefore al-

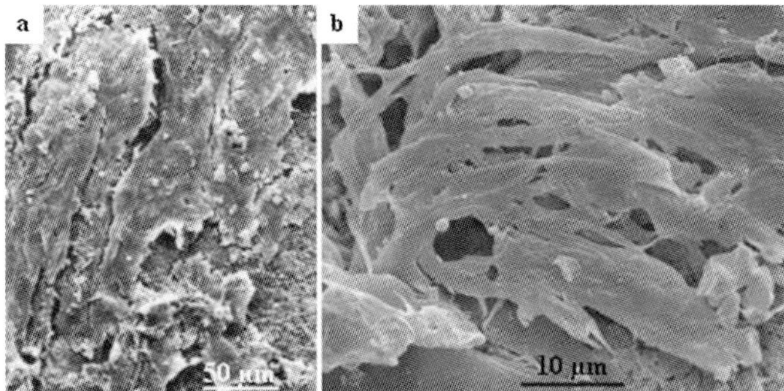

Fig. 25 Silicified microbial mat formed on a beach surface under evaporitic conditions (Westall et al. 2006b). (**a**) Plan view of the evaporite mineral-encrusted and desiccated filamentous mat. (**b**) Portions of the mat exhibit overturned filaments and bundles of filaments indicating initial formation under flowing water

low greater biological productivity than chemolithotrophic metabolisms, it still takes a long time for a mat to be developed.

In comparison, the meter-high stromatolites found today in the protected habitat of Shark Bay, Western Australia, formed by highly efficient oxygenic photosynthesising primary producers (plus associates), are estimated to be several thousands of years old. This means that, for any serious microbial mat development, the environmental conditions need to be stable over long periods of time. The Early-Mid Archean environment, with its very active tectonic cycle and rapidly changing basin edge scenarios, was not particularly conducive to the construction of layers upon layers of slowly accumulating microbial mats.

There are some exceptions however and they are important also for the implications regarding the changing tectonic regime of the geologically evolving Earth. Small stromatolites do occur in the Pilbara and in Barberton (Byerly et al. 1986; Hofmann et al. 1999; Allwood et al. 2006). Averaging about 10 cm or less in height, these bio-constructions are far smaller than their larger, Later Archean to recent counterparts that testify to the effectiveness of oxygenic photosynthesis. Despite their small size, the Early-Mid Archean stromatolites probably required equally as long to grow, given the limitations of anoxygenic photosynthesis. However, the important implication is that, at least in certain small areas of the Earth at this period, stabilized continents with stable, shallow water continental platforms were starting to come into existence. It was only through this geological evolution that suitable habitats started to appear that provided the breeding grounds for opportunistic anoxygenic photosynthesiers first and then the oxygenic photosynthesisers that made our planet what it is today.

18 Sustaining Habitability: The Long Term Scale—Is There an Outlook?

Is Mars a lesson? Does habitability end? Will Earth become Mars-like?—or, for that matter, Venus-like? Why has Earth, uniquely, sustained habitability?

The habitable zone is usually defined as the range of distances from the Sun where liquid water is stable on the planet's surface. This definition has both parochial and operational aspects. The latter in particular focus attention on the kinds of extrasolar planets we might plausibly detect within the horizons of our lives. The conventional definition excludes planets or moons for which liquid water is only present in clouds, or only present underground

or under ice cover. It is widely thought that life on such extrasolar planets or moons would be undetectable to telescopic inspection of extrasolar planets; in any event, if such life is present now in our own Solar System, it has remained hidden from us.

The inner and outer limits of the habitable zone are determined by the evaporation and freezing of water, respectively. Neither has been as clearly defined as we should like. A hard inner boundary is set by the runaway greenhouse limit (see chapter by Zahnle et al.). If the planet were any closer to the Sun, all liquid water on the surface would evaporate. There is some uncertainty in the precise location of the inner bound that depends on how much sunlight the planet reflects, compared to how much it absorbs. For a cloudy ocean world the reflectivity would probably be around 40%.

Kasting et al. (1993) defined the inner limit to habitability as the threshold for hydrogen escape. This is, however, not necessarily the crucial limiting factor. On Earth today hydrogen escape is tiny because the top of the troposphere is cold enough that very little water reaches the stratosphere or above. With very little hydrogen at high altitudes there is very little hydrogen escape. If the Earth were somewhat closer to the Sun (0.95 AU) this atmospheric cold trap would disappear and the limit on hydrogen escape would not exist. Kasting et al. called this the "moist greenhouse". However, such a planet would have liquid water oceans—possibly for a very long time albeit not forever. Thus the hydrogen-threshold is not obviously the limiting factor, especially if life can appear and evolve complexity faster than it has done on Earth—say in a few hundred million years.

A second issue arises in the difference between wet and dry planets. The runaway greenhouse limit applies only if water condenses in the atmosphere or at the surface. If a planet is dry enough its tropical atmosphere can be truly cloud-free. If so, it can be sited closer to the Sun than a wet planet and still have condensed water near the poles (Abe and Abe-Ouchi 2005). Because hydrogen escapes before the ocean evaporates, Kasting et al.'s (1993) moist greenhouse could evolve into a dry planet without ever becoming uninhabitable. Indeed, it is an interesting speculation if Venus could remain habitable by aerophile acidophiles floating in the upper atmospheric smog.

The outer limit to the habitable zone is determined by the strength of the greenhouse effect. In principle there isn't an outer limit because, given the right gases and the will to use them, there is no upper limit on the greenhouse effect. Stevenson (1999) pointed out that a thick hydrogen atmosphere could even make a planet lost in space habitable. But would such a planet be recognizable as habitable? For specificity Kasting et al. (1993) took the conservative view that CO_2 is responsible for the greenhouse effect. CO_2 condenzation then sets the outer boundary at ~ 1.7 AU today in our Solar System.

How long can an Earth-like planet remain habitable? Figure 27 explores the bounds of habitability for an Earth-like planet at a given current distance. Habitable planets have been defined as receiving an insolation S between 0.7 and 1.4 times what Earth receives today. These bounds correspond to the bounds for the habitable zone as defined in Fig. 26. The more restrictive habitable zone for insolation $0.75 < S < 1.1$ corresponds to the habitable zone originally defined by Kasting et al. (1993); the inner bound the "moist" greenhouse of the other figure. The less restrictive $0.4 < S < 1.4$ sets the inner boundary by the runaway greenhouse effect and the outer boundary includes a rough upper limit on the greenhouse warming that can be obtained from CO_2 clouds (Kasting 1996, 1997). The figure was prepared using standard models of solar evolution applied over the full lifetime of our Sun. Late in the Sun's life, when it is a red giant, it becomes much brighter than it is now. During this time the habitable zone leaves the realm of the terrestrial planets, but extends into the realm of the giant planets. The structure in the curve arises from stationary points in the Sun's evolution.

 Springer

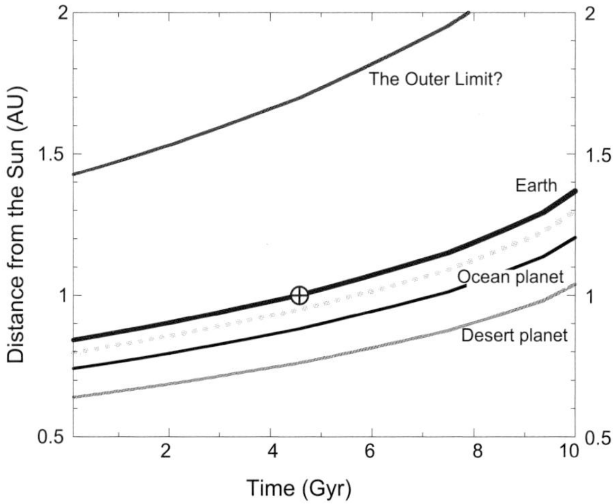

Fig. 26 The habitable zone over the first ten Ga of solar evolution. Here we consider the long-term habitability of an Earth-like planet as the Sun evolves. Solar evolution is from Sackmann et al. (1993). The inner edge of the habitable zone is set by the runaway greenhouse effect. This is more permissive for a desert planet (like Mars) than it is for a planet with oceans of water (like Earth). The less permissive *dashed curve* refers to the "moist" greenhouse, which Kasting et al. (1993) identify with the inner edge of the habitable zone. The moist greenhouse describes a perfectly habitable ocean planet that suffers rapid hydrogen escape; the importance of this is debatable. The outer edge of the habitable zone as drawn here is set by the condenzation of CO_2 clouds (Kasting et al. 1993). This is conservative. If the CO_2 clouds themselves have a strong greenhouse warming effect (Forget and Pierrehumbert 1997), or if other less volatile greenhouse gases are abundant, the habitable zone expands. Stevenson (1999) argued that, because there is no upper bound on the greenhouse effect, there is no outer limit to the habitable zone

Fig. 27 The wave of quickening. As the Sun evolves, the habitable zone sweeps through the solar system (assuming the standard solar evolution from Sackmann et al. 1993). Figure 27 shows how long any location in our Solar System remains in the habitable zone. More and less generous assessments of the habitable zone are shown. It is perhaps no accident that Earth-bound natural philosophers find that Earth is in the best place in the Solar system

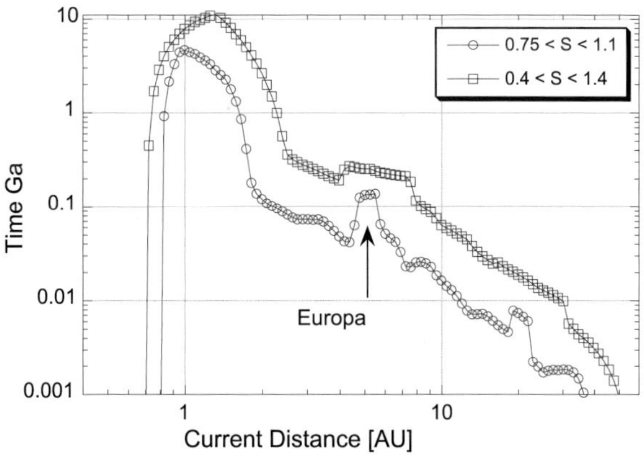

The main story is one of steady brightening during the roughly 11 Ga that the Sun spends on the main sequence. This is the longest stage in the Sun's life, when it fuses hydrogen into helium in its core. The core is only 10% of the Sun's mass. When hydrogen in the core is exhausted, the Sun begins to fuse the rest of its hydrogen into helium, but to do so the Sun must get hotter inside and therefore gets brighter. Thus begins the red giant phase, which lasts about 1 Ga and consumes much of the remaining hydrogen. As it runs out of hydrogen, the Sun becomes very big and bright and the habitable zone sweeps to Uranus and beyond. At a certain point the helium in the core begins to fuse into carbon. When this happens the

Sun enters a 100 million year stage of relative constancy that happens to put Jupiter and its moons in the habitable zone. This is the reason why there is a local maximum, marked "Europa", in Fig. 27.

There, perhaps, lies our last home.

Acknowledgements Many thanks to Norman Sleep, Dave Des Marais and other reviewers, and Dave Waltham for very useful comments, to Ernst Hauber for help with figure 2.12, and especially to Kathryn Fishbaugh for her help and control, and for very helpful detailed editing. K.Z. thanks the NASA Exobiology Program for support. R.J. and J.H. thank the DLR Space Research Program for support.

References

Y. Abe, A. Abe-Ouchi, Stability of liquid water on a land planet: Wider habitable zone for a less water planet than an aqua planet. American Geophysical Union, Fall Meeting 2005, abstract #P51D-0965, 2005

C.M. Aitken, D.M. Jones, S.R. Larter, Nature **431**, 291–294 (2004)

A.C. Allwood, M.R. Walter, B.S. Kamber, C.P. Marshall, I.W. Burch, Nature **441**, 714–718 (2006)

R.E. Arvidson, E.A. Guinness, S. Lee, Nature **278**, 533–535 (1979)

R.E. Arvidson, F. Poulet, J.-P. Bibring, M. Wolff, A. Gendrin, R.V. Morris, J.J. Freeman, Y. Langevin, N. Mangold, G. Bellucci, Science **307**, 1591–1593 (2005)

V.R. Baker, R.G. Strom, S.K. Croft, V.C. Gulick, J.S. Kargel, G. Komatsu, Nature **352**, 589–594 (1991)

V.A. Baker, M.H. Carr, V.C. Gulick, C.R. Williams, M.S. Marley, in *Mars*, ed. by H.H. Kieffer, B.M. Jakosky, C.W. Snyder, M.S. Matthews (University of Arizona Press, Tucson, 1992), pp. 493–522.

V.A. Baker, Nature **412**, 228–236 (2001)

L.G. Benning, V.R. Phoenix, N. Yee, M.J. Tobin, Geochim. Cosmochim. Acta **68**, 729–741 (2004)

K. Benzerara, N. Menguy, F. Guyot, F. Skouri, G. de Lucca, T. Heulin, Earth Planet. Sci. Lett. **228**, 439–449 (2004a)

K. Benzerara, T.-H. Yoon, T. Tyliszczak, B. Constantz, A.M. Spormann, G.E. Brown Jr., Geobiology **2**, 249–259 (2004b)

K. Benzerara, T.-H. Yoon, N. Menguy, T. Tyliszczak, G.E. Brown Jr., Proc. Natl. Acad. Sci. USA **102**, 979–982 (2005)

J.-P. Bibring, Y. Langevin, A. Gendrin, B. Gondet, F. Poulet, M. Berthé, A. Soufflot, R. Arvidson, Mangold, J. Mustard, P. Drossart, the OMEGA team, Science **307**, 1576–1581 (2005)

H. Bieman, J. Oró, P. Toulmin, L.E. Orgel III, A.O. Nier, D.M. Anderson, P.G. Simmonds, D. Flory, A.V. Diaz, D.R. Rushneck, J.E. Biller, A.L. LaFleur, J. Geophys. Res. **82**, 4641–4658 (1977)

P.J. Boston, M.N. Spilde, D.E. Northup, L.A. Melim, D.S. Soroka, L.G. Kleina, K.H. Lavoie, L.D. Hose, L.M. Mallory, C.N. Dahm, L.J. Crossey, R.T. Schelble, Astrobiology **1**, 25–55 (2001)

W.V. Boynton, W.C. Feldman, S.W. Squyres, T. Prettyman, J. Brückner, L.G. Evans, R.C. Reedy, R. Starr, J.R. Arnold, D.M. Drake, P.J.A. Englert, A.E. Metzger, I. Mitrofanov, J.I. Trombka, C. d'Uston, H. Wänke, O. Gasnault, D.K. Hamara, D.M. Janes, R.L. Marcialis, S. Maurice, I. Mikheeva, R. Torkar, C. Shinohara, Science **297**, 81–85 (2002)

S.A. Bowring, T. Housh, Science **269**, 1535–1540 (1995)

A. Brack, in *Astrobiology: the Quest for the Conditions of Life*, ed. by G. Horneck, C. Baumstark-Khan (Springer, Berlin, 2002), pp. 79–88.

M.D. Brasier, O.R. Green, J.F. Lindsay, N. McLoughlin, A. Steele, C. Stoakes, Precambr. Res. **140**, 55–102 (2005)

D. Breuer, T. Spohn, J. Geophys. Res. **108**, E/ (2003). doi:10.1029/2002JE001999

J.C. Bridges, P.H. Warren, J. Geol. Soc. London **163**(2), 229–251 (2006)

D.M. Burr, J.A. Grier, A.S. Mcewen, L.P. Keszthelyi, Icarus **159**, 53–73 (2002)

G.R. Byerly, M.M. Walsh, D.L. Lowe, Nature **319**, 489–491 (1986)

D.E. Caldwell, S.J. Caldwell, Microb. Ecol. **47**, 252–265 (2004)

R.A. Carddock, T.A. Maxwell, J. Geophys. Res. **98**, 3453–3468 (1993)

M.H. Carr, Nature **326**, 30–35 (1987)

M.H. Carr, H. Wänke, Icarus **98**, 61–71 (1992)

M.H. Carr, J. Geophys. Res. **100**, 7479–7507 (1995)

M.H. Carr, *Water on Mars* (Oxford University Press, 1996), 229 pp.

M.H. Carr, M.C. Malin, Icarus **146**, 366–386 (2000)

S. Castanier, G. Le Metayer-Levrel, J.P. Perthuisot, Sediment. Geol. **126**, 9–23 (1999)

D.C. Catling, K.J. Zahnle, C.P. McKay, Science **293**, 839–843 (2001)

 Springer

C.S. Chan, G. De Stasio, S.A. Welch, M. Girasole, B.H. Frazer, M.V. Nesterova, S. Fakra, J.F. Banfield, Science **303**, 1656–1658 (2004)

F.H. Chapelle, K. O'neill, P.M. Bradley, B.A. Methé, S.A. Ciufo, L.L. Knobel, D.R. Lovley, Nature **415**, 312–315 (2002)

A.E. Chichibabin, *Basic Principals of Organic Chemistry*, vol. 1 (Goskhimizdat, Leningrad, 1953) 796 pp. (in Russian)

P.R. Christensen, B.M. Jakosky, H.H. Kieffer, M.C. Malin, H.Y. McSween Jr., K. Nealson, G.L. Mehall, S.H. Silverman, S. Ferry, M. Caplinger, M. Ravine, Space Sci. Rev. **110**, 85–130 (2004)

S.M. Clifford, J. Geophys. Res. **98**, 10973–11016 (1993)

S.M. Clifford, T.J. Parker, Icarus **154**, 40–79 (2001)

C.S. Cockell, Planet. Space Sci. **4**, 203–214 (2000)

J.E.P. Connerney, M.H. Acuña, P. Wasilevski, N.F. Ness, H. Rèmw, C. Mazelle, D. Vignes, R.P. Lin, D. Mitchell, P. Cloutier, Science **284**, 794–798 (1999)

J.E.P. Connerney, M.-H. Acuna, N.F. Ness, G. Kletetschka, D.L. Mitchell, R.P. Lin, H. Reme, PNAS **102**(42), 14970–14975 (2005)

J.C. Dann, A.H. Holzheid, T.L. Grove, H.Y. McSween, Meteorit. Planet. Sci. **36**, 793–806 (2001)

S.T. De Vries, Early Archean sedimentary basins: depositional environment and hydrothermal systems. Examples from the Barberton and Coppin Gap greenstone belts. Geologica Ultraiectina, University of Utrecht, 2004, 159 pp.

S. D'Hondt, B.B. Jørgensen, J. Miller, A. Batzke, R. Blake, B.A. Cragg, H. Cypionka, G.R. Dickens, T. Ferdelman, K.-U. Hinrichs, N. G. Holm, R. Mitterer, A. Spivack, G. Wang, B. Bekins, B. Engelen, K. Ford, G. Gettemy, S.D. Rutherford, H. Sass, C.G. Skilbeck, I.W. Aiello, G. Gue'Rin, C.H. House, F. Inagaki, P. Meister, T. Naehr, S. Niitsuma, R.J. Parkes, A. Schippers, D.C. Smith, A. Teske, J. Wiegel, C. Naranjo Padilla, J.L. Solis Acosta, Science **306**, 2216–2221 (2004)

S. D'Hondt, S. Rutherford, A.J. Spivack, Science **295**, 2067–2070 (2002)

G. Dreibus, H. Wänke, Icarus **71**, 225–240 (1987)

G. Dreibus, H. Wänke, in *Origin and Evolution of Planetary and Satellite Atmospheres*, ed. by S.K. Atria, J.B. Pollack, M.S. Matthews (Univ. of Arizona Press, Tucson, 1989), pp. 268–288

D. Emerson, W.C. Ghiorse, Appl. Env. Microbiol. **58**, 4001–4010 (1992)

B. Fegley Jr., R.G. Prinn, H. Hartman, G.H. Watkins, Nature **319**(6051), 305–308 (1986)

W.C. Feldman, W.V. Boynton, R.L. Torkar, T.H. Prettyman, O. Gasnault, S.W. Squyres, R.C. Elphic, D.J. Lawrence, S.L. Lawson, S. Maurice, G.W. McKinney, K.R. Moore, R.C. Reedy, Science **297**, 75–78 (2002)

F. Forget, R. Pierrehumbert, Science **278**, 1273–1276 (1997)

V. Formisano, S. Atreya, T. Encrenaz, N. Ignatiev, M. Giuranna, Science **306**, 1758–1761 (2004)

V. Formisano, F. Angrilli, G. Arnold, S. Atreya, G. Bianchini, D. Biondi, A. Blanco, M.I. Blecka, A. Coradini, L. Colangeli, A. Ekonomov, F. Esposito, S. Fonti, M. Giuranna, D. Grassi, V. Gnedykh, A. Grigoriev, G. Hansen, H. Hirsh, I. Khatuntsev, A. Kiselev, N. Ignatiev, A. Jurewicz, E. Lellouch, J. Lopez Moreno, A. Marten, A. Mattana, A. Maturilli, E. Mencarelli, M. Michalska, V. Moroz, B. Moshkin, F. Nespoli, Y. Nikolsky, R. Orfei, P. Orleanski, V. Orofino, E. Palomba, D. Patsaev, G. Piccioni, M. Rataj, R. Rodrigo, J. Rodriguez, M. Rossi, B. Saggin, D. Titov, L. Zasova, Planet. Space Sci. **53**(10), 963–974 (2005)

R.B. Frankel, D.A. Bazylinski, Rev. Mineral. Geochem. **54**, 95–114 (2003)

H.V. Frey, J.H. Roark, K.M. Shockey, E.L. Frey, S.E.H. Sakimoto, Geophys. Res. Lett. **29**(10), 1384 (2002). doi:10.1029/2001.GL013832

H. Furnes, K. Muehlenbachs, O. Tumyr, T. Torsvik, I.H. Thorseth, Terra Nova **11**, 228–233 (1999)

H. Furnes, H. Staudigel, Earth Planet. Sci. Lett. **166**, 97–103 (1999)

H. Furnes, N.R. Banerjee, K. Muehlenbachs, H. Staudigel, M. de Wit, Science **304**, 578–581 (2004)

H. Furnes, N.R. Banerjee, K. Muehlenbachs, A. Kontinen, Precambr. Res. **136**, 125–137 (2005)

S.J.G. Galer, K. Mezger, Precambr. Res. **92**, 389–412 (1998)

R. Gellert, R. Rieder, R.C. Anderson, J. Brückner, B.C. Clark, G. Dreibus, T. Economou, G. Klingelhöfer, G.W. Lugmair, D.W. Ming, S.W. Squyres, C. d'Uston, H. Wänke, A. Yen, J. Zipfel, Science **305**, 829–832 (2004)

A. Gendrin, N. Mangold, J.-P. Bibring, Y. Langevin, B. Gondet, F. Poulet, G. Bonello, C. Quantin, J. Mustard, R. Arvidson, S. LeMouélic, Science **307**, 1587–1590 (2005)

M.V. Gerasimov, in *Catastrophic Events and Mass Extinctions: Impacts and Beyond*, ed. by C. Koeberl, K.G. MacLeod. Special Paper No 356 (Geological Society of America, Boulder, 2002), pp. 705–716

M.V. Gerasimov, L.M. Mukhin, M.D. Nussinov, Doklady Akademii Nauk of the USSR **275**(3), 646–650 (1984)

M.V. Gerasimov, Yu.P. Dikov, O.I. Yakovlev, F. Wlotzka, in *Lunar and Planetary Science XXV, Lunar and Planet. Inst.* (Abstracts) (Houston, 1994a), pp. 413–414

M.V. Gerasimov, Yu.P. Dikov, O.I. Yakovlev, F. Wlotzka, in *Lunar and Planetary Science XXV, Lunar and Planet. Inst.* (Abstracts) (Houston, 1994b), pp. 415–416

M.V. Gerasimov, Yu.P. Dikov, O.I. Yakovlev, F. Wlotzka, in *Large Meteorite Impacts and Planetary Evolution* (abstract), Sudbery, LPI Contr. No 922, 1997, pp. 15–16

M.V. Gerasimov, B.A. Ivanov, O.I. Yakovlev, Yu.P. Dikov, in *Laboratory Astrophysics and Space Research*, ed. by P. Ehrenfreund, K. Krafft, H. Kochan, V. Pirronello. Astrophysics and Space Science Library, vol. 236 (Kluwer, Dortrecht, 1999), pp. 279–330

M.V. Gerasimov, Yu.P. Dikov, O.I. Yakovlev, F. Wlotzka, Deep-sea research Part II. Top. Stud. Oceanogr. **49**(6) 995–1009 (2002)

W.C. Ghiorse, Annu. Rev. Microbiol. **38**, 515–550 (1984)

T.M. Gihring, P.L. Bond, S. Peters, J.F. Banfield, Extremophiles **7**, 123–130 (2003)

M.P. Golombek, N.T. Bridges, J. Geophys. Res. **105**, 1841–1853 (2000)

N.V. Grassineau, P. Abell, P.W.U. Appell, D. Lowry, E.G. Nisbet, in *Evolution of Early Earth's Atmosphere, Hydrosphere and Biosphere – Constraints from Ore Deposits*, ed. by S.E. Kesler, H. Ohmoto (Geological Society of America, 2006), pp. 33–52

L.A. Haskin, A. Wang, B.L. Jolliff, H.Y. McSween, B.C. Clark, D.J. DesMarais, S.M. McLennan, N.J. Tosca, J.A. Hurowitz, J.D. Farmer, A. Yen, S.W. Squyres, R.E. Arvidson, G. Klingelhöfer, C. Schröder, P.A. de Souza Jr., D.W. Ming, R. Gellert, J. Zipfel, J. Brückner, J.F. Bell III, K. Herkenhoff, P.R. Christensen, S. Ruff, D. Blaney, S. Gorevan, N.A. Cabrol, L. Crumpler, J. Grant, L. Soderblom, Nature **436**, 66–69 (2005)

E. Hauber, K. Gwinner, D. Reiss, F. Scholten, G.G. Michael, R. Jaumann, G.G. Ori, L. Marinangeli, G. Neukum, The HRSC Co-Investigator Team, Lunar Planet. Sci. **36** (2005a), abstract # 1661

E. Hauber, S. van Gasselt, B. Ivanov, S.C. Werner, J.W. Head, G. Neukum, R. Jaumann, R. Greeley, M. Mitchell, P. Muller, the HRSC Co-Investigator Team, Nature **434**, 356–360 (2005b)

J.M. Hayes, J.R. Waldbauer, Philos. Trans. Roy. Soc. Lond. B Biol. Sci. **361**, 931–950 (2006)

J.W. Head, R. Greeley, M.P. Golombek, W.K. Hartmann, E. Hauber, R. Jaumann, P. Masson, G. Neukum, L.E. Nyquist, M.H. Carr, in *Chronology and Evolution on Mars*, ed. by R. Kallenbach, J. Geiss, W.K. Hartmann. Space Science Series of ISSI. Space Sci. Rev. **96**, 263–292 (2001)

J.W. Head, J.F. Mustard, M.A. Kreslavsky, R.E. Milliken, D.R. Marchant, Nature **426**, 797–802 (2003)

J.W. Head, G. Neukum, R. Jaumann, H. Hiesinger, E. Hauber, M. Carr, N. Mangold, M. Kreslavsky, S. Milkovich, the HRSC Co-Investigator Team, Nature **434**, 346–351 (2005)

J. Helbert, D. Reiss, E. Hauber, J. Benkhoff, Geophys. Res. Lett. **32**(17), L17201 (2005). CiteID

J.L. Heldman, M.T. Mellon, Icarus **168**, 285–304 (2004)

A.M. Hessler, D.R. Lowe, R.L. Jones, D.K. Bird, Nature **428**, 736–8 (2004)

B.A. Hofmann, J.D. Farmer, Planet. Space Sci. **48**, 1077–1086 (2000)

B.A. Hofmann, Y. Krüger, A.E. Fallick, M. Eggimann, M.J. Van Kranendonk, Proc. 6th European Workshop on Astrobiology, Lyon, 16–18 October 2006, Ecole Normale Superieure, Lyon, France, 2006

H.J. Hofmann, K. Grey, A.H. Hickman, R.I. Thorpe, Geol. Soc. Am. Bull. **111**, 1256–1262 (1999)

N. Hoffman, P.R. Kyle, The ice towers of Mt. Erebus as analogues of biological refuges on Mars. Presented at 6th International Mars Conference, #3105 (2003)

N. Hoffman, Modern geothermal gradients on Mars and implications for subsurface liquids. Conference on the Geophysical Detection of Subsurface Water on Mars (2001)

N.G. Holm, M. Dumont, M. Ivarsson, C. Konn, Geochem. Trans. **7**, 7 (2004). doi:10.1186/1467-4866-7-7

A. Hynes, Earth Planet. Sci. Lett. **185**, 161–172 (2001)

B.A. Ivanov, D.D. Badukov, O.I. Yakovlev, M.V. Gerasimov, Yu.P. Dikov, K.O. Pope, A.C. Ocampo, in *The Cretaceous-Tertiary Event and other Catastrophes in Earth History*, ed. by G. Ryder, D. Fastovsky and S. Gartner. Special paper No 307 (Geological Society of America, 1996), pp. 125–139

R. Jaumann, Die Erosionsmorphologie des Mars: Genese, Verteilung und Stratigraphie von Erosionsformen und deren klimatische Bedeutung, DLR Forschungsbericht **2003–20**, pp. 261 (2003)

R. Jaumann, E. Hauber, J. Lanz, H. Hoffmann, G. Neukum, in *Astrobiology*, ed. by G. Horneck, C. Baumstark-Kahn (Springer, Berlin, 2002), pp. 89–109.

R. Jaumann, D. Reiss, S. Frei, G. Neukum, F. Scholten, K. Gwinner, T. Roatsch, K.-M. Matz., E. Hauber, U. Köhler, J.W. Head, H. Hiesinger, M. Carr, Geophys. Res. Lett. **32**, L16203 (2005). doi:10.1029/2002GL023415

A. Kappler, B. Schink, D.K. Newman, Geobiology **3**, 235–245 (2005)

J.S. Kargel, R.G. Storm, Anc. Glaciat. Mars Geol. **20**, 3–7 (1992)

K. Kashevi, D.R. Lovley, Science **301**, 934 (2003)

J.F. Kasting, Astrophys. Space Sci. **241**, 3–24 (1996)

J.F. Kasting, Origins of Life **27**, 291–307 (1997)

J.F. Kasting, D.P. Whitmire, R.T. Reynolds, Icarus **101**, 108–128 (1993)

P. Kharecha, J. Kasting, J. Siefert, Geobiology **3**, 53–76 (2005)

H.H. Kieffer, P. Zent, in *Mars*, ed. by H.H. Kieffer, B.M. Jakosky, C.W. Snyder, M.S. Matthews (University of Arizona Press, Tucson, 1992), pp. 1180–1220

G. Klingelhöfer, R.V. Morris, B. Bernhardt, C. Schröder, D.S. Rodionov, P.A. de Souza Jr., A. Yen, R. Gellert, E.N. Evlanov, B. Zubkov, J. Foh, U. Bonnes, E. Kankeleit, P. Gütlich, D.W. Ming, F. Renz, T. Wdowiak, S.W. Squyres, R.E. Arvidson, Science **306**, 1740–1746 (2004)

K.O. Konhauser, V.R. Phoenix, S.H. Bottrell, D.G. Adams, I.M. Head, Sedimentology **48**, 415–433 (2001)

M. Kretzschmar, Facies **7**, 237–260 (1982)

Y. Langevin, F. Poulet, J.-P. Bibring, B. Gondet, Science **307**, 1584–1586 (2005)

A. Lenardice, F. Nimmo, L. Moresi, J. Geophys. Res. **109**, E02003 (2004). doi:1029/2003JE002172

D.R. Lowe, G.R. Byerly, F.T. Kyte, A. Shukulyukov, F. Asaro, A. Krull, Astrobiology **3**, 7–48 (2003)

A.M. Macgregor, South Afr. J. Sci. **24**, 155–172 (1927)

M.C. Malin, K.S. Edgett, Science **290**, 1927–1937 (2000a)

M.C. Malin, K.S. Edgett, Science **288**, 2330–2335 (2000b)

M.C. Malin, K.S. Edgett, L.V. Posiolova, S.M. McColley, E.Z. Noe Dobrea, Science **314**, 1573–1577 (2006)

P. Masson, M.H. Carr, F. Costard, R. Greeley, E. Hauber, R. Jaumann, in *Chronology and Evolution on Mars* ed. by R. Kallenbach, J. Geiss, W.K. Hartmann. Space Science Series of ISSI. Space Sci. Rev. **96**, 333–364 (2001)

J.P. Mckinley, T.O. Stevens, F. Westall, Geomicrobiol. J. **17**, 43–54 (2000)

H.Y. McSween Jr., Meteoritics **29**, 757–779 (1994)

H.Y. McSween Jr., S.L. Murchie, J.A. Crisp, N.T. Bridges, R.C. Anderson, J.F. Bell III, D.T. Britt, J. Brückner, G. Dreibus, T. Economou, A. Ghosh, M.P. Golombek, J.P. Greenwood, R.J. Johnson, H.J. Moore, R.V. Morris, T.J. Parker, R. Rieder, R. Singer, H. Wänke, J. Geophys. Res. **104**(E4), 8679–8715 (1999)

H.Y. McSween, R.E. Arvidson, J.F. Bell III, D. Blaney, N.A. Cabrol, P.R. Christensen, B.C. Clark, J.A. Crisp, L.S. Crumpler, D.J. DesMarais, J.D. Farmer, R. Gellert, A. Ghosh, A. Gorevan, T. Graff, J. Grant, L.A. Haskin, K.E. Herkenhoff, J.R. Johnson, B.L. Jolliff, G. Klingelhoefer, A.T. Knudson, S. McLennan, K.A. Milam, J.E. Moersch, R.V. Morris, R. Rieder, S.W. Ruff, P.A. de Souza Jr., S.W. Squyres, H. Wänke, A. Wang, M.B. Wyatt, A. Yen, J. Zipfel, Science **305**, 842–845 (2004)

C.E. Mire, J.A. Tourjee, W.F. O'Brien, K.V. Ramanujachary, G.B. Hecht, Appl. Environ. Microbiol. **70**, 855–864 (2004)

D. Möhlmann, Icarus **168**, 318–323 (2004)

R.V. Morris, G. Klingelhöfer, B. Bernhardt, C. Schröder, D.S. Rodionov, P.A. de Souza Jr., A. Yen, R. Gellert, E.N. Evlanov, J. Foh, E. Kankeleit, P. Gütlich, D.W. Ming, J.F. Mustard, F. Poulet, A. Gendrin, J.-P. Bibring, Y. Langevin, B. Gondet, N. Mangold, G. Belucci, F. Altieri, Science **307**, 1594–1597 (2005)

D.P. Moser, T.M. Gihring, F.J. Brockman, J.K. Fredrickson, D.L. Balkwill, M.E. Dollhopf, B.S. Lollar, L.M. Pratt, E. Boice, G. Southam, G. Wanger, B.J. Baker, S.M. Pfiffner, L.-H. Lin, T.C. Onstott, Appl. Environ. Microbiol. **71**, 8773–8783 (2005)

L.M. Mukhin, M.V. Gerasimov, E.N. Safonova, Nature **340**, 46–48 (1989)

J.F. Mustard, F. Poulet, A. Gendrin, J.-P. Bibring, Y. Langevin, B. Gondet, N. Mangold, G. Belucci, F. Altieri, Science **307**, 1594–1597 (2005)

G. Neukum, R. Jaumann, H. Hoffmann, E. Hauber, J.W. Head, A.T. Basilevsky, B.A. Ivanov, S.C. Werner, S. van Gasselt, J.B. Murray, T. McCord, the HRSC Co-Investigator Team, Nature **432**, 971–979 (2004)

F. Nimmo, D. Stevenson, J. Geophys. Res. **105**, 11969–11979 (2000)

E.G. Nisbet, *The Young Earth* (G. Allen and Unwin, London, 1987) 402 pp.

E.G. Nisbet, M.J. Cheadle, N.T. Arndt, M.J. Bickle, Lithos **30**, 291–307 (1993)

E.G. Nisbet, N.H. Sleep, Nature **409**, 1083–1091 (2001)

B. Orberger, V. Rouchon, F. Westall, S.T. de Vries, C. Wagner, D.L. Pinti, in *Processes on the Early Earth*, vol. 405, ed. by W.U. Reimold, R. Gibson (Geol. Soc. Amer. Spec. Pub., 2006), pp. 133–152

R.J. Parkes, B.A. Cragg, S.J. Bale, J.M. Getliff, K. Goodman, P.A. Rochelle, J.C. Fry, A.J. Weightman, S.M. Harvey, Nature **371**, 410–413 (1994)

R.J. Parkes, G. Webster, B.A. Cragg, A.J. Weightman, C.J. Newberry, T.G. Ferdelman, J. Kallmeyer, B.B. Jørgensen, I.W. Aiello, J. Fry, Nature **436**, 390–394 (2005)

A.A. Pavlov, J.F. Kasting, L.L. Brown, K.A. Rages, R. Freedman, J. Geophys. Res. **105**, 11981–11990 (2001)

K. Pedersen, FEMS Microbiol. Lett. **185**, 9–16 (2000)

V.R. Phoenix, D.G. Adams, K.O. Konhauser, Chem. Geol. **169**, 329–338 (2000)

V.R. Phoenix, K.O. Konhauser, D.G. Adams, Geology **29**, 823–826 (2001)

E. Pierazzo, H.J. Melosh, Meteorit. Planet. Sci. **35**, 117–130 (2000)

D. Pieri, Science **210**, 895–897 (1980)

D.L. Pinti, K. Hashizume, Precambr. Res. **105**, 85–88 (2001)

D.L. Pinti, K. Hashizume, J. Matsuda, Geochim. Cosmochim. Acta **65**, 2301–2315 (2001)

R.G. Prinn, B. Fegley Jr., Earth Planet. Sci. Lett. **83**, 1–15 (1987)

N.E. Putzig, M.T. Mellon, K.A. Kretke, R.E. Arvidson, Icarus **173**, 325–341 (2005)

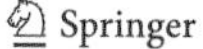

D. Reiss, R. Jaumann, Geophys. Res. Lett. **30**(6) (2003). doi:1029/2002GL016704

A. Ricardo, M.A. Carrigan, A.N. Olcott, S.A. Benner, Science **303**, 196 (2004)

R. Rieder, H. Wänke, T. Economou, A. Turkevich, J. Geophys. Res. **102**, 4027–4044 (1997a)

R. Rieder, T. Economou, H. Wänke, A. Turkevich, J. Crisp, J. Brückner, G. Dreibus, H.Y. McSween Jr., Science **278**, 1771–1774 (1997b)

R. Rieder, R. Gellert, R.C. Anderson, J. Brückner, B.C. Clark, G. Dreibus, T. Economou, G. Klingelhöfer, G.W. Lugmair, D.W. Ming, S.W. Squyres, C. d'Uston, H. Wänke, A. Yen, J. Zipfel, Science **306**, 1746–1749 (2004)

J.R. Rogers, P.C. Bennett, Chem. Geol. **203**, 91–108 (2004)

J. Sackmann, A.I. Boothroyd, K.E. Kraemer, Astrophys. J. **418**, 457–468 (1993)

M.A. Saito, D.M. Sigman, F.M.M. Morel, Inorganica Chimica Acta **356**, 308–318 (2003)

S. Schultze-Lam, G. Harauz, T.J. Beveridge, J. Bacteriol. **174**, 7971–7981 (1992)

D.H. Scott, J.M. Dohm, Martian highland channels: An age reassessment, Lunar Planet. Sci. Conf. XXIII, LPI, Houston, 1992, pp. 1251–1252

Y. Sekine, S. Sugita, T. Kadono, T. Matsui, J. Geophys. Res. **108**(E7), 5070 (2003). doi:10.1029/2002JE002034

R.P. Sharp, M.C. Malin, Geol. Soc. Am. Bull. **86**, 593–609 (1975)

C. Siebert, J.D. Kramers, T. Meisel, P. Morel, T.F. Nagler, Geochimica Cosmochimica Acta **69**, 1787–1801 (2005)

N.H. Sleep, J. Geophys. Res. **99**, 5639–5655 (1994)

N.H. Sleep, Annu. Rev. Earth Planet. Sci. **33**, 369–393 (2005)

N.H. Sleep, K.J. Zahnle, J. Geophys. Res. **103**, 28529–28544 (1998)

N.H. Sleep, K. Zahnle, P.S. Neuhoff, Proc. Nat. Acad. Sci. **98**, 3666–3672 (2001)

A.P. Sommer, D.S. McKay, N. Ciftcioglu, U. Oron, A.R. Mester, E.O. Kajander, J. Proteome Res. **2**, 441–443 (2003)

T. Spohn, M.H. Acuña, D. Breuer, M. Golombek, A. Greeley, A. Halliday, E. Hauber, R. Jaumann, F. Sohl, in *Chronology and Evolution on Mars*, ed. by R. Kallenbach, J. Geiss, W.K. Hartmann. Space Science Series of ISSI. Space Sci. Rev. **96**, 231–261 (2001)

S.W. Squyres, M.H. Carr, Science **231**, 249–252 (1986)

S.W. Squyres, J. Grotzinger, R.E. Arvidson, J.F. Bell III, W. Calvin, P.R. Christensen, B.C. Clark, J.A. Crisp, W.H. Farrand, K.E. Herkenhoff, J.R. Johnson, G. Klingelhöfer, A.H. Knoll, S.M. McLennan, H.Y. McSween Jr., R.V. Morris, J.W. Rise Jr., R. Rieder, L.A. Soderblom, Science **306**, 1709–1714 (2004)

L.J. Stahl, in *Biostabilization of sediments*, ed. by W.E. Krumbein (Biblioteks und Informationssystem der Carl von Ossietzky Universität, Oldenburg, 1994), pp. 41–54

T. Stevens, FEMS Microbiol. Rev. **20**, 327–337 (1997)

T.O. Stevens, J.P. Mckinley, Science **270**, 450–454 (1995)

T.O. Stevens, J.P. Mckinley, J.K. Fredrickson, Microb. Ecol. **25**, 35–50 (1993)

D.J. Stevenson, Nature **400**, 32 (1999)

B.M. Tebo, J.R. Bargar, B.G. Clement, G.J. Dick, K.J. Murray, D. Parker, R. Verity, S.M. Webb, Annu. Rev. Earth Planet. Sci. **32**, 287–328 (2004)

M.M. Tice, D.R. Lowe, Nature **431**, 549–552 (2004)

J.K.W. Toporski, A. Steele, F. Westall, K.L. Thomas-Keprta, D.S. McKay, Astrobiology **2**, 1–26 (2002)

N.H. Trewin, A.H. Knoll, Palaios **14**, 288–294 (1999)

Y. Ueno, H. Yoshioka, S. Maruyama, Y. Isozaki, Geochim. Cosmochim. Acta **68**, 573–589 (2004)

J.W. Valley, W.H. Peck, E.M. King, S.A. Wilde, Geology **30**, 351–354 (2002)

P. Van Thienen, N.J. Vlaar, A.P. van der Berg, Phys. Earth Planet. Int. **142**, 61–74 (2004)

M. van Zuilen, A. Lepland, G. Arrhenius, Nature **418**, 627–630 (2002)

M.A. van Zuilen, K. Mathew, B. Wopenka, A. Lepland, K. Marti, A. Arrhenius, Geochim. Cosmochim. Acta **69**, 1241–1252 (2005)

J.C.G. Walker, Nature **329**, 710–712 (1987)

M.M. Walsh, Precambr. Res. **54**, 271–293 (1992)

M.M. Walsh, Astrobiology **4**, 429–437 (2004)

P.H. Warren, J. Geophys. Res. **98**, 5335–5345 (1993)

A.J. Watson, T.M. Donahue, W.R. Kuhn, Earth Planet. Sci. Lett. **68**, 1–6 (1984)

E.B. Watson, T.M. Harrison, Science **308**, 841–844 (2005)

B.P. Weiss, Y.L. Yung, K.H. Nealson, Proc. Natl. Acad. Sci. **97**, 1395–1399 (2000)

S.C. Werner, Major aspects of the chronostratigraphy and geologic evolutionary history of Mars. PhD Thesis, Free University Berlin, 2005

F. Westall, Palevol **2**, 485–501 (2003)

F. Westall, G. Southam, in *Archean Geodynamics and Environments*, ed. by K. Benn et al. AGU Geophys. Monogr., vol. 164 (2006), pp. 283–304

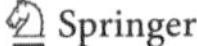

F. Westall, L. Boni, E. Guerzoni, Paleontology **38**, 495–528 (1995)

F. Westall, S.T. de Vries, W. Nijman, V. Rouchon, B. Orberger, V. Pearson, J. Watson, A. Verchovsky, I. Wright, J.-N. Rouzaud, D. Marchesini, S. Anne, in *Processes on the Early Earth*, ed. by W.U. Reimold, R. Gibson, Geol. Soc. Amer. Spec. Pub., vol. 405 (2006a), pp. 105–131

F. Westall, C.E.J. de Ronde, G. Southam, N. Grassineau, M. Colas, C. Cockell, H. Lammer, Phil. Trans. Roy. Soc. B **185**, 1857–1875 (2006b)

F. Westall, L. Lemelle, M. Salomé, A. Simionovici, G. Southam, C.E.J. Ronde, N. Grassineau, M. Colas, B. Orberger, Geochemical profiling of an Early Archean, littoral environment, photosynthetic microbial mat from the Barberton greenstone belt, South Africa. EANA, Lyon 16–18 October 2006c, Abstr.

J. Wierzchos, L.G. Sancho, C. Ascaso, Environ. Microbiol. **7**, 566–575 (2005)

S.A. Wilde, J.W. Valley, W.H. Peck, C.M. Graham, Nature **409**, 175–178 (2001)

O.I. Yakovlev, Yu.P. Dikov, M.V. Gerasimov, Geochem. Int. **30**(7), 1–10 (1993)

O.I. Yakovlev, Yu.P. Dikov, M.V. Gerasimov, F. Wlotzka, J. Huth, in *Lunar and Planetary Science XXXIII*, Abstract #1271 (Lunar and Planet. Inst., Houston, 2002) (CD-ROM)

A.S. Yen, R. Gellert, C. Schröder, R.V. Morris, J.F. Bell III, A.T. Knudson, B.C. Clark, D.W. Ming, J.A. Crisp, R.E. Arvidson, D. Blaney, J. Brückner, P.R. Christensen, D.J. DesMarais, P.A. de Souza Jr., T.E. Economou, A. Ghosh, B.C. Hahn, K.E. Herkenhoff, L.A. Haskin, J.A. Hurowitz, B.L. Jolliff, J.R. Johnson, G. Klingelhöfer, M.B. Madsen, S.M. McLennan, H.Y. McSween, L. Richter, R. Rieder, D. Rodionov, L. Soderblom, S.W. Squyres, N.J. Tosca, A. Wang, M. Wyatt, J. Zipfel, Nature **436**, 49–54 (2005)

K.J. Zahnle, N.H. Sleep, in *Comets and the Origin and Evolution of Life*, 1st edn., ed. by P.J. Thomas, C.F. Chyba, C.P. McKay (Springer, New York, 1997), pp. 175–208

K.J. Zahnle, N.H. Sleep, in *Comets and the Origin and Evolution of Life*, 2nd edn., ed. by P.J. Thomas, R.D. Hicks, C.F. Chyba, C.P. McKay (Springer, Berlin, 2006), pp. 207–252

M.Y. Zolotov, E.L. Shock, Meteorit. Planet. Sci. **35**, 629–638 (2000)

Space Sci Rev (2007) 129: 123–165
DOI 10.1007/s11214-007-9193-3

Conversations on the Habitability of Worlds:
The Importance of Volatiles

J.-L. Bertaux · M. Carr · D.J. Des Marais · E. Gaidos

Received: 17 February 2006 / Accepted: 10 April 2007 /
Published online: 26 June 2007
© Springer Science+Business Media B.V. 2007

Abstract Our scientific forefathers discuss the interrelationships between water, climate, the atmosphere, and life on Earth and other terrestrial planets at a workshop in Nichtchâtel, Switzerland.

Keywords Origin of life · Habitability · Planetary volatiles · Mars · Biogeochemical cycles · Planetary atmospheres · Biological evolution

Guillaume: Good morning everyone. My name is Charles Guillaume and I am very honored to chair this session on "Planetary Volatiles, Atmospheres, and Habitability". I look forward to what our distinguished session speakers have to say about planetary habitability. I must confess that I know nothing of the subject, having spent my entire career working on the properties of nickel-steel alloys that have application in precision instruments. I thought this was a worthwhile endeavor, and evidently so did the Norwegian Nobel Committee in 1920. A reminder that our format will consist first of four talks, followed by a panel discussion involving our speakers. We will first hear from Charles Messier, who will speak on "The Origin of Earth's Water", followed by William Herschel, who will discuss "The History of Water and Climate on Mars". We then have Charles Lyell, who will present "Volatile

J.-L. Bertaux
Service d'Aéronomie, CNRS, Paris, France
e-mail: jean-loup.bertaux@aerovre.jussieu.fr

M. Carr
USGS, Menlo Park, CA, USA
e-mail: carr@usgs.gov

D.J. Des Marais
NASA Ames Research Center, Moffett Field, CA, USA
e-mail: david.j.desmarais@nasa.gov

E. Gaidos (✉)
University of Hawaii, Honolulu, HI, USA
e-mail: gaidos@hawaii.edu

 Springer

Reservoirs and Exchange Processes on Earth and Mars", and finally Charles Darwin, who will finish with "Co-evolution of the Atmosphere and Life on Earth". Well, we must stay on schedule; this is Switzerland, after all, so I would like to introduce the first speaker, Charles Messier. Mssr. Messier was a self-educated astronomer until he began working for the French navy in 1751. Six years later he began searching for Comet Halley, which was predicted to return in 1758. His independent recovery of that famous object in early 1759 launched his career as a comet hunter, but he is best known for the catalog of *nebulae* that he developed while searching for comets. He was elected to the Royal Society of London in 1764, and to the Royal Academy of Sciences of Paris in 1770. He was appointed Astronomer of the Navy the following year. Charles?

1 The Origin of Earth's Water

Messier: Merçi. May I have the first slide? All life on Earth requires water; it seems to be an immutable requirement of planetary habitability, at least for life as we know it. Therefore, what is the origin of Earth's water? Because Earth is not alone in this region of the solar system, we should also consider the case of Venus and Mars, where liquid water is or could have been present and where life might have emerged independently. Also, if we find a good theory for the origin of water on Earth, it must also be compatible with the current content of water on the two other planets.

In planetary environments, water typically exists in the following four forms: solid ice, liquid, vapor, and as water molecules that are bound in a great variety of minerals, for example, gypsum: $CaSO_4^{2-} \cdot 2(H_2O)$. These minerals can liberate their water when they are heated, which indeed occurred during the formation of the planets. Later, radioactive heating and outgassing from the interiors of planets transferred this water to the oceans and atmosphere.

1.1 Planetary Inventories of H_2O

The present inventory of water on the three planets is as follows. On Venus, where the crust is hot and degassed, the only reservoir is in the atmosphere as vapor. If transformed into liquid, it would represent a layer of only 3 cm. This estimate may be calculated from the H_2O vertical profile data obtained by Moroz et al. (1983), re-interpreted by Ignatiev et al. (1997) as a constant mixing ratio of 3 ppmv. Let us take 3 cm as representative for Venus, which makes it the driest, by far, of the three planets today. The Earth has 2.8 km of liquid water, if spread uniformly over the surface, and an additional amount of chemically and physically bound water in the crust and mantle of about 4 km equivalent thickness. It is interesting that, if the Earth had more liquid water, a total of about 11 km or more equivalent thickness, then there would be no land. The Himalaya summits would be submerged.

What is the quantity of H_2O on Mars now? The main measurable reservoir is the permanent North polar cap; its volume was measured by Mars Global Surveyor Laser Altimeter (MOLA) (Zuber et al. 1998). This quantity of ice corresponds to a layer of about 9 m, if spread uniformly as liquid on the surface. The Mars Express orbiter discovered that the permanent south polar cap is made mainly of H_2O ice, with a layer of CO_2 ice on top (e.g., Bibring et al. 2005). The visible permanent south polar cap is much smaller than its northern counterpart, but overlays ice-rich layered deposits that rival its northern counterpart. Neutron and gamma ray measurements on Mars Odyssey (Boynton et al. 2002; Mitrofanov et al. 2002; Feldman et al. 2004) also discovered a large fraction of H in the surface at high latitudes, but these techniques can probe only the first meter of the subsurface.

 Springer

Therefore, the total amount of *measured* H_2O reservoir today is about 12 m (± 2 m). This corresponds to a volume of 1.7×10^6 km^3 of water, a number to remember when William Herschel will present later the geological evidence for erosion on the surface, and estimate the amount of water necessary to create this erosion.

1.2 Evolution of Planetary Volatiles

Coming back to Earth, do we know the origin of its plentiful water? Well, to know better, we have to journey backward in time to the formation of the solar system, 4.55 billion years ago. And we have to combine two scientific disciplines that fortunately are communicating with each other better and better, although each of them requires very specialized skills. One discipline performs ever more refined measurements of elemental and isotopic compositions of Earth, Moon, and meteorites, including meteorites derived from Mars and dust from comets. This allows us to reconstitute a remarkably detailed and accurate chronology of a number of events in the early history of the solar system. The other scientific discipline develops theories of the formation of planets through accretion of material from the proto-solar nebula. Numerical simulations of mutual interactions between planet-forming objects play an important role in model development.

In the early descriptions by Safronov (1969), small dust particles aggregated into km-size bodies, called planetesimals, that eventually assembled into the planetary bodies, such as the terrestrial planets, that assumed more or less their present orbital location and present sizes. The giant planets accreted an additional huge amount of primitive nebula gas. One problem with this picture is that the study of meteorites indicates that material formed inward of 2.5–3 AU (Astronomical Unit, equal to the Sun–Earth distance) was too dry to supply the quantity of water in Earth's oceans (Lunine et al. 2003).

However, the theory of accretion has progressed since the early picture of Safronov. After the first stage of formation of planetesimals, dynamical simulations indicate that these planetesimals combined to form Mercury- to Mars-mass planetary embryos in quasi-circular and coplanar orbits, from 0.3 to $\sim$4 AU (Petit et al. 2001). These orbits extended outward to the asteroid belt, where the original material contained minerals with bound water. Minerals such as these have survived in a special class of meteorites, the carbonaceous chondrites.

Then, Jupiter accreted its gaseous envelope and reached its present mass, with enormous consequences for the planetary embryos. In particular, the ones orbiting at more than 2 AU and nearest the orbit of Jupiter, were either ejected from the solar system, sent to the Sun, or sent to orbits crossing the orbits of the growing terrestrial planets, provoking gigantic collisions, but providing a good supply of water to the otherwise dry planetary embryos (Lunine et al. 2003). Interestingly, these and other simulations do not usually produce a Mars-mass planet in an orbit similar to the Mars orbit; they are either much more massive in a gas-free system, or, with gas drag, a number of very small planets form.

Venus and Earth were therefore growing through these gigantic collisions with embryos sent by Jupiter from orbits beyond that of Mars. The development of this scenario was stimulated by the work of Cameron and Benz (1991), who described a theory that the Moon was formed through a gigantic impact of the Earth with a Mars-size body, now called Theia in the literature. Though, by principle, we do not like to invoke catastrophic events to explain what we see (and in particular our own existence), it seems that such a catastrophe is the only way to explain the present composition of Earth and Moon, as inferred from samples returned by the Apollo missions. This theory is now widely accepted, and the collision event is now dated to have occurred rather precisely, at 40 to 50 million years after the formation of the oldest meteorites, 4.567 billion years ago.

But what happened to water during this terrible collision? It is easy to imagine that the oceans definitively could have been lost to space. Detailed simulations of collisions show a more complicated picture, however. If the Earth had only an atmosphere and no oceans (water present only as steam), then 90% of the atmosphere would survive a moderate velocity collision (Genda and Abe 2004). It is reasonable to assume that the collision occurred at a moderate velocity because, otherwise, the Earth would have decreased in size and the Moon would not have formed after the collision with Theia. But if there was an ocean (if water had already condensed before 40 million years after accretion), then the mechanics of the shock are entirely different. The transmission of the shock through the liquid phase is easier, the resulting velocity of the shocked atmosphere is larger, even at antipode of the shock, and the ocean might have been partially lost to space and the atmosphere totally lost (Genda and Abe 2005).

If such an event indeed occurred, it could have lowered the content of ^{36}Ar (a noble, non-radiogenic gas) in Earth's atmosphere to a quantity 50 times less than that on Venus, as is found today. This difference has been a long standing-puzzle. If water existing only as steam on Venus early in its history, then its primitive atmosphere, including original inventory of ^{36}Ar, would have mostly survived the large collisions. Pursuing this scenario further, we could very well imagine that, indeed, the gigantic collisions with the Earth could have also eliminated a substantial fraction of the ocean. Some scientists believe that the challenge to understanding the history of water on Earth is not to find an adequate source, because the sources are large, but rather to find ways to eliminate a good fraction of its water. Maybe the collision with Theia, or another large collision that preceded it, could have eliminated 80% of the water. Without this enormous loss of water, the ocean would have inundated the Himalaya. Possibly life could still emerge on this Planet Ocean, but we would not be here to speculate on the habitability of terrestrial planets.

Studies of deuterium to hydrogen ratios (D/H) in the atmospheres (and in the ocean) of terrestrial planets and in meteorites and comets also offer clues about the origin of water and its evolution through the aeons (Robert 2001). The terrestrial D/H value is 149 parts per million, similar to that of seawater that is referred to as SMOW (Standard Mean Ocean Water). It is very near the values commonly found in carbonaceous chondrites, whereas D/H values found in the atmospheres of three long-period comets are twice as great (Balsiger et al. 1995; Bockelée-Morvan et al. 1998; Meier et al. 1998). Such differences in D/H do not favor comets as the main source of water on Earth. One may, however, claim that these three comets are not representative, or invoke another, ad hoc source of water with less deuterium that would combine to yield the present D/H ratio in the oceans (Owen and Bar-Nun 2000).

Over time the atmosphere of a terrestrial planet may be enriched in D through differential escape. Water vapor is transported upward and photo-dissociated by solar UV radiation to yield H and D atoms that diffuse up to the exosphere, the most external part of an atmosphere, where the density is so low that essentially no collisions occur. Atoms having velocities larger than the escape velocity are sent on hyperbolic trajectories and lost from the planet. Clearly the higher mass of D atoms makes their escape less likely, with a subsequent enrichment of HDO, relative to H_2O, in the atmosphere. The escape of D atoms may be further reduced by preferential condensation of HDO as ice, preventing much of the HDO from attaining higher altitudes where it is photodissociated by solar UV (Bertaux and Montmessin 2001).

The atmosphere of Venus is enormously enriched in HDO; its HDO/H_2O value is $\sim$150 times that of SMOW (de Bergh et al. 1991; Bjoraker et al. 1992; Donahue et al. 1997). If no D atoms escaped to space during the entire history of Venus after the end of big collisions that provoked undifferentiated hydrodynamic escape to space, the atmosphere's present-day

 Springer

HDO content would indicate that there was never more than 3×130 cm, or 4 m of water. This is a very small quantity indeed, compared to Earth. But this is an absolute lower limit; if there is now some escape of D atoms (and we will know with the forthcoming ESA Venus Express mission), then some scaling to the present escape of H can be done and extrapolated back in time for a more accurate estimate of past water inventories on Venus. For Earth, we assume that the escape of hydrogen to space caused little or no enrichment in the D/H values of the oceans and atmosphere. As a matter of fact, the present measured escape rate of H of 2×10^8 atoms cm^{-2} sec^{-1} represents a loss of only 4 m of liquid water during all of Earth's history (Bertaux 1974). On Mars, the HDO enrichment is found to be $\sim$5.5 SMOW (Owen et al. 1988). Applying the same line of reasoning as for Venus, we infer that Mars once had $\sim$70 m of water, or 10^7 km^3. Is this quantity sufficient to explain all of the water erosion features that we observe on Mars? This inventory seems to be sufficient, especially when considering that liquid water may have flowed many times, either through seasonal cycles or through climatic cycles controlled by the rotation axis inclination and orbit eccentricity. In this case, there might be no significant amount of water ice buried below the surface of Mars. But William Herschel will now give the facts about what we have actually observed on Mars.

Guillame: Thank you, Charles. Our next speaker is William Herschel. Sir Herschel began his career as a musician—an organist, I believe—first in Germany and then in England. In 1773, At the age of 35 he became interested in astronomy and started constructing the world's most advanced telescopes. Eight years later he discovered the planet Uranus and was soon after elected to the Royal Society as well as appointed Court Astronomer. Besides his work on *nebulae*, he also discovered two moons of Uranus, two moons of Saturn, and he first proposed that the polar caps of Mars were made of ice and snow. William?

2 The History of Water and Climate on Mars

Herschel: Thanks. I will discuss the hydrologic history of Mars and what it means for climate on that planet. Although the origin of the valleys and channels on Mars has been debated for over thirty years their origin remains almost as puzzling as it was when they were first observed in 1972 during the Mariner 9 mission. For liquid water to be thermodynamically stable, temperatures must exceed 273 K and the atmospheric pressure must exceed 6.1 mbar. For most of the planet the total pressure is less than 6.1 mbar and liquid water at the surface would rapidly boil and freeze. Ice, on heating, would sublimate rather than pass through an intervening liquid phase. Liquid water is stable at the pressures and temperatures found at low elevations at midday in summer. However, the partial pressure of water vapor in the atmosphere typically falls 2–3 orders of magnitude short of 6.1 mbar and thus any liquid water would eventually evaporate. Liquid water could exist transiently in the upper centimeter of an ice-rich soil heated by the Sun, if water vapor were inhibited from diffusing into the atmosphere, but the amount of water involved would be minute. These conditions were mostly understood when the valleys and channels were first discovered, so that an ice streams idea (Lucchitta et al. 1981), and origins other than water erosion were explored. Included were faulting, mass wasting, and erosion by wind, lava, liquid carbon dioxide, and other exotic fluids. Despite these possibilities, the broad consensus is that most of the channels and valleys were cut by water. The consensus has been reinforced recently by detection of water soluble minerals at the surface in several places (Gendrin et al. 2005), by discovery of evaporites in water-lain sediments at

Meridiani (Squyres et al. 2004), by extensive aqueous alteration of rocks in the Columbia Hills of Gusev Crater (Squyres et al. 2006), and by confirmation that water–ice is widespread just below the surface at high latitudes (Boynton et al. 2002; Mitrofanov et al. 2002; Feldman et al. 2004). In the following discussion we will assume that, except for lava channels in volcanic regions and some local mass-wasting and ice sculpted features, most of the valleys, channels, and gullies are water-worn.

2.1 Stability of Liquid Water

With mean annual temperatures close to 215 K at the equator and 160 K at the poles, the ground is everywhere frozen to kilometer depths. The exact depth at which liquid water might be found depends on the heat flow, the thermal conductivity of the crustal rocks, and the salinity of the water. Taking plausible ranges for these parameters Clifford (1993) and Clifford and Parker (2001) estimated the mean thickness of the cryosphere (permanently frozen crust) to be in the 2–20 km range. In volcanically active areas, the thickness of the cryosphere could be reduced to zero as hydrothermal fluids reach the surface, but no such areas, either fossil or active, have been detected. The thickness range applies not only to today but probably to much of post-Noachian times since cold climatic conditions appear to have prevailed for much of Mars' history, and the increase in heat flow back in time (McGovern et al. 2002) is unlikely to result in values that fall outside the range considered by Clifford and co-workers. (The Noachian is the earliest epoch in Mars' history ending sometime between 3.8 and 3.5 billion years ago.)

Under present conditions, not only is liquid water unstable everywhere on the surface, but so is ice, since the surface temperature everywhere exceeds the frost point at some time during the year. It is unknown whether the polar layered deposits are currently in a state of net deposition or net sublimation. At latitudes higher than 40 degrees, ice, while unstable on the surface, is stable at depths greater than roughly one meter where the mean temperature of the ground never exceeds the frost point. Consistent with these relations, hydrogen in amounts equivalent to several tens of percent ice, has been detected below a thin dehydrated layer at high latitudes (Boynton et al. 2002; Mitrofanov et al. 2002; Feldman et al. 2004). In contrast, at low latitudes temperatures exceed the frost point at all depths so that ground ice is unstable: Any ice present will eventually sublimate and the water vapor lost to the atmosphere. Detection of several percent water at low latitudes by neutron and gamma-ray techniques is, therefore, inconsistent with the stability relations. The water detected may be chemically bound water, or water inherited from an earlier era when conditions were different, and has yet to completely diffuse from the regolith.

The conditions just described are for the present day. The stability of water near the surface is sensitive to the obliquity, and Mars may have recently emerged from an era when its obliquity was higher (Laskar et al. 2002). At an obliquity lower than today's, equatorial temperatures rise and the amount of water in the atmosphere probably falls as the water–ice polar deposits grow. As a result the latitude belt over which ice is unstable at all depths widens (Mellon and Jakosky 1995). At a higher obliquity the reverse occurs. Equatorial temperatures fall and the amount of water present in the atmosphere increases and with it the frost point temperature. Ice may then be stable at shallow depths below the surface at all latitudes. At the highest values of obliquity water ice may be driven from the poles and accumulate on the surface at low to mid latitudes.

Given the stability conditions just described, a major issue with respect to the seemingly water worn features is whether climate changes were required to form them and, if so, when they occurred, what their magnitude was, how sustained the climatic changes were, and what

 Springer

caused them. Coupled to these issues are others such as the former presence of oceans and the fate of the missing water discussed by Mssr. Messier. These issues will be addressed by examining the different types of water worn features and what they might imply about the hydrologic history of the planet.

Three classes of likely water worn features are recognized. Outflow channels are linear swaths of scoured ground, tens to hundreds of kilometers across that commonly contain streamlined remnants of the pre-existing terrain. Most start full size and have no tributaries. Because of their close resemblance to terrestrial flood features, they are widely, although not universally, believed to have formed by huge floods. Valley networks are comprised of valleys that are typically only 1–5 km across, although they may be hundreds even thousands of kilometers long. The networks they form resemble terrestrial river systems in plan. They are thought to have formed mostly by slow erosion of running water. Gullies are smaller still. They are restricted to steep slopes and are only several to tens of meters wide and hundreds of meters long.

2.2 Outflow Channels

Outflow channels vary greatly in size. They occur in several regions of the planet and start in several different types of geologic terrains (Fig. 1). They clearly do not all originate in the same way. The most prominent concentration of outflow channels is around the Chryse basin. The largest outflow channel on the planet, Kasei Vallis, enters the Chryse basin from the west, and several large channels emerge from the cratered uplands to the south of the basin and affect an 800 km wide swath of terrain clearly marked by longitudinal scour and numerous teardrop shaped islands. The southern channels extend northward across the Chryse basin to 25 N where they curl northwestward and merge with the scoured ground from Kasei Vallis, and continue to roughly 40 N where traces of the channels become lost in the northern plains. Crater counts and translation relations suggest that most of the circum-Chryse channels are from the Hesperian (the middle epoch in Mars history ending sometime between 3.5 and 1.8 billion years ago). Various estimates have been made of the discharge Q implied by the large size of the channels. The estimates are based on relations observed for terrestrial rivers, after correcting for the lower gravity on Mars (Komar 1979). The basic equation is

$$Q = A\sqrt{\frac{g_m S R^{4/3}}{g_e n^2}},\tag{1}$$

where A is the cross-sectional area of the flow, g_m and g_e are gravity on Mars and Earth, S is the local slope, R is the hydraulic radius (ratio of cross-sectional area to wetted perimeter and n is the Manning roughness coefficient (Williams et al. 2000). The roughness coefficient takes into account factors such as the roughness of the stream bed and the sinuosity of the channel and is determined empirically from terrestrial rivers. Applying this equation to Mars is uncertain because what value is appropriate for the Manning coefficient is uncertain, the depth of the floods is usually unknown and (1) was determined from terrestrial rivers, which are orders of magnitude smaller than the martian channels. For Kasei Vallis estimates for the peak discharge range from 10^4 m^{-3} s^{-1} (Williams et al. 2000) to 10^9 m^{-3} s^{-1} (Robinson and Tanaka 1990). The total amount of water involved is even more uncertain. Roughly 6×10^5 km^3 was removed to form Kasei Vallis, suggesting that at least 10^6 km^3 of water was needed and possibly considerably more.

Many of the valleys that extend into the Chyse basin (e.g., Maja Vallis, Salbatana Vallis, Ravi Vallis, and Ares Vallis) start full size in rubble filled hollows (Fig. 2a). The relations

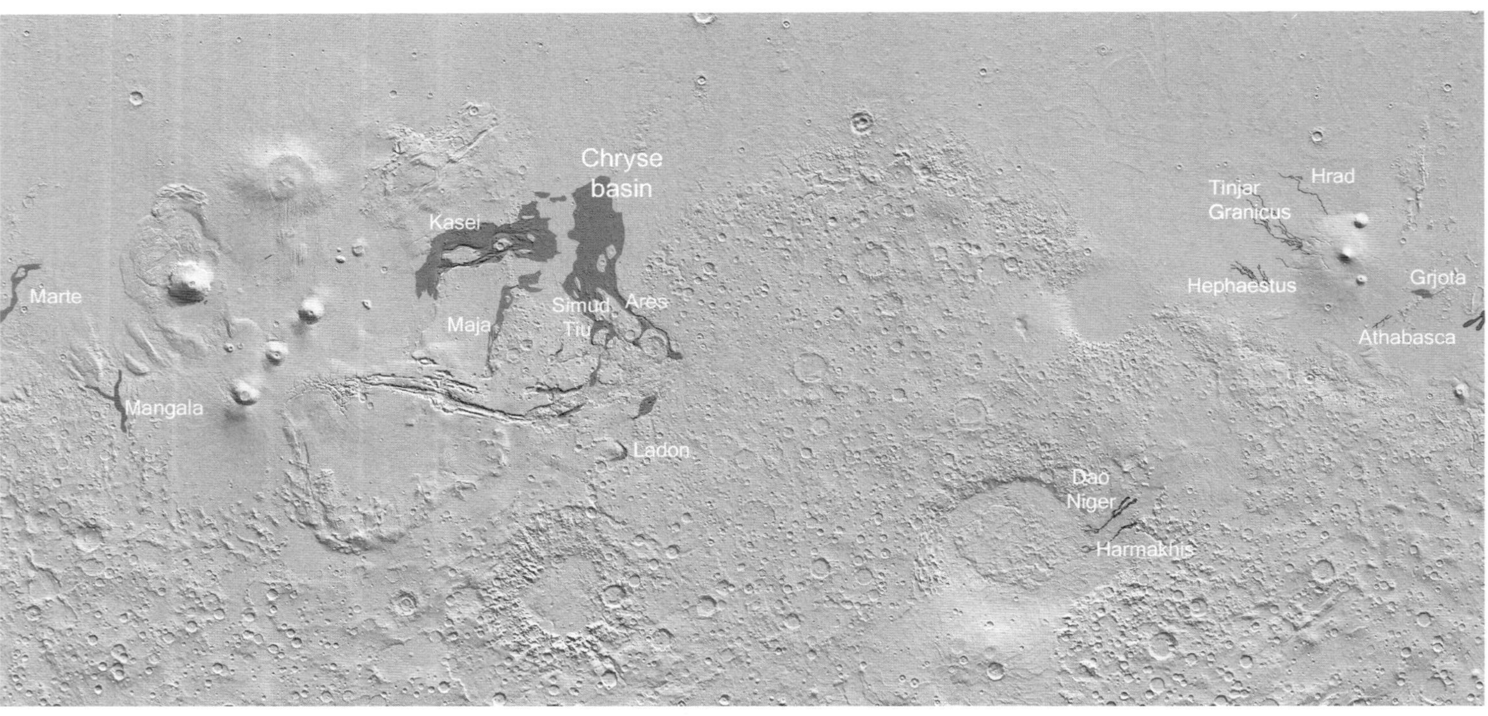

Fig. 1 *Left*: Global map of outflow channels. Around the Chryse basin only the largest channels are named. Base map is MOLA/MGS shaded relief

suggest that the floods were caused by catastrophic eruptions of groundwater followed by collapse of the source areas (Carr 1979). To achieve the discharges needed the groundwater must have been under high pressure, possibly as a result of being trapped under a thick cryosphere. Others valleys (Tiu Vallis, Simud Vallis, and Kasei Vallis) can be traced upstream into substantial canyons. In these cases, the relations suggest catastrophic drainage of former lakes within the canyons. The presence of former lakes within Vallis Marineris (and in Juventae Chasma) is supported by the presence of extensive internal layered deposits (McCauley 1978; Malin and Edgett 2001) containing gypsum and kieserite (Gendrin et al. 2005). The lakes may have been ice-covered. Crater counts and superposition relations indicate that most of the outflow channels in the Chryse region are upper Hesperian in age (Scott and Tanaka 1986), probably 2.0–3.8 billion years old (Hartmann and Neukum 2001).

Several outflow channels in the Amazonis-Elysium region appear to have a different origin from those around Chryse since they start at graben. The largest is Mangala Vallis (Fig. 2b). It starts abruptly at a 10 km wide notch in a graben wall then extends over 1000 km to the north, being in places up to 100 km wide. Like the Chryse channels it has a rich array of teardrop-shaped islands, convergent and divergent striations and other streamlined forms. Other smaller channels, Athabasca Vallis (10°N, 158°E) and Grjota Vallis (16°N, 163°E) also start at graben among and to the west of the hills that separate Amazonis Planitia from Elysium. The channels can be traced for hundreds of kilometers before they disappear in low-lying Cerberus plains just north of the plains-upland boundary at 170°E. From these plains emerges yet another large channel, Marte Vallis that extends northeastward into Amazonis Planitia. These are the youngest outflow channels on the planet. Burr et al. (2002), using the crater chronology of Hartmann and Neukum (2001), estimate that the ages of Athabasca Vallis, Grjota Vallis and Marte Vallis are 2–8 million years, 10–40 million years and 35–140 million years, respectively. If only approximately correct, these ages are so young as to suggest that outflow channels such as these could form today. Other channels in Tharsis, such as the Olympica Fossae and the Gordii Fossae, also start at graben. As with Athabasca Vallis, channels with streamlined forms clearly formed from fluids that erupted from the graben, but in these cases, whether the fluid was lava or water is more uncertain.

Eruption of water from graben could result from several causes. Faulting may simply have disrupted the cryosphere seal over a deep aquifer, thereby allowing water from the underlying cryosphere to reach the surface. The water may have been under high pressure because of the regional topography. Alternatively, tectonic forces may have pressurized the aquifer causing the water to flood to the surface when faults broke the seal (Hanna and Phillips 2005). Another possibility is that pressurization resulted from the injection of dikes that accompanied formation of the graben (Head et al. 2003).

Several valleys, including the Granicus, Tinjar and Hrad Valles emerge from graben on the western flank of the Elysium dome. At higher elevations, the graben are typical of those elsewhere on the planet, being narrow, steep-walled depressions. To the west, at lower elevations, however, they transition into channels with streamlined walls, teardrop-shaped islands and scoured floors. Many have broad textured rims, with lobate outer margins, as though the channels had overflowed and left a deposit on the rim. The latter observations suggested to Christiansen (1989) that the channels were cut by or at least were utilized by lahars or mudflows. Fluvial landforms are generally restricted to elevations less than −3400 m (relative to the 6.1 mbar datum), which suggests that this was the elevation of the local water table when the valleys formed (Russell and Head 2003). As with the channels discussed in the previous section, formation of the valleys likely resulted from tectonic disruption of the cyosphere, possibly accompanied by dike injection.

 Springer

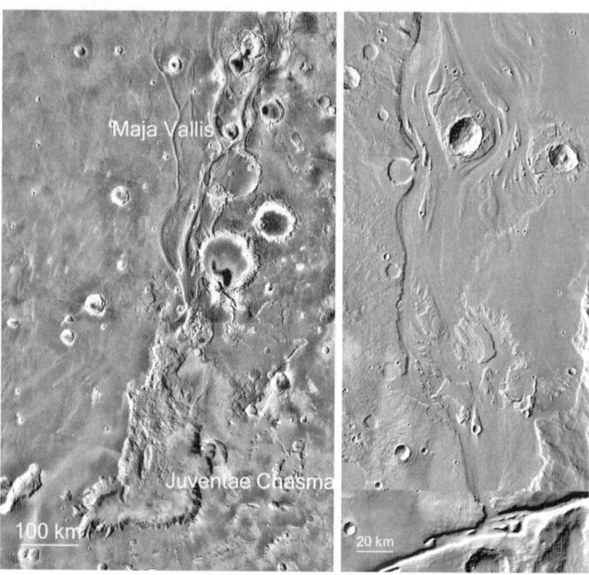

Fig. 2 *Left*: Maja Vallis and Juventae Chasma. Maja Vallis emerges from a 5 km deep rubble-filled depression. The outlet is at an elevation 4 km above the floor of the depression, so after Maja Vallis formed a lake must have been left in the depression, consistent with detection of sulfates there (Gendrin et al. 2005). The relations indicate that Maja Vallis formed by a massive eruption of groundwater (Viking Orbiter camera MDIM2). *Right*: Mangala Vallis. The channel starts at a notch in a graben, which suggests eruption of groundwater as a result of faulting and/or dike injection. (Mars Thermal Emission Imaging System, THEMIS mosaic, Mars Odyssey)

The most puzzling features in this area are the Hephaestus Fossae and Hebrus Vallis (Fig. 4a). The upper portions of both these valleys are fluvial in appearance but the lower portions are distinctly non-fluvial. They consist of linear segments that meet at high angles to form a pattern like a cracked pavement, except that the individual segments commonly consist of lines of unconnected depressions. The origin of the networks is unknown. Lava tubes form lines of depressions but not linear networks such as we see here. The pattern more resembles karst. They may indicate subsurface drainage through soluble rocks, such as carbonates, or through ice rich rocks.

Three large valleys, the Dao, Niger and Harmakhis Valles, start on the east rim of Hellas and extend for over 1000 km down into the floor of the Hellas basin. The Dao and Niger Valles both start near the volcano Hadriaca Patera, suggesting that the volcano was somehow connected with their origin, possibly causing local melting of ground ice or explosive release of groundwater following injection of hot magma into the hydrosphere and cryosphere.

Thus outflow channels occur widely across the planet with ages that range from over 3.5 billion years almost to the present. Most can plausibly be interpreted as the result of flooding by release of groundwater that followed tectonic, volcanic, or other disruptions of the cryosphere. High hydrostatic pressures in the underlying hydrosphere could have resulted from a variety of causes, including regional topography, tectonic deformation and volcanic intrusions. Another possible cause of flooding, particularly adjacent to the Valles Marineris, is catastrophic release of water from lakes. The lakes may have formed also by release of groundwater, possibly accompanying the massive faulting that caused much of the canyon relief. If a cryosphere was present when the lakes formed they would have quickly become ice covered. Most of the floods likely occurred under cold climatic conditions when a thick cryosphere was present. This conclusion is consistent with formation by massive eruptions of groundwater which had to be contained prior to eruption, with the presence of geologically recent outflow channels such as Athabasca, and with low erosion rates (Golombek and Bridges 2000) and low weathering rates (Haskins et al. 2005) of the basalts on the Hesperian plains in Gusev, which suggests that if the surface experienced warm episodes during the Hesperian and Amazonian (the latest epoch in Mars' history), then these episodes were short. The larger channels must have left large bodies of water at their termini, which would have rapidly frozen, given the likely presence of a thick cryosphere.

2.3 Valley Networks

The second class of water-worn feature is the valley network. Most of the cratered uplands are dissected by networks of branching valleys, no more than a few kilometers wide but up to hundreds, even thousands of kilometers long. Most are readily distinguishable from the outflow channels. They occur mostly in the cratered uplands but are distributed unevenly. Northwest Arabia and large areas to the west and southeast of Hellas, for example, are sparsely dissected whereas the broad swath of terrain just south of the equator from 340°E eastward to 180°E is highly dissected (Fig. 3). Most valleys are short and drain into local lows. However, several valleys extending down the regional slope through Terra Meridiani toward the Chryse basin are over 1000 km long. Valleys are typically 1–4 km wide, have cross sectional shapes that range from V-shaped in the upper reaches to U-shaped or rectangular in the lower reaches (Fig. 4b). In planimetric form they resemble terrestrial river systems. Drainage densities range widely up to values at the low end of the terrestrial range (Hynek and Williams 2001). Some of the most densely dissected surfaces are on the flanks of volcanoes.

The ages of the valleys are difficult to determine. The vast majority of the valleys are in Noachian terrain, which is mostly dissected with some exceptions, as noted above. In contrast, Hesperian plains such as Syria, Solis, Hesperia and Launae Plana are largely undissected. While these generalizations are valid, there are numerous exceptions. Some of the largest and freshest-appearing valleys such as Nanedi and Nirgal Valles cut Hesperian terrain. Well-developed, dense networks cut Hesperian plains at the southern end of Echus Chasma (Mangold et al. 2004) and on some volcanoes such as Alba Patera, Amazonian surfaces are dissected. It appears that the dominant period of valley formation was in the Noachian and that the rate fell dramatically at the end of the Noachian but formation continued either episodically or at a very low rate for much of Mars history. Quantitative measures of stream profiles and basin shapes suggest that most martian valley networks are less well developed than their terrestrial counterparts (Stepinski and O'Hara 2003). In most terrestrial

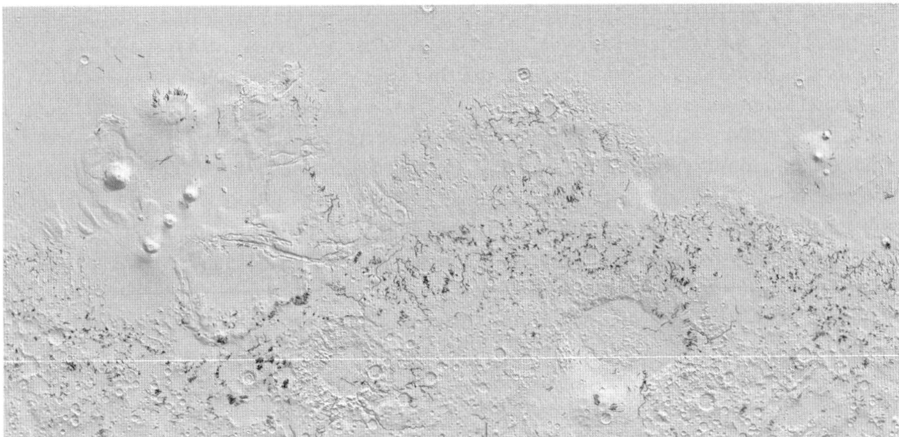

Fig. 3 Global map of valley networks. Most are in the southern uplands, although western Arabia and the uplands between Argyre and Hellas are only poorly dissected. The plains are largely undissected (MOLA)

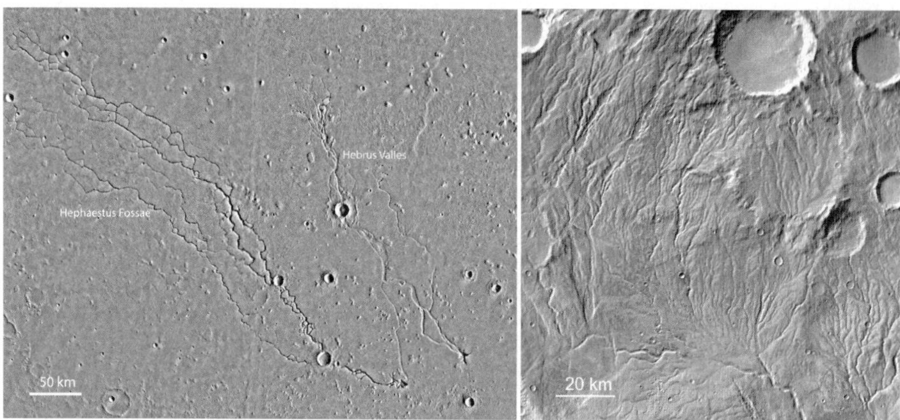

Fig. 4 *Left*: Utopia Channels. Hephaestus Fossa and Hebrus Valles have both fluvial and tectonic characteristics. Both start at irregular depressions, from which emerge fluvial-like channels. The Hephaestus Fossae change downstream into an array of discontinuous, intersecting linear segments. The Hebrus Valles similarly terminate in linear segments. The cause of the tectonic-like patterns are unknown. (MDIM2). *Right*: Warrego Valles. The terrain here is densely dissected by small valleys. Such area-filling dense dissection is characteristic of fluvial regimes in which surface runoff dominates and precipitation was widespread (THEMIS mosaic)

basins the local slope $S \sim A^{-\phi}$ where A is the area upstream at the given point and ϕ is called the concavity exponent, which is a measure of how concave the basin is. The values for martian basins (0.2–0.3) are consistently less than terrestrial values (0.3–0.7) indicating poor basin development. Another indicator of basin development is how the circularity of the basin varies with elevation. The higher the elevation slice through a typical terrestrial basin, the more circular the basin outline. This tendency is significantly less with martian basins.

Although the valleys were almost certainly cut by water, the source of the water and the conditions under which formation took place remain controversial. Water could have been introduced onto the surface in three ways; as groundwater seepage, as rainfall or as snowfall. That groundwater seepage played a prominent role in development of some of the valley networks was recognized early from the open networks, the amphitheater-like terminations of tributaries (Fig. 5a), rectangular cross sections, and other properties (Sharp and Malin 1975; Pieri 1980). Some authors have argued that the valley networks could form exclusively by groundwater sapping under conditions similar to those that prevail today (Squyres and Kasting 1994; Gaidos and Marion 2003). However, formation of valley networks by groundwater sapping alone seems unlikely. Dense area-filling networks (Fig. 4) are common and normally do not form where seepage dominates (Craddock and Howard 2002; Hynek and Williams 2001). In addition, many valleys start at local highs such as crater rims and central peaks where groundwater seepage is unlikely. Moreover, seepage draws down the local water table so some form of recharge is needed to sustain erosion. Although Clifford (1987) suggested that the global hydrosphere could be recharged by basal melting of ice at the poles, his concern was mainly for post-Noachian times, not for the Noachian era when most of the valleys formed. Many of the valleys are also at higher elevations than the base of the south polar layered terrain so could not provide the hydrostatic head needed to enable seepage (Carr 2002).

Arguments against rainfall are several: (1) It is difficult to warm early Mars when most of the valleys formed because of the faint, young Sun (Kasting 1991); (2) the early Mars

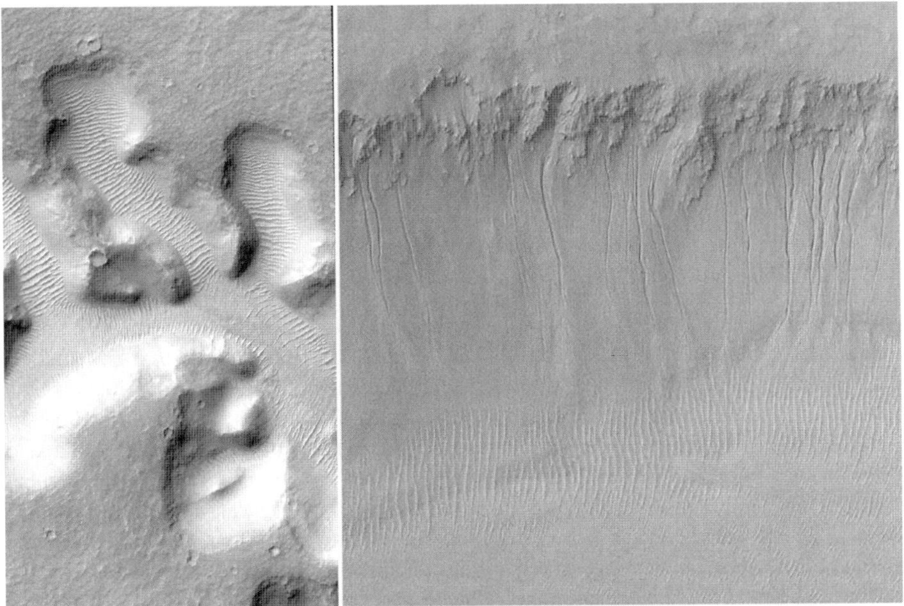

Fig. 5 *Left*: Nirgal Vallis at 27.8°S, 316.6°E. The tributaries have alcove like terminations. They do not divide into ever smaller valleys as in the previous picture. The relations suggest origin by groundwater sapping. The scene is 2.8 km across. Nirgal Vallis is Hesperian in age (Mars Observer camera MOC E0202651). *Right*: Gullies on the south-facing wall of Nirgal Vallis at 29.7°S, 321.4°E. The scene is 2.3 km across. The gullies are mostly incised into talus below a bedrock outcrop at the top of the slope. Fans of debris eroded by the gullies appear to be superimposed on dunes in the floor of the main valley (MOC M03-22990)

atmosphere is vulnerable to blow-off by large impacts; (3) until recently, hydrated minerals, evaporites and carbonates had not been detected from orbit; and (4) easily weathered olivine has been detected from orbit. Craddock and Howard (2002) argue, however, that the geomorphic evidence for surface runoff and precipitation is so compelling that some assumptions in the modeling studies must be wrong. They also suggest that observational artifacts are hindering our ability to detect weathered minerals from orbit. The arguments against precipitation have been recently undermined by the finding of evaporites and water lain sediments in Meridiani (Squyres and Knoll 2005) and by detection of evaporite minerals and phyllosilicates from orbit by the near-infrared spectrometer OMEGA on Mars Express (Bibring et al. 2005).

Precipitation may have been rain or snow. Snowfall can occur during the present epoch. The obliquity may have been as high as 45° within the last 10 million years (Laskar et al. 2002). At a high obliquity, water would be driven from the poles and deposited as ice (snow) at lower latitudes. Melting of such snow could, in principle, provide meltwater to cut valleys. However, unless these changes are accompanied by significantly warmer surface conditions, a snow cover is unlikely to produce sufficient meltwater to cut valleys (Clow 1987), although there may be sufficient water to cut gullies on steep, poleward-facing slopes as discussed below. Basal melting of snow from internal heat flow is also unlikely to produce meltwater in sufficient quantities because of the low internal heat flow. Climate change appears to be needed to form valleys. This conclusion is consistent with the patterns of dissection, the high Noachian erosion rates, detection of phyllosilicates produced by chemical weathering and finding of waterline sediments at Meridiani.

Greenhouse warming by a CO_2–H_2O atmosphere, however, may not have been sufficient to allow rainfall (Kasting 1991; Haberle 1998). Other greenhouse gases may have been involved or greenhouse warming may not have been the primary mechanism for warming the planet. An alternative is that the warming resulted from the injection of large amounts of hot rock and water into the atmosphere during large impacts (Segura et al. 2002). The proposal is attractive because it is consistent with formation of the valleys mainly in the Noachian and continuing at a low rate subsequently.

2.4 Gullies

Finally, there are the gullies. Gully is the term applied to small, linear, seemingly young erosion features incised into steep, mostly poleward-facing slopes at mid to high latitudes. They typically consist of an upper theater-shaped alcove that tapers downward to converge on one or more channels that are mostly meters to tens of meters wide and hundreds of meters long (Fig. 5b). They are much smaller than the valleys just discussed and confined to steep slopes. Malin and Edgett (2000) originally ascribed them to groundwater sapping, but there is considerable uncertainty about how they formed. With mean annual temperatures close to 215 K the ground is frozen to kilometer depths for all plausible heat flows and thermal conductivities. Moreover many gullies occur on central peaks, or reach to the lip of craters where groundwater is unlikely. They may have formed during periods of high obliquity from the summer melting of snow that accumulated on slopes in winter (Costard et al. 2002; Christensen 2003). According to this model, they form preferentially on steep poleward facing slopes because these are constantly illuminated in summer at high obliquity.

Regarding ice, neutron and gamma-ray data indicate that ice is abundant at high latitudes at depths of roughly 1 meter, the detection limit from orbit. Geologic evidence suggests that ground ice is also abundant to depths of at least hundreds of meters. One indicator is a general softening of the terrain at mid to high latitudes where ice is stable (Squyres and Carr 1986). The softening has been attributed to ice-abetted creep of near-surface materials. Another indicator of ground ice is the abundance of features at mid to high latitudes that indicate glacier-like flow of materials shed from slopes. In addition, a wide array of landforms in the northern plains, Argyre, and Hellas have been attributed to glaciers on the surface, some of which likely formed by the freezing of lakes fed by large floods (Kargel et al. 1995). Glacial features have also been identified on and adjacent to volcanoes in Tharsis, but these had a different origin, having possibly formed by accumulation of ice during periods of high obliquity (Head et al. 2005) (Fig. 6a).

2.5 Lakes and Oceans

What about lakes and oceans? Numerous standing bodies of water of widely different sizes must have accompanied formation of the outflow channels and valley networks. In the cratered uplands valley networks commonly converge on local lows where water likely ponded. On a larger scale lakes must have formed at the ends of outflow channels. A common relation in the uplands is the breaching of crater walls by valleys, both incoming and outgoing. In either case, the crater must have formerly contained a lake. Breaks in slope on crater walls that might be shorelines internal to craters are rare, although occasionally present. More common are deltas deposited from streams entering craters (Fig. 6b) (Malin and Edgett 2003; Hauber et al. 2005). In addition, most large upland craters have shallow flat floors. Where the floors are eroded, they are commonly revealed to be underlain by finely layered deposits. While not proof of lacustrine deposition, they are consistent

 Springer

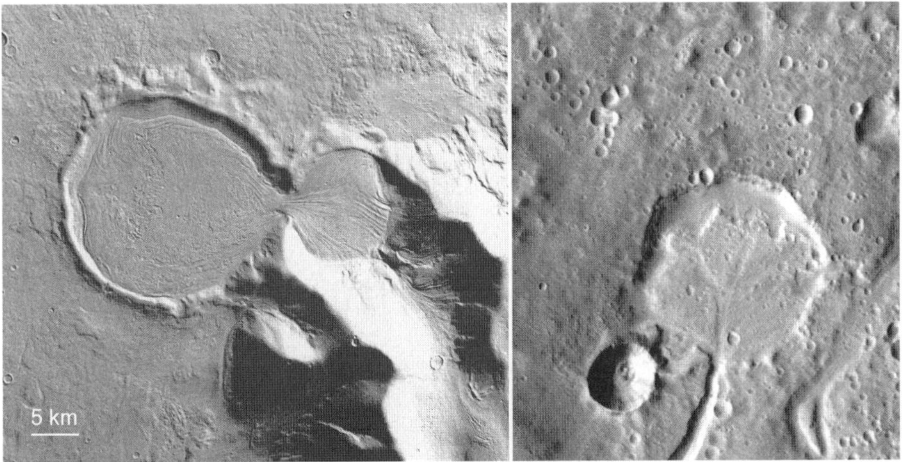

Fig. 6 *Left*: Possible glacier east of Hellas at 38°S, 104°E. Material has flowed from an alcove in a massif in the upper right into a nearby crater, and from there through a gap into another larger crater 500 m lower in elevation (HRSC SEM80HRMDGE). *Right*: Former crater lake at 8.5°N, 312°E. A delta formed within the 6.4 km diameter crater from material brought down the channel that enters through southern rim. Such must have been common throughout the uplands in the Noachian when most of the valley networks formed (THEMIS V0184911)

with it. In some areas, such as around Meridiani, mounds of sediments are common within craters. They appear to be remnants of former regional deposits such as were sampled by "Opportunity", and may contain lacustrine deposits as do those at the Meridiani landing site.

The former presence of much larger bodies of water has been suggested for the northern plains (Parker et al. 1989, 1993; Clifford and Parker 2001), for Hellas (Moore and Wilhelms 2001) and Argyre (Parker et al. 2000). Their former presence is not controversial, although their sizes are. Estimates for the size of the northern ocean range from roughly 2×10^7 km^3 for the Deuteronilus shoreline of Clifford and Parker (2001) (Fig. 7) to 3×10^8 km^3, based on the assumption that the Olympus Mons cliff was wave cut (Baker 2001). The Deuteronilus shoreline, as mapped by breaks in slope and textural changes encloses an area where deposits of upper Hesperian age, interpreted as effluent from large floods, have partly buried craters and ridges (Head et al. 2002). Thus the shoreline mapped from breaks in slope and similar features is supported by the regional geology. Evidence for larger bodies of water is primarily based on alignment of terraces, benches, escarpments and so forth. Their validity as shorelines is difficult to assess because of the multiple ways that such features can form. Moreover, bodies of water larger than that outlined by the Deuteronilus shoreline are not required by the dimensions of the channels entering the basin.

As indicated above the large outflow channels around the Chryse basin are mostly upper Hesperian in age as are the sediments identified as effluent by Head et al. (2002). Most of the shoreline features identified by Parker and his co-workers are also post-Noachian. Thus a plausible case can be made for large bodies of water in the northern plains in the upper Hesperian. Evidence for earlier oceans in the Noachian, for which there is stronger evidence for warm conditions, is much weaker. Yet if warm conditions prevailed, as strongly suggested by the valley networks, then large bodies of water must have been present in low areas such as the northern plains. The evidence has probably been largely erased by subsequent events. The proposed shorelines in Hellas are more continuous than those around

Fig. 7 Possible shoreline. This view, looking down on the north poles, shows the −3760 m contour, which roughly corresponds to the lowest shoreline proposed by Parker et al. (1989, 1993). The volume below the contour is 2×10^7 km^3, and is indicative of the size of a body of water that would be left after one of the largest floods

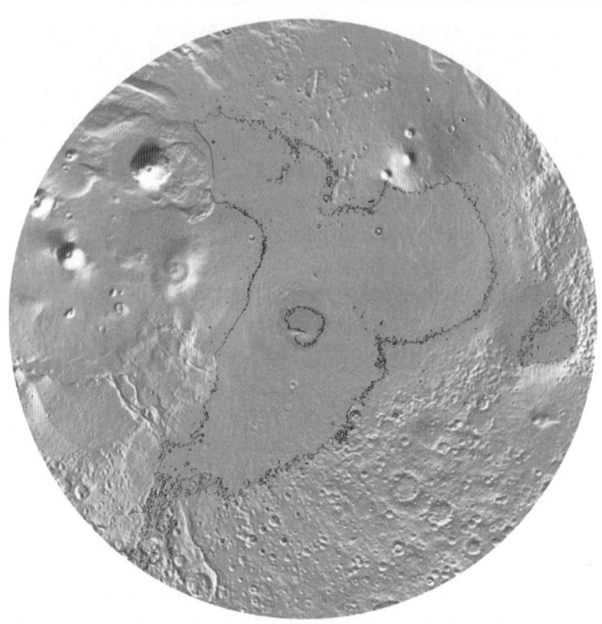

the northern basin. Moore and Wilhelms (2001) recognized two strandlines, one at −5800 m and one at −3100 m. Both can readily be traced at a constant elevation around a substantial fraction of the basin.

The presence of young outflow channels such as Athabasca, Grjota and Marte Valles implies that lakes were present at least transiently, and ice-covered in the recent geologic past. Interpretation of some platy features as ice flows (Murray et al. 2005) where such lakes may have been present is, however, controversial.

2.6 Surface Observations

We now have observations from the surface that pertain to the hydrologic history of Mars. The findings of the rover "Opportunity" in Meridiani strongly support the presence of liquid water at the surface in the late Noachian. The rover landed on a sequence of rocks that unconformably overlie cratered and dissected Noachian rocks. Crater counts suggest that the sequence dates back to close to the Noachian–Hesperian boundary, or roughly 3.5–3.7 billion years old. While most of the sequence examined was deposited by the wind, the fragmental debris is comprised of roughly 50% silicic material and 50% sulfates, and was likely derived from a nearby playa-like source. Moreover, in the upper part of the sequence are beds that were deposited from liquid water (Grotzinger et al. 2005). In addition, the dissolution and recrystallization of the soluble salts within the sequence indicates vertical movement of the local water table (McLennan et al. 2005). The entire sequence appears to have been deposited in a dune field with interdunal ponds and a fluctuating water table.

The findings of the rover "Spirit" at Gusev are more ambiguous. The rocks on the Hesperian age floor of Gusev are basalts that are almost pristine except for a millimeter thick, sulfate-rich rind. The rocks of the Columbia Hills are, however, very different. They vary greatly in type from dunites to sulfate-rich rocks, and their silicic components range from pristine primary minerals, to highly oxidized and hydrated minerals. Many of these rocks

 Springer

have clearly been altered under warm aqueous conditions, but whether these conditions occurred at the surface or deep below the surface is uncertain in most cases, as are the climatic conditions, if any, that are implied (Squyres et al. 2006).

Finally, there is evidence based on erosion rates. Evidence for a dramatic change in erosion rates at the end of the Noachian is unambiguous. On Hesperian and Amazonian terrains most impact craters are almost perfectly preserved, including the most subtle textures on their ejecta blankets. In contrast, in Noachian terrains almost all the larger older craters are highly eroded, many being detectable only by subtle topographic signatures (Schultz and Frey 1990). Golombek and Bridges (2000) estimate that the erosion rates decreased by 3–6 orders of magnitude at the end of the Noachian.

2.7 Hydrologic and Climatic Change on Mars

To summarize what we know about hydrologic history: That a dramatic change in the hydrologic regime occurred at the end of the Noachian cannot be doubted. The rate of valley formation declined dramatically, erosion rates fell by orders of magnitude, and weathering rates as indicated by production of phyllosillicates also fell. Widespread, dense dissection of the Noachian terrains, implies an active Noachian hydrologic cycle with precipitation, surface runoff, infiltration, movement of groundwater and accumulation of standing bodies of water, all consistent with warm climatic conditions. In contrast, the scarcity of valley networks, low erosion rates, lack of weathering products, and presence of outflow channels that appear to have formed mostly by massive eruptions of groundwater, all suggest that most post-Noachian times were characterized by cold surface conditions, a thick cryosphere, and the occasional large floods that created bodies of water that rapidly froze.

Despite the likely warm conditions and an active hydrologic cycle during the Noachian, large integrated drainage basins comparable to those of the Mississippi, Amazon, and Nile did not develop. Most of the valleys in the Noachian terrain drain into local lows. They typically do not merge with other networks to form large basins thousands of kilometers across, as has happened in many areas on Earth. The immaturity of the drainage system is also reflected in the poor concavity and relations between circularity and elevation within the basins as described above. These indications of immaturity may simply reflect the balance between the rate of terrain creation, which in the Noachian was largely by impacts and volcanism, and the rate of terrain degradation by fluvial erosion. On Noachian Mars the balance appears to have been more in favor of terrain creation than on present-day Earth. Whether this was the result a higher rate of terrain formation or a lower rate of terrain degradation is unclear.

The extent of any Noachian oceans is also unclear. However, if there was an active hydrologic cycle as appears likely, then there must have been large bodies of water in low areas. Much of the near-surface inventory of water on Mars today is likely sequestered as ice in the kilometers thick cryosphere or trapped in the hydrosphere below (Clifford 1993). If the surface of Mars was warm and wet in the Noachian then much of the water now locked in or under the cryosphere would have been able to percolate into the low-lying basins such as Hellas and the northern plains and so participate in the global hydrologic cycle. The Meridiani shoreline around the northern basin, proposed by Clifford and Parker (2001), if real, would be Noachian in age. It encloses the equivalent of a global layer 1.5 km thick. If this was truly a Noachian shoreline, then approximately half the planet would have been covered with water. If the upper strandline in Hellas was close to the global sea level then the volume of water would have been 0.5 km spread evenly over the whole planet, and 20% of the planet would have been under water.

In addition to large bodies of water in basins such as Hellas and Isidis, there would have been numerous small bodies of water in the hollows of the poorly graded landscape. The presence of lakes within many Noachian craters is indicated by valleys entering or leaving the craters. Many of these local lows, including craters, now contain finely layered deposits of yet to be determined origin.

The Noachian rocks of the Columbia Hills in Gusev Crater are interbedded volcanic rocks and impact breccias that show a wide range of aqueous alteration, as would be expected from an era of high rates of volcanism and impacts and abundant near-surface water. High rates of hydrothermal activity are also expected although no unambiguous hydrothermal deposits have been detected.

At the end of the Noachian conditions changed dramatically. The most characteristic hydrologic feature of the Hesperian is the outflow channel. Valley formation became highly localized, sapping characteristics became more common, erosion rates fell dramatically, and phyllosilicates production declined although sulfates deposits continued to form. The change from a regime dominated by surface runoff to one dominated by outflow channels, many of which appear to be eruptions of groundwater, could be explained by a change in the global climate at the end of the Noachian and development of a thick cryosphere. However, while such a change likely did occur, the story is more complicated. There are local, highly dissected post-Noachian surfaces. Many of the most densely dissected on the planet are on post-Noachian volcanoes such as Ceraunius Tholus, and Hecates Tholus. In addition, there are rare, local, highly dissected surfaces away from volcanoes, such as the Hesperian plains adjacent to the southern end of Echus Chasma (Mangold et al. 2004).

Most of the fluvial features of the post-Noachian era are consistent with the presence of a thick cryosphere. Eruption of water from below the crysophere as a result of tectonic activity, volcanism, or impacts may have been catastrophic as in the case of the large outflow channels, or more gentle as with valleys with sapping characteristics such as Nirgal Vallis and Nanedi Vallis. Ouflow channels continued to form throughout much of Mars' history, although the younger outflow channels such as the Athabasca and Gordi Valles are much smaller than those that formed in the late Hesperian. Water brought to the surface would have pooled wherever it encountered hollows, either at the start of the channels (Juventae Chasma) or at the ends of the channels (the northern plains). Finding of sulfates within the canyons and other hollows is consistent with sublimation of lakes formed by groundwater brought to the surface. Lack of hydrous silicate minerals in post-Noachian terrains is consistent with a dominantly cold climate and a thick cryosphere.

While this simple picture explains much of what we see, there are anomalies as previously noted. Various proposals have been made to explain young valley systems: (1) The young valleys on volcanoes result from circulation of groundwater onto the surface as a result of hydrothermal activity (Gulick 1998). (2) Formation of large outflow channels by CO_2-charged floods episodically caused temporary changes in the global climate (Baker 2001). (3) The global climate was episodically and temporarily changed by large impacts (Segura et al. 2002) or massive volcanic eruptions. (4) The valleys formed by melting of ice during periods of high obliquity (Jakosky and Carr 1985). There may be other possibilities. However, despite these possible excursions, for most of the Hesperian and Amazonian, Mars appears to have been cold, with a thick cryosphere, and extremely low rates of erosion and weathering.

Guillame: Thank you, William. Now we have Charles Lyell. Sir Lyell received both his B.A. and M.A. from Exeter College. After a brief legal career he devoted himself to geology and quickly became an established proponent of what was later known as the uniformatarianism

view in which geologic phenomena were explained by the cumulative effect of processes acting gradually over an immense period of time. He was elected a fellow of the Royal Society in 1826 and began publishing his well-known *Principles of Geology* in 1830. Charles?

3 Volatile Reservoirs and Exchange Processes on Earth and Mars

Lyell: Cheers. Geological processes that sustain reservoirs of volatiles and their geochemical cycling in planets are part of the "life support system" that can make planets habitable. To sustain life a planet must provide key chemical constituents as well as environmental conditions and sources of energy that can fuel biological processes. Igneous rocks that are abundant in the crusts of silicate-rich planets in our inner Solar System offer many of the needed elemental ingredients. Silicate rocks provide phosphorous and essential metals such as iron, magnesium, calcium and key trace elements. However organisms consist principally of water and compounds of the elements C, N, S, and P, and silicate rocks typically have relatively minor amounts of these. Although the Earth's atmosphere, hydrosphere and sedimentary rocks contain abundant water, oxidized C compounds, N, Cl, and various S compounds, these substances clearly did not derive principally from crustal igneous rocks. W. Rubey proposed that such "excess volatiles", including water, derived from volcanic outgassing and other thermal processes that have been sustained by Earth's geologic activity (Rubey 1951).

These "excess volatiles" are significant also because they helped to create and maintain habitable environmental conditions at and near Earth's surface. They warm and moderate Earth's climate and thereby create temperatures and pressures necessary to stabilize liquid water, one of the key ingredients essential for life as we know it. These volatiles also interact with the solid planet in ways that can buffer environmental perturbations caused by volcanism, large impacts and long-term changes in solar luminosity.

Habitable planetary environments also must provide sources of energy that living systems can utilize to drive their metabolism. Today's biosphere is dominated by photosynthetic biota that can harvest abundant solar energy. However many other organisms are non-photosynthetic and can obtain useful chemical energy by reacting oxidized and reduced chemical compounds in so-called "redox" reactions. Processes of volatile exchange in the Earth's crust have delivered both oxidized and reduced chemical compounds to habitable environments.

A proper understanding of the roles played by volatile exchange processes requires that they be considered in the context of a network of volatile reservoirs in the atmosphere, hydrosphere, crust and mantle that are linked by physical and chemical processes. Such networks are called geochemical cycles because a key dynamic is that volatile constituents can move back and forth between two or more reservoirs over time. On Earth such networks are called biogeochemical cycles because life itself is an important crustal process. Global biogeochemical cycles actually consist of multiple nested cyclic pathways that differ with respect to their reservoirs and processes. However, all pathways ultimately pass through reservoirs of volatiles in the hydrosphere and atmosphere, and these shared reservoirs unite all of the sub-cycles and allow even their most remote constituents to influence the biosphere.

3.1 Biogeochemical Cycling of Volatiles on Earth

The present-day biogeochemical cycles of volatiles on the Earth are represented graphically in Fig. 8 as an integrated system of reservoirs and processes. Reservoirs and processes

Fig. 8 Schematic diagram of Earth's biogeochemical cycles of volatiles, showing reservoirs (*boxes*) in the mantle, crust, oceans and atmosphere, and showing the processes (*arrows*) that unite these reservoirs. The *vertical bars at right* indicate the timeframes within which a volatile element or compound typically completely traverses each of the four sub-cycles (the HAB, SED, MET and MAN sub-cycles, see text). For example, C can traverse the hydrosphere–atmosphere–biosphere (HAB) sub-cycle typically in the time scale between 0 to 1000 years

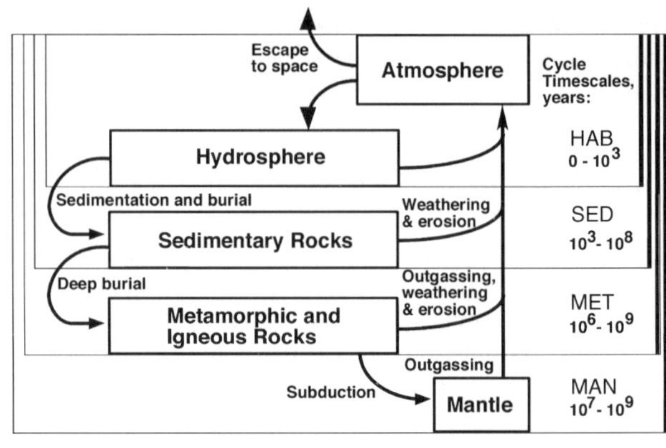

are shown as boxes and labeled arrows, respectively, that delineate the various sub-cycles, denoted in the figure by the labels "HAB", "SED", "MET" and "MAN". The range of timescales typically needed for a particular chemical constituent to traverse each of these sub-cycles is indicated along the right margin, below their corresponding labels. Of course the actual physical boundaries between these sub-cycles are not so sharply delineated in nature. But the sub-cycles depicted here do represent characteristic domains along the continuum of reservoirs and processes that collectively constitute the Earth system.

Now I will discuss each individual "sub-cycle": The first includes the hydrosphere, atmosphere, and biosphere (HAB). The HAB sub-cycle includes only the volatile reservoirs in the hydrosphere, atmosphere and biosphere (Fig. 8) and is characterized by typically geologically high rates of exchange between these reservoirs. Volatile constituents can traverse the HAB sub-cycle within time scales of minutes to hours, for example, during the rapid cycling of constituents between biota and their environment or across the air-sea interface. Alternatively, volatiles in seawater can require a thousand years or more to be cycled due to the circulation rate of the global ocean (Broecker and Peng 1982). The ocean indeed exerts a dominant role in the HAB sub-cycle due to its great size and enormous stored content of thermal energy from the Sun.

The next sub-cycle in terms of timescale is the sedimentary (SED) sub-cycle which includes the entire HAB sub-cycle plus reservoirs of volatile constituents in sedimentary rocks (Fig. 8). The SED sub-cycle strongly influences the HAB sub-cycle. For example, sedimentation limits global productivity by removing nutrients, phosphorus in particular. The balance between sedimentation of oxidized (carbonate, sulfate, Fe^{3+}) versus reduced (organic C, sulfides, Fe^{2+}) species determines the abundances of O_2 and sulfate in the atmosphere and hydrosphere (Holland 1984). The weathering and transport of igneous and sedimentary rocks deliver nutrients to the oceans, thereby modulating global biological productivity. Organisms can substantially enhance weathering rates. The rates of weathering and erosion of sediments and rocks, together with the contents of reduced species in those deposits, determine their rates of consumption of O_2 and other chemically reactive constituents.

The rate at which an atom or chemical constituent traverses the SED sub-cycle is determined principally by tectonic controls upon rates of formation and destruction of sedimentary rocks. Reduced C, N and S compounds are buried principally in deltaic-shelf sediments and in sediments beneath highly productive open-ocean regions. The global net burial rates

of C and S species are controlled ultimately by the rates of delivery of those species to the oceans by weathering and transport and by global rates of their reaction with submarine basalts (Garrels and Perry 1974). Sedimentary reservoirs of C and S are much larger than C and S reservoirs in the HAB sub-cycle (Des Marais 2001). The average sedimentary rock survives for about 200 million years (Derry et al. 1992). Therefore key biogeochemical properties of the oceans and atmosphere, including their nutrient inventories and oxidation states, are modulated by the SED sub-cycle over timescales between tens of thousands to hundreds of millions of years (Fig. 8).

Next is the metamorphic (MET) sub-cycle (Fig. 8), which includes more deeply buried sedimentary and igneous rocks that are altered (metamorphosed) by elevated temperatures and/or pressures. Volatiles in the MET sub-cycle include those in rocks that enter subduction zones but ultimately escape injection into the mantle either because they are degassed or because their host rocks also escape subduction, for example, by lateral accretion into continental crust. The mass of rocks in the MET sub-cycle greatly exceeds those in the SED sub-cycle (Lowe 1992). However inventories of reduced C in the MET sub-cycle are smaller (Hunt 1972), due both to losses during metamorphism of sedimentary rocks and to the typically much lower reduced C contents of crustal igneous rocks. Carbonate C is also lost during thermal metamorphism as CO_2. Volatile constituents require typically tens of millions to billions of years to traverse the MET sub-cycle. These cycle times are typically longer than those for the SED sub-cycle and reflect the longer lifetimes of more deeply buried continental rocks.

Finally, there is the deepest mantle-crust or MAN sub-cycle which includes the mantle volatile reservoirs (Fig. 8) and the processes of subduction and mantle outgassing. The modern global rates of outgassing of key volatile species from the mantle are summarized in Table 1. An atom of C requires between tens of millions of years to as long as billions of years to traverse the MAN sub-cycle (Des Marais 2001). Over timescales of tens of millions to billions of years, the processes of mantle-crust exchange probably modulated both the sizes and the overall oxidation states of the much smaller C reservoirs in the HAB, SED and MET sub-cycles. In addition, thermal emanations of other reduced species (principally

Table 1 Midocean ridge hydrothermal fluxes of reduced species to the ocean and atmosphere. Modern fluxes are from Elderfield and Schultz (1996)

Species	Flux [10^{12} mol yr^{-1}]	O_2 consumed [10^{12} mol yr^{-1}]
$S_{reduced}$	0.085–0.96	0.18–1.92
Fe^{2+}	0.023–0.19	0.02–0.19
Mn^{2+}	0.011–0.034	0.01–0.034
H_2	0.003–0.015	0.002–0.008
CH_4	0.007–0.024	0.014–0.048
Modern total		0.2–2.1
Ancient total	(2.1 Ga[a])	0.74–7.7
	(2.2 Ga[a])	0.77–8.0
	(3.0 Ga[b])	0.97–10

[a]Linear interpolation between fluxes estimated for today and for 3.0 Ga

[b]Assumes that thermal fluxes of reduced chemical species have scaled linearly to mid-ocean ridge spreading rates, and that these spreading rates have scaled with the square of the heat flow (Sleep 1979). At 3.0 Ga, heat flow is estimated to have been approximately 2.2 times the modern value (Turcotte 1980)

sulfides, H_2, and Fe^{2+}) consume O_2 and also contribute reducing power for biosynthesis by microbes that derive energy from redox reactions. Thus the balance between biological productivity and decomposition, sediment cycling, and thermal processes has modulated the overall oxidation state of the surface environment and the crustal reservoirs of key redox-sensitive elements such as C, N, S and Fe.

Biota that dwell at or below the sea floor along the mid-ocean ridges obtain their energy principally from oxidation–reduction reactions involving reduced hydrothermal emanations (Jannasch and Wirsen 1979). They utilize reduced S, H_2 (derived from water–rock reactions), and Fe^{2+}. Today this total flux of reduced constituents, expressed as O_2 equivalents, is in the range $0.2–2.1 \times 10^{12}$ mol yr^{-1} (Table 1) (Elderfield and Schultz 1996).

3.2 Changes in Biogeochemical Cycling over Time

Several processes have changed rates of volatile exchange during early Earth history. One such process is the Sun itself. Most current models of stellar evolution predict that the Sun was less luminous when it first entered the main sequence about 4.6 billion years ago. As the Sun burned H to He, its core became denser and therefore hotter. This increased the rate of thermonuclear burning, which, in turn, increased the Sun's luminosity over time. According to one estimate (Gough 1981), the sun was only 70% as luminous 4.6 billion years ago as it is today. This "faint early sun" presents a paradox for Earth's early climate (Sagan and Mullen 1972) because, if one assumes that parameters such as atmospheric composition and planetary albedo had been the same as today, then Earth's mean surface temperature would have been below freezing during the first 2 to 2.5 billion years of its history. However, very ancient (ca. 3.7 billion-year-old) metasedimentary rocks indicate that large standing bodies of water were abundant (e.g., Schopf 1983).

Another is impacts. Large impacts should have released substantial quantities of volatiles (Sleep and Zahnle 2001). Impacts on Mars have been credited with releasing crustal volatiles (CO_2, H_2O, etc.), which helped to enhance an ancient greenhouse (Newsom 1980; Segura et al. 2002). However impacts also created dust clouds that, in the aftermath of the impacts, might have weathered quickly, thus rapidly removing atmospheric CO_2 and perhaps triggering periodic profound cooling of the Hadean climate (Sleep and Zahnle 2001). More recent impacts probably affected surface volatiles, at least over shorter timescales. For example, the impact at the Cretaceous–Paleogene boundary probably released large quantities of C and S from sedimentary carbonates and sulfates and upper mantle materials that were within the Yucatan target zone (Toon 1997). Impacts of this size very likely affected the HAB subcycle for as long as thousands of years. Such impacts might have exerted more profound and permanent effects upon the surface environment, but interpretations of the specific details of these effects have been considerably more controversial.

A third is mantle-outgassing. Over timescales of tens of millions to billions of years, processes that govern the exchange of volatile species between Earth's surface, deep crust and upper mantle probably affected volatile inventories in the crust, oceans and atmosphere. Mid-ocean ridge volcanism is quantitatively the most important source of mantle volatiles. The heat flow from Earth's interior was substantially greater during the earlier Precambrian (e.g., Lambert 1976). During the past 3.0 billion years, decay of the radioisotopes ^{238}U, ^{235}U, ^{232}Th and ^{40}K has been the principal source of this heat, therefore their decay over time has caused global heat flow to decline. Thermal fluxes of volatiles, including CO_2, can be scaled linearly to mid-ocean ridge spreading rates, and these rates vary with the square of heat flow (Sleep 1979). Thus, for example, heat flow 3.0 billion years ago has been estimated to have

been 2.2 times its modern value (Turcotte 1980), therefore the mid-ocean ridge mantle CO_2 flux was perhaps approximately 5 times its present-day value.

A fourth process is subduction of the crust at plate margins returning volatiles to the mantle. For example, if the evolution of temperature and pressure regimes of subducting slabs of modern oceanic crust are considered together with the stability of C species in such slabs, then a substantial fraction of the subducted C could escape dissociation and melting and be carried to considerable depths, possibly to be retained in the mantle for very long periods of time (Huang et al. 1980). However, three billion years ago, subducted C would, at any particular pressure, have experienced considerably greater temperatures (McCulloch 1993). These greater temperatures enhanced the likelihood that subducted C reacted to form mobile phases that migrated upward and therefore escaped injection into the mantle (Des Marais 1985).

A fifth is metamorphism. If the total C inventory in the HAB sub-cycle has been maintained over time at near-steady state, then net losses of C from the HAB sub-cycle during sedimentation and burial must have been balanced by the release of CO_2 during thermal metamorphism of sediments (Berner et al. 1983). For example, because carbonates decompose to CO_2 during the subduction of sediments, rates of CO_2 outgassing should vary with global mean spreading rates (which correlate with global mean subduction rates). Also, greater outgassing of CO_2 from deep continental interiors should correlate with greater worldwide tectonism that should correlate with faster mean global spreading rates (Berner et al. 1983). Therefore the mean global rate of CO_2 release from rock metamorphism should be proportional to global heat flow. Consequently, due to the long-term decline in global heat flow, the rates of transfer of CO_2 by thermal processes from the crustal sedimentary rocks (SED sub-cycle) to reservoirs in the HAB sub-cycle should have declined.

What are the roles played by continents? Continents are important because; (1) subaerial weathering is a key sink for volatiles; (2) rivers strongly affect seawater chemistry; and (3) continents have been much more stable repositories of sedimentary volatile reservoirs than have ocean basins. Accordingly any substantial long-term changes in the total area, thickness, stability, and subaerial exposure of continents would have contributed to major long-term changes in the cycling of volatile elements.

The large continents as we know them today required a considerable interval of geologic time during early Earth history to grow and become stabilized (thicker and less dense) by anatexis (partial melting and chemical fractionation), metamorphism, and under-plating (Lowman Jr. 1989). Due to higher heat flow on early Earth, virtually all of the ancestral pre-stabilized crust was destroyed, largely by recycling into the mantle. These factors probably shortened the lifetimes of sedimentary reservoirs of volatiles. Higher heat flow led to higher sea floor spreading rates (Sleep 1979) and lower mean ages of oceanic crust. Therefore typical Archean oceanic crust was younger, hotter and thus more buoyant than typical modern seafloor. An overall greater buoyancy of oceanic crust on early Earth created shallower ocean basins that, in turn, probably displaced more seawater onto the continents (Hays and Pitman 1973). Thus even if the mass of ancient continental crust had been similar to that of modern crust, the land area on the young Earth probably was less extensive.

Volatile exchange processes have co-evolved with the global environment. The Archean rock record supports the view that the basic architecture of biogeochemical cycles, namely the nested HAB, SED, MET and MAN cycles (Fig. 8) was in place before 3.5 billion years ago. The major changes over time occurred principally in the relative sizes of the various reservoirs and the fluxes between them. For example, the most direct solution to the "faint early sun problem" (Sagan and Mullen 1972) is to invoke a stronger greenhouse effect in the early atmosphere that was sustained by substantially higher CO_2 levels. One-dimensional

 Springer

climate models have been used to estimate the CO_2 levels required (e.g., Kasting 1987). For example, if the global mean temperatures 4.5 and 2.5 billion years ago were equal to the modern global mean temperature, pure CO_2 atmospheres of 1 and 0.1 bar, respectively, would have been required. Therefore, CO_2 declined perhaps by a factor of 1000 or more, from 4.5 billion years ago to the present. This large CO_2 decline required that the HAB and SED sub-cycles changed over time. These changes could have occurred as an expression of a self-regulating climate control system (Walker et al. 1981). For example, low solar luminosity would have favored low global temperatures that, in turn, would have reduced rates of water evaporation, precipitation and therefore lowered rates of chemical erosion of silicate rocks. Low erosion rates lowered the rate of CO_2 removal from the atmosphere, which would have allowed thermal CO_2 sources to increase atmospheric CO_2 levels. This would have raised surface temperatures until the rate of CO_2 removal by weathering achieved a balance with the thermal sources. As solar luminosity increased slowly over time, the CO_2 levels needed to maintain the temperature at which erosional (CO_2 sink) and thermal (CO_2 source) processes balanced would have slowly declined. If the pH of the global ocean remained reasonably constant over time, then seawater HCO_3^- and CO_3^{2-} levels would have declined in close parallel with the decline in atmospheric CO_2 levels. Such a substantial decline in seawater HCO_3^- concentrations (and, therefore, CO_3^{2-} concentrations) undoubtedly affected processes of carbonate precipitation (Grotzinger and Kasting 1993).

Recently, atmospheric CH_4 has gained favor as potentially a very significant source of greenhouse warming during the Archean and early Proterozoic Eons (e.g., Pavlov et al. 2001a). The production and accumulation of abundant CH_4 would have been favored by hydrothermal activity, which was greater during the Archean and that produced CH_4 and H_2, and by methane-producing microbes that occupied widespread anoxic environments and utilized H_2 and other sources of reducing power.

The hotter early mantle must have influenced significantly the inventories of volatiles in the crust, oceans and atmosphere. Higher rates of crustal production were accompanied by higher rates of mantle outgassing (e.g., Des Marais 2001). A hotter mantle retained subducted volatiles with greater difficulty (McCulloch 1993). These considerations are consistent with an early Earth in which the crustal C inventory might even have exceeded the modern inventory (Des Marais 1985; Zhang and Zindler 1993), and the cycling of volatiles between the mantle and crust was more vigorous than today.

Well-preserved sedimentary rocks are not abundant within the relatively few provinces of lithosphere older than 2.7 billion years. The best-preserved Archean sediments occur in the 3.5 to 3.2 billion-year-old Kapvaal Craton of South Africa and the Pilbara Block of Western Australia (Lowe 1992). These deposits are associated with episodes of greenstone activity and intrusive events that created stable microcontinents or cratons. These cratons later became the nuclei of full-sized modern continents. Most of the Archean continental crust had yet to become stabilized by cratonization (Rogers 1996), therefore a greater fraction of continental sediments experienced relatively higher rates of instability and thermal alteration. The tectonically more active regime within the Archean marine basins favored rapid destruction by continental collisions, partial melting and mantle/crust exchange (Windley 1984).

Chemical weathering was very effective during the Archean, consistent with high CO_2 levels (Walker 1985) and a warm climate (Lowe 1992). Weathering of a typical uplifted rock sequence produced coarse clastic sediments that became enriched in the most chemically resistant components such as cherts and silicified komatiitic and dacitic tuffs (Nocita and Lowe 1990). Despite the rapid uplift and transport of these rocks and their debris, their less chemically resistant components were efficiently degraded. The apparently highly effective

weathering is consistent both with relatively warm, moist conditions and with elevated atmospheric CO_2 levels (Lowe 1994). Altered evaporites also occur in greenstone sequences between 3.5 and 3.2 billion years ago (e.g., Buick and Dunlop 1990). Their formation in such tectonically unstable environments is consistent with high rates of evaporation.

In the late Archean and early Proterozoic, following the inevitable decay of radioactive nuclides in the mantle, the heat flow from Earth's interior declined (Turcotte 1980). This decreased the rates of both sea floor hydrothermal circulation and volcanic outgassing of reduced species. The style of subduction also changed (McCulloch 1993). In the early-to mid-Archean, subducted slabs were dehydrated, sustained partial melting, and largely disaggregated in the upper 200 km of the mantle. Later, the reduced heat flow and lower temperatures permitted colder, stronger oceanic lithosphere to develop. Subducting slabs thus sustained perhaps only partial dehydration and, together with volatiles such as CO_2 and H_2O, penetrated to depths exceeding 600 km (McCulloch 1993). It has been proposed (Kasting et al. 1993) that the upper mantle was oxidized by the subduction of water, followed by the escape of reduced gases. A progressive oxidation of the upper mantle has not yet been demonstrated (Delano 2001), but, if it had occurred, its effect upon the redox balance of volatiles would have been substantial.

The reworking of Archean continental crust by tectonism, igneous activity and metamorphism also had important consequences for the processing of volatiles. Marine carbonates of this age record a substantial increase in $^{87}Sr/^{86}Sr$ values, indicating greater continental erosion and runoff (Mirota and Veizer 1994). New and extensive stable shallow water platforms became sites for the deposition and long-term preservation of carbonates (Grotzinger 1989) and organic C (Des Marais 1994). Increased global continental erosion rates also would have accelerated the rate of decline of atmospheric CO_2 (Walker 1990). Increased subaerial weathering would have enhanced the delivery of nutrients to coastal waters, enhancing biological productivity (Betts and Holland 1991). Greater productivity would have removed more CO_2 from surface seawater, but, given the still-higher-than-present oceanic and atmospheric inorganic C contents, the effect of this productivity on the atmosphere should have been minor.

Patterns of carbonate deposition, as well as the presence or absence of gypsum and anhydrite in associated evaporites, indicate that seawater concentrations of HCO_3^- and CO_3^{2-} have declined and SO_4^{2-} has increased since the late Archean (Grotzinger and Kasting 1993). Late Archean platform sequences include relatively abundant evidence of abiotic carbonate precipitation as tidal flat tufas and marine cements. Evaporite sequences often proceed directly from carbonate to halite deposition, thus excluding gypsum/anhydrite deposition. These observations are consistent with significantly lower seawater SO_4^{2-} concentrations and/or considerably greater HCO_3^- concentrations. In either case, the ratio of HCO_3^- to Ca^{2+} was sufficiently large to prevent deposition of gypsum/anhydrite in marine or marginal-marine environments. These observations are consistent with the view that inorganic C reservoirs within the HAB subcycle were much higher during the Archean and Paleoproterozoic (Walker 1985).

During most of the Archean Eon (prior to ca. 2.7 billion years ago), SO_4^{2-} was locally present in shallow seawater at concentrations well below 1 mM (Canfield et al. 2000), considerably less than modern seawater concentrations ($\sim$27 mM). Locally high SO_4^{2-} concentrations were precipitated as evaporitic sulfate minerals. The deposition of gypsum rather than anhydrite (Lowe 1983) indicates that temperatures very likely were below 58°C. Sulfate concentrations increased between 2.5 and 2.1 billion years ago, perhaps in response to extensive oxidation of sulfur due to widespread oxygen-producing photosynthetic biota.

Molecular oxygen emanating from these communities might have hastened the rate of destruction of atmospheric CH_4, leading to global cooling and the onset of glaciation about 2.5 billion years ago (Kasting and Catling 2003). Still, the deposition of banded iron chemical sediments persisted until ca 1.8 billion years ago and indicates that the deep oceans prior to that time contained appreciable concentrations of dissolved Fe^{2+}. After 1.8 billion years ago, levels of seawater SO_4^{2-} were sufficient to enable sulfate-reducing microbes to produce sulfide at rates that exceeded the rate of supply of Fe^{2+} to the global oceans (Canfield and Raiswell 1999). Thereafter, banded iron formations disappeared and the deep oceans remained anoxic and sulfide-rich until perhaps sometime after 800 million years ago (Canfield and Raiswell 1999). Thereafter, a series of global climate perturbations ensued and included episodic widespread glaciations. These ended just prior to the dawn of the Phanerozoic Eon (543 million years ago), during which the deep ocean became more oxidized as it approached its modern state.

3.3 The Geochemistry of Volatiles on Mars

That is our current picture of Earth volatiles. I'll now discuss cycling of volatiles on Mars. Mars is perhaps the most likely planet to provide a second example in our solar system of a habitable planet. Mars is the only major planet other than Earth where evidence indicates that liquid water contributed substantially to the development of its crust. But there are important differences. For example, today, Mars clearly has no ocean or smaller standing bodies of water. Because Mars is smaller and has lower a gravitational field than Earth, and because it lacks an active magnetic dynamo, losses of volatiles to space by various mechanisms (Jakosky and Jones 1994) have been far more extensive than on Earth. Mars' smaller size and greater distance from the Sun can help us begin to understand how processes that exchange volatiles have influenced the habitability of diverse planets. It is important to determine whether the exchange of volatiles on Mars could ever have provided the key chemical constituents, environmental conditions and sources of energy that could have sustained life.

Volatiles in the Martian atmosphere and surface have exchanged with reservoirs elsewhere in the planet, some volatiles have reacted irreversibly with crustal materials, and some have been lost to space permanently (Fig. 9). Martian meteorites and spacecraft observations have provided clues about reservoirs of volatiles in the atmosphere and crust as

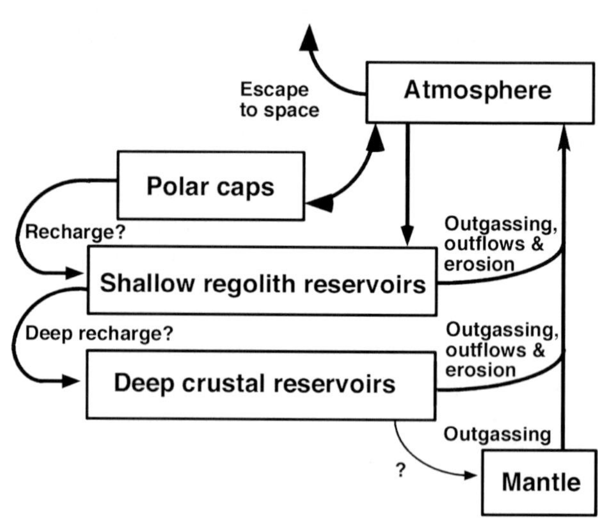

Fig. 9 Schematic diagram of geochemical cycles of volatiles on Mars, showing reservoirs (*boxes*) in the mantle, crust, oceans and atmosphere, and showing the processes (*arrows*) that unite these reservoirs. Such cycles on Mars differ from those on Earth in several respects, including the following: Mars lacks standing bodies of water, atmospheric escape processes probably have exerted relatively larger effects, and subduction into the Martian mantle has been less important, if it occurs at all

well as the processes that linked them. Hydrothermal systems sustained by volcanism should have been widespread on Mars (Gulick 1998). Volcanic aerosols have long been proposed as the source of the high S and Cl ("excess volatiles") in the Martian soil (Clark and Baird 1979) and the abundant sulfate found at both of the Mars Exploration Rover landing sites. Volcanic hydrothermal fluids might have transported mobile elements to the soil (Newsom and Hagerty 1999). Large impacts played several key roles. They added volatiles but also removed them by impact erosion of atmosphere to space. Throughout much of Martian history, impacts might have episodically released volatiles to the surface and atmosphere (Segura et al. 2002) and created local hydrothermal systems (Newsom et al. 1996) that probably sustained habitable environments, much as volcanogenic hydrothermal systems have done throughout Earth history.

The Martian meteorites probably represent conditions of formation that are somewhat different from those that are presently found at the surface. Their composition is most consistent with alteration in an alkaline and reduced environment controlled by water–rock interactions rather than in the oxidized acidic environment at the surface (Zolotov and Shock 2004). The carbonates in the ALH 84001 meteorite very likely derived from CO_2-rich fluids (Golden et al. 2001). However carbonates have not been documented to exist on the surface. Thermal infrared spectra obtained from orbit provide evidence that basalts in some areas have experienced widespread alteration (Wyatt and McSween 2003). Near-IR spectra from Mars Express indicate that clay minerals exist in the very ancient highlands (Bibring et al. 2005).

The Mars rover "Opportunity" has found clear evidence that aqueous processes at and near the Martian surface created chemical sediments (sulfates) and precipitates (hematite) in the Meridiani Planum region (Squyres and Knoll 2005). The Spirit rover found evidence of aqueous basalt alteration and chemical precipitates in rocks on the floor of Gusev crater and in the nearby Columbia Hills (Ming et al. 2006).

Measurements of stable isotopes can reveal important aspects of the histories of volatiles in Martian materials. Each major volatile element has more than one stable isotope. Stable isotopes exhibit ranges of abundance ratios (e.g., D/H, $^{18}O/^{16}O$, $^{13}C/^{12}C$, $^{34}S/^{32}S$, $^{38}Ar/^{36}Ar$, etc.) in natural materials because isotopes having different masses can respond differently to chemical and physical processes. For example, water molecules consisting of H and ^{16}O have a higher vapor pressure than molecules having the heavier isotopes. When oxidized and reduced compounds of C (or S) coexist in chemical equilibrium, their lighter isotopes are relatively more abundant in the more reduced compounds. When organic matter is thermally decomposed, ^{12}C–^{12}C bonds are broken more frequently, forming ^{12}C-enriched gases.

The Martian atmosphere has been profoundly altered by the loss of volatiles to space due to a variety of processes (Kulikov et al. 2007). Hydrogen is lost by thermal escape. Photochemical reactions can deliver sufficient energy to heavier elements such as C, O and N to promote their escape. The loss to space of volatile compounds of these elements is expected to enrich the remaining atmosphere in the heavier isotopes D, ^{13}C, ^{18}O, and ^{15}N (Jakosky and Jones 1997). Gases in the upper atmosphere are not well mixed and so the lighter isotopes become relatively more abundant at higher altitudes where the probability of loss to space is greater.

As discussed by Messier, the Martian atmosphere is strongly enriched in deuterium with respect to hydrogen. Mass-spectroscopy measurements on Viking orbiters revealed relatively high values of also $^{13}C/^{12}C$ and $^{15}N/^{14}N$ (Nier and McElroy 1977), and these high values are interpreted to reflect substantial losses to space of volatiles that were relatively enriched in H, ^{12}C and ^{14}N (e.g., Jakosky and Jones 1997). These losses probably affected

the long-term evolution of Martian climate and have contributed substantially to its present-day cold, dry state.

Martian meteorites have allowed geochemists to extend the measurements of isotopic patterns into the Martian crust and its potential reservoirs of volatiles. The isotopes ^{130}Xe, ^{132}Xe and ^{136}Xe in gas trapped by these samples display a range of relative abundances indicating that mixing has occurred between two end member reservoirs. The most likely explanation is that such mixing occurred on Mars; one of these end members is an atmospheric component enriched in the heavier isotopes. Meteorite mineral phases enriched in the lighter Xe isotopes are phases that probably formed earliest and that also retain Xe most tightly (Ott et al. 1988). Distributions of H isotopes in the Martian meteorites also indicate that multiple volatiles reservoirs exist (Watson et al. 1994). Elevated D/H values occur in minerals such as kaersutite and apatite that might have been influenced by exchange with D-enriched hydrogen in the atmosphere and in surface materials. The hydrogen reservoir having lower D/H values might consist of water and hydrated minerals from the crust and mantle that have not exchanged substantially with the atmosphere. Thus the lower D/H values might represent hydrogen from the interior of Mars that represents its primordial endowment of hydrogen, whereas elevated D/H values represent hydrogen reservoirs in surface materials and atmosphere that have been fractionated isotopically by processes of hydrogen escape to space that have been active for virtually all of Martian history. Values of ^{13}C/^{12}C also provide evidence for multiple components of volatiles. Carbon components released at high temperatures from Martian meteorites during heating experiments have relatively low ^{13}C/^{12}C values that are interpreted to represent the reservoir of C in the Martian mantle (Carr et al. 1985). In contrast, carbonates in these meteorites have much higher ^{13}C/^{12}C values that have been interpreted to reflect exchange of C in the atmosphere and shallow deposits that has been fractionated isotopically by processes of atmospheric escape to space.

3.4 Volatiles and Life on Ancient Mars

Cycling of these volatiles has evolved during Martian history. Although Mars and Earth are quite different today, their early evolution might have been similar because they shared similar processes that evolved in similar ways. For example, volcanism has been important on both planets and its intensity has declined over time. Both planets experienced early episodes of large impacts that were more substantial than later in history. And, of course, both planets share the same star and the effects of its long-term evolution. These processes had multiple consequences for the exchange of volatiles. Figure 10 depicts the processes that influenced the inventories of Martian volatiles and their exchange between the surface environment over time. A substantial endowment of volatiles on early Mars might have helped to maintain a denser atmosphere that, in turn, allowed liquid water to exist at the surface at least episodically and create the features described earlier by Sir Herschel. However over time these volatiles were removed either by substantial losses to space or by sequestration in the subsurface (Fig. 3 in Jakosky and Jones 1997).

Under present harsh conditions at the Martian surface, the deep subsurface has probably provided the most widespread and stable environment for liquid water during Martian history (Clifford 1993). Diverse populations of microorganisms (chemoautotrophs) that obtain their energy from redox reactions can populate subsurface environments on Earth (e.g., Kelley et al. 2005). Chemoautotrophs have been documented across a broad range of environmental extremes of temperature, pH and salinity (Rothschild and Mancinelli 2001). Accordingly, chemoautotrophs that utilize H, S and CH$_4$ might represent useful analogs for

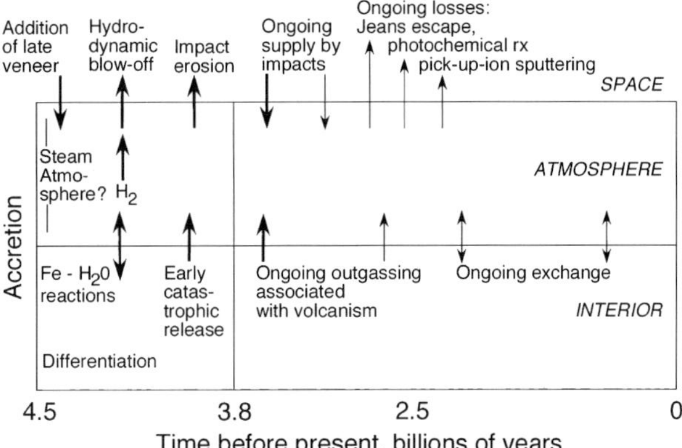

Fig. 10 Schematic diagram of the history of key events and processes on Mars that have affected the cycling of volatiles between the interior, surface and atmosphere and their gains from and losses to space. Modified after Jakosky and Jones (1997)

potential Martian microbiota (Boston et al. 1992; Jakosky and Shock 1998). The subsurface holds the greatest potential for having hosted the development of complex microbial ecosystems.

What does Mars tell us about the potential for habitable environments on other planets with masses different than that of Earth? Mars already has indicated that important geologic processes such as volcanism decline more rapidly on smaller planets than on Earth, leading perhaps to the more rapid deterioration of their surface environments. But how do planets more massive than Earth evolve over time? One might expect that a larger planet could maintain for longer periods of time the levels of geologic activity that were characteristic of Earth's earliest history. Hydrothermal processes would be more widespread and processes of atmospheric escape would be less important on a planet larger than Earth. Future space exploration indeed promises new insights about the full diversity of habitable planetary environments and their potential to sustain life.

Guillame: Thank you Charles. Our final speaker is Charles Darwin. Mr Darwin's inauspicious beginnings as a scholar included brief stints studying medicine and theology in Edinburgh and Cambridge, respectively. At Cambridge, his interest in the natural world was encouraged by John Henslow and Adam Sedgwick. The former arranged for his fateful voyage on the *Beagle* in 1831, and the rest, as they say, is natural history. Charles?

4 Co-evolution of the Atmosphere and Life on Earth

Darwin: Thank you, Charles. I will discuss how living things modify the geochemistry of the surface of the Earth, and how that affects atmospheric composition. Why the atmosphere? Earth's atmosphere influences surface temperatures, and hence the availability of liquid water, by the greenhouse effect and by scattering sunlight back to space. It protects Earth's biota from harmful radiation, it serves as a medium through which organisms exchange gases such as oxygen and carbon dioxide, and it is the principal reservoir of nitrogen at the surface of our planet. That the composition of the life-giving atmosphere is itself modified by biological activity gives rise to interesting feedbacks, as I shall discuss later. That modification is also an important signature for the presence of life, one that is potentially detectable from distances as great as the nearest stars around which humans will search for, and may find, other Earth-like planets.

 Springer

4.1 Evolution of the Atmosphere and the Evolution of Life

Science is certain that the composition of our atmosphere has dramatically evolved over our planet's 4.5 billion year history, but past compositions are difficult to determine with any precision. The composition of the present atmosphere was only finalized after the discovery of argon by William Ramsay in 1894. The composition of Earth's earliest atmosphere is the subject of much debate, as there is no rock record from this time, and the atmosphere may have been involved in the prebiotic chemistry that led to life (Miller and Urey 1959). Perhaps the most significant change occurred within the first one hundred million years when any primitive atmosphere captured from the primordial solar nebula, or degassed during the impact of accreting planetary material was lost to space. Evidence for such atmospheric escape consists of the isotopic and elemental abundance of rare gases in the modern atmosphere (Porcelli and Pepin 2000). These abundances are consistent with the pattern produced when heavier elements such as the rare gases are carried away by rapidly escaping hydrogen from the uppermost layers of the atmosphere. The hydrogen itself is heated to temperatures energetically equivalent to gravitational escape by elevated ultraviolet flux from the more active young Sun (Zahnle and Walker 1982; Ribas et al. 2005) or by impacts.

The primordial atmosphere was replaced by one whose composition was determined by the chemistry of gases from volcanoes and fumaroles. If these gases exsolved from silicate melts that were in chemical equilibrium with the mantle, and if mantle chemistry has evolved little (Canil 2002; Li and Lee 2004), then these gases were primarily H_2O, CO_2, and N_2, with small amounts of CH_4 and H_2. Atmospheric composition continued to evolve, however, the most notable event being the appearance of persistent atmospheric oxygen about 2400 million years ago (Kasting 1993). Evidence for anoxic conditions prior to that time now seems, I hesitate to say, airtight: That evidence consists of leaching of soluble ferrous iron (Fe^{2+}) from paleosols (ancient soils), the presence of (insoluble) detridal pyrite (FeS_2) and uraninite (UO_2) in stream deposits, and mass-independent fractionation of stable isotopes of sulfur (Canfield 2005). The last indicates the absence of the modern sulfur cycle (and, indirectly, the absence of atmospheric oxygen) in which sulfur dioxide from volcanoes or pyrite from sedimentary rocks is oxidized to sulfate in the oceans, and then re-reduced by sulfate-reducing microorganisms in marine sediments or the water column (Farquhar et al. 2000). Once oxygen appeared, we can say relatively little about its concentration, except that for the next billion years it was probably lower than today because the deep oceans remained anoxic (Kasting 1987). Sometime in the late Precambrian, perhaps 700 million years ago, the level of oxygen rose to near its present value (Canfield and Teske 1996) and has remained within a factor of two of that level since that time (Berner et al. 2000; Berner 2001).

There is substantial, although not uncontroversial, evidence that atmospheric carbon dioxide concentrations were higher in the past (Rye et al. 1995; Pearson and Palmer 2000; Kaufman and Xiao 2003; Hessler et al. 2004). Decreasing carbon dioxide levels over geologic time might be symptomatic of a negative feedback that stabilizes climate against the increasing luminosity of the Sun on the main sequence (Walker et al. 1981). It is not known if the concentration of the primary component of the atmosphere, dinitrogen, has changed. The atmosphere is the principle reservoir of nitrogen on the surface of the Earth and modern nitrogen transformations are mediated by life, particularly bacteria. Fixation of nitrogen from the atmosphere into organic form, and its return to the atmosphere by biotic and abiotic processes cycles the entire atmospheric reservoir through living matter and sedimentary nitrogen in about 20 million years. In principle, the partial pressure of nitrogen could have

evolved if a significant fraction of nitrogen had been shunted to these other reservoirs, but there is good reason to think this never occurred: The amount of nitrogen in living matter is limited by the total size of the biosphere and this has never been much larger in the past, as indicated by the relatively constant isotopic composition of carbon in marine carbonates (Schidlowski 1988). The size of the sedimentary nitrogen budget has probably also has been limited by the propensity for life to efficiently scavenge sources of single nitrogen.

To discuss the effect of life on the chemistry of the atmosphere, one must discuss life, and to discuss life one must not only consider biochemistry, but history. Life has evolved not only according to absolute physiochemical limitations, but also, it seems, according to events that impose conditions on its subsequent evolution. Such contingencies are sometimes called 'frozen accidents'. The small and large subunits of the ribosome, an ancient holoenzyme responsible for protein synthesis and critical to all known life, is one example of a possible frozen accident. The ribonucleic acid (RNA) polymer moiety of the ribosome, a possible relic from an early 'RNA world' that preceded the advent of proteins (Gilbert 1986), has not been replaced by more efficient proteins as is presumably the case for other enzymes.

Speciation and the divergence of lineages of organisms is another example of a frozen accident. Once two species diverge sufficiently under the influence of natural selection they become genetically isolated and the process becomes irreversible, although it is now well established that even genetically distant microorganisms can exchange functional genetic material over evolutionary time through a process called horizontal gene transfer (Gogarten and Townsend 2005). Despite this complication, the common ancestry of lineages, and the degree to which they have diverged, can be described in simplified terms as an evolutionary tree—a Tree of Life. This slide (Fig. 11a) is a conceptual evolutionary tree that I drew in my journal. The next slide (Fig. 11b) is a schematic of a phylogenetic tree constructed from the analysis of the sequences of conserved molecules in extant life. Figure 11b can be used to illustrate several other 'frozen accidents' in the history of life: Most obvious is the existence of three "domains", Bacteria, Archaea, and Eukarya (Woese and Fox 1981), not two, or five, as previously thought (Whittaker 1969). Another "accident" is the presence of the photosynthetic apparatus uniquely in the Bacteria. (Eukaryotic plants carry out photosynthesis because of the presence of chloroplasts—the extremely derived descendants of bacterial endosymbionts.) Speaking of plants, all known algae are thought to originate from

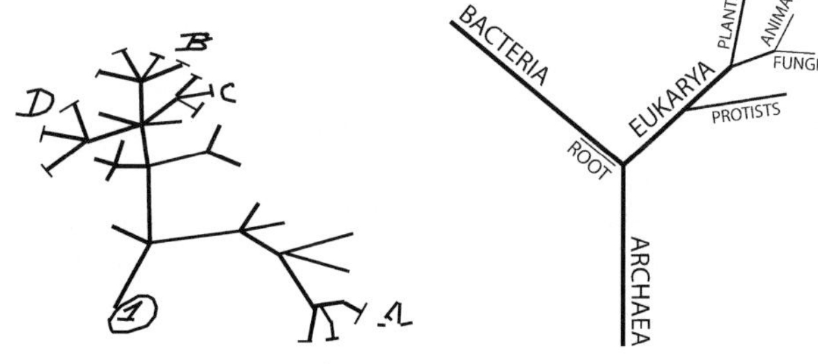

Fig. 11 (**a**) A reproduction of a phylogenetic tree from Charles Darwin's "First Notebook on Transmutation of Species" (1837). (**b**) A schematic tree of life showing the phylogenetic relationships between some of the major kingdoms of organisms based on sequences of conserved molecules. The Bacteria and Archaea each comprise multiple, diverse groups that are not shown here for clarity. The location of the "root" of the tree in the Bacteria, inferred from ancient genes that duplicated before the appearance of the three domains, means that the Bacteria emerged as a distinct lineage before either the Archaea or Eukarya

one or more endosymbiotic events in which an engulfed or infectious bacteria came to reside obligately in a eukaryotic host cell (McFadden 2001). The eukaryotic cell may itself be the result of a fusion between two cell types, one of which gave rise to the nucleus (Gupta and Golding 1996). The formation of the nucleus had consequences for the environmental range of the eukaryotic cell: The need to selectively transport messenger RNA out of the nucleus and proteins into the nucleus drove the evolution of large nuclear pore complexes. These large assemblies of proteins compromise the thermal stability of the nuclear membrane. As a result of this and other factors, eukaryotic cells have a much lower maximum known growth temperature, e.g., 62°C for thermophilic fungi (Maheshwari et al. 2000) compared to as high as 121°C in the nucleus-lacking prokaryotes (Kashefi and Lovley 2003). I will discuss a conjectural consequence of this limitation for the atmosphere and global climate at the end of my talk.

4.2 The Impact of Life on the Atmosphere

Life affects the atmosphere in three ways: First, organisms carry out chemical reactions directly involving atmospheric gases. Secondly, they mediate chemical reactions that indirectly involve atmospheric gases. Third, they create structures that alter the diffusion of gases between the atmosphere and surface of the Earth. Organisms interact directly with the atmosphere by producing and consuming gases, including N_2, O_2, CO_2, CH_4, and N_2O. The last three molecules are strong greenhouse gases. Carbon dioxide is fixed by oxygenic phototrophs into complex organic molecules such as glucose ($C_6H_{12}O_6$); oxygen is evolved as a by-product. The reverse reaction occurs when sugars are respired aerobically for energy (Table 2). Carbon dioxide is also fixed or produced during chemoautotrophic growth and catabolism of energy-rich organic molecules such as glucose under anaerobic conditions (fermentation), some reactions of which yield CO_2 and/or H_2. Methane is produced either by combining CO_2 and H_2 or decomposing acetate ($C_2H_4O_2$) (Table 2). Methane is consumed during aerobic or anaerobic methane oxidation. The latter takes place concomitantly with the reduction of sulfate (SO_4^{2-}) to sulfide (HS^-) in marine environments. (Sulfide evolves to the gas H_2S under low pH.) Nitrous oxide is produced during denitrification by the reduction of nitrate (NO_3^-) to nitrite (NO_2^-) and nitric oxide (NO). Dinitrogen (N_2), the

Table 2 Metabolisms involving atmospheric gases

Metabolism	Reaction	Domains
Photosynthesis	$6CO_2 + 6H_2O \rightarrow C_6H_{12}O_6 + 6O_2$	B, E
Aerobic respiration	$C_6H_{12}O_6 + 6O_2 \rightarrow 6CO_2 + 6H_2O$	B, A, E
Sugar fermentation	$C_6H_{12}O_6 \rightarrow 2C_2H_5OH + 2CO_2$	B, E
Methanogenesis	$CO_2 + 4H_2 \rightarrow CH_4 + 2H_2O$	A
Acetoclastic methanogenesis	$CH_3COOH \rightarrow CO_2 + CH_4$	A
Methane oxidation	$CH_4 + 2O_2 \rightarrow CO_2 + 2H_2O$	B
Anaerobic methane oxidation	$CH_4 + SO_4^{2-} + 2H^+ \rightarrow H_2S + CO_2 + 2H_2O$	A
Nitrogen fixation	$N_2 \rightarrow N_{org}$	B, A?
Denitrification	$2NO_3^- + 12H^+ \rightarrow N_2 + 6H_2O + 6CO_2$	B, A
Ammonium oxidation	$2NH_4^+ + 4O_2 \rightarrow 2NO_3^- + 4H^+ + 2H_2O$	B
Anammox	$NH_4^+ + NO_2^- \rightarrow N_2 + 2H_2O$	B
Sulfate reduction	$SO_4^{2-} + 2H^+ + 2CH_2O \rightarrow H_2S + 2CO_2 + 2H_2O$	B, A

dominant constituent of Earth's modern atmosphere, is converted into organic form by the process of nitrogen fixation and also released during denitrification.

The distribution of these metabolisms in the tree of life is very uneven (Table 2). Although organisms capable of aerobic respiration are present in all three domains of life (Bacteria, Archaea, and Eukarya), only a single group of the Bacteria—the Cyanobacteria—are known to be capable of oxygenic photosynthesis. Likewise, methanogens are found only amongst the Archaea. The genes for proteins involved in nitrogen fixation have been found in both the Bacteria and Archaea (particulary methanogens) (Zehr et al. 2004). By comparison, eukaryotic cells are not very metabolically facile, and so far shown only to be capable of aerobic respiration, oxygenic photosynthesis, and fermentation. In fact, the first two metabolisms were imported long ago when oxygen-respiring α-proteobacteria and cyanobacteria came to permanently occupy the eukaryotic cell and the eukaryotic plant cell, eventually becoming the mitochondrian and the chloroplast, respectively. Methanogenic symbionts have been found in certain protists, but it is not known if these benefit the eukaryotic host cell (Fenchel and Finlay 1991).

Life also alters atmospheric composition indirectly by mediating chemical reactions that affect atmospheric gases. One example is biological control over weathering reactions of silicate minerals at the surface of the Earth. Many silicate minerals created at high pressures and temperatures within the Earth are unstable under conditions at the Earth's surface. In the presence of water and an acid they transform to more thermodynamically favored phyllosilicates, clays and oxides by a variety of weathering reactions, e.g.:

$$CaAl_2Si_2O_8 + 8H^+ \rightarrow Ca^{2+} + 2Al^{3+} + 2SiO_2 + 4H_2O. \tag{2}$$

Protons can be supplied by carbonic acid, HCO_3^-, a ubiquitous acid in waters under CO_2-containing atmospheres:

$$CO_2(aq) + H_2O \rightarrow H_2CO_3 \rightarrow HCO_3^- + H^+. \tag{3}$$

Abiotic weathering of silicates and its role in atmospheric CO_2 and climate have been reviewed by Kump et al. (2000). Plants and fungi produce a variety of organic acids such as oxalic acid that accelerate the weathering process. Vegetation may also increase the relative mobility and release of Ca, Mg, and Sr, thereby favoring the right-hand side of (2) (Gislason et al. 1996). (Vascular plants also influence weathering in other ways, as I shall talk about in a moment.) Acceleration of weathering by organisms may have counteracted warming by a brighter, evolving Sun (Schwartzman and Volk 1989; Berner 1992).

The complementary reaction to weathering is the biogenic precipitation of aragonite or calcite in the oceans:

$$Ca^{2+} + 2HCO_3^- \rightarrow CaCO_3 + CO_2 + H_2O. \tag{4}$$

In the inorganic cycling of carbon, net formation of sedimentary carbonates balances CO_2 produced by volcanoes and metamorphism. Photosynthetic uptake of carbon dioxide favors carbonate predicipation, however the *rate* will still be limited by the supply of calcium ions to the ocean. For billions of years before the emergence of multicellular life, precipitation was abiotic or driven by photosynthesis. Biomineralization by animals involves actively maintaining and controlling the precipitation process. Active pumping of calcium ions and bicarbonate to biomineralization sites means that ambient concentrations, and thus the calcite saturation state of the ocean, can fall. Equation (3) tells us then that ambient CO_2 concentrations will also fall until equilibrium is

re-established. This has obvious consequences for climate, and the suggestion that animals were somehow responsible for a dramatic cooling of the planet at the end of the Precambrian is not new. However, the oldest fossils of biomineralizing animals, the "small shelly fauna" including *Cloudina* and *Namacalathus*, appear 550 million years ago (Grant 1990; Grotzinger et al. 2000), and only *after* multiple episodes of extreme glaciation and the decline of massive limestone formations.

Biotic manipulation of atmospheric composition can also affect the fraction of sunlight that is reflected back into space, a quantity called the planetary albedo. Biogenic aerosols influence climate by scattering incoming radiation and serving as cloud condensation nuclei (CCN). Natural aerosols are primarily composed of sulfate and carbonaceous compounds (Pöschl 2005) and, in the marine boundary layer, salts (Murphy et al. 1998). Biogenic sulfate derives from oxidation of dimethyl sulfide (DMS) from marine organisms and carbonyl sulfide (COS) from terrestrial vegetation (Andreae and Crutzen 1997). An elevated atmospheric CCN content decreases average cloud droplet size, increasing cloud albedo and cloud lifetime against precipitation, thereby cooling the Earth (Ackerman et al. 2000). Aerosols may have been more abundant on the Archean Earth when oxygen was absent from the atmosphere (Pavlov et al. 2001b).

The final effect of living organisms is what one might call a "compartmentalization" of the atmosphere by biogenic structures. One example of this type of effect is a soil: Soils are an admixture of the weathering products of rocks and decomposing organic matter. Physically, a soil is a permeable structure, maintained in place by vascular plant roots and its own cohesiveness. It contains significant pore space containing gases and aqueous solutions and because their transport is limited by diffusion, their composition can depart radically from that of the atmosphere. These differences can affect the pathways and kinetics of the weathering reactions of the parent rock and the decomposition of organic matter. It will also affect biological activity through, for example, the abundance of oxygen. More specifically, aerobic respiration of organic matter by the roots and soil microorganisms can increase the concentration of CO_2 in soil pore space to as much as 10,000 ppm. There is also a concomitant decrease in the concentration of O_2. Because there is much more O_2 in the atmosphere compared to CO_2, the decrease in O_2 will in general not be significant. The exceptions are soils in which a significant amount of the pore volume is occupied by water, through which oxygen diffuses a factor of 10^4 times more slowly than in air, and in which conditions can be hypoxic or even anoxic. High soil CO_2 leads to an increase in the concentration of hydrogen ions in porewater (3), hence favoring chemical weathering (2). In the absence of soil, atmospheric CO_2 (and, through the greenhouse effect, temperatures) would be much higher (Berner 1992). Vascular plans also actively pump soil water into their tissues and transpire it into the atmosphere, thereby changing relative humidity, cloud formation, and planetary albedo in unknown but potentially significant ways.

A second example is the digestive tract of animals. Rather than simply filtering particles from seawater (as sponges do) or absorbing dissolved molecules (which the enigmatic placozoans do), animals with guts process their food in internal compartments that are isolated from the atmosphere. Again, as in the case of soils or sediments, biological activity by the microorganisms, in addition to the host animal itself, alters the chemistry of the immediate environment compared to surface conditions at the Earth or in the oceans (Plante and Jumars 1992). To illustrate how dramatic the difference is, I show here a table (Table 3) giving median concentrations of gases in human flatus (Suarez et al. 1997) compared with the composition of the modern atmosphere. Organic matter passing through guts is processed under conditions quite dissimilar to those at the surface of the modern Earth: Metabolisms such as methanogenesis that require abundant H_2, for example, are favored. The consequences

Table 3 Median composition (% number) of human colonic flatus (Suarez et al. 1997) and the modern atmosphere

Chemical species	Flatus	Present atmosphere
CO_2	29.5%	0.038
H_2	32.2	0.00005
N_2	18.9	78
CH_4	0.006	0.00019
O_2	2.95	21
H_2S	1.1	0

for the global carbon cycle of the appearance with animals with guts are not understood but may have been dramatic (Rothman et al. 2003).

Compartmentalization of the atmosphere did not stop with the origin of animal guts: New forms emerged as animals evolved, including burrows, reefs, and termite mounds. We humans continue a time-honored tradition when we build and inhabit edifices, heating or cooling their interiors by burning fossil fuels, releasing CO_2, and altering the outside atmosphere and climate.

4.3 Feedbacks and Planetary Homeostasis

I have described three ways in which living things alter the atmospheric composition of the Earth, and thereby the climate and environment which they experience. I have also described in previous work (Darwin 1859) how the environment acts through natural selection on the variation between individuals in a species, thereby causing the species to evolve, or even new species to emerge, over geologic time. Does this combination of forces constitute a feedback between the organic and inorganic worlds? Margulis and Lovelock (1974) proposed that surface conditions are regulated by life so as to permit maximum growth of the biosphere and that planetary homeostasis emerges naturally. This extraordinary claim (the "Gaia hypothesis") has met with great interest, and perhaps even greater incredulity by the scientific community.

Watson and Lovelock (1983) illustrated planetary homeostasis by a simple model involving oppositely-pigmented members of the *Asterales* order of the Plantae (black and white daisies). Both strains have the same growth response to temperature, but because they absorb sunlight to a different degree (black completely, white not at all), they experience different local temperatures. As a result, the growth of one type of daisy will be favored until that growth has changed the planet's albedo and has heated or cooled it sufficiently to eliminate that advantage. As a consequence, the evolving brightness of the planet's parent star will be accompanied by a concomitant change in the relative proportions of black and white daisies. Black daisies will be favored at lower illumination levels; white daisies at higher. The surface temperature of Daisyworld remains constant as an initially pure black daisy population is replaced by white daisies (Fig. 12a). Subsequent versions of the Daisyworld model included intermediate strains of "grey" daisies and predation (Harding and Lovelock 1996).

Evolutionary biologists have been reluctant to take such models seriously because they seem to imply phenomenae that would not emerge from natural selection acting on individual organisms. In particular, the idea that natural selection can act on groups of organisms ("group selection"), rather than individuals, has fallen out of favor, with the exception of close relatives that share alleles ("kin selection"). Another objection is that adaptation of species to their environment should lead to an uncontrolled "drift" of the system and loss

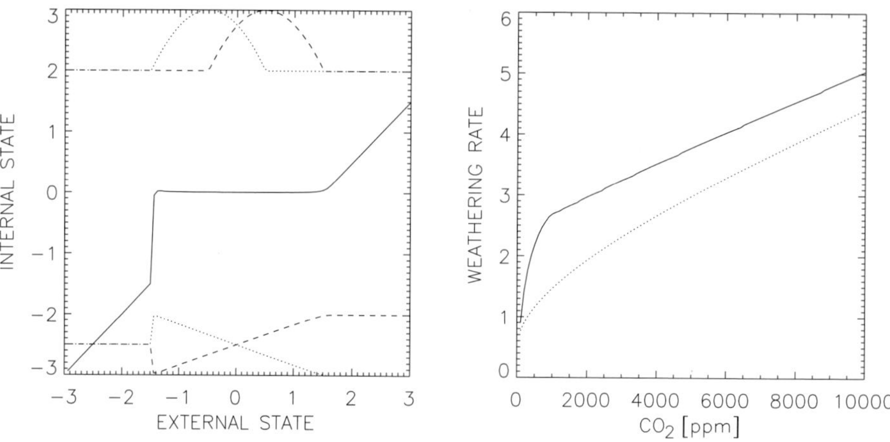

Fig. 12 (**a**) Biological homeostasis ("Daisyworld") in the abstract: As an external parameter (e.g., stellar radiation) varies, the relative growth of two species (black and white daisies), which affect the internal state (e.g., temperature) in opposite ways, causes the internal state to remain constant over a certain range (*solid curve*). *Curves* of growth rate as a function of the internal parameter value are at the *top*; the relative population sizes are the *bottom curves*. Regulation fails once one species dominates. (**b**) One possible example of climate regulation by life: Relative rate of weathering and removal of CO_2 from the atmosphere as carbonates as a function of atmospheric CO_2, based on a coupled climate–vegetation–soil model (Gaidos, in prep). The *solid curve* is for the Earth with the current configuration of continents, orbital parameters, and fixed vegetation biomass; the *dotted curve* is for an unvegetated world. The steeper slope of the vegetated world is due to migration of vegetation to, and soil development at, high latitudes with increasing temperature as well as enhanced gross primary productivity under a CO_2-rich atmosphere. A steeper slope implies a stronger negative feedback

of self-regulation (Robertson and Robinson 1998; Lenton 1998). Indeed, it would seem that the essence of Daisyworld is lack of or incomplete adaptation to the environment due to some physiological barrier, for example. In the case of the original model, it is the finite temperature range of growth (Lenton and Lovelock 2000).

Does Earth's atmosphere reflect any Daisyworld-like self-regulation? Charlson et al. (1987) have proposed that the production of dimethylsulfide by marine algae, the incorporation of that sulfur into cloud condensation nuclei and the effect of clouds on the global albedo is one such atmospherically-mediated climate regulation. There are many other potential climate-biosphere feedbacks (Lashof et al. 1997). One such feedback may involve the maximum growth temperatures of lichens and plants involved in enhancing weathering of silicate rocks. The upper temperature range for growth is about 62°C for fungi (Maheshwari et al. 2000) and 40–50°C for plants. (Lower limits are set by the unavailability of significant liquid water below 0°C.) As a result, the latitude range of plants shifts in response to climate (Davis and Shaw 2001). Thus, if the concentration of atmospheric carbon dioxide decreases, temperatures fall, plants retreat from high latitudes and weathering at those latitudes decreases. As a consequence, carbon dioxide levels increase, warming the Earth. Conversely, if the Earth warms, plants occupy higher latitudes, increasing the development of soil and rates of weathering there, and lowering atmospheric CO_2. This feedback is in addition to the abiotic thermostat proposed by Walker et al. (1981). The vegetation migration effect is stronger than the abiotic effect alone at low atmospheric CO_2 levels and temperatures, but becomes weaker with increasing CO_2 and mean surface temperature because high latitudes constitute a smaller fraction of Earth's weatherable surface area (Fig. 12b). A potential instability point exists when temperatures at the equator approach the limit of plant

 Springer

growth. Beyond that point, plants die back at low latitudes, leading to decreased weathering, increased CO_2, and still higher temperatures. A similar collapse of regulation occurs in a 2-D Daisyworld model when increasing equatorial temperatures drive desertification and an increase in albedo at low latitudes (Ackland et al. 2003).

The Gaia hypothesis emerged from the proposition that life on Mars should have already been detected by its effect on atmospheric composition, and therefore did not exist in abundance (Hitchcock and Lovelock 1966). Regardless of what one thinks of the Gaia conjecture, the notion that Earth's atmosphere bears the imprint of life continues to be accepted as a practical approach to detecting life on another planet and recent studies have addressed this possibility (Des Marais et al. 2002). The simultaneous presence of oxygen and methane in an atmosphere, originally considered by Hitchcock and Lovelock (1966), is considered to be a strong indicator of biological activity and in an interesting experiment, they were detected in Earth's atmosphere by a passing spacecraft (Sagan et al. 1993).

Since my publication of *Origin of Species*, one species in particular has increased its influence on the atmosphere. The atmospheric concentration of CO_2 is now at the highest level it has ever been in the past 600,000 years (Siegenthaler et al. 2005) and the annual increase shows no signs of abating (Keeling et al. 1989). The concentration of methane has also dramatically increased since the expansion of major anthropogenic sources, particularly rice paddies and ruminant livestock. Our influence on the atmosphere would be detectable from far away, if an observer knew what to look for. Our engineering of the atmosphere is reckless and without a plan. Intelligent species on other planets might be more deliberate and constructive in their work. For example, they could introduce a suite of long-lived "super" greenhouse gases such as fluorine compounds that would block most of the infrared radiation escaping from the surface, creating an intense greenhouse effect that would warm planets like Mars that are too distant from their parent star and otherwise too cold to support liquid water at their surfaces (Gerstell et al. 2001). The unambiguous spectroscopic signature of such a feat—strong absorption features at many infrared wavelengths—might be detected from a great distance. Thank you.

5 Discussion

Guillaume: Thank you, Charles. We now move into the panel discussion part of our session. Charles, I see you already have your hand up, so please go ahead.

Lyell: Are there examples of destabilizing, or "anti-Gaia" biotic feedbacks? The geologic activity of our planet is also thought to have slowly decreased with time as the radiogenic heat produced in the Earth decays and the mantle convection needed to reject that heat slows. What are the implications of this for climate stabilization by biology?

Darwin: Excellent question. Lovelock and Kump (1994) addressed some aspects of this issue, arguing that above a global mean temperature of 20°C, both marine and terrestrial ecosystems supply a destabilizing *positive* climate feedback by taking up less CO_2 or releasing more to the atmosphere. Clearly, long-term brightening of the Sun will eventually defeat any stabilization. Kasting and Caldeira (1992) showed that, well before the increase in surface temperature and atmospheric water vapor content create a runaway "wet" greenhouse, the atmospheric carbon dioxide concentration will fall below the level where C4 plants can carry out photosynthesis. Of course, strong selection pressure may produce plants capable of growing at even *lower* CO_2 concentrations but eventually levels will reach zero and a biotic

crisis will ensue. Lower rates of geologic activity and CO_2 output in the future (Franck et al. 2000) will only hasten that day.

Darwin: I have a question for Sir Herschel: You mention that greenhouse gases other than carbon dioxide may have been important in determining the climate and of early Mars. What are those gases?

Herschel: The most likely candidate is methane. Per molecule, methane is twenty times more effective as a greenhouse gas. Methane is an attractive candidate because it absorbs at different wavelengths than CO_2. Kasting (1997) and Pavlov et al. (2001a) argued that an atmosphere with about one bar of CO_2 and 1% methane could keep Mars surface temperatures above freezing early in its history when the Sun was fainter. The real difficulty with this scenario is that it requires a source of CH_4 that is comparable with the present-day biological source on Earth (much of which is anthropogenic in nature). Purely abiotic geologic activity produces methane on the Earth, i.e., through production of H_2 by the serpentinization of mafic rocks and reaction of hydrogen with CO_2, but in relatively small (and uncertain) amounts compared to biogenic methane. Ongoing fluid-rock chemistry in the crust may be responsible for the purported few tens of ppb of CH_4 detected on Mars (Lyons et al. 2005). Alternatively, methane may only now be escaping from the destabilization of hydrates formed during an earlier, more geologically active phase (Oze and Sharma 2005). However, the level of geologic activity on early Mars is not known and it is not at all clear that even a very active Mars could have produced the necessary methane to warm the surface without the intervention of biology. Furthermore, the photochemical lifetime on an ancient, wet Mars would be much shorter than its present value of a few centuries. So perhaps this all implies that life—methanogens—*did* exist on Mars!

Darwin: My question is for Sir Lyell: After the "Opportunity" and "Spirit" rovers, which question about the past or present habitability of Mars now looms largest? What are the next "opportunities", if I may turn a phrase, to answer that question?

Lyell: The twin Mars Exploration rovers have demonstrated that liquid water once existed at and near the surface and that it chemically altered the rocks. But spacecrafts orbiting Mars have revealed other extensively altered terrains where environments might have been even more hospitable to life and persisted for longer periods. The presence of methane in the Martian atmosphere might indicate that liquid water exists even today in the subsurface. Every mission reveals that the Martian crust and atmosphere are even more complex than we had previously suspected. Therefore the next question looms largest: Did liquid water ever co-exist with sources of energy and environmental conditions that could have allowed life to develop and persist? Additional questions immediately follow. What has been the full diversity of habitable environments on Mars and can spacecraft explore them and/or their geologic records? How extensive geographically were such environments and how long did they persist? We must identify geologic deposits that preserved evidence of ancient habitable environments and also, potentially, signatures of life. Examples of such deposits include carbonates, aqueously deposited silica, phosphates, and evaporite minerals such as well-cemented sulfates and halides.

Regarding upcoming opportunities, both the Mars Express and the Mars Reconnaissance orbiters will probably continue to identify minerals that indicate where volatile species such as water and CO_2 have interacted with crustal rocks and perhaps created both habitable environments and aqueous precipitates. The orbiters will also probably discover sites that are

 Springer

even more promising than those visited by the MER landers. In the nearest future NASA's Phoenix lander (2007 launch) and Mars Science Laboratory (MSL) rover (2009 launch) will analyze both minerals and volatile species. The MSL rover, another "robotic field geologist", like "Spirit" and "Opportunity", might someday explore a site that represents an ancient lake, hydrothermal system, or some other kind of ancient habitable environment that, once upon a time, indeed fulfilled all of life's requirements.

Guillaume: This is indeed a fascinating discussion, but my highly accurate Swiss watch tells me that we are out of time. Let's thank our speakers again and enjoy the rest of this day in this especially habitable part of Earth.

Acknowledgements DJD acknowledges support from the NASA Astrobiology Institute and the NASA Exobiology program. EG acknowledges travel support by the NASA Terrestrial Planet Finder Foundation Science program and the NASA Astrobiology Institute. This chapter was improved by reviews by S. Rodin, L. Ksanfomality, and K. Fishbaugh.

References

A.S. Ackerman, O.B. Toon, D.E. Stevens, A.J. Heymsfield, V. Ramanathan, E.J. Welton, Science **288**, 1042 (2000)

G.J. Ackland, M.A. Clark, T.M. Lenton, J. Theor. Biol. **223**, 39 (2003)

M.O. Andreae, P.J. Crutzen, Science **276**, 1052 (1997)

V.R. Baker, Nature **412**, 228 (2001)

H. Balsiger, K. Altwegg, J. Geiss, J. Geophys. Res. **100**, 5827 (1995)

R.A. Berner, Geochim. Cosmochim. Acta **56**, 3225 (1992)

R.A. Berner, Geochim. Cosmochim. Acta **65**, 685 (2001)

R.A. Berner, A.C. Lasaga, R.M. Garrels, Am. J. Sci. **283**, 641 (1983)

R.A. Berner, S.T. Petsch, J.A. Lake, D.J. Beerling, B.N. Popp, R.S. Lane, E.A. Laws, M.B. Westley, N. Cassar, F.I. Woodward, W.P. Quick, Science **287**, 1630 (2000)

J.L. Bertaux, Thèse de Doctorat d'Etat, University of Paris (UPMC), 1974

J.L. Bertaux, F. Montmessin, J. Geophys. Res. **106**, 32,879 (2001)

J.N. Betts, H.D. Holland, Palaeogeogr. Palaeoclimatol. Palaeoecol. **97**, 5 (1991)

J.-P. Bibring, Y. Langevin, A. Gendrin, B. Gondet, F. Poulet, M. Berth'e et al., Science **307**, 1576 (2005)

G.L. Bjoraker, H.P. Larson, M.J. Mumma, R. Timmermann, J.L. Montani, Bull. Am. Astron. Soc. **24**, 995 (1992)

D. Bockelée-Morvan, D. Gautier, D.C. Lis, K. Young, J. Keene, T.G. Phillips et al., Icarus **133**, 147 (1998)

P.J. Boston, M.V. Ivanov, C.P. McKay, Icarus **95**, 300 (1992)

W.V. Boynton, W.C. Feldman, S.W. Squyres, T. Prettyman, J. Brückner, L.G. Evans et al., Science **297**, 75 (2002)

W.S. Broecker, T.H. Peng, *Tracers in the Sea* (Eldigio Press, Palisades, New York, 1982)

R. Buick, J.S.R. Dunlop, Sedimentology **37**, 247 (1990)

D.M. Burr, J.A. Grier, A.S. McEwen, L.P. Kesztherlyi, Icarus **159**, 53 (2002)

A.G.W. Cameron, W. Benz, Icarus **92**, 204 (1991)

D.E. Canfield, Annu. Rev. Earth Planet. Sci. **33**, 1 (2005)

D.E. Canfield, A.P. Teske, Nature **382**, 127 (1996)

D.E. Canfield, R. Raiswell, Am. J. Sci. **299**, 697 (1999)

D.E. Canfield, K. Habicht, B. Thamdrup, Science **288**, 658 (2000)

D. Canil, Earth Planet. Sci. Lett. **195**, 75 (2002)

M.H. Carr, J. Geophys. Res. **84**, 2995 (1979)

M.H. Carr, J. Geophys. Res. **107**, 5131 (2002)

R.H. Carr, M.M. Grady, I.P. Wright, C.T. Pillinger, Nature **314**, 248 (1985)

R.J. Charlson, J.E. Lovelock, M.O. Andreae, S.G. Warren, Nature **326**, 655 (1987)

P.R. Christensen, Nature **422**, 45 (2003)

E.H. Christiansen, Geology **17**, 203 (1989)

B.C. Clark, A.K. Baird, Geophys. Res. Lett. **6**, 811 (1979)

S.M. Clifford, J. Geophys. Res. **92**, 9135 (1987)

S.M. Clifford, J. Geophys. Rev. **98**, 10,973 (1993)

S.M. Clifford, T.J. Parker, Icarus **154**, 40 (2001)

G.D. Clow, Icarus **72**, 95 (1987)

F. Costard, F. Forget, N. Mangold, J.P. Peulvast, Science **295**, 110 (2002)

R.A. Craddock, A.D. Howard, J. Geophys. Res. **107**, 5111 (2002)

C. Darwin, *On the Origin of Species by Means of Natural Selection, or, The Preservation of Favoured Races in the Struggle for Life* (John Murray, London, 1859)

M.B. Davis, R.G. Shaw, Science **292**, 673 (2001)

J.W. Delano, Orig. Life Evol. Biosphere **31**, 311 (2001)

C. de Bergh, B. Bezard, T. Owen, D. Crisp, J.P. Maillard, B.L. Lutz, Science **251**, 547 (1991)

L. Derry, A.J. Kaufman, S.B. Jacobsen, Geochim. Cosmochim. Acta **56**, 1317 (1992)

D.J. Des Marais, in *The Carbon Cycle and Atmospheric CO_2: Natural Variations Archean to Present*, vol. 32, ed. by E.T. Sundquist, W.S. Broecker (Am. Geophys. U., Washington, 1985), p. 602

D.J. Des Marais, Chem. Geol. **114**, 303 (1994)

D.J. Des Marais, in *Stable Isotope Geochemistry, Rev. Mineral. Geochem.* vol. 32, ed. by J.W. Valley, D.R. Cole (Mineral. Soc. Amer., 2001), p. 555

D.J. Des Marais, M.O. Harwit, K.W. Jucks, J.F. Kasting, D.N.C. Lin, J.I. Lunine, J. Schneider, S. Seager, W.A. Traub, N.J. Woolf, Astrobiology **2**, 153 (2002)

T.M. Donahue, D.H. Grinspoon, R.E. Hartle, R.R. Hodges, in *Venus II*, ed. by S.W. Bougher, D.M. Hunten, R.J. Phillips (Univ. of Arizona Press, Tucson, 1997), p. 385

H. Elderfield, A. Schultz, Annu. Rev. Earth Planet. Sci. **24**, 191 (1996)

J. Farquhar, H. Bao, M. Thiemens, Science **289**, 756 (2000)

W.C. Feldman, T.H. Prettyman, S. Maurice, J.J. Plaut, D.L. Bish, D.T. Vaniman et al., J. Geophys. Res. **109**, E09006 (2004)

T. Fenchel, B.J. Finlay, Eur. J. Protistol. **26**, 200 (1991)

S. Franck, A. Block, W. von Bloh, J.J. Schellnhuber, Y. Svirezhev, Tellus B **52**, 94 (2000)

E. Gaidos, G. Marion, J. Geophys. Res. **108**, 5055 (2003)

R.M. Garrels, E.A. Perry Jr., in *The Sea*, vol. 5, ed. by E.D. Goldberg (Wiley, New York, 1974), p. 303

H. Genda, Y. Abe, Icarus **164**, 149 (2004)

H. Genda, Y. Abe, Nature **433**, 842 (2005)

A. Gendrin, N. Mangold, J.-P. Bibrin, Y. Langevin, B. Gondet, F. Poulet et al., Science **302**, 1587 (2005)

M.F. Gerstell, J.S. Francisco, Y.L. Yung, C. Boxe, E.T. Aaltonee, Proc. Nat. Acad. Sci. USA **98**, 2154 (2001)

W. Gilbert, Nature **319**, 618 (1986)

S.R. Gislason, S. Arnorsson, H. Armannson, Am. J. Sci. **296**, 837 (1996)

J.P. Gogarten, J.P. Townsend, Nat. Rev. Microbiol. **3**, 679 (2005)

D.C. Golden, D.W. Ming, C.S. Schwandt, H.V. Lauer, R.A. Socki, R.V. Morris et al., Am. Mineral. **86**, 370 (2001)

M.P. Golombek, N.T. Bridges, J. Geophys. Res. **105**, 1841 (2000)

D.O. Gough, Sol. Phys. **74**, 21 (1981)

S.W.F. Grant, Am. J. Sci. **290-A**, 261 (1990)

J.P. Grotzinger, in *Controls on Carbonate Platform and Basin Development*, ed. by P.D. Crevello, J.L. Wilson, J.F. Sarg, J.F. Read, Special Publication 44. Tulsa, OK, 1989, p. 79

J.P. Grotzinger, J.F. Kasting, J. Geol. **101**, 235 (1993)

J.P. Grotzinger, W.A. Watters, A.H. Knoll, Paleobiology **26**, 334 (2000)

J.P. Grotzinger, R.E. Arvidson, J.F. Bell III, W. Calvin, B.C. Clark, D.A. Fike et al., Earth Planet. Sci. Lett. **240**, 11 (2005)

V.C. Gulick, J. Geophys. Res. **103**, 19,365 (1998)

R.S. Gupta, G.B. Golding, Trends Biochem. Sci. **21**, 166 (1996)

R.M. Haberle, J. Geophys. Res. **103**, 28,467 (1998)

J.C. Hanna, R.J. Phillips, Lunar Planet. Sci. Conf. **36**, 2261 (2005)

S.P. Harding, J.E. Lovelock, J. Theor. Biol. **182**, 109 (1996)

W.K. Hartmann, G. Neukum, in *Chronology and Evolution of Mars*, ed. by R. Kallenbach, J. Geiss, W.K. Hartmann (Springer, 2001), p. 165

L.A. Haskins, A. Wang, H.Y. McSween, B.C. Clark, D.J. Des Marais, S.M. McClennan et al., Nature **436**, 66 (2005)

E. Hauber, K. Gwinner, D. Reiss, F. Scholten, G. Michael, R. Jaumann et al., Lunar Planet. Sci. **36**, 1641 (2005)

J.D. Hays, W.C. Pitman, Nature **246**, 18 (1973)

J.W. Head, G. Neukum, R. Jaumann, H. Hiesinger, E. Hauber, M. Carr et al., Nature **434**, 346 (2005)

J.W. Head, L. Wilson, K.L. Mitchel, Geophys. Res. Lett. **30**, 1577 (2003)

J.W. Head, M.A. Kreslavsky, S. Pratt, J. Geophys. Res. **107**, 5003 (2002)

A.M. Hessler, D.R. Lowe, R.L. Jones, D.K. Bird, Nature **428**, 736 (2004)

D.R. Hitchcock, J.E. Lovelock, Icarus **7**, 149 (1966)

H.D. Holland, *The Chemical Evolution of the Atmosphere and Oceans* (Princeton, 1984), 582 pp

W.-L. Huang, P.J. Wyllie, C.E. Nehru, Am. Mineral. **65**, 285 (1980)

J.M. Hunt, AAPG Bull. **56**, 2273 (1972)

B.M. Hynek, R.J. Williams, Geology **29**, 407 (2001)

V.I. Ignatiev, N.I. Moroz, B.E. Moshkin, A.P. Ekonomov, V.I. Gnedykh, A.V. Grigoriev, I.V. Khatuntsev, Planet. Space Sci. **45**, 427 (1997)

B.M. Jakosky, M.H. Carr, Nature **315**, 559 (1985)

B.M. Jakosky, J.H. Jones, Nature **370**, 328 (1994)

B.M. Jakosky, J.H. Jones, Rev. Geophys. **35**, 1 (1997)

B.M. Jakosky, E.L. Shock, J. Geophys. Res. **103**, 19,359 (1998)

H.W. Jannasch, C.O. Wirsen, Bioscience **29**, 592 (1979)

J.S. Kargel, V.R. Baker, J.E. Beget, J.F. Lockwood, T.L. Pewe, J.S. Shaw, R.G. Strom, J. Geophys. Res. **100**, 5351 (1995)

K. Kashefi, D.R. Lovley, Science **301**, 934 (2003)

J.F. Kasting, Precambr. Res. **34**, 205 (1987)

J.F. Kasting, Icarus **94**, 1 (1991)

J.F. Kasting, Science **259**, 920 (1993)

J.F. Kasting, Science **276**, 1213 (1997)

J.F. Kasting, K. Caldeira, Nature **360**, 721 (1992)

J.F. Kasting, D. Catling, Annu. Rev. Astron. Astrophys. **41**, 429 (2003)

J.F. Kasting, D.H. Eggler, S.P. Raeburn, J. Geol. **101**, 245 (1993)

A.J. Kaufman, S. Xiao, Nature **425**, 279 (2003)

C.D. Keeling, R.B. Bacastow, A.F. Carter, S.C. Piper, T.P. Whorf, M. Heimann et al., in *Aspects of Climate Variability in the Pacific and the Western Americas*, ed. by D.H. Peterson. Geophys. Monogr., vol. 55, (1989), p. 165

D.S. Kelley, J.A. Karson, G.L. Fruh-Green, D.R. Yoerger, T.M. Shank, D.A. Butterfield et al., Science **307**, 1428 (2005)

P.D. Komar, Icarus **42**, 317 (1979)

Y.N. Kulikov, H. Lammer, H.I.M. Lichtenegger, T. Penz, D. Breuer, T. Spohn, R. Lundin, H.K. Biernat, 2007 (this issue). doi:10.1007/s11214-007-9192-4

L.R. Kump, S.L. Brantley, M.A. Arthur, Annu. Rev. Earth Planet. Sci. **28**, 611 (2000)

R.S.J. Lambert, in *The Early History of the Earth*, ed. by B.F. Windley (Wiley, New York, 1976), p. 363

D.A. Lashof, B.J. DeAngelo, S.R. Saleska, J. Harte, Annu. Rev. Energy Environ. **22**, 75 (1997)

J. Laskar, B. Levrard, J.F. Mustard, Nature **419**, 375 (2002)

T.M. Lenton, J.E. Lovelock, J. Theor. Biol. **206**, 109 (2000)

T.M. Lenton, Nature **394**, 439 (1998)

Z.-X.A. Li, C.-T. Lee, Earth Planet. Sci. Lett. **228**, 483 (2004)

J.E. Lovelock, L.R. Kump, Nature **369**, 732 (1994)

D.R. Lowe, Precambr. Res. **19**, 239 (1983)

D.R. Lowe, in *The Proterozoic Biosphere: A Multidisciplinary Study*, ed. by J.W. Schopf, C. Klein (Cambridge, 1992), p. 67

D.R. Lowe, in *Early Life on Earth. Nobel Symposium 84*, ed. by S. Bengtson (Columbia Univ., New York, 1994), p. 24

P.D. Lowman Jr., Precambr. Res. **44**, 171 (1989)

B. Lucchitta, D. Anderson, H. Shoji, Nature **290**, 759 (1981)

J.I. Lunine, A. Chambers, J. Morbidelli, L.A. Leshin, Icarus **165**, 1 (2003)

J.R. Lyons, C. Manning, F. Nimmo, Geophys. Res. Lett. **32**, L13201 (2005)

R. Maheshwari, G. Bharawaj, M.K. Bhat, Microb. Mol. Biol. Rev. **64**, 461 (2000)

M.C. Malin, K.S. Edgett, Science **302**, 1931 (2003)

M.C. Malin, K.S. Edgett, J. Geophys. Res. **106**, 23,429 (2001)

M.C. Malin, K.S. Edgett, Science **288**, 2330 (2000)

N. Mangold, C. Quantin, V. Anson, C. Delacourt, P. Allemand, Science **305**, 78 (2004)

L. Margulis, J.E. Lovelock, Icarus **21**, 471 (1974)

J.F. McCauley, Geologic map of the Coprates quadrangle of Mars, USGS Misc. Inv. Map I-897, 1978

M.T. McCulloch, Earth Planet. Sci. Lett. **115**, 890 (1993)

G.I. McFadden, J. Phycol. **37**, 1 (2001)

P.J. McGovern, S.C. Solomon, D.E. Smith, M.T. Zuber, M. Simons, M.A. Wieczorek et al., J. Geophys. Res. **107**, 5136 (2002)

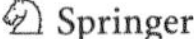

S.M. McLennan, J.F. Bell III, W.M. Calvin, P.R. Christensen, B.C. Clark, P.A. de Souza et al., Earth Planet. Sci. Lett. **240**, 95 (2005)

R. Meier, H.E. Owen, T.C. Matthews, D.C. Jewitt, D. Bockelée-Morvan, N. Biver et al., Science **279**, 842 (1998)

M.T. Mellon, B.M. Jakosky, J. Geophys. Res. **100**, 11,781 (1995)

S.L. Miller, H.C. Urey, Science **130**, 245 (1959)

D.W. Ming, D.W. Mittlefehldt, R.V. Morris, D.C. Golden, R. Gellert, A. Yen et al., J. Geophys. Res. **111**, E02S12 (2006)

M.D. Mirota, J. Veizer, Geochim. Cosmochim. Acta **58**, 1735 (1994)

I. Mitrofanov, D. Amfimov, A. Kozyrev, M. Litvak, A. Sanin, V. Tret'yakov et al., Science **297**, 78 (2002)

J.B. Murray, J.-P. Muller, G. Neukum, S.C. Werner, S. van Gasselt, E. Hauber et al., Nature **434**, 352 (2005)

J.M. Moore, D.E. Wilhelms, Icarus **154**, 258 (2001)

V.I. Moroz, B.E. Moshkin, A.P. Ekonomov, A.V. Grigoriev, V. Grigoriev, V.I. Gnedykh, Y.M. Golovin, Kosm. Issled. **21**, 246 (1983)

D.M. Murphy, J.R. Anderson, P.K. Quinn, L.M. McInnes, F.J. Brechtel, S.M. Kreidenwies et al., Nature **392**, 62 (1998)

H.E. Newsom, Icarus **44**, 207 (1980)

H.E. Newsom, G.E. Brittelle, C.A. Hibbitts, L.J. Crossey, A.M. Kudo, J. Geophys. Res. Planets **101**, 14,951 (1996)

H.E. Newsom, J.J. Hagerty, J. Geophys. Res. **104**, 88,717 (1999)

A.O. Nier, M.B. McElroy, J. Geophys. Res. **82**, 4341 (1977)

B.W. Nocita, D.R. Lowe, Precambr. Res. **48**, 375 (1990)

U. Ott, H.P. Lohr, F. Begemann, Meteoritics **23**, 295 (1988)

C. Oze, M. Sharma, Geophys. Res. Lett. **32**, L10203 (2005)

T. Owen, J.P. Maillard, C. de Bergh, B.L. Lutz, Science **240**, 1767 (1988)

T. Owen, A. Bar-Nun, in *Origin of the Earth and Moon*, ed. by R.M. Canup, K. Richter (Univ. Ariz. Press, Tucson, 2000), p. 459

T.J. Parker, R.S. Saunders, D.M. Schneeberger, Icarus **82**, 111 (1989)

T.J. Parker, D.S. Gorsline, R.S. Saunders, D. Pieri, D.M. Schneedberger, J. Geophys. Res. **98**, 11,061 (1993)

T.J. Parker, S.M. Clifford, W.B. Banerdt, Lunar Planet. Sci. Conf. **31**, 2033 (2000)

A.A. Pavlov, J.F. Kasting, L.L. Brown, K.A. Rages, R. Freedman, J. Geophys. Res. **105**, 11,981 (2001a)

A.A. Pavlov, J.F. Kasting, J.L. Eigenbrode, K.H. Freeman, Geology **29**, 1003 (2001b)

P.N. Pearson, M.R. Palmer, Nature **406**, 695 (2000)

J.M. Petit, A. Morbidelli, J. Chambers, Icarus **153**, 338 (2001)

D.C. Pieri, Science **210**, 895 (1980)

C. Plante, P. Jumars, Microb. Ecol. **23**, 257 (1992)

D. Porcelli, R.O. Pepin, in *Origin of the Earth and Moon*, ed. by R.M. Canup, K. Richter (Univ. Ariz. Press, Tucson, 2000), p. 435

U. Pöschl, Angewandte Chemie **44**, 7520 (2005)

I. Ribas, E.F. Guinan, M. Güdel, M. Audard, Astrophys. J. **622**, 680 (2005)

F. Robert, Science **293**, 1056 (2001)

D. Robertson, J. Robinson, J. Theor. Biol. **195**, 129 (1998)

M.S. Robinson, K.L. Tanaka, Geology **18**, 902 (1990)

J.J.W. Rogers, J. Geol. **104**, 91 (1996)

D.H. Rothman, J.M. Hays, R.E. Summons, Proc. Nat. Acad. Sci. USA **100**, 8124 (2003)

L.J. Rothschild, R.L. Mancinelli, Nature **409**, 1092 (2001)

W.W. Rubey, GSA Bull. **62**, 1111 (1951)

P.S. Russell, J.W. Head, J. Geophys. Res. **108**, 5064 (2003)

R. Rye, P.H. Kuo, H.D. Holland, Nature **378**, 603 (1995)

V.S. Safronov, *Evolution of the Protoplanetary Cloud and Formation of the Earth and Planets* (Nauka, Moscow, 1969)

C. Sagan, G. Mullen, Science **177**, 52 (1972)

C. Sagan, W.R. Thompson, R. Carlson, D. Gurnett, C. Hord, Nature **365**, 715 (1993)

J.W. Schopf, *Earth' s Earliest Biosphere* (Princeton University Press, Princeton, 1983)

D.W. Schwartzman, T. Volk, Nature **340**, 457 (1989)

M. Schidlowski, Nature **333**, 313 (1988)

R.A. Schultz, H.V. Frey, J. Geophys. Res. **95**, 14,175 (1990)

D.H. Scott, K.L. Tanaka, Geologic map of the western equatorial region of Mars, U.S. Geol. Survey Misc. Inv. Map I1802-A, 1986

T.L. Segura, O.B. Toon, A. Colaprete, K. Zahnle, Science **298**, 1977 (2002)

R.P. Sharp, M.C. Malin, Geol. Soc. Am. Bull. **86**, 593 (1975)

U. Siegenthaler, T.F. Stocker, E. Monnin, D. Lüthi, J. Schwander, B. Stauffer et al., Science **310**, 1313 (2005)

N.H. Sleep, J. Geol. **87**, 671 (1979)

N.H. Sleep, K. Zahnle, J. Geophys. Res. Planets **106**, 1371 (2001)

S.W. Squyres, M.H. Carr, Science **231**, 249 (1986)

S.W. Squyres, J.F. Kasting, Science **265**, 744 (1994)

S.W. Squyres, A.H. Knoll, Earth Planet. Sci. Lett. **240**, 1 (2005)

S.W. Squyres, R.E. Arvidson, J.F. Bell III, J. Brückner, N.A. Cabrol, W. Calvin et al., Science **305**, 794 (2004)

S.W. Squyres, R.E. Arvidson, B.C. Blaney, D.L. Clark, L. Crumpler, W.H. Farrand et al., J. Geophys. Res. **111**, E02S17 (2006)

T.F. Stepinski, W.J. O'Hara, Lunar Planet. Sci. Conf. **34**, 1659 (2003)

F. Suarez, J. Furne, J. Springfield, M. Levitt, Am. J. Physiol. Gastrointest. Liver Physiol. **272**, 1028 (1997)

O.B. Toon, Rev. Geophys. **35**, 41 (1997)

D.L. Turcotte, Earth Planet. Sci. Lett. **48**, 53 (1980)

J.C.G. Walker, Orig. Life **16**, 117 (1985)

J.C.G. Walker, Palaeogeogr. Palaeoclimatol. Palaeoecol. **82**, 261 (1990)

J.C.G. Walker, P.B. Hays, J.F. Kasting, J. Geophys. Res. **86**, 9776 (1981)

A.J. Watson, J.E. Lovelock, Tellus **35B**, 286 (1983)

L.L. Watson, I.D. Hutcheon, S. Epstein, E.M. Stolper, Science **265**, 86 (1994)

R.H. Whittaker, Science **163**, 150 (1969)

R.M. Williams, R.J. Phillips, M.C. Malin, Geophys. Res. Lett. **27**, 1073 (2000)

B.F. Windley, *The Evolving Continents* (Wiley, New York, 1984)

C.R. Woese, G.E. Fox, Proc. Nat. Acad. Sci. USA **74**, 5088 (1981)

M.B. Wyatt, H.Y. McSween, Nature **421**, 712 (2003)

K.J. Zahnle, J.C.G. Walker, Rev. Geophys. Space Phys. **20**, 280 (1982)

J.P. Zehr, B.D. Jenkins, S.M. Short, G.F. Steward, Environ. Microbiol. **5**, 539 (2004)

Y. Zhang, A. Zindler, Earth Planet. Sci. Lett. **117**, 331 (1993)

M.Y. Zolotov, E.L. Shock, Meteorit. Planet. Sci. **39**, A119 (2004)

M.T. Zuber, D.E. Smith, S.C. Solomon, J.B. Abshire, R.S. Afzal, O. Aharonson et al., Science **282**, 2053 (1998)

Space Sci Rev (2007) 129: 167–203
DOI 10.1007/s11214-007-9149-7

Water, Life, and Planetary Geodynamical Evolution

**P. van Thienen · K. Benzerara · D. Breuer ·
C. Gillmann · S. Labrosse · P. Lognonné · T. Spohn**

Received: 30 October 2006 / Accepted: 18 January 2007 /
Published online: 5 May 2007
© Springer Science+Business Media, Inc. 2007

Abstract In our search for life on other planets over the past decades, we have come to
understand that the solid terrestrial planets provide much more than merely a substrate on
which life may develop. Large-scale exchange of heat and volatile species between planetary
interiors and hydrospheres/atmospheres, as well as the presence of a magnetic field, are

P. van Thienen (✉)
Faculty of Geosciences, Utrecht University, P.O. Box 80.021, 3508 TA Utrecht, The Netherlands
e-mail: thienen@geo.uu.nl

K. Benzerara
Institut de Minéralogie et de Physique des Milieux Condensés, UMR 7590 and Institut de Physique
du Globe de Paris, 140 Rue de Lourmel, 75015 Paris, France
e-mail: karim.benzerara@impmc.jussieu.fr

D. Breuer
German Aerospace Center (DLR), Rutherfordstrasse 2, 12489 Berlin, Germany
e-mail: Doris.Breuer@dlr.de

C. Gillmann · P. Lognonné
Équipe Études spatiales et planétologie, Institut de Physique du Globe de Paris, 4 Avenue de
Neptune, 94107 Saint-Maur-des-Fossés, France

C. Gillmann
e-mail: gillmann@ipgp.jussieu.fr

P. Lognonné
e-mail: lognonne@ipgp.jussieu.fr

S. Labrosse
Laboratoire des sciences de la Terre, École Normale Supérieure de Lyon, 46 Allée d'Italie, 69364
Lyon Cedex 07, France
e-mail: stephane.labrosse@ens-lyon.fr

T. Spohn
German Aerospace Center (DLR), Rutherfordstrasse 2, 12489 Berlin, Germany
e-mail: Tilman.Spohn@dlr.de

important factors contributing to the habitability of a planet. This chapter reviews these processes, their mutual interactions, and the role life plays in regulating or modulating them.

Keywords Mantle dynamics · Habitability · Magnetic field · Volatiles · Thermal evolution

1 Introduction

Although in our current knowledge life appears to be limited to a relatively narrow and superficial zone extending no more than a few kilometers from Earth's surface in either direction, the vast volume of mostly solid rock and solid or liquid iron alloy which is inside this envelope provides several functions that are important if not crucial for planetary habitability. It appears logical that these functions may also contribute to the habitability of other planets (or their absence to hostility towards life). Life has not been detected on any other astronomical body, but there have been inferences and speculations for Mars (e.g. Formisano et al. 2004) and Europa (see Chyba and Phillips 2001). Titan has also attracted the attention of exobiologists (Raulin and Owen 2002).

This chapter aims to provide an overview of the many processes and interactions of the interior of terrestrial planets which may help sustain life. Obviously this is a field of study which is only starting to be uncovered, and on many subjects our knowledge and understanding is far from complete. Therefore we also try to discuss some important open questions which need to be answered.

Figure 1 gives a general overview of the direct and indirect interactions of planetary core, mantle, crust, hydrosphere/cryosphere and atmosphere processes with the biosphere for planets with plate tectonics (Earth, and possibly early Mars and early Venus) and for one-plate planets (Mars, Venus). The transfer of heat from the core through the mantle and out into the atmosphere and space is shared by both types (though in a one-plate planet, the core and the mantle are not necessarily cooling down), as are degassing and heat transfer by volcanism. Both show erosion of the atmosphere by the solar wind, but the presence of a magnetic field may mitigate this in the case of plate tectonics. However, only the plate tectonics scenario includes a means of reintroducing material back into the planet's interior through subduction, generating a cycle between interior and exterior reservoirs, though a mechanism has been proposed for one-plate planets as well (Elkins-Tanton 2007a).

It is immediately clear that several interactions consist of the transfer of volatile species like water and CO_2 between different reservoirs. This is very important since both species are the essential materials of life (nitrogen, phosphorus and sulfur are other important basic raw materials), and also main constituents of the hydrosphere/cryosphere and atmosphere.

Less obvious, but not less important, is the effect that volatiles, and specifically water, may have on the dynamics of the planetary interior. The presence of water is thought to play a crucial role in the generation of plate tectonics through weakening the lithosphere (Regenauer-Lieb et al. 2001). The presence or absence of plate tectonics significantly affects the thermal evolution of a planet, which in turn determines whether a magnetic field can be generated to protect the atmosphere against the solar wind (see Chap. 6). In addition to this, volatile species play a significant role in determining the surface temperature of a planet, thus affecting the temperature difference between interior and exterior which drives the dynamics of the mantle.

In this chapter, these interactions are reviewed in more detail, combining two approaches to habitability. On the one hand, we stay on firm ground by looking at Earth and what makes it habitable. On the other hand, we look at other planets, specifically Mars and Venus, and

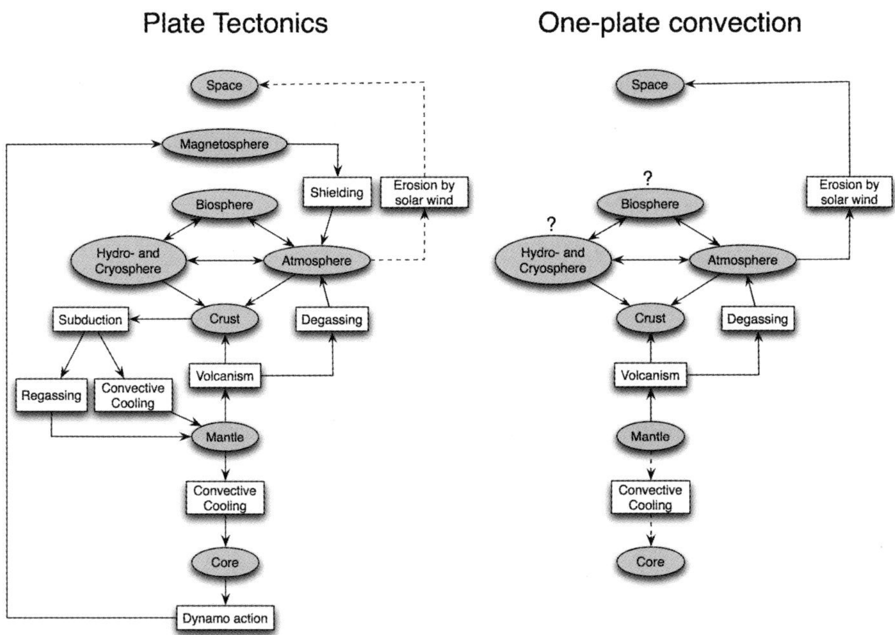

Fig. 1 Sketch comparing the possible relationships between interior, surface and atmosphere of a planet with plate tectonics on the one hand, and a planet with convection underneath a single plate on the other hand. *Dashed arrows* denote low efficiency mechanisms. By promoting very complex and global geological cycles, plate tectonics tends to generate conditions favorable for life. In the case of a one-plate planet (no subduction) the geologic interactions between the different subsets are reduced and more localized. Such a planet may not be able to sustain its internal magnetic field and recycle volatiles (extracted from the interior by magmatism) into the mantle; finally it evolves toward a planet with either an atmosphere that is too thin (Mars) or too thick (Venus), being quite probably uninhabitable at the present time

consider what makes these planets less suitable for life. We start by discussing the inventories and fluxes of volatiles in terrestrial planets (focusing on Earth, which is the only planet for which we have sufficient information) in Sect. 2. In Sect. 3, we discuss several processes involving these volatile species which are important for maintaining planetary habitability. The dynamical evolution of Earth, Mars and Venus in the context of these processes is presented in Sect. 4, and finally a synthesis discussing the implications for planetary habitability and open questions is presented in Sect. 5.

2 Origins, Inventories and Cycles

In the light of the central position volatiles appear to play in the direct and indirect interaction between planetary interiors and biospheres, the following sections discuss the present-day inventories and fluxes of volatiles in Earth's interior and exterior and the Martian and Venusian atmospheres, and the origin and survival of these inventories. The focus will be on H_2O and CO_2, as these appear to be the most important and best studied. As N and S are crucial components of amino-acids, these elements will be treated as well. We finish by discussing the oxidation state of the mantle and its implications for habitability.

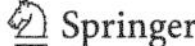

2.1 Earth's Volatile Inventories

Even for planet Earth, which is the best studied of the terrestrial planets, there remains considerable uncertainty about distribution of inventories between different reservoirs, the core remaining the most elusive. Table 1 shows a compilation of recent reservoir size estimates for Earth's core, mantle, and hydrosphere/atmosphere.

Most mantle minerals, including nominally anhydrous minerals, may carry several hundred ppm's of water. Xenolith samples have shown the following ranges of water contents for upper mantle minerals (Withers et al. 1998; Ingrin and Skogby 2000; Koga et al. 2003; Mosenfelder et al. 2006): 100–1300 ppm in clinopyroxene, 60–650 ppm in orthopyroxene, up to several thousand ppm in olivine, and up to 1000 ppm in garnet. In the transition zone, $(Mg,Fe)_2SiO_4$—γ-spinel can contain up to 2.7 wt.% of H_2O (Inoue et al. 1998; Williams and Hemley 2001). This may make the transition zone an important reservoir for the Earth's interior water (Bercovici and Karato 2003). Mg-perovskite may contain up to 700 atoms of H per 10^6 atoms of Si at 27 GPa and 1830°C, but increasing the pressure and adding Al may increase this solubility (Williams and Hemley 2001). Murakami et al. (2002) report that magnesiowüstite and $MgSiO_4$-rich perovskite may contain 0.2 wt.% water, and $CaSiO_4$-rich perovskite may contain 0.4 wt.%. These numbers suggest that the lower mantle may contain up to 5 ocean masses. The Earth's core may be the most important of the hydrogen reservoirs (Williams and Hemley 2001; Hirao et al. 2004), as shown in Table 1, storing hydrogen in the form of iron hydrides. High pressure experiments of Saxena et al. (2004) confirm the stability of iron hydride at core pressures.

Most of the Earth's carbon is in its interior. However, the solubility of CO_2 in olivine, pyroxenes, garnet and spinel is very low, about 0.1–1 ppm by weight. Keppler et al. (2003) conclude from this that there must be a separate carbon-bearing phase in the deep upper mantle, possibly carbonates. The carbon content of the atmosphere and oceans is orders of magnitude smaller. The Earth's core may contain 2–4 wt.% C (Wood 1993). Core compo-

Table 1 Inventories of important volatile species in the Earth's interior and exterior, and the martian and venusian atmospheres

Element/ compound	Reservoir size (g)			Mars atmosphere[o]	Venus atmosphere[p]
	Earth				
	Core	Mantle/crust	Hydrosphere/atmosphere		
H_2O	$1 \cdot 10^{25} -$ $1.4 \cdot 10^{26}$ [a,b,n]	$3.6 \cdot 10^{24} -$ $8.3 \cdot 10^{24}$ [a,n]	$1.4 \cdot 10^{24}$ [m]	$2.3 \cdot 10^{15}$	$4.8 \cdot 10^{18}$
C	$4 \cdot 10^{24} -$ $8 \cdot 10^{25}$ [e,n]	$3 \cdot 10^{22} -$ $4.9 \cdot 10^{23}$ [c,d,n]	$3.6 \cdot 10^{19}$ [l]	$6.5 \cdot 10^{18}$	$1.3 \cdot 10^{23}$
N	$1.4 \cdot 10^{23}$ [n]	$3.8 \cdot 10^{20} -$ $8.1 \cdot 10^{21}$ [d,h,n]	$4 \cdot 10^{21}$ [f]	$4.4 \cdot 10^{17}$	$1.1 \cdot 10^{22}$
S	$1.9 \cdot 10^{25} -$ $3.3 \cdot 10^{25}$ [g,i,n]	$1.0 \cdot 10^{24}$ [h,n]	$1.3 \cdot 10^{21}$ [k]		$5.3 \cdot 10^{19}$

References: a) Williams and Hemley (2001), b) Hirao et al. (2004), c) Coltice et al. (2004), d) Zhang and Zindler (1993), e) Wood (1993), f) Galloway (2001), g) Allègre et al. (1995), h) McDonough and Sun (1995), i) Dreibus and Palme (1996), j) Jahnke (1992), k) Charlson et al. (1992), l) Holmen (1992), m) Murray (1992), n) McDonough (2003), o) Williams (2004), p) Williams (2005)

 Springer

sitions of 0.3 wt.% carbon or more, in the presence of another light element like S, results in the formation of Fe_3C as the first crystallizing phase, which is more consistent with the inner core density than a pure Fe or FeNi composition.

It is unclear whether the core reservoir has significant interaction with the mantle, and thus ultimately with the biosphere. Tungsten isotope data have been applied to infer the presence or absence of core material in the source material of oceanic island basalts, but the results are equivocal and remain controversial (Scherstén et al. 2004; Jellinek and Manga 2004; Humayun et al. 2004), with both geochemical and geophysical evidence arguing against it (Lassiter 2006).

2.2 Volatile Inventories of Mars and Venus

Obviously, the volatile inventories of Mars and Venus are much more difficult to estimate than those for Earth. Most information is available on the atmospheres, which is listed in Table 1. Note that on Mars, significant exchange (20% of atmosphere mass) of volatiles between the atmosphere and the surface, the seasonal ice caps, and the polar layered deposits takes place on a seasonal basis (Mischna et al. 2003; Aharonson et al. 2004). The north polar cap primarily consists of water, and has an estimated permanent water inventory of $1.4 \cdot 10^{21}$ g (Zuber et al. 1998). The south polar cap may contain, when assuming it to consist of water only, about $2.5 \cdot 10^{21}$ g (Smith et al. 1999). In addition to this, there appear to be significant volumes of ice in the Martian subsurface (Mitrofanov et al. 2002; Feldman et al. 2002; Boynton et al. 2002).

Dreibus and Wänke (1987) estimated the water content of the Martian mantle to be 36 ppm, i.e. a relatively dry mantle. This corresponds to a mantle inventory on the order of 10^{22} g, or two orders of magnitude smaller than in Earth's mantle (see Table 1). Note however that the uncertainty in this number is significant (see Lodders and Fegley 1997), and a homogeneous distribution in the Martian mantle is not to be expected. To illustrate this point, we mention that Dann et al. (2001) and McSween et al. (2001) infer pre-eruptive water contents of Shergottite magmas of 1.8%, though this is disputed by Jones (2004). The Venusian mantle is generally assumed to be dry (see Nimmo and McKenzie 1998) like its atmosphere (Grinspoon 1993).

2.3 Origin and Survival of the Volatile Inventories

Oceanic island basalts samples around the globe show a large range of $^3He/^4He$ ratios. Whereas the latter of the two helium isotopes is the product of α-decay of several radioactive elements, the former is not. Since noble gases do not easily bind to anything and, as a consequence, there is nearly no subduction of noble gases (Staudacher and Allègre 1988), the presence of ^3He in the mantle suggests that ^3He was added to our growing planet during accretion, and that some of this primordial reservoir is still present. Similarly, the uniformity of the isotope composition of C in different types of basalts but also diamonds and carbonitites has been interpreted as evidence for a primitive C source in the mantle (Tingle 1998).

A relatively small amount of water in the original building blocks of the terrestrial planets, which could later have progressively degassed, would suffice to explain the present-day water budget of planet Earth. But contrary to carbon, there has been significant escape of hydrogen to space (see below). This may demand additional post-accretional sources. Possible sources are (see Gaidos et al. 2005 and references therein): (1) material from e.g. asteroids;

(2) comets; (3) ice particles from the outer region of the primordial solar system; (4) 'wet' planetary embryos.

During core segregation, significant amounts of hydrogen may have been transported into the core resulting from reactions between liquid metallic iron and water in the mantle (Ohtani et al. 2005):

$$3Fe + H_2O \rightarrow 2FeH_x + FeO. \tag{1}$$

Ohtani et al. observed that the reaction can take place between 6 and 84 GPa, thus being able to scavenge water from a large part of the primordial mantle. Okuchi (1997) suggests that more than 95 mole% of the water present in a primordial terrestrial magma ocean may have reacted for form FeH_x and moved into the core.

SNC meteorite studies by Dreibus and Wänke (1987) suggest that the Martian mantle is very poor in water, containing about 36 ppm. This appears to be in conflict with the general trend of increasing water availability which is observed when moving away from the sun in the solar system. However, other volatile elements like Na and K are more abundant in Mars than in Earth (references in Kuramoto 1997). It is likely that a lot of water was lost from Mars by the oxidation of metallic iron in the Martian mantle:

$$H_2O + Fe \rightarrow FeO + H_2. \tag{2}$$

The hydrogen which is released in this way can either dissolve into the core (Zharkov and Gudkova 2000), or be segregated into the atmosphere, from where it escapes into space (Hunten et al. 1987). However, an equilibrium may already have been reached during the accretion process in the planetesimals, possibly eliminating further loss after accretion.

Kuramoto (1997) proposed an alternative reducing agent: reduced C species. He argues that, whereas the Earth's core may contain significant amounts of carbon, the high sulfur content of the Martian core may have prevented the uptake of large amounts of C into the core, leaving it in the mantle. The idea of the core as a C reservoir is supported by the fact that graphite and carbites are common in iron meteorites, the geochemistry of these minerals suggests a planetesimal core origin of these meteorites, and the carbon isotope composition is similar to that of Earth's mantle (Tingle 1998, and references therein). Furthermore, Wood (1993) has shown that C is quite soluble in iron liquids at core pressure, and that dissolution of C in iron liquids during planetesimal differentiation may be expected.

2.4 Earth Volatile Cycles

Table 2 summarizes the fluxes of the species listed in Table 1 for Earth.

2.4.1 Volatile Fluxes

The subduction system as a whole is an important agent in the recycling of volatile species into the mantle. However, significant fractions of these species may not enter the deep mantle but may be reinjected into the atmosphere by arc volcanism.

Water is carried into the mantle in the form of sediment and igneous rock pore and structural water (Jarrard 2002). The most important water carrying minerals in subducting crust are probably lawsonite ($CaAl_2Si_2O_7(OH)_2 \cdot H_2O$, 11 wt.% water) and serpentine ($Mg_3Si_2O_5(OH)_4$, up to 13 wt.% water) (Williams and Hemley 2001). Serpentine is stable up to halfway down the upper mantle, and lawsonite down to the deep upper mantle. Water-carrying phases of lesser or unknown importance are zoisite ($CaAl_3Si_3O_{12}(OH)$, significantly lower water content), magnesian pumpellyite ($Mg_5Al_5Si_6O_{21}(OH)_7$), K-amphibole,

alkali-rich hydrous phases, humite ($nMg_2SiO_4 \cdot Mg(OH,F)_2$), and Dense Hydrous Magnesium Silicate, or the so-called alphabet phases (Williams and Hemley 2001). Some of these minerals are stable down to great pressures (e.g. DHMS phase B 12–24 GPa, DHMS phase D down to 50 GPa) and may play a role in transporting subducted hydrogen down into the deep mantle. Apart from these mafic minerals, sediments also appear to play a role in transporting volatiles into the mantle. Plank and Langmuir (1998) calculated the composition of the average subducted sediment, and from this estimated global subduction fluxes of sediments and their components. Their estimate for the flux of H_2O carried by sediments during subduction is $9.49 \cdot 10^{13}$ g/yr, which is about 1/9 of the total H_2O subduction flux.

The fraction of the amount of subducted water which is released through arc volcanism is sometimes called the recycling ratio R_{rec} (Franck and Bounama 2001). Its complementary value, i.e. the fraction of water which is actually subducted into the deep mantle is usually expressed as the regassing ration R_{H_2O}. The geophysical modeling of Franck and Bounama (2001) gives a preferred value for R_{H_2O} of 0.3 to 0.4. Numerical mantle convection models confirm that parts of the subducting slab remain sufficiently cool to prevent complete dehydration (Van Keken et al. 2002). Rüpke et al. (2004) found that specifically serpentinized lithospheric mantle may be an important agent in transporting up to 40% of it's initial water content into the mantle, corresponding to $R = 0.6$. In contrast, geochemical based estimates for R_{H_2O} are much higher, around 0.9 (e.g. Ito et al. 1983), though the value of Peacock (1990) of about 0.16 is at the other end of the spectrum.

A common way to estimate the flux of CO_2 at mid-ocean ridges is by determining volatile contents of glasses sampled at mid-ocean ridges and calibrating C to ^{3}He (Marty and Tolstikhin 1998). Alternatively, pre-eruptive volatile contents of normal mid-ocean ridge (MOR) basalts can be determined (Saal et al. 2002), which results in similar flux estimates. Aubaud et al. (2004) looked at kinetic desequilibrium during degassing of MOR magmas, and reconstructed the original carbon content of the magma. They extrapolated this value to the global ridge system and estimated a carbon flux of $1.6^{+0.6}_{-0.3} \cdot 10^{14}$ g/yr of carbon, which is equivalent to $5.9^{+2}_{-0.7} \cdot 10^{14}$ g of CO_2 per year, i.e. much higher than previous estimates.

Carbon is subducted mostly in the form of carbonates added within the basaltic crust through hydrothermal circulation (Jarrard 2002) and carbonate sediments deposited at the top of the oceanic crustal column (Plank and Langmuir 1998; Kerrick and Connolly 2001). Plank and Langmuir (1998) report carbonate sequence thicknesses of up to several hundred meters at subduction zones in the subducting crust. Because of a limited number of direct measurements of arc volcanic CO_2 fluxes and uncertainty about representivity of present-day fluxes over geological history, it is better to estimate carbon fluxes based on magma production rates (Marty and Tolstikhin 1998). This requires knowledge of relative amounts of slab-derived and mantle-derived carbon. Based on $CO_2/^3$He ratios and carbon isotopes in arc gases, Varekamp et al. (1992) concluded that more than 80% of arc volcanic CO_2 originates from subducted carbon species. Kerrick and Connolly (2001) studied metamorphic devolatilization of sediments and found that release of CO_2 requires specific conditions of pressure and temperature and also the presence of H_2O rich fluids, which explains the large amount of carbon that is not released.

Nitrogen is subducted in the form of ammonium and nitrates in the sediment cover of the oceanic crust. Ammonium is produced by life, and although nitrate can be produced inorganically by lightning, this requires the presence of oxygen, which in turn is produced by photosynthetic life (Boyd 2001). Much of this sedimentary nitrogen is returned into the atmosphere through arc volcanism. Thus, about 85% of arc volcanic nitrogen is of sedimentary origin (Sano et al. 2001a). The total net (i.e. atmospheric nitrogen subtracted) nitrogen

subduction zone flux is $6.0 \cdot 10^8$ mol/yr, including both arc volcanism and backarc spreading centers (Sano et al. 2001b).

Sulphur is fixated in the oceanic crust and its sediment cover in the form of metal sulfides like pyrite. Thus, these are subducted into the mantle with the downgoing slab. Additionally, sulfates in pore water in the subducting slab and occasional evaporitic sulfate deposits may be subducted (Canfield 2004).

2.4.2 Non-steady State Contributions

Although Table 2 shows the present-day fluxes which can be considered to represent most of the past 180 My, during which the plate creation and subduction rates were probably relatively constant (Rowley 2002), there are punctuated events which form a large additional source of volatiles: the eruption of flood basalts. For example, the eruption of the Deccan flood basalts, around 65 Myr ago, resulted in the eruption of about $6 \cdot 10^{18}$ g carbon into the atmosphere over a period of 1.36 Myr (Wignall 2001), corresponding to an additional flux of $4 \cdot 10^{12}$ g/yr, which is similar to the lower end of the estimates for the present-day mid-ocean ridge flux. In addition to this, and more importantly, about $6 \cdot 10^{18}$ g of S was also injected into the atmosphere, also corresponding to an annual flux of $4 \cdot 10^{12}$ g/yr (Wignall 2001), which is significantly more than the present-day flux of arcs, ridges and hotspots combined. The magmas associated with a flood basalt event are not only a direct source but also an indirect source of carbon, since interaction of voluminous melts with carbon bearing sediments in the crust may cause additional release of carbon into the atmosphere (e.g. Svensen et al. 2004). On the early Earth, flood basalt eruptions were both larger and more frequent (Abbott and Isley 2002). Therefore, they were probably a more important factor in the atmosphere/hydrosphere volatile balance.

2.4.3 Flux Balances

Rüpke et al. (2004) investigated the evolution of subduction zone water transport in the course of Earth's history. Assuming a high spreading rate and young average age of the oceanic crust for the early Earth, and the inverse for the present-day Earth, they use a simple model to predict a high rate of degassing and low rate of reinjection of water in the early Earth. A flipping of the balance between degassing and regassing would take place somewhere between 2.5 Ga and 900 Ma, depending on assumptions of the amount of serpentinization of oceanic lithospheric mantle. Since then, the total amount of hydrosphere and atmosphere water has been slowly declining at the expense of mantle water. The models of Rüpke et al. (2004) predict a sea level drop of several hundred meters over the last 600 Myr.

Edmond and Huh (2003) argue that at the present day there is an imbalance between deposition of CO_2 in the form of $CaCO_3$ at ocean bottoms and subduction of this material, as the Atlantic and Indian oceans form significant regions of $CaCO_3$ deposition but have no significant subduction of this material. This illustrates the possibility of changes in the carbon cycle on time scales of the Wilson cycle (periodic opening and closing of ocean basins), and a weak direct coupling between rates of CO_2 consumption by weathering and recycling of fixated carbon through subduction zone degassing (Edmond and Huh 2003).

For carbon, nitrogen, and sulfur, Table 2 shows a net flux out of the mantle. The sulfur flux has probably been positive for the past 700 Myr, due to the oxic nature of the oceans. A significantly larger subduction flux of sulfur has been inferred for the Archean and Proterozoic, when the ocean water was anoxic, allowing massive deposition and subduction of pyrite (Canfield 2004).

 Springer

Table 2 Fluxes of important volatile species between the Earth's interior and exterior

Element/compound	Fluxes (g/yr)				Balance
	Subduction ↓	Arc and backarc ↑	MOR ↑	Hotspots ↑	
H_2O	$8.7 \cdot 10^{14} -$ $1.83 \cdot 10^{15}$ [a,n]	$1.4 \cdot 10^{14}$ [a]	o	o	↓ [j,k]
C	$1.1 \cdot 10^{13} -$ $4.2 \cdot 10^{13}$ [b,d,e,n]	$3.0 \cdot 10^{13}$ [c]	$0.6 \cdot 10^{13} -$ $2.2 \cdot 10^{14}$ [c,g,h,i]	$3.5 \cdot 10^{11} -$ $4 \cdot 10^{13}$ [c,h]	↑
N	$9.0 \cdot 10^{9}$ [f]	$1.8 \cdot 10^{10}$ [f]	$1.1 \cdot 10^{10}$ [f]	$6.2 \cdot 10^{7}$ [f]	↑
S	$1.2 \cdot 10^{13} -$ $2.4 \cdot 10^{13}$ [m]	——————— $2.8 \cdot 10^{13}$ [l] ———————			↑

References: a) Peacock (1990), b) Varekamp et al. (1992), c) Marty and Tolstikhin (1998), d) Plank and Langmuir (1998), e) Kerrick and Connolly (2001), f) Sano et al. (2001b), g) Resing et al. (2004), h) Sano and Williams (1996), i) Aubaud et al. (2004), j) Rüpke et al. (2004), k) Wallmann (2001), l) Charlson et al. (1992), m) Canfield (2004), n) Jarrard (2002), o) no estimate found in the literature

2.5 Volatiles and the Oxidation State of Earth's Mantle and Surface

As all life is driven by redox gradients, and significant exchange of chemical species between planetary interiors and atmospheres may take place, it is important to consider the mantle oxidation state when talking about habitability.

The initial oxidation state of a planetary mantle is probably already determined during accretion and early magma ocean stages. One possibly important process is the oxidation of ferrous iron by hydrogen with subsequent escape of hydrogen into space or sequestration in the core (Hunten et al. 1987; Ohtani et al. 2005), though this is difficult to quantify, and fails to explain the relatively reduced state of Martian basalts where this planet should be more volatile-rich than Earth (Wade and Wood 2005). Wade and Wood (2005) proposed that a layer of magnesium silicate perovskite may act as an 'oxygen pump' during accretion. In the perovskite, $Mg^{2+}Si^{4+}$ may be substituted by $Fe^{3+}Al^{3+}$ during fractional crystallization at the bottom of a deep magma ocean. This substitution is independent of the redox state and a disproportionation may take place of the form $3Fe^{2+} = 2Fe^{3+} + Fe^{0}$ (Frost et al. 2004). As the metallic iron which is formed in this way may be transported into the core by sinking iron diapirs, and exchange of silicate material between the perovskite and the magma ocean (and later the overlying peridotite) may take place, this results in the effective oxidation of the mantle. As Mars is probably not big enough to have significant amounts of perovskite in its mantle (the pressure in the lowermost mantle barely reaching that of the phase transition), this may also explain the apparently relatively reduced state of its mantle (Wade and Wood 2005).

The geochemistry of old and recent rocks may provide evidence for the evolution of the redox state of the Earth's mantle. A suitable set of tracers is V and Sc, since these elements behave in a similar fashion during partial melting of the mantle, and only V is sensitive to redox conditions during a differentiation event (Li and Lee 2004). Recently, iron isotopes ($^{57}Fe/^{54}Fe$) were proposed as a proxy for the mantle oxidation state, since ^{57}Fe preferentially fractionates into bonds with Fe^{3+}, whereas ^{54}Fe prefers to associate itself with Fe^{2+} (Williams et al. 2004). The heterogeneity of this isotope ratio in the mantle is larger than can be explained by metasomatism and partial melting. Therefore, this suggests that

part of the variation is caused by a secular variation in the mantle oxidation state (Williams et al. 2005). However, other workers, using combinations of V, Cr and/or Sc as a proxy for redox state, find that the mantle oxygen fugacity has been relatively constant since the Archean (Canil 1997; Li and Lee 2004), possibly since ~4 Ga (Delano 2001). The upper mantle may have undergone a progressive oxidation due to an imbalance between the effects of crustal recycling and introduction of carbon from the deep mantle (Kadik 1997), but so far this appears to be ill constrained by the available data.

The advent of an oxygen-rich atmosphere, also called the Great Oxidation Event (GOE, around 2.3 Ga, Holland 2002), has obviously had an enormous influence on the evolution of life on Earth. The relative constancy of the mantle oxidation state before and after this event suggests that the great oxidation event was not related to changes in the redox state of the mantle (Li and Lee 2004). However, Holland (2002) states that the GOE may have been related to "a gradual increase in the oxidation state of volcanic gases and hydrothermal inputs". A coupled change in the the oxidation states of the mantle and atmosphere was originally proposed by Kasting et al. (1993). Holland (2002) determined that the required change in the oxidation state is sufficiently small to be indiscernible in the present record of Cr and V data. Kump et al. (2001) proposed the following scenario: During the pre-GOE era, reaction of water with ferric iron in basalts produces hydrogen, which may escape into space, and ferrous iron. The operation of this process and subsequent subduction of oxidized crust into the deep mantle over a long period of time results in the formation of an oxidized reservoir in the deep mantle. During two periods of intense volcanism in the late Archean and early Proterozoic, the oxidized reservoir provides source material for the volcanism, and produces more oxidized associated volcanic gasses. This change in the oxidation state of (reduced) volcanic gasses, which form an atmospheric oxygen sink, could have changed the balance with photosynthetic oxygen production (cyanobacteria existed long before the GOE, see e.g. Brocks et al. 1999), resulting in an oxidizing instead of a reducing atmosphere (Kump et al. 2001).

Although there are several uncertain or even speculative aspects to this scenario, it does illustrate the possible significance of endogenic processes for the oxidation state at a planet's surface and therefore the conditions under which life may form and evolve.

3 Mechanisms and Feedbacks

3.1 Geodynamical Regimes

As will be discussed below, several processes characteristic of one or more geodyamical regimes play significant roles in creating and maintaining the habitability of a planet. Before discussing these, we first present an overview of geodynamical regimes.

3.1.1 Overview

Four distinct geodynamical regimes can be distinguished, three of which are depicted explicitly in Fig. 2, and the fourth, an episodic switching between two of the former regimes, implicitly. The most obvious is plate tectonics, which is active on present-day Earth. From an instantaneous point of view, characteristics of plate tectonics include spreading ridges where oceanic crust is created, and generally subparallel to these ridges, subduction zones where oceanic lithosphere is returned into the mantle. On longer time scales, the opening, growth, shrinking and closure of oceanic basins (Wilson cycles) have been inferred (Wilson 1966;

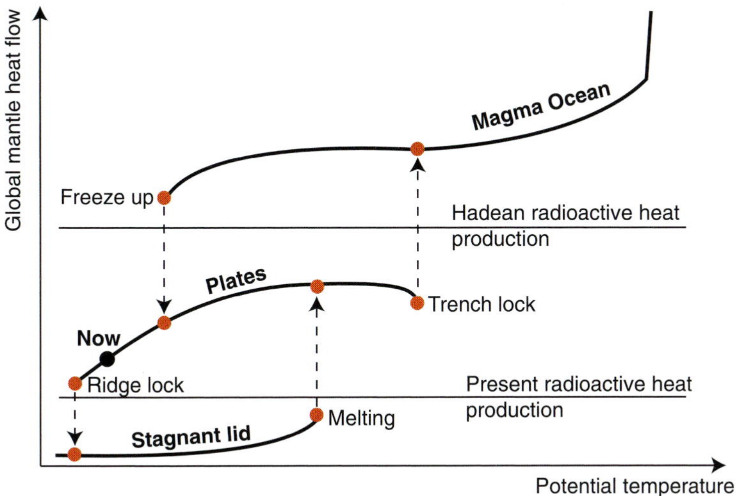

Fig. 2 Different branches of the relation between heat flow and temperature corresponding to different regimes in which the Earth and any other terrestrial planet could have operated. Adapted from Sleep (2000)

Turcotte and Schubert 2002). The main driving force of plate tectonics is *slab pull*, which is exerted by the subducting slab. In general, this pull is transmitted through the slab in its shallower parts, whereas deeper parts tend to exert their contribution through viscous drag of the mantle (Conrad and Lithgow-Bertelloni 2002). Mid-ocean ridge topography exerts a *ridge push*, which also contributes to the driving force (Turcotte and Schubert 2002). Additionally, *trench suction* associated with arc rollback (e.g. Wilson 1993) may play a significant role. The main opposing forces are lithosphere buoyancy (Oxburgh and Parmentier 1977; Sleep and Windley 1982; Vlaar and Van den Berg 1991), depending on the age of the slab and mantle temperature, and bending forces at the subduction zone (Conrad and Hager 2001).

The stagnant lid regime is well known from numerical and laboratory studies of fluids with a strongly temperature dependent viscosity (e.g. Davaille and Jaupart 1993, 1994; Reese et al. 1998), which results in a thick *lid* being formed on top of a convecting fluid layer. The efficiency of heat transport of this mechanism is significantly smaller than that of plate tectonics. Mars is thought to be in the stagnant lid regime (Spohn et al. 2001), showing a very old surface (Zuber 2001) with no evidence of subduction (though the hypothesis of an early phase of plate tectonics has been proposed, Sleep 1994; Connerney et al. 1999, 2005).

An episodic regime, with alternating phases of lid stagnation and lid mobilization, has been observed in numerical convection models (e.g. Stein et al. 2004; Van Thienen et al. 2004b). It has been suggested that Venus may presently be in this mode of convection (Parmentier and Hess 1992; Turcotte 1995), since crater counting studies indicate a relatively uniform surface age between about 300 Ma and 1 Ga (Schaber et al. 1992; McKinnon et al. 1997), which suggests a period of large scale resurfacing followed by a period of quiescence.

During the earliest phases of planetary evolution, sufficient energy was probably available to melt entire sections of planetary mantles, resulting in the magma ocean regime. The main sources of energy were the release of gravitational energy during accretion and core segregation (Horedt 1980), and the decay of short-lived and longer-lived radioactive

isotopes. Some large impacts during the final stages of accretion were also sufficiently energetic to produce a magma ocean. Abe (1997) computed that the survival time of magma oceans generated in this way are on the order of 100–200 Myr for Earth.

Systematic studies of the conditions under which stagnant lid convection, active lid convection, and the intermediate episodic regime take place have been done by several workers (Solomatov 1995; Moresi 1998; Kameyama and Ogawa 2000; Stein et al. 2004). In general, the balance between the viscous and/or brittle strength of the lithosphere and convective stresses in the mantle determines the style of convection. In these numerical studies, a stagnant lid regime is found for situations with a high lithospheric yield stress and relatively low vigor of convection (Rayleigh number), and for high viscosity contrasts between the lid and the interior. A mobile lid regime is observed for a relatively low viscosity contrast, high Rayleigh number, and relatively small yield stress. Parameters for the transitional episodic regime are between these extremes.

3.1.2 Role of Volatiles in the Operation of Geodynamic Regimes

Volatiles, principally water, affect the operation of any regime primarily due to their weakening effect on the rheology of mantle rocks (Chopra and Paterson 1981; Karato 1986), lowering the viscosity by up to two orders of magnitude (Hirth and Kohlstedt 1996). In addition to this, they appear to be particularly important for the operation of plate tectonics. The absence of plate tectonics on Venus has been ascribed to the dryness of this planet relative to Earth (e.g. Nimmo and McKenzie 1998). One important effect of water is the reduction of the strength of the lithosphere. Regenauer-Lieb et al. (2001) produced numerical models of the initiation of subduction by sediment loading. Using a complex, laboratory experiment based rheology which depends on the water content, they found that the presence of water promotes the localization of stresses and the formation of a shear zone through the entire lithosphere. It is clear that the breaking of the lithosphere is an important first step for starting subduction.

Water also reduces the solidus and liquidus temperatures of mantle peridotite by up to several hundred Kelvin (e.g. Mysen and Boettcher 1975; Katz et al. 2003). As a result, a wet mantle undergoes more chemical differentiation through partial melting than a dry mantle. The generally basaltic melt products may find their way to the (near) surface to add material to the crust. The depleted residue which remains in the mantle has a lower density than the starting material (Jordan 1979), and may act to stabilize the mantle (Parmentier and Hess 1992; De Smet et al. 2000; Zaranek and Parmentier 2004). On the other hand, basaltic crustal material may, when reaching a sufficiently high pressure, transform into dense eclogite (Hacker et al. 2003), which can destabilize the crust and lithosphere (Dupeyrat and Sotin 1995; Van Thienen et al. 2004b). The balance between these two effects remains to be investigated in detail.

Combined, the effects of mantle weakening and increased melt production by volatiles result in an increased capacity for cooling of the mantle in the plate tectonics scenario. The former effect causes faster advection of heat in the mantle. The latter results in more advection of heat by melts to the (near) surface. For the stagnant lid scenario, the net effect is less clear, as increased heat transport may be mitigated by a decreased advective and conductive mantle heat flow due to a stabilizing, depleted buoyant layer in the upper part of the mantle (Schott et al. 2002; Zaranek and Parmentier 2004).

3.1.3 Regime Transitions

A relatively simple way to consider regime transitions is by considering the energetics of convection. Stevenson (2003b) wrote this energy balance down as follows for an active lid

(plate tectonics) scenario:

$$\text{gravitational energy release} = \text{viscous dissipation of the internal flow}$$
$$+ \text{work done deforming the plates.} \qquad (3)$$

The second term of the right hand side takes place primarily at subduction zones (Conrad and Hager 2001). This term obviously depends on the mechanical properties of the lithosphere, which in turn depend on temperature and water content. From energetic considerations, one may predict a negative correlation between the magnitude of this term and the number of plates, with a single plate planet as an end member case (Stevenson 2003b).

Sleep (2000) studied the transitions between the three major modes of mantle convection, magma ocean, plate tectonics and stagnant lid, in the context of secular cooling of terrestrial planets, primarily considering the effects and degree of melting (see Fig. 2). Note that in this context, the term magma ocean refers to a partially molten region (mush) within the planetary interior, in contrast to an accretion-related initial magma ocean which extends to the surface. The following presents an overview of these melting-related and additional processes which may cause or be part of a regime transition.

When the mantle temperature falls below the level which allows partial melting to take place underneath the mid-ocean ridges, ridge lock occurs. In the absence of axial magma chambers and associated systems the ridges now become mechanically strong and no longer easily support spreading (Sleep 2000). In this case, plate tectonics stops and a stagnant lid begins.

A transition from plate tectonics to a magma ocean takes place when the amount of heat removed from the mantle by plate tectonics is less than the amount of heat produced in the mantle. The mantle heats up and at some point, the crust which is formed at MORs does not completely solidify to its base anymore (Sleep 2000), resulting in a trench lock as well. Note that thin lids on top of a magma ocean tend to founder, causing the magma ocean to extend to the surface. A sufficiently high crystal fraction in the mush may mitigate this (Abe 1997; Elkins-Tanton et al. 2005).

Alternatively, a locking at the trenches may also occur when oceanic lithosphere becomes insubductable due to its inherent low density. This results in a transition to the stagnant lid regime. Generally speaking, a thicker, less easily subductable lithosphere is created for higher mantle temperatures and/or a lower gravitational acceleration (Sleep and Windley 1982; Vlaar and Van den Berg 1991; Davies 1980). Results of 1D melting models by Van Thienen et al. (2004c) show that the upper limits of potential mantle temperature allowing plate tectonics are about 1500°C for Earth, 1450°C for Venus, and 1300–1400° for Mars, based on lithosphere buoyancy considerations at the subduction zone (which is likely an oversimplification, Conrad and Lithgow-Bertelloni 2002; Labrosse and Jaupart 2007).

Mechanical instability of the lithosphere in a stagnant lid regime may cause the initiation of plate tectonics (Sleep 2000). Small-scale sublithospheric convection may also initiate subduction (Solomatov 2004), possibly starting the plate tectonics cycle. However, the initiation of subduction remains poorly understood (Stern 2004).

When the amount of heat generated in a planet's interior is greater than the amount which can be removed by conduction through a stagnant lid, the mantle may heat up to a point where large-scale partial melting starts to take place (Reese et al. 1998). The crust which is produced from this melt can either start subducting, thus initiating plate tectonics, or remain stable, thus retaining a low surface heat flow and continuing the heating up of the interior, and result in a magma ocean (Sleep 2000). In general, big planets produce thinner, more easily subductable crust than small planets (Van Thienen et al. 2004c), so the former would

tend to evolve to the plate tectonics regime whereas a magma ocean would be a more likely outcome in the latter case (Sleep 2000).

In the model of Sleep (2000), the magma ocean freezes by upward solidification of the mush up to the base of the lid. This causes a transition to plate tectonics if part of the lid can subduct or to a stagnant lid regime if not. Note, however, that differentiation during the solidification of a magma ocean results in an unstable density stratification which may overturn to produce a very stable density stratification (Elkins-Tanton et al. 2003, 2005), thus inhibiting the onset of convection in the mantle.

A mechanism for the cessation of a hypothetical early phase of plate tectonics on Mars (e.g. Sleep 1994) was proposed by Lenardic et al. (2004). They suggest that the thick southern hemisphere crust may have grown from a small original size to such an extent that it reduced the mantle heat flow to a point where the mantle stopped cooling down and started heating up again. As the mantle heated up, the viscosity dropped and convective stresses dropped below a critical value required to mobilize the lithosphere.

A geodynamical regime transition in the other direction would also be possible, when cooling of the planetary interior results in an increase in convective stresses. Lenardic et al. (2004) suggest on the basis of similar arguments that Earth may have had a stagnant lid phase before plate tectonics became active. Based on thermal calculations, one would expect Mars to have switched back to plate tectonics as well. Possible explanations why this has not happened include the high rigidity of the Martian lithosphere and chemical buoyancy of the thick crust stabilizing the lithosphere.

Many of the changes in geodynamical regime discussed above are related to melting, as discussed by Sleep (2000), and thus changes in the internal temperature of a planet. However, recent work by Loddoch et al. (2006) demonstrates that, at least for the stagnant lid case, this is not a necessity. Their 3D numerical convection models show incidental mobilization of a stagnant lid in thermally equilibrated models in an apparent statistical steady state.

3.2 Magnetic Field Generation, Requirements and Effects

The existence of a global magnetic field is important for the habitability on terrestrial planets as it can protect life from severe radiation. Without the magnetic field, life will have to retreat to places covered by enough rock to be shielded from damaging radiation effects, a circumstance that could limit biosphere diversification. In addition to the direct influence of the magnetic field on life, there exists an important cycle, which relates the dynamo generation to the stability of the atmosphere, the surface conditions and the interior dynamics. Before the description of these feedback mechanisms, we introduce the general concept of magnetic field generation and its requirements.

The global magnetic fields of terrestrial planets are generated in their iron-rich and liquid cores. Although the dynamo mechanism is imperfectly understood, it is generally accepted that core convection is needed, as well as a finite rotation rate of the body, though the slow rate of Venus already suffices (Stevenson 2003a). Thus, a necessary condition for a dynamo is the presence of convection, which can be either by thermal or compositional convection.

Thermal convection in the core, like thermal convection in the mantle, is driven by a sufficiently large super-adiabatic temperature difference between the core and the mantle. It occurs if the core heat flow supersedes that conducted along the core adiabat. The latter heat flow, therefore, serves as a criterion for the existence of thermally driven convection in the core. If the mantle removes heat from the core at a rate that exceeds the critical heat flow, then the core will convect. If the mantle removes heat at a rate less than the critical heat flow,

the core is thermally stably stratified; dynamo action by thermal convection would not be possible. To cool the core above the threshold is difficult to achieve and it is suggested that a purely thermal dynamo may exist only for a short time of a planets evolution. The most likely period is just after accretion, in particular if the core is superheated with respect to the mantle. The initially strong cooling rates decline rapidly, possibly within a few thousands to a few hundred million years, to values where the heat flow out of the core can be accommodated by conduction alone. How long thermal convection can be sustained depends on the efficiency of mantle cooling. The more efficient the mantle cooling, the longer the existence of a thermal dynamo (e.g. Breuer and Spohn 2006). The existence of a thermal dynamo can be further prolonged in case of radioactive heating in the core. It is usually assumed that the terrestrial cores do not contain a significant amount of any radioactive elements, but it has been suggested that a larger concentration of potassium could have been incorporated at high pressure in the core during its formation (e.g. Lee and Jeanloz 2003).

Compositional density differences are 100% efficient in driving convection, in contrast to temperature differences (described by the Carnot efficiency factor, which is between 0.06–0.11, e.g. Braginsky 1964; Stevenson et al. 1983; Braginsky and Roberts 1995; Lister and Buffett 1995). The mechanical work provided by convection equals the viscous and ohmic dissipation (Hewitt et al. 1975; Backus 1975; Braginsky and Roberts 1995; Lister and Buffett 1995; Labrosse 2003), the former of which can be neglected in the core (Braginsky and Roberts 1995), and the latter of which is the expression of the magnetic field. Compositional convection can occur due to the release of positively buoyant light material during the process of solid inner core freezing from a fluid core with non-eutectic composition (Braginsky 1964). The dominant light element is suggested to be sulphur but also silicon, oxygen or hydrogen are possible candidates. For Earth, oxygen may be the most likely (Alfè et al. 2002).

Compositional convection and the associated generation of a magnetic field in the core occur if the temperature in the fluid (outer) core ranges between the solidus and the liquidus of the core material. As a consequence of the formation of an inner, nearly pure iron-rich core, the outer fluid becomes enriched in the lighter elements resulting in a decrease of the melting temperature. The melting temperature of the eutectic composition is in general very low (e.g. Fei et al. 1997) and it is very difficult to totally freeze the core of a terrestrial planet. This circumstance suggests that once a chemical dynamo has started—which is presumably late in the evolution—its existence is likely to be longstanding and it is difficult to stop as long as the core is cooling. Whether an inner core can grow depends on the initial amount of light elements in the core and on the efficiency to cool the interior below the liquidus. The more light elements in the core the lower is the melting temperature and the more difficult to cool the core below that temperature.

For Mercury it is suggested that the sulphur content is only a few percent and core freezing starts with respectively high core temperatures. A different situation is given for Mars, where a sulphur content of about 14% is suggested from SNC meteorites (Wänke and Dreibus 1994). Here, the melting temperature is comparatively low. In such a case, the interior needs to cool efficiently to initiate inner core formation and a chemical dynamo. Planets with one-plate tectonics cool inefficiently whereas planets with plate tectonics cool comparative efficiently. Thus, for the latter, inner core growth is much more likely, e.g. for the Earth where the inner core growth has initiated about 1 to 3.5 Ga ago (e.g. Labrosse et al. 2001; Gubbins et al. 2003). Note however, that the probability for a one-plate planet to cool below the core liquidus is enhanced for a wet and weak mantle rheology and/or for a thin mantle as it might be the case for Ganymede and Mercury, respectively.

 Springer

3.3 Interior Atmosphere Interaction

Surface conditions such as atmospheric pressure, temperature and the presence of water are key parameters for planetary habitability. These may strongly depend on interactions between the outer and inner layers of the planets. Most of the publications concerning mantle dynamics and volatile exchange focus on Earth (e.g. McGovern and Schubert 1989; Franck and Bounama 1995, 1997, 2001) and few involve complete two-way coupling between interior and exterior (to some extent in Franck et al. 1999; Phillips et al. 2001a). Here we mainly focus on Mars and Venus since the interactions are simpler for these planets, but nevertheless illustrate the most important processes.

The planetary science data available mostly consist of atmospheric measurements and surface observations, since these are most easily obtained, and less of data pertaining to planetary interiors. The data suggest both planets are currently uninhabitable. They lack stable liquid water at the surface, and have either a tenuous atmosphere (Mars) or one that is much too hot and dense (Venus). Despite these observations, we can tell that in the past, things were certainly different.

3.3.1 Mars

Mars in particular shows remnants of water-related features such as outflow channels and valley networks which are evidence of the existence of liquid water (stable or not) on the surface at various times in Mars' past for unknown periods of time (Carr 1996; Jakosky and Phillips 2001; and Chaps. 3 and 4 of this volume). Moreover, missions such as Mars Express with its High Resolution Stereo Camera and the OMEGA spectrometer have provided us with new data on the morphology and composition of certain areas of Mars. Hydrated minerals have been found and also sulfate-rich layered deposits in Valles Marineris (Bibring et al. 2005), as well as in the north polar region (Squyres et al. 2004; Langevin et al. 2005). These sulfates imply that liquid water was present during the formation of the layers, though not necessarily as stable bodies of water. Other observations (e.g. ASPERA data) have allowed estimates for the present atmospheric escape rate of Mars, suggesting a value of about 1 kg/s (see Carlsson et al. 2006). This implies an evaporation of the present atmosphere in some 200 Myr at present conditions. The HRSC has revealed recent episodic volcanic activity on Mars (Neukum et al. 2004). Calderas on major volcanoes present signs of activation and resurfacing during the last 800 Myr with events that may be as young as 4 Myr. Recent detection of methane, with a photochemical lifetime of around 300 years, in Mars' atmosphere (Krasnopolsky et al. 2004; Formisano et al. 2004) suggests a presently active source. Both a biogenic origin (Formisano et al. 2004) and a geological origin (Lyons et al. 2004) have been proposed. Lyons et al. (2004) show that limited recent dike-like activity and hydrothermal alteration can account for the observations, which would support continued addition of volatile species to the Martian atmosphere from its interior. As discussed in Sect. 2.2, the polar layered depositits and ice caps are major agents in the present-day surface-atmosphere interactions of water and CO_2.

These observations suggest that in the past and in the present, volatiles have been and still are released into the atmosphere of Mars and have likely been essential to the evolution of the planet. The primary interaction between the atmosphere of a terrestrial planet and its mantle is indeed the degassing of volatiles (water and CO_2 in particular). It is commonly admitted that early hydrodynamic escape and heavy bombardment occurring during the first hundreds of million years of the planets history efficiently remove the primordial gaseous layer leaving up to 1% of the early CO_2-rich atmosphere (see Carr 1996). Thus volcanism

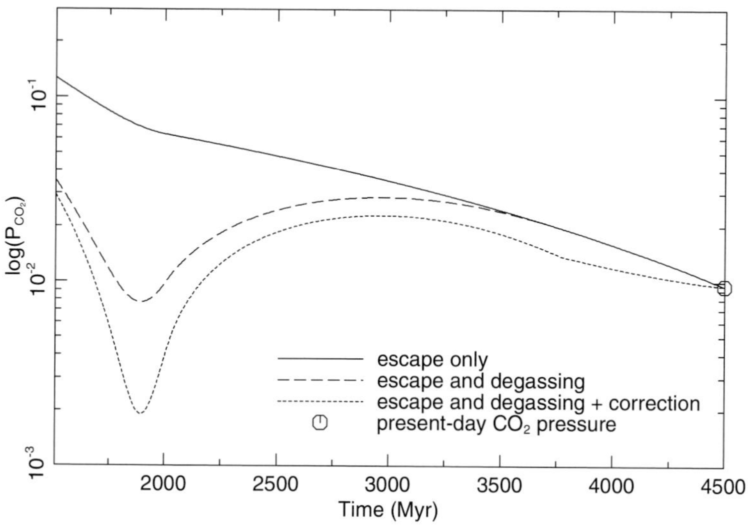

Fig. 3 Evolution of the maximum CO_2 pressure in the atmosphere of Mars over the last three billion years. The *solid curve* shows only atmospheric escape as described in Chassefière et al. (2007). The *dashed curve* shows escape and includes degassing of CO_2 by crust production using results from Breuer and Spohn (2006). The *dotted curve* uses the same model as the *dashed one* but adds a corrective factor for late activity which is not present in the crust production model

can have a strong influence over medium to late atmospheric history. It is however difficult to quantify the amount of degassing for two reasons. First, data is scarce and melt production rate estimates can be obtained either by observation of resurfacing (Hartmann and Neukum 2001) or with numerical modeling by looking at the crust production. Both methods lack the desired accuracy and provide only an order of magnitude estimate at best. The second difficulty originates from uncertainties in the composition of the planet's interior, which, for Mars, is rather poorly constrained. It is nonetheless clear that inner dynamics determine the amount of degassing by controlling partial melting through the temperature of the mantle and therefore influence the evolution of the atmosphere. A simple numerical model taking into account only degassing (from crust production data, see Breuer and Spohn 2006) and non-thermal atmospheric escape (Chassefière et al. 2007) can be used to provide an estimate of the history of CO_2 pressure on Mars that fits with present conditions (Fig. 3).

In these models, most, if not all, of the present day CO_2 has been released by volcanism, as the primordial atmosphere is blown away during the first hundreds of million years by bombardment and hydrodynamic escape. Only low pressures which are insufficient to create enough greenhouse effect to obtain a warm climate during the past three billion years are observed. This is consistent with the current consensus that a dense (> 1 bar CO_2) atmosphere could not have persisted more than some tens of millions of years before collapsing, with CO_2 forming carbonates that no one has been able to detect so far (Catling and Leovy 2006).

Phillips et al. (2001b) calculated that the emplacement of Tharsis may have resulted in the degassing of $1.7 \cdot 10^{22}$ g water and $1.9 \cdot 10^{21}$ g carbon in the form of CO_2. Compared to the present-day atmosphere inventory (see Table 1), these are vary large quantities.

3.3.2 Venus

For Mars, most of the interaction between the interior and the atmosphere appears to take place in the form of mass and heat transport from the former towards the latter reservoir. However, in the opposite direction, the atmosphere or hydrosphere of a terrestrial planet may influence the dynamics of the solid interior, primarily through the action of water (see Sect. 3.1.2). In addition to this, surface conditions provide the upper thermal boundary condition for the solid planet. Thus, a dense and hot atmosphere can delay the cooling of the planetary mantle, affecting its thermal and also its magmatic/volcanic evolution. Phillips et al. (2001a) investigated such a coupled evolution for Venus. They used a simple parameterized mantle convection model (including partial melting) and a radiative-convective atmospheric model. Feedback was incorporated by release of water associated with melt extraction into the Venusian atmosphere which adds to the greenhouse effect and thus affects the surface temperature. After two billion years of evolution the coupled model was hotter by several tens of K and its extrusive volcanic flux was four times higher than the case without coupling. Their study shows a complex interplay between Venus convective evolution, volcanic activity and atmospheric state even with such a simple model featuring only basic processes. More complex and elaborate models are required to obtain a more detailed understanding of the interactions.

3.4 Life, CO_2 Fixation and Climate

CO_2 is a major greenhouse gas impacting habitability and Earth global climate. On geological time scales, the atmospheric CO_2 budget is mostly controlled by volcanic degassing, and consumption through the coupling of silicate weathering and carbonate precipitation (Berner et al. 1983), known as the Urey reaction (see reactions (4)–(6)):

$$(Ca, Mg, Fe)SiO_3 + 2H_2CO_3 = (Ca^{2+}, Mg^{2+}, Fe^{2+}) + 2HCO_3^- + SiO_2 + H_2O \quad (4)$$

(weathering of silicates, mostly on continents)

$$Ca^{2+}, Mg^{2+}, Fe^{2+} + 2HCO_3^- = (Ca, Mg, Fe)CO_3 + H_2CO_3 \quad (5)$$

(cations and HCO_3^- are consumed by carbonate precipitation in oceans)

$$(Ca, Mg, Fe)SiO_3 + CO_2 = (Ca, Mg, Fe)CO_3 + SiO_2 \quad (6)$$

(net budget showing that C is transferred from the atmosphere to sediments through silicate weathering).

Silicate weathering and carbonate precipitation are thus two important processes to consider, and we discuss here the impact of life on them.

3.4.1 Diversity of Factors Affecting Silicate Weathering

Many studies have focused on river geochemistry, including small watersheds or at a larger scale using the main rivers worldwide to infer parameters governing weathering (for a review, see Dupre et al. 2003). Weathering rates can also be determined in laboratory experiments conducted over short periods of time or from analysis of soil profiles formed over geological timescales. Simple parametric laws describing the dependency of weathering on diverse environmental factors are then established and introduced into numerical models

 Springer

calculating the evolution of atmospheric CO_2 over geological times. The environmental factors impacting silicate weathering that are classically considered are: (1) climate (Chadwick et al. 2003), silicate weathering increasing with temperature (White et al. 1999) and runoff, both enhanced under high CO_2 values; (2) Lithology of the weathered rocks, basalt weathering acting in particular as a major atmospheric CO_2 sink (Dessert et al. 2003); (3) the age of the rock substrates (White and Brantley 2003); (4) the intensity of physical denudation with the important role of soils on chemical weathering (Gaillardet et al. 1999); (5) Configuration of continents, etc. . . . Life is an additional parameter, whose quantitative role on silicate weathering is however still poorly constrained.

3.4.2 Life as an Enhancer or Inhibitor of Silicate Weathering

There are numerous mechanisms mediated by life that may potentially enhance silicate weathering (e.g. Schwartzman and Volk 1991): e.g. release of CO_2 inducing a weak amplification of weathering rate (Von Bloh et al. 2003), lowering of local pH (Barker et al. 1997), trapping of nutrients freed by weathering, physical weathering through microfracturing of mineral grains, release of organic acids and ion-complexing organic ligands (organic molecules which may act as ion carriers) like siderophores (e.g. Liermann et al. 2000), modification of Fe redox modifying Fe-containing silicates dissolution (for a review, see Barker et al. 1997). The debate over which effects are most significant is important as some mechanisms may be a passive result for other processes, whereas others involve an active input of energy providing them a different status from an evolutionary perspective.

Several studies have evidenced a biological enhancement of silicate weathering either in laboratory experiments or in the field, at the watershed scale (e.g. Aghamiri and Schwartzman 2002; Moulton and Berner 1998; Lucas 2001; Hutchens et al. 2003; Rogers and Bennett 2004). However, several processes mediated by life could also inhibit silicate dissolution: e.g. production of organic polymers reducing silicate surface reactivity and/or the water/rock ratio, enhancement of secondary phase precipitation passivating silicate surfaces, and modification of iron redox, lowering iron-containing silicate reactivity (e.g. Santelli et al. 2001). Several studies have evidenced a biological inhibition of silicate weathering both in laboratory and field observations (e.g. Santelli et al. 2001; Ullman et al. 1996; Benzerara et al. 2004; 2005a). A study of Hawaiian basalt flows of known age indicates that there is an acceleration of chemical weathering on the order of at least 10 to 100 times for lichen covered surfaces relative to bare rock (Jackson and Keller 1970). Moulton and Berner (1998) indicated an enhancement factor of two to five in a Western Iceland catchment. Aghamiri and Schwartzman (2002) evidenced problems of elemental incongruency as enhancement factors were found to be 16 considering Mg^{2+} and only 4 considering Si.

Considering the high number of mechanisms potentially controlling silicate weathering, there is probably no single value for a biological enhancement factor, which might vary depending on biological communities and external physico-chemical conditions (Kelly et al. 1998). Hence, many more studies at the watershed scale should be systematically conducted to find potential constants in the biological enhancement/inhibition of silicate weathering.

3.4.3 Impact of Vascular Plants

An important question is whether vascular land plants have a higher effect on silicate weathering than microbes, as a significant difference between the impact of microbes and plants on silicate weathering would imply a qualitative and quantitative shift of the modalities of silicate weathering over Earth history as vascular plants emerged in the Silurian (440 Ma ago).

The point is not to present microbes and plantes as intrinsically opposite agents, as most of plants nowadays live in close association with microbes. Moulton and Berner (1998) concluded from their study that trees enhance silicate weathering more than mosses and lichen partially covering rocks. However, this kind of study should be conducted more systematically and pertinent comparisons should be made whenever possible.

3.4.4 Impact of Life on Carbonate Precipitation

The precipitation of carbonates (reaction (5)) is an additional step of the Urey reaction where life can have a dramatic impact. There is no unequivocal evidence that calcium carbonate can precipitate abiotically from seawater (e.g. Wright and Oren 2005).

Although calcium carbonate and dolomite should precipitate spontaneously from supersaturated solutions in seawater based on thermodynamic considerations, they do not (Lippmann 1973), and it has been suggested that kinetic inhibitors to carbonate precipitation are effective. On the other hand, there is increasing evidence that these inhibitors can be overcome through microbial activity by e.g. modifications of solution chemistry, control of pH at the microscale, removal of kinetic inhibitors (for a review, see Wright and Oren 2005). In marine environments throughout the Phanerozoic, carbonate precipitation has been considered almost exclusively a biological process mediated by skeleton-forming invertebrates, algae or protozoa. Could the formation of Precambrian carbonates have been mediated by microbiological processes? Or might they have been simply chemical precipitates? Would carbonates precipitate in the absence of life? This is one of the major questions in the present discussion and remains as a key issue in geobiology (e.g. Arp et al. 2001; Wright and Oren 2005).

Numerous experimental studies have shown that microorganisms can promote precipitation of carbonates (e.g. Castanier 1999; Morita 1980), but their importance in the natural environment at the global scale has yet to be quantified. Different mechanisms can be involved: e.g. removal of CO_2 from the medium by autotrophic pathways (e.g. Riding 2000), active ionic exchanges of Ca^{2+}/Mg^{2+} through the cell membrane, production of carbonate ions and pH increase by ammonification, dissimilatory reduction of nitrate or degradation of urea, production of carbonates, pH increase and removal of sulphates which are inhibitors for calcium carbonate precipitation by dissimilatory reduction of sulphates (e.g. Vasconcelos et al. 1995). Microbes can also provide preferential nucleation sites. It is still much debated whether the polymers they excrete are enhancers or inhibitors of calcium carbonate precipitation (e.g. Arp et al. 1999; DeFarge et al. 1996). Recent observations show that those polymers are widespread in many biological carbonate deposits and suggest that they do have an impact on carbonate precipitation (e.g. Gautret et al. 2004; Benzerara et al. 2006).

3.4.5 Microbial Urey Reaction at a Much Shorter Time Scale?

The Urey reaction is usually seen as a two-step process on a time scale of hundreds of thousands of years, with silicate weathering occurring on continents and carbonates precipitating in oceans (Berner 1995). It is interesting to note however that few studies have shown the possibility that the two steps occur in a much shorter time and at the same location when catalyzed by microbes (e.g. Ferris et al. 1994; Rogers and Bennett 2004; Benzerara et al. 2005b). The importance of such a microbial promoted Urey reaction needs to be evaluated.

 Springer

3.4.6 Modeling of the Quantitative Effect of Life on Climate Regulation

The Urey reaction can provide a purely abiotic negative feedback on atmospheric CO_2 content regulation. Indeed, an increase of atmospheric CO_2 raises global temperatures which enhance silicate weathering and hence CO_2 trapping. If life does significantly enhance silicate weathering, then the efficiency of this self-regulating process is greatly amplified and becomes more sensitive to any perturbation. This has been considered as the best established Gaian negative feedback: if global temperatures increase, biological growth increases and rock weathering increases, which ultimately reduces carbon dioxide in the atmosphere. As temperature decreases due to the reduction in atmospheric CO_2 (a greenhouse gas), life-mediated silicate weathering decreases. It is noticed that it is a non-selective feedback on growth: if any studies have suggested that biological weathering strategies have evolved because they provide a source of limiting nutrients like phosphorus or iron (e.g. Lenton 1998; Bennett 2001; Rogers and Bennett 2004), they were not selected for their effect on global carbon dioxide content. Changes in atmospheric carbon dioxide content or temperature do likely not alter the selective advantage of a strategy for nutrient search. Hence, this feedback does not drive Earth to a state intrinsically dependent on the presence of life and responding only passively to external forcing.

Several studies have incorporated the biological amplification of the Urey reaction in models of the global carbon cycle over geological time scales (e.g. Lenton 1998; Lenton and Lovelock 2001; Von Bloh et al. 2003). A stronger biotic amplification of weathering implies a stronger negative feedback. It moreover provides a responsive regulation against perturbations that tend to increase carbon dioxide and/or temperature (Von Bloh et al. 2003). Von Bloh et al. (2003) show that specific values of the biotic amplification on silicate weathering extend the life span of the biosphere. Refined models taking in account other processes as well as the measure of the biotic effect on silicate weathering throughout Earth history would improve our knowledge of the stability of the Earth system.

4 Planetary Evolution

As we have treated the processes relevant for life and habitability in the previous sections, we will now move on to discuss these in the context of the (geodynamical) evolution of planets Earth, Mars and Venus. Whereas Earth appears to have had clement surface conditions for several billion years, there is evidence for significant changes in the conditions at the Martian and Venusian surfaces during their evolution. Along the same vein, Mars and Venus show signs of possible changes in the geodynamical regime, the former very early in its history and the latter relatively recently. Earth shows evidence for the continued operation of plate tectonics over a long period of time, starting 1 up to 4 billion years ago (De Wit 1998; Hamilton 1998; Stern 2005). Changes in the thermal evolution of these planets are implied by the changes in the dynamical evolution.

4.1 Thermal Evolution Time Scales

The thermal evolution of a planet is controlled by the heat balance equation,

$$MC\frac{\partial(T - T_s)}{\partial t} = -Q + H(t),$$
(7)

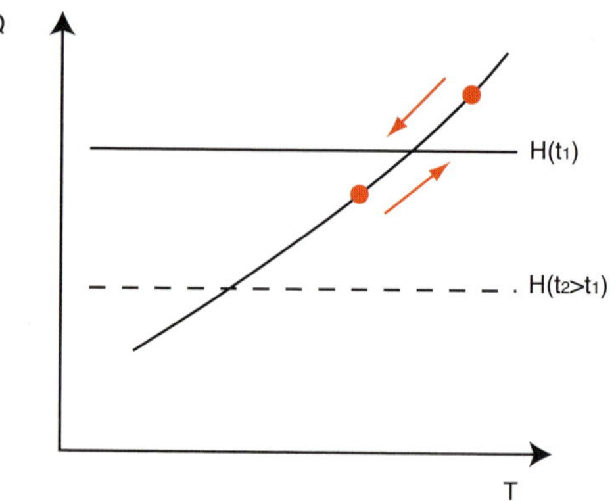

Fig. 4 Schematic thermal evolution of a planet. The direction of the evolution depends on the balance between the heating term and the heat flow term in (7). Scaling of mantle convection provides a relationship between the heat flow and the potential temperature of the planet. The monotonic decay of radiogenic heating eventually leads to cooling of the planets

with M the mass of the planet, C its averaged heat capacity, T the potential temperature, T_s the surface temperature, Q the heat flow at the surface and H the time decaying radiogenic heat production. Several isotopes contribute to $H(t)$ and each has a separate time-scale, ranging from less than 1 Gyr to 20 Gyr. From an initial temperature T_0, the planet can either cool or heat up, depending on the sign of the right-hand side of (7), but the decay of heat production will eventually make this sign negative and lead the planet to cooling (Fig. 4). The time scale with which the temperature decay follows the decay of heat production is given by comparing the left-hand side of (7) and the first term on the right-hand side:

$$\tau = \frac{MC(T - T_s)}{Q}. \tag{8}$$

The heat flow can be measured at the surface of the planet (it is rather well known at the surface of the present Earth, as will be discussed below) but is subject to important evolution with time. The first order effect comes from the evolution of temperature itself. Indeed, the heat flow per unit surface of a planet can be written as $q = k(T - T_s)/\delta$, with k the thermal conductivity, and δ the thickness of the boundary layer at the top of the mantle. When the interior of the planet cools, the heat flow decreases, linearly if the boundary layer thickness is kept constant. This is however usually not the case, since this parameter is the result of dynamical balances in the mantle, in which for example the temperature dependent viscosity plays an important role.

Most models of the thermal evolution of planets use scaling laws that relate the heat flow at the surface to the potential temperature (e.g. Turcotte and Oxburgh 1967; Schubert et al. 1980; Christensen 1985; Solomatov 1995; Reese et al. 1998). These scaling laws are obtained using numerical or analog models of thermal convection and simple dynamical arguments that balance the dominant stresses in the system. The difficulty in this exercise comes from the identification of the dominant balance in the considered planet which, as discussed above (Sect. 3.1) and citing Sleep (2000), can evolve on different regimes, each having a different $Q(T)$ scaling (Fig. 2). Some of these scalings are rather well established, like for example the stagnant lid convection (e.g. Davaille and Jaupart 1993; Solomatov 1995) but the conditions for transition between regimes are still poorly understood (see Sect. 3.1.3). Sleep (2000) emphasizes the role of melting which is clearly a key unresolved issue in the problem. Melting of course depends on temperature but also on composition, particularly in volatile elements, and influences volatile extraction and rheology. This means that using the potential temperature as unique parameter controlling the heat flow is probably not correct.

 Springer

One effect that has not been sufficiently investigated yet and might be important in the context of habitability is the evolution of the surface temperature T_s which is controlled by the radiative balance of the atmosphere (see Sect. 3.3). The amount and nature of the chemical species present in the atmosphere depend in part on the degassing history of the internal part. Small variations of T_s have a tremendous effect on life but very large variations and disparities among planets can also occur, as exemplified by the difference between the Earth and Venus.

4.2 Thermal and Dynamical Evolution of Earth

The Earth is the best place to start constructing models of thermal evolution since the amount of available data places strong constraints on the general procedures developed in these models. The history of this topic shows that the issue is far from obvious and the points of view of geochemistry have traditionally been orthogonal to that of geophysical modeling, due to the difficulty in satisfying constraints concerning both the present heat production in the mantle and the heat flow at the surface. The reason for this difficulty can be understood by getting a closer look at the way most geophysical models work.

The thermal evolution of the Earth is modeled using (7) and usually a scaling relationship of the form

$$Q = AT^{1+\beta}\nu^{-\beta} \tag{9}$$

with A a scaling constant, ν the temperature dependent viscosity of the mantle and β an exponent close to $1/3$ in usual boundary layer theory (e.g. Schubert 1979). The temperature dependence of the viscosity is responsible for a short adjustment time scale τ. Following Christensen (1985), one can assume a viscosity law of the form $\nu = \nu_0 (T/T_0)^{-n}$ and linearly develop (7) around T_0 to obtain as adjustment time scale:

$$\tau_a = \frac{MCT_0}{(1+\beta+\beta n)Q_0}. \tag{10}$$

Using the classical value $\beta = 1/3$ and $n \simeq 35$ that approximate best the Arrhenius law of mantle materials (Davies 1980; Christensen 1985) gives $\tau_a \sim 800$ Ma. This leads to a heat flow at the mantle surface that closely follows the decay of heat production, and at least 70% of the present surface heat flow is radiogenic (the Urey number is 0.7). On the other hand, the geochemical models of the Earth favor a present radiogenic heat production of about 20 TW or less (Javoy 1999) to be compared to the present total heat loss of 46 TW (see Jaupart et al. 2007, for a comprehensive discussion of the energetics of Earth's mantle).

A classical way to solve the issue is to change the exponent β to a lower value, as was first done by Christensen (1985). Indeed, (10) shows that the thermal adjustment time scale is then increased and the heat flow decay with time can lag much longer after the decay of radiogenic heating. Several models have been proposed to lead to a smaller β: effect of temperature dependent viscosity (Christensen 1985), resistance of plate to bending (Conrad and Hager 1999), chemical buoyancy of a thicker crust associated with deeper melting at high temperature (Sleep 2000; Korenaga et al. 2003). All these effects may well be important but no evidence has been shown to support the Earth being in the regime where they actually are.

Another view of this problem has been recently proposed by Labrosse and Jaupart (2007) who use the observed distribution of surface heat flux as input to a thermal evolution model, therefore not relying on any scaling assumption. Because of the difficulties associated with

 Springer

the measurements of seafloor heat flux, the heat loss of the Earth is usually computed using the distribution of seafloor ages, that is linked to the heat flux using a half-space cooling model that give

$$q = \frac{kT}{\sqrt{\pi \kappa a}},$$ (11)

where a is the age of the seafloor and κ is the thermal diffusivity. Everywhere a good sediment coverage allows reliable heat flux measurement, the data agrees very well with this model (Jaupart et al. 2007).

Using maps of seafloor ages (Sclater et al. 1980; Müller et al. 1997) one can obtain a distribution of seafloor ages (Sclater et al. 1980; Rowley 2002; Cogné and Humler 2004) which can be used to obtain the total heat loss through the oceans in the form

$$Q_{\text{oc}} = \lambda(f) \frac{A_{\text{oc}} kT}{\sqrt{\pi \kappa a_{\text{max}}}},$$ (12)

$\lambda(f)$ being a geometrical factor depending on the distribution of seafloor ages, presently equal to $8/3$, $A_{\text{oc}} = 3.09 \cdot 10^{14}$ m^2 the total oceanic surface and $a_{\text{max}} = 180$ Ma the maximum age of the oceanic lithosphere.

The continental contribution to the heat loss of the Earth is more difficult to measure because the continental heat flux varies greatly, particularly on short length scales. These variations cannot be attributed simply to geologic ages but essentially reflect the variability in concentration in radioactive elements of crustal rocks. In fact, the heat flux at the surface can be attributed in a large part to heat production in the crust and the flux from the mantle has been estimated to values of the order of 7–15 mW m^{-2} (Pinet et al. 1991; Jaupart et al. 2007), to be compared to the average of 100 mW m^{-2} of the oceans. It is then reasonable to assume the continents to be perfect insulators when modeling the thermal evolution of the Earth (Grigné and Labrosse 2001). Equation (12) can then be used in the heat balance (7), after removing the heat flow at the surface of continents from the radiogenic heat production term.

Assuming that $\lambda(f)$, A_{oc} and a_{max} have been constant through time, the heat balance (7) can be solved (Labrosse and Jaupart 2007) and the resulting evolution of temperature and heat flow are shown in Fig. 5. The main outcome of this model is that the heat flow from the seafloor has been essentially constant through time and the total change of temperature is of order 150 K. Such a mild thermal evolution could have been anticipated by estimating the Earth's thermal adjustment time scale according to (8) which, using (12) gives

$$\tau_E = \frac{MC \sqrt{\pi \kappa a_{\text{max}}}}{A_{\text{oc}} k \lambda(f)} = 10 \text{ Gyr.}$$ (13)

Another interesting feature of this model is that equilibrium between radiogenic heat production and heat loss is reached about 3 Gyr ago. Whether the warming of the Earth for times prior to that actually occurred depends on the validity of the assumptions made. In particular, the total oceanic surface may not have been constant before that (Collerson and Kamber 1999).

The assumption that could be subject to most criticism is that of a_{max} being constant. Another assumption that could be made is that its value is set by the stability of the lithosphere, that is to say that the Rayleigh number computed with the lithosphere thickness at that age,

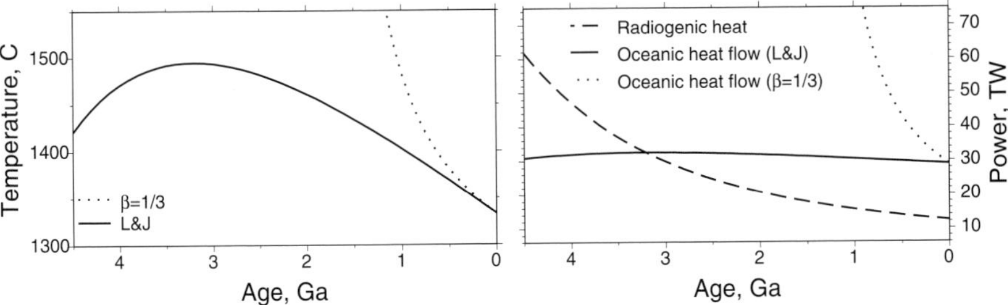

Fig. 5 Thermal evolution of the Earth based on the observed distribution of seafloor ages (*solid lines*, labeled L&J) or from a boundary layer stability hypothesis (*dotted lines*, labeled $\beta = 1/3$). After Labrosse and Jaupart (2007)

$\sqrt{\pi \kappa a_{\max}}$, is equal to the critical one R_c, that is

$$\sqrt{\pi \kappa a_{\max}} = \left(\frac{R_c \kappa \nu}{\alpha g T} \right)^{1/3}. \tag{14}$$

With this assumption we recover the classical scaling given by (9) with $\beta = 1/3$ and the numerical solution to the heat balance equation is shown as dotted lines in Fig. 5. This shows that the assumption that the maximum age of the oceanic lithosphere is set by its stability cannot be accepted. The distribution of seafloor ages actually shows that the Earth is able to subduct a lithosphere of any age with a more or less equal probability (Rowley 2002) and this justifies our assumption of a constant maximum age. Alternatively, the assumption that the mechanism resulting in (11) and (12) has been active since Earth's early history may be wrong. Geologists are still divided on when plate tectonics started to operate (see e.g. Stern 2005), and the other terrestrial planets in our solar system warn that plate tectonics should not be taken for granted.

On the other hand, the geometrical coefficient $\lambda(f)$ clearly depends on the distribution of continents at the surface of the Earth and has a large effect on the heat flow at the surface. If the distribution of seafloor ages was flat, the value of λ would be 2, instead of the present 8/3, which would give a 30% decrease of the oceanic heat flow compared to the present value. This means that large fluctuations of the heat flow on the time scale of 500 Myr are to be expected and their magnitude overwhelms its long term evolution (Grigné et al. 2005).

4.3 Thermal and Dynamical Evolution of Mars and Venus

The surface morphologies of Mars and Venus differ significantly from that of Earth most likely mirroring fundamental differences in the present-day tectonic regimes and heat-loss mechanisms of these planets. The Earth has a dichotomy between oceanic and continental regions and features mid-oceanic ridges, orogenic mountain belts and island arc systems, features that are commonly linked to plate tectonics. No significant plate tectonic features have been identified on Mars and Venus although on the former the linearity of the Tharsis volcano chain has been likened to island arc volcanic chains on Earth (Sleep 1994). More recently, crustal remnant magnetization patterns have been used to suggest some form of plate tectonics for early Mars (see Connerney et al. 2004 for a recent review and Connerney et al. 2005 for most recent data). Early plate tectonics have also been proposed (Nimmo and Stevenson 2000; Breuer and Spohn 2003) to allow for a better cooling of the core thereby

facilitating the explanation of an early magnetic field. For Venus, early plate tectonics have been suggested (e.g. Phillips and Hansen 1998) but as for Mars there is little, if any, factual evidence for it. It is commonly held that plate tectonics requires fluid water on the surface to operate, the water modifying the rheology of the cold plates such that subduction becomes feasible. The high D/H ratio in the Venusian atmosphere strongly suggests but does not prove that it had substantial water in the past (e.g. Fegley Jr. 2004). Today Venus is desiccated as far as the available spacecraft data can tell.

4.3.1 Volcanism

Mars has a major volcanic center, Tharsis, that covers almost a hemisphere and rises up to 20 km above the datum. There is a second much smaller center, Elysium. Other volcanoes are distributed more randomly across the surface. The surface of the planet shows a dichotomy between roughly the southern hemisphere and the northern hemisphere. While the southern hemisphere is old (about 4 to 4.5 Ga) the northern hemisphere is about 1 Ga younger (Zuber 2001). Whether or not the northern hemisphere has been resurfaced by volcanic activity or by sedimentation is debated. Mars Global Surveyor gravity data show (Smith et al. 1999) that an old cratered surface exists buried underneath the northern hemisphere. Tharsis is centered almost at the border between these two hemispheres. Volcanic activity on Tharsis has apparently continued until the very recent past, albeit at a small rate (Neukum et al. 2004). The recent knowledge on Mars volcanism has been summarized by Solomon et al. (2005).

Venus has no distinct volcanic center but a prominent rift system, the Beta, Atla and Themis Regiones. Volcanic features include a large number of comparatively small shield volcanoes that are apparently randomly distributed, volcanic plains and about 500 coronae—circular large scale volcano-tectonic features. It is widely accepted that coronae form above upwelling plumes (Smrekar and Stofan 1999), though downwellings rather than upwellings have been suggested as well (Hoogenboom and Houseman 2006). Johnson and Richards (2003) have recently reanalyzed the coronae distribution and have concluded that the coronae activity at least tended to concentrate in the Beta, Atla and Themis rift system over time. The paucity of impact craters on the surface has been used to suggest that Venus was resurfaced about 500–700 million years ago by a volcanic event of global scale (Schaber et al. 1992; Bullock et al. 1993; McKinnon et al. 1997). This event was followed, as the model suggests, by almost no volcanic activity. The suggestion prompted a number of papers that speculated on an episodic thermal history of Venus with pulses of global activity followed by epochs of relative tranquillity like the present one (e.g. Turcotte 1995; Phillips and Hansen 1998; Reese et al. 1999; Ogawa 2000; Nimmo 2002). The early conclusions from the cratering record have been challenged by e.g. Hauck et al. (1998) and Campbell (1999) and most recently by Bond and Warner (2006). According to the latter authors, the cratering record allows a variety of interpretations in terms of volcanic resurfacing including a global decrease in time in the rate of an otherwise statistically distributed volcanic activity. Previous workers had still concluded that Venus underwent a major transition in tectonic style, albeit more gradual than the previously postulated sudden global resurfacing event. The surface geology also seems to indicate that modifications of the surface are planet wide and and gradual over long periods of time rather than episodic and locally confined (Ivanov and Head 2006).

Whether or not there is present day volcanic activity on Venus is unclear. Some sulphuric clouds have been interpreted as being evidence of recent volcanism (Fegley and Prinn 1989). The recent data of the infrared spectrometer Virtis on Venus Express (e.g. Helbert and

 Springer

Benkhoffm 2006) suggest differences in surface temperature that can be linked to morphologic features on the surface but it is too early for more far reaching conclusions.

4.3.2 Thermal and Dynamical Evolution

Since planets can be regarded as heat engines that convert thermal energy to mechanical work, part of which is necessary to maintain the magnetic field against losses by ohmic dissipation, the evolution of the planet subsequent to formation and early differentiation is strongly coupled to its thermal evolution or cooling history. Also, as noted above, the cooling history is strongly related to the history of the magnetic field. In the context of the discussion of the evolution of habitability of a planet and even life, the outgassing history matters. The latter is again linked to the thermal history of the planet (e.g. McGovern and Schubert 1989; Phillips et al. 2001a). Phillips et al. (2001a) have presented models where the evolutions of the Venusian atmosphere greenhouse and the interior are coupled. These models suggest that the coupling can be quite effective for Venus.

Early models of the thermal histories of Mars and Venus (e.g. Schubert 1979; Stevenson et al. 1983; Schubert and Spohn 1990; Spohn 1990) used simple parameterized convection schemes that did not include effects of the temperature dependence of the viscosity on the convection. But Turcotte (1995) had already noted the difficulty for Venus to remove its internal heat if it lacked plate tectonics as the surface geology suggests. Following the work of Davaille and Jaupart (1993), Solomatov (1995), Moresi and Solomatov (1995) and Grasset and Parmentier (1998), stagnant lid models of planetary convection and thermal history were developed and successfully applied to Mars (e.g. Hauck and Phillips 2002; Breuer and Spohn 2003: Breuer and Spohn 2006; see also Spohn et al. 2001 and Connerney et al. 2004 for recent reviews). These models suggest that Mars is continuously cooling at a rate of about 30 K/Ga. Volcanic activity and thereby outgassing decreased rapidly during the first few 100 million years and gradually from there on to non-zero but almost negligible (volume wise) values during the past Ga. On Mars, outgassing would therefore have contributed to or generated an atmosphere early on that was subsequently lost due to atmospheric loss processes in the absence of the protecting magnetic field. The thermal history calculations would principally allow for a wet early Mars. But the geologic record as revealed by the HRSC camera on Mars Express and the MER suggests a desert planet that had wet spots with surface water in places of volcanic activity that melted the permafrost (e.g. Head et al. 2005).

Although the continuously cooling thermal history models of Mars are successful in explaining the observations (e.g. crustal production, magnetic field), it is not excluded that the Martian thermal evolution could be episodic. The Rayleigh number for Martian mantle convection is comparatively small (10^5 to 10^6), a fact that allows for quasi stationary convection but the interaction of the flow with possible mantle phase transitions such as the olivine to spinel and the spinel to perovskite phase transitions may introduce time dependence. For instance, since the spinel-perovskite transition is strongly temperature dependent an early perovskite layer at the bottom of the lower mantle of Mars may thin and eventually disappear (Spohn et al. 1998) and allow for increased heat flow from the core. However, presently available data do not constrain the present-day presence or absence of the perovskite layer (Van Thienen et al. 2006). The interaction of Martian mantle convection with phase transitions has been used before to explain the singularity as of Tharsis as volcanic center (Harder and Christensen 1996; Breuer et al. 1996, 1997, 1998) but most recently Schumacher and Breuer (2006) have criticized the hypothesis of Tharsis being maintained by a mantle superplume and have introduced a model of thermal blanketing and mantle melting, instead.

The thermal history of Venus appears to be more complicated (see Schubert et al. 1997, for a review). Although it must in principle be possible to remove the heat generated by radiogenic elements in a stagnant lid convection regime the resulting thermal lithosphere may be too thin to be consistent with the available gravity data (Anderson and Smrekar 2006). The latter data suggest an elastic lithosphere of up to 100 km thickness and a crust of 90 km thickness. The necessity of removing the heat in the absence of plate tectonics leads to episodic scenarios as have been proposed by Turcotte (1993), Turcotte (1995), Turcotte et al. (1999), Reese et al. (1999), and Ogawa (2000). Some authors (Turcotte 1995; Phillips and Hansen 1998; Nimmo 2002; and most recently Van Thienen et al. 2005) suggest that present Venus may actually be heating up. This would offer another explanation of the absence of a dynamo magnetic field on the planet. Stevenson et al. (1983) had attributed the absence of a field to the absence of an inner core and a cooling rate of the core that was too small for convection and dynamo action to occur. If instead Venus had an inner core but a mantle the temperature of which would be increasing in time, then the inner core would have ceased to grow (it may actually be shrinking in radius) and the dynamo could not be driven by chemical buoyancy released upon inner core growth. Chemical buoyancy is believed to be the main driver of the Earths dynamo (e.g. Labrosse 2003; Stevenson 2003a).

The D/H ratio on Venus has been used to infer that Venus had a wet past with water on the surface with the equivalent of a global ocean of up to about 500 m depth (Donahue and Russell 1997). It is not known when the transition from a more habitable climate to the present desiccated state occurred, but see Grinspoon and Bullock (2005) for a recent discussion of the water history of the planet. It can be speculated that the removal of water from the atmosphere was accompanied by a transition from plate tectonics to the present tectonic state. The transition may have been accompanied by the loss of the magnetic field. Whether or not remnant crustal magnetization may have survived is speculative. If observed by future space missions, it would suggest a more habitable past of the planet.

5 Discussion and Important Open Questions

5.1 Discussion

Looking at Tables 1 and 2, it is clear that the amounts of volatiles present in Earth's interior are greater by far than the amounts available to the biosphere. The fluxes of the listed species are such that the entire atmospheric inventory is processed through the mantle on time scales significantly shorter than the age of planet: on the order of 10^6 yr (C), 10^8 yr (S), and 10^9 yr (H_2O), respectively (Note, however, that the errors in these estimates are large due to the large uncertainties in the fluxes). Therefore, the internal dynamics have the potential to significantly affect the availability of these species to the biosphere. Indeed, the balances listed in Table 2 show that the internal dynamics are constantly changing the reservoir sizes. Nevertheless, life's involvement in the fixation of volatile species, specifically C and N, allowing them to be subducted eventually, is also important; not so much for the mantle reservoir, but certainly for the hydrosphere and atmosphere reservoirs. In addition to this, life may contribute to determining the amount which is subducted. More specific, in the case of sulfur the oxidation state of the ocean, depending on the availability of oxygen, determines the amount of sulfur which is deposited in the form of pyrite in oceanic sediments, which may eventually be subducted. For carbon, the total volcanic flux is only three times as large as the subduction flux. This means that the biosphere, through its capability to precipitate carbonates, has the capacity to noticeably influence the amount of carbon (see Sect. 3.4)

 Springer

available in the hydrosphere/atmosphere on epoch time scales (this is apart from the fixation of carbon in organic matter on much shorter time scales). It is hard to quantify this due to uncertainties in the numbers of Tables 1 and 2, variations of these values with time, and also the partial decoupling of carbonate deposition in the ocean basins and subduction of these carbonates (see Sect. 2.4.3).

Because of the size of the solid planet, these biological processes do not significantly influence the composition of the mantle. However, through their regulation of the atmospheric CO_2 concentration, they strongly influence the greenhouse effect and thus surface conditions in terms of temperature and availability of water. As we have shown in Sect. 4.1, the surface temperature affects the rate of convection and cooling of the mantle. The availability of water may determine whether the lithosphere is weak enough to break (Sect. 3.1) and allow subduction zones, an essential ingredient of plate tectonics, to be formed.

Such considerations also illustrate the importance of plate tectonics compared to one-plate convection. As illustrated in Fig. 1, the latter regime is generally thought not to allow the return of volatiles into the planet's interior. As a consequence, it is more difficult for any fixated volatile to be made available again to the biosphere. When plate tectonics case is active, elevated temperatures in the mantle contribute to releasing at least a part of the fixated material again in volatile form. However, some return of volatiles into the mantle may be possible, even in the stagnant lid case, by delamination of lower crustal material (e.g. Dupeyrat and Sotin 1995; Van Thienen et al. 2004a, 2004b; Elkins-Tanton 2007a, 2007b), assuming the sinking material contains volatile species. On Venus, the high atmospheric pressure may prevent exsolution of volatile species from lavas at the surface. In this case, return of the volatiles into the atmosphere is also possible by devolatilization or partial melting of sinking material.

Plate tectonics with volcanism at ridges and subduction zones is a continuous regime both in time and space, whereas the stagnant lid regime with plume and local tectonics-related volcanism is not. This is a fundamental difference. Earth's mid-ocean ridges form a single connected grid (Ricou 2004). At these ridges, production of new crust is more or less continuous. A recent compilation of sea floor ages suggests more or less constant spreading rates over the past 180 Myr (Rowley 2002). Subduction related volcanism is neither spatially nor temporally continuous, but the discontinuities both in space and time are limited. Reorganization, creation, and destruction of ridges and subduction zones take place as part of the Wilson cycle (see Turcotte and Schubert 2002). However, this does not interfere with the continuous existence of spreading centers and subduction zones.

On the other hand, volcanism related to plume activity (whether caused by a plume or an alternative mechanism, see e.g. King and Anderson 1995) is by definition limited to individual isolated localities. Also from a temporal point of view, activity is limited, starting with the eruption of flood basalts, and ending with the termination of the plume tail. Geological evidence for an increase in size and frequency of flood basalt eruptions for the younger Earth has been presented by Abbott and Isley (2002), corresponding to a decrease in distance in both space and time. However, Ernst et al. (2005) note that there are still many uncertainties which hinder such trend analyses.

As volcanism provides a source of nutrient elements, and contributes to the cycling of volatile species which play a role in the atmospheric energy balance (see Sect. 3.3), the distinction between temporal continuity and episodicity corresponds to a distinction between: (a) continuous and episodic supply of nutrient elements and (b) a stabilizing versus a destabilizing effect on planetary climate. The extent of the first effect also depends on the characteristic residence time of the nutrient elements in the biosphere before burial in a geological reservoir. The latter effect is illustrated by modeling studies. For example, Phillips et al.

(2001b) showed that the injection of CO_2 and H_2O into the Martian atmosphere by volcanism associated with the formation of Tharsis resulted in a warmer climate, which declined afterwards.

Episodic clement climate conditions and nutrient availability require a form of life which is capable of surviving intermediate periods, which may be very long (on the order of millions of years or more). Moreover, it may require some mobility in case the availability of nutrients is limited to the location of volcanic activity and these locations change with time.

On the other hand, flood volcanism events may act to significantly disrupt an already established biosphere by affecting climate. The eruption of flood basalts coincides with several large mass extinctions (e.g. Courtillot et al. 1986; Renne et al. 1995; White 2002; Courtillot 2003). Though the largest mass extinctions may have required the combined effects of flood volcanism and a bolide impact (White and Saunders 2005), it is clear that massive volcanic eruptions may cause a cooling of the climate on short time scales through silicate and sulphate aerosols (McCormick et al. 1995) and warming on a longer time scale through the injection of massive amounts of carbon into the atmosphere as described in Sect. 2.4.2 which results in a moderate increase of atmospheric CO_2 levels (Berner 2002).

As mentioned above, the existence of a dynamo can be important for the habitability on terrestrial planets due to the feedback mechanisms between the interior and the atmosphere (Fig. 1): The magnetic field protects the atmosphere from erosion of light elements, primarily hydrogen, due to sputtering and/or hydrodynamic effects (see Chap. 6). A sufficiently thick atmosphere with the associated greenhouse effect is necessary to allow fluid water to be present on the surface. However, water will not be stable in a liquid state if the atmosphere is too thick and the greenhouse effect is too strong. Water affects the evolution of a planetary mantle and the planetary tectonic engine. First, it makes the lithosphere deformable enough for subduction of the crust to occur. Second, water recycled to the mantle by plate tectonics reduces both the activation energy for creep and the solidus temperature of mantle rock, thereby enhancing the cooling of the interior. At the end of the cycle, the sufficient cooling by plate tectonics and/or a wet mantle rheology may allow the formation of an inner core and thus, a longstanding magnetic field. In the long term, an interruption or a change of this cycle may result in conditions that are not favorable for life. For example, the cessation of plate tectonics results in the heating of the interior. As a consequence the core cannot cool, the inner core growth stops and the magnetic field vanishes. This starts a runaway process through which atmosphere erosion and loss is promoted; the surface conditions change and water may become unstable at the surface. A similar scenario could be observed in case of one-plate tectonics throughout the evolution. Due to inefficient cooling of the interior, inner core growth never starts. An early thermal dynamo, which exists only in case of a superheated core as a consequence of core formation, vanishes rapidly after few hundred million years. Thus, an early atmosphere can be eroded efficiently due to the lack of a global protecting magnetic field. Whether and how long the surface conditions allow the existence of water during the planets evolution may depend strongly on the density and composition of the early atmosphere.

5.2 Important Open Questions

An important question which needs to be investigated is to what extent life is essential for the long-term stability of water at a planet's surface. Would the Earth's atmosphere be hot like Venus' atmosphere, boiling away all the water, if life would not have arisen in the Archean and started fixating CO_2 from the atmosphere? Would the absence of oceans in such a scenario have prevented the operation (or even onset) of plate tectonics on Earth?

Would this, and the higher surface temperature, have reduced the heat flow from the mantle to such an extent that the core dynamo could not be sustained and Earth's protection against the solar wind would be removed?

At the moment we can only speculate on the answers to these questions, but they may hold the key to understanding why the Earth and its sister Venus and brother Mars are so alike in terms of composition and size, yet so different in terms of habitability.

Although it was primarily defined by astrophysicists, habitability is a central notion to which geobiologists and even geodynamicists can contribute. Life itself may modify habitability. This idea was central to the Gaia theory (e.g. Lovelock and Watson 1982). Although this theory has been largely criticized there is potentially an important role of life in the global carbon dioxide cycle at geological time scales. A better understanding of the mechanisms involved in this process, as well as a much better quantitative view of the biological role have yet to be obtained. Ironically, while some effort is done to look for extraterrestrial life, we need for our purpose to envision the evolution of a purely abiotic Earth which seems so far a remote aim.

Acknowledgements   We would like to thank Linda Elkins-Tanton for a very constructive and helpful review which led to a significantly improved chapter (and also for providing preprints), Kathryn Fishbaugh for helpful comments, particularly concerning language, and Jeroen van Hunen for constructive observations on the manuscript. Peter van Thienen acknowledges the financial support provided through the European Community's Human Potential Programme under contract RTN2-2001-00414, MAGE. This is IPGP contribution number 2195.

References

D.H. Abbott, A.E. Isley, J. Geodyn. **34**, 265–307 (2002)

Y. Abe, Phys. Earth Planet. Inter. **100**, 27–39 (1997)

R. Aghamiri, D.W. Schwartzman, Chem. Geol. **188**, 249–259 (2002)

O. Aharonson, M.T. Zuber, D.E. Smith, G.A. Neumann, W.C. Feldman, T.H. Prettyman, J. Geophys. Res. **109**(E05004) (2004). doi:10.1029/2003JE002223

D. Alfè, M.J. Gillan, G.D. Price, Earth Planet. Sci. Lett. **195**, 91–98 (2002)

C.J. Allègre, J. Poirier, E. Humler, A.W. Hofmann, Earth Planet. Sci. Lett. **134**, 515–526 (1995)

F.S. Anderson, S.E. Smrekar, J. Geophys. Res. **111**(E8), E08–009 (2006)

G. Arp, A. Reimer, J. Reitner, Eur. J. Phycol. **34**, 393–403 (1999)

G. Arp, A. Reimer, J. Reitner, Science **292**, 1701–1704 (2001)

C. Aubaud, F. Pineau, A. Jambon, M. Javoy, Earth Planet. Sci. Lett. **222**, 391–406 (2004)

G.E. Backus, Proc. Natl. Acad. Sci. **72**(4), 1555–1558 (1975)

W.W. Barker, S.A. Welch, J.F. Banfield, Rev. Mineral. **35**, 391–428 (1997)

P.C. Bennett, Geomicrobiol. J. **18**, 3–19 (2001)

K. Benzerara, N. Menguy, F. Guyot, G. De Luca, T. Heulin, C. Audrain, Geomicrobiol. J. **21**, 341–349 (2004)

K. Benzerara, N. Menguy, F. Guyot, C. Vanni, P. Gillet, Geochimica et Cosmochimica Acta **69**, 1413–1422 (2005a)

K. Benzerara, T.H. Yoon, N. Menguy, T. Tyliszczak, G. Brown, Proc. Natl. Acad. Sci. USA **102**, 979–982 (2005b)

K. Benzerara, N. Menguy, P. López-García, T.H. Yoon, J. Kazmierczak, T. Tyliszczak, G.E.J. Brown, Proc. Natl. Acad. Sci. USA **103**, 9440–9445 (2006)

D. Bercovici, S. Karato, Nature **425**, 39–44 (2003)

R. Berner, Rev. Mineral. **31**, 565–583 (1995)

R. Berner, A.C. Lasaga, R. Garrels, Am. J. Sci. **283**, 641–683 (1983)

R.A. Berner, Proc. Natl. Acad. Sci. **99**(7), 4172–417710 (2002). doi:1073/pnas.032095199

J. Bibring, Y. Langevin, A. Gendrin, B. Gondet, F. Poulet, M. Berth, A. Soufflot, R. Arvidson, N. Mangold, J. Mustard, P. Drossart, the OMEGA team Science **307**, 1576–1581 (2005)

T.M. Bond, M.R. Warner, Dating Venus: Statistical models of magmatic activity and impact cratering. 37th Annual Lunar and Planetary Science Conference, Abstr. No 1957 (2006)

S.R. Boyd, Precambrian Res. **108**, 158–173 (2001)
W.V. Boynton, W.C. Feldman, S.W. Squyres, T.H. Prettyman, J. Brockner, L.G. Evans, R.C. Reedy, R. Starr, J.R. Arnold, D.M. Drake, P.A.J. Englert, A.E. Metzger, I. Mitrofanov, J.I. Trombka, C. d'Uston, H. Wnke, O. Gasnault, D.K. Hamara, D.M. Janes, R.L. Marcialis, S. Maurice, I. Mikheeva, G.J. Taylor, R. Tokar, C. Shinohara, Science **297**(5578), 81–85 (2002)
S. Braginsky, Geomagn. Aeron. **4**, 698–712 (1964)
S.I. Braginsky, P.H. Roberts, Geophys. Astrophys. Fluid. Dyn. **79**, 1–97 (1995)
D. Breuer, T. Spohn, J. Geophys. Res. **108**(E7) (2003). doi:10.1029/2002JE001999
D. Breuer, T. Spohn, Planet. Space Sci. **54**(2), 153–169 (2006)
D. Breuer, H. Zhou, D.A. Yuen, T. Spohn, J. Geophys. Res. **101**(E3), 7531–542 (1996)
D. Breuer, A.D. Yuen, T. Spohn, Earth Planet. Sci. Lett. **148**, 457–469 (1997)
D. Breuer, D.A. Yuen, T. Spohn, S. Zhang, Geophys. Res. Lett. **25**(3), 229–232 (1998)
J.J. Brocks, G.A. Logan, R. Buick, R.E. Summons, Science **285**, 1033–1036 (1999)
M.A. Bullock, D.H. Grinspoon, J.W. Head, Venus resurfacing rates: Constraints provided by 3-d Monte Carlo simulations. Twenty-fourth Lunar and Planetary Science Conference, 1993
B.A. Campbell, J. Geophys. Res. **104**(E9), 21951–21955 (1999)
D.E. Canfield, Am. J. Sci. **304**, 839–861 (2004)
D. Canil, Nature **389**, 842–845 (1997)
E. Carlsson, A. Fedorov, S. Barabash, E. Budnik, A. Grigoriev, H. Gunell, H. Nilsson, J. Sauvaud, R. Lundin, Y. Futaana, Icarus **182**(2), 320–328 (2006)
M.H. Carr, *Water on Mars* (Oxford University Press, 1996)
S. Castanier, Sediment. Geolog. **126** (1999)
C. Catling, C. Leovy, in *Encyclopedia of the Solar System*, ed. by L.A. McFadden, P. Weissman, T. Johnson (Academic, San Diego, 2006)
O.A. Chadwick, R.T. Gavenda, E.F. Kelly, K. Ziegler, C.G. Olson, W.C. Elliott, D.M. Hendricks, Chem. Geol. **202**, 195–223 (2003)
R.J. Charlson, T.L. Anderson, R.E. McDuff, in *Global Biogeochemical Cycles*, ed. by S.S. Butcher, R.J. Charlson, G.H. Orians, G.V. Wolfe (Academic, San Diego, 1992)
E. Chassefière, F. Leblanc, B. Langlais, Planet. Space Sci. **55**(3), 343–353 (2007)
P.N. Chopra, M.S. Paterson, Tectonophysics **78**, 453–473 (1981)
U.R. Christensen, J. Geophys. Res. **90**(B4), 2995–3007 (1985)
C.F. Chyba, C.B. Phillips, Proc. Natl. Acad. Sci. **98**, 801–804 (2001). doi:10.1073/pnas.98.3.801
J.P. Cogné, E. Humler, Earth Planet. Sci. Lett. **227**, 427–439 (2004)
K.D. Collerson, B.S. Kamber, Science **283**, 1519–1522 (1999)
N. Coltice, L. Simon, C. Lécuyer, Geophys. Res. Lett. **31** (2004). doi:10.1029/2003GL018873
J.E.P. Connerney, M.H. Acuña, P.J. Wasilewski, N.F. Ness, H. Rème, C. Mazelle, D. Vignes, R.P. Lin, D.L. Mitchell, P.A. Cloutier, Science **284**, 794–798 (1999)
J.E.P. Connerney, M.H. Acuña, N.F. Ness, T. Spohn, G. Schubert, Space Sci. Rev. **111**(1–2), 1–32 (2004)
J.E.P. Connerney, M.H. Acuña, N.F. Ness, G. Klettetscha, D.L.D. Mitchell, R.P. Lin, H. Reme, Proc. Natl. Acad. Sci. **102**(42), 14970–14975 (2005)
C.P. Conrad, B.H. Hager, Geophys. Res. Lett. **26**(19), 3041–3044 (1999)
C.P. Conrad, B.H. Hager, Geophys. J. Int. **144**, 271–288 (2001)
C.P. Conrad, C. Lithgow-Bertelloni, Science **298**, 207–209 (2002)
V. Courtillot, J. Besse, D. Vandamme, R. Montigny, J. Jaeger, H. Cappetta, Earth Planet. Sci. Lett. **80**, 361–374 (1986)
V.P.R. Courtillot, Comptes Rendus Geosci. **335**, 113–140 (2003)
J.C. Dann, A.H. Holzheid, T.L. Grove, H.Y. McSween Jr., Meteorit. Planet. Sci. **36**, 793–806 (2001)
A. Davaille, C. Jaupart, J. Fluid. Mech. **253**, 141 (1993)
A. Davaille, C. Jaupart, J. Geophys. Res. **99**(B10), 19853–19866 (1994)
G.F. Davies, J. Geophys. Res. **85**, 2517–2530 (1980)
G.F. Davies, Geology **20**, 963–966 (1992)
J.H. De Smet, A.P. Van den Berg, N.J. Vlaar, Tectonophysics **322**, 19–33 (2000)
M.J. De Wit, Precambrian Res. **91**, 181–226 (1998)
C. DeFarge, J. Trichet, A. Jaunet, M. Robert, J. Tribble, F.J. Sansone, J. Sediment. Res. **66**, 935–947 (1996)
J.W. Delano, Orig. Life Evol. Biosphere **31**, 311–341 (2001)
C. Dessert, B. Dupre, J. Gaillardet, L.M. Francois, C.J. Allègre, Chem. Geol. **202**, 257–273 (2003)
T.M. Donahue, C.T. Russell, in *Venus II: Geology, Geophysics, Atmosphere, and Solar Wind Environment*, ed. by S.W. Bougher, D.M. Hunten, R.J. Philips (University of Arisona Press, Tucson, 1997), p. 3
G. Dreibus, H. Palme, Geochimica et Cosmochimica Acta **60**, 1125–1130 (1996)
G. Dreibus, H. Wänke, Icarus **71**, 225–240 (1987)
L. Dupeyrat, C. Sotin, Planet. Space Sci. **43**(7), 909–921 (1995)

B. Dupre, C. Dessert, P. Oliva, Y. Godderis, J. Viers, L. Francois, R. Millot, J. Gaillardet, Comptes Rendus Geosci. **335**, 1141–1160 (2003)

J.M. Edmond, Y. Huh, Earth Planet. Sci. Lett. **216**, 125–139 (2003)

L.T. Elkins-Tanton, J. Geophys. Res. **112**, B03405 (2007a). doi:10.1029/2005JB004072

L.T. Elkins-Tanton, J. Geophys. Res. (2007b accepted)

L.T. Elkins-Tanton, E.M. Parmentier, P.C. Hess, Meteorit. Planet. Sci. **38**(12), 1753–1771 (2003)

L.T. Elkins-Tanton, S.E. Zaranek, E.M. Parmentier, P.C. Hess, Earth Planet. Sci. Lett. **236**, 1–12 (2005)

R.E. Ernst, K.L. Buchan, I.H. Campbell, Lithos **79**, 271–297 (2005)

B.J. Fegley, R.G. Prinn, in *The Formation and Evolution of Planetary Systems*, ed. by H.A. Weaver, L. Danly (Elsevier, Amsterdam, 1989), pp. 171–205

B. Fegley Jr., in *Treatise on Geochemistry*, ed. by A.M. Davis (Cambridge University Press, Cambridge, 2004)

Y. Fei, C.M. Bertka, L.W. Finger, Science **275**, 1621–1623 (1997)

W.C. Feldman, W.V. Boynton, R.L. Tokar, T.H. Prettyman, O. Gasnault, S.W. Squyres, R.C. Elphic, D.J. Lawrence, S.L. Lawson, S. Maurice, G.W. McKinney, K.R. Moore, R.C. Reedy, Science **297**(5578), 75–78 (2002)

F.G. Ferris, R.G. Wiese, W.S. Fyfe, Geomicrobiol. J. **12**, 1–13 (1994)

V. Formisano, S. Atreya, T. Encrenaz, N. Ignatiev, M. Giuranna, Science **306**(5702), 1758–1761 (2004)

S. Franck, C. Bounama, Phys. Earth Planet. Inter. **92**, 57–65 (1995)

S. Franck, C. Bounama, Phys. Earth Planet. Inter. **100**, 189–196 (1997)

S. Franck, C. Bounama, J. Geodyn. **32**, 231–246 (2001)

S. Franck, K. Kossacki, C. Bounama, Chem. Geol. **159**, 305–317 (1999)

D. Frost, C. Liebske, F. Langenhorst, C.A. McCammon, R.G. Tronnes, D.C. Rubie, Nature **428**, 409–412 (2004)

E. Gaidos, B. Deschenes, L. Dundon, K. Fagan, C. Mcnaughton, L. Menviel-Hessler, N. Moskovitz, M. Workman, Astrobiology **5**(2), 100–126 (2005)

J. Gaillardet, B. Dupre, P. Louvat, C.J. Allegre, Chem. Geol. **159**, 3–30 (1999)

J.N. Galloway, in *Biogeochemistry*, ed. by H.D. Holland, K.K. Turekian, vol. 8 (Elsevier–Pergamon, Oxford, 2001)

P. Gautret, G. Camoin, S. Golubic, S. Sprachta, J. Sediment. Res. **74**, 462–478 (2004)

O. Grasset, E.M. Parmentier, J. Geophys. Res. **103**(B8), 18171–18181 (1998)

C. Grigné, S. Labrosse, Geophys. Res. Lett. **28**(14), 2707–2710 (2001)

C. Grigné, S. Labrosse, P.J. Tackley, J. Geophys Res. **110**(B3), B03–409 (2005). doi:10.1029/2004JB003376

D.H. Grinspoon, Nature **363**, 428–431 (1993)

D.H. Grinspoon, M.A. Bullock, Geochimica et Cosmochimica Acta **69**(10), A750 (2005)

D. Gubbins, D. Alfè, G. Masters, G.D. Price, M.J. Gillan, Geophys. J. Int. **155**, 609622 (2003)

B.R. Hacker, G.A. Abers, S.M. Peacock, J. Geophys. Res. **107** (2003). doi:10.1029/2001JB001127

W.B. Hamilton, Precambrian Res. **91**, 143–179 (1998)

H. Harder, U. Christensen, Nature **280**, 507–509 (1996)

W.K. Hartmann, G. Neukum, Space Sci. Rev. **96**, 165–194 (2001)

S. Hauck II, R.J. Phillips, J. Geophys. Res. **107**(E7), (2002). doi:10.1029/2001JE001801

S.A. Hauck II, R. Phillips, M.H. Price, J. Geophys. Res. **103**(E6), 13635 (1998)

J. Head, G. Neukum, R. Jaumann, H. Hiesinger, E. Hauber, et al., Nature **434**, 346–351 (2005)

J. Helbert, J. Benkhoffm, Planet. Space Sci. **54**(4), 331–336 (2006)

J.M. Hewitt, D.P. McKenzie, N.O. Weiss, J. Fluid. Mech. **68**, 721–738 (1975)

N. Hirao, T. Kondo, E. Ohtani, K. Takemura, T. Kikegawa, Geophys. Res. Lett. **31** (2004). doi:0.1029/2003GL019380

G. Hirth, D.L. Kohlstedt, Earth Planet. Sci. Lett. **144**, 93–108 (1996)

H.D. Holland, Geochimica et Cosmochimica Acta **66**(21), 3811–3826 (2002)

K. Holmen, in *Global Biogeochemical Cycles*, ed. by S.S. Butcher, R.J. Charlson, G.H. Orians, G.V. Wolfe (Academic, San Diego, 1992)

T. Hoogenboom, G.A. Houseman, Icarus **180**, 292–307 (2006)

G.P. Horedt, Phys. Earth Planet. Inter. **21**, 22–30 (1980)

M. Humayun, L. Qin, M.D. Norman, Science **306**, 91–94 (2004)

D.M. Hunten, R.O. Pepin, J.C.G. Walker, Icarus **69**, 532–549 (1987)

E. Hutchens, E. Valsami-Jones, S. McEldowney, W. Gaze, J. McLean, Mineral. Mag. **67**, 1157–1170 (2003)

J. Ingrin, H. Skogby, Eur. J. Mineral. **12**, 543–570 (2000)

T. Inoue, D.J. Weidner, P.A. Northrup, J.B. Parise, Earth Planet. Sci. Lett. **160**, 106–113 (1998)

D.M. Ito, E. Harris, A.T.J. Anderson, Geochimica et Cosmochimica Acta **47**, 1613–1624 (1983)

M.A. Ivanov, J.W. Head, Testing directional (evolutionary) and non-directional models of the geologic history of venus: Results of mapping in a geotraverse along the equator of Venus. 37th Annual Lunar and Planetary Science Conference, Abstr. No 1366 (2006)

T. Jackson, W. Keller, Am. J. Sci. **269**, 446–466 (1970)

R.A. Jahnke, in *Global Biogeochemical Cycles*, ed. by S.S. Butcher, R.J. Charlson, G.H. Orians, G.V. Wolfe (Academic, San Diego, 1992)

B.M. Jakosky, R.J. Phillips, Nature **412**, 237–244 (2001)

R.D. Jarrard, Geochem. Geophys. Geosyst. **4**(5) (2002). doi:10.1029/2002GC000392

C. Jaupart, S. Labrosse, J.C. Mareschal, in *Treatise of Geophysics. Mantle*, ed. by D. Bercovici (Elsevier, 2007, in press)

M. Javoy, Comptes Rendus Acad. Sci. Paris **329**, 537–555 (1999)

A.M. Jellinek, M. Manga, Rev. Geophys. **42**(RG3002) (2004). doi:10.1029/2003RG000144

C.L. Johnson, M.A. Richards, J. Geophys. Res. **108**(E6), 5058 (2003)

J.H. Jones, The edge of wetness: the case for dry magmatism on Mars. Lunar and Planetary Science Conference XXXV, Abstr. No 1798 (2004)

T.H. Jordan, in *The Mantle Sample: Inclusions in Kimberlites and Other Volcanics* (American Geophysical Union, 1979), pp. 1–14

A. Kadik, Phys. Earth Planet. Inter. **100**, 157–166 (1997)

M. Kameyama, M. Ogawa, Earth Planet. Sci. Lett. **180**, 355–367 (2000)

S. Karato, Nature **319**, 309–310 (1986)

J.F. Kasting, D.H. Eggler, S.P. Raeburn, J. Geol. **101**, 245–257 (1993)

R.F. Katz, M. Spiegelman, C.H. Langmuir, Geochem. Geophys. Geosyst. **4**(9) (2003). doi:10.1029/2002GC000433

E.F. Kelly, O.A. Chadwick, T.E. Hilinski, Biogeochemistry **42**, 21–53 (1998)

H. Keppler, M. Wiedenbeck, S.S. Shcheka, Nature **424**, 414–416 (2003)

D.M. Kerrick, J.A.D. Connolly, Nature **411**, 293–296 (2001)

S.D. King, D.L. Anderson, Earth Planet. Sci. Lett. **136**(3–4), 269–279 (1995)

K. Koga, E. Hauri, M. Hirschmann, D. Bell, Geochem. Geophys. Geosyst. **4**(2) (2003). doi:10.1029/2002GC000378

J. Korenaga, Geophys. Res. Lett. **30**(8) (2003). doi:10.1029/2003GL016982

V.A. Krasnopolsky, J.P. Maillard, T.C. Owen, Icarus **172**, 537–547 (2004)

L.R. Kump, J.F. Kasting, M.E. Barley, Geochem. Geophys. Geosyst. **2** (2001). doi:10.1029/2000GC000114

K. Kuramoto, Phys. Earth Planet. Inter. **100**, 3–20 (1997)

S. Labrosse, Phys. Earth Planet. Inter. **140**(1–3), 127–143 (2003)

S. Labrosse, C. Jaupart, Earth Planet. Sci. Lett. (2007, accepted for publication)

S. Labrosse, J. Poirier, J. Le Mouël, Earth Planet. Sci. Lett. **190**, 111123 (2001)

Y. Langevin, P. Fran, J.P. Bibring, B. Gondet, Science **307**(5715), 1584–1586 (2005)

J.C. Lassiter, Earth Planet. Sci. Lett. **250**, 306–317 (2006)

K.K.M. Lee, R. Jeanloz, Geophys. Res. Lett. **30**(23), 2212 (2003)

A. Lenardic, F. Nimmo, L. Moresi, J. Geophys. Res. **109**(E02003) (2004). doi:10.1029/2003JE002172

T.M. Lenton, Nature **394**, 439–447 (1998)

T.M. Lenton, J.E. Lovelock, Tellus Ser. B – Chem. Phys. Meteorol. **53**, 288–305 (2001)

Z.A. Li, C.A. Lee, Earth Planet. Sci. Lett. **228**, 483–493 (2004)

L.J. Liermann, B.E. Kalinowski, S.L. Brantley, J.G. Ferry, Geochimica et Cosmochimica Acta **64**, 587–602 (2000)

F. Lippmann, *Sedimentary Carbonate Minerals* (Springer, Berlin, 1973) p. 228

J. Lister, B.A. Buffett, Phys. Earth Planet. Inter. **91**, 17–30 (1995)

K. Lodders, B. Fegley Jr., Icarus **126**, 373–394 (1997)

A. Loddoch, C. Stein, U. Hansen, Earth Planet. Sci. Lett. **251**, 79–89 (2006)

J.E. Lovelock, A.J. Watson, Planet. Space Sci. **30**, 795–802 (1982)

Y. Lucas, Annu. Rev. Earth Planet. Sci. **29**, 135–163 (2001)

J.R. Lyons, C. Manning, F. Nimmo, Geophys. Res. Lett. **32** (2004). doi:10.1029/2004GL022161

B. Marty, I.N. Tolstikhin, Chem. Geol. **145**, 233–248 (1998)

M.P. McCormick, L.W. Thomason, C.R. Trepte, Nature **373**, 399–404 (1995)

W.F. McDonough, in *Treatise on Geochemistry*, ed. by R.W. Carlson. The Mantle and Core, vol. 2 (Elsevier–Pergamon, Oxford, 2003), pp. 547–568

W.F. McDonough, S. Sun, Chem. Geol. **120**, 223–253 (1995)

P.J. McGovern, G. Schubert, Earth Planet. Sci. Lett. **96**, 27–37 (1989)

W.B. McKinnon, K.J. Zahnle, B.A. Ivanov, H.J. Melosh, in *Venus II*, ed. by S.W. Bougher, D.M. Hunten, R.J. Phillips (University of Arizona Press, Arizona, 1997)

H. McSween Jr., T.L. Grove, R.C.F. Lentz, J.C. Dann, A.H. Holzheld, L.R. Riciputi, J.G. Ryan, Nature **409**, 487–490 (2001)

M.A. Mischna, M.I. Richardson, R.J. Wilson, D.J. McCleese, J. Geophys. Res. **108**(E6), 5062 (2003). doi:10.1029/2003JE002051

I. Mitrofanov, D. Anfimov, A. Kozyrev, M. Litvak, A. Sanin, V. Tret'yakov, A. Krylov, V. Shvetsov, W. Boynton, C. Shinohara, D. Hamara, R.S. Saunders, Science **297**(5578), 78–81 (2002)

L. Moresi, V. Solomatov, Geophys. J. Int. **133**, 669–682 (1998)

L. Moresi, V.S. Solomatov, Phys. Fluids **7**, 2154–2162 (1995)

R.Y. Morita, Geomicrobiol. J. **2**(63) (1980)

J.L. Mosenfelder, N.I. Deligne, P.D. Asimow, G.R. Rossman, Am. Mineral. **91**, 285–294 (2006)

K.L. Moulton, R.A. Berner, Geology **26**, 895–898 (1998)

R.D. Müller, W.R. Roest, J.Y. Royer, L.M. Gahagan, J.G. Sclater, J. Geophys. Res. **102**(B2), 3211–3214 (1997)

M. Murakami, K. Hirose, H. Yurimoto, S. Nakashima, N. Takafuji, Science **295**, 1885–1887 (2002)

J.W. Murray, in *Global Biogeochemical Cycles*, ed. by S.S. Butcher, R.J. Charlson, G.H. Orians, G.V. Wolfe (Academic, San Diego, 1992),

B.O. Mysen, A.L. Boettcher, J. Petrol. **16**, 520–548 (1975)

G. Neukum, R. Jaumann, H. Hoffmann, E. Hauber, J.W. Head, A.T. Basilevsky, B.A. Ivanov, S.C. Werner, S. van Gasselt, J.B. Murray, T. McCord, Nature **432**, 971–979 (2004)

F. Nimmo, Geology **30**(11), 987–990 (2002)

F. Nimmo, D. McKenzie, Annu. Rev. Earth Planet. Sci. **26**, 23–51 (1998)

F. Nimmo, D.J. Stevenson, J. Geophys. Res. **105**(E5), 11969–11979 (2000)

M. Ogawa, J. Geophys. Res. **105**(E3), 6997–7012 (2000)

E. Ohtani, N. Hirao, T. Kondo, M. Ito, T. Kikegawa, Phys. Chem. Miner. **32**, 77–82 (2005). doi:10.1007/s00269-004-0443-6

T. Okuchi, Science **278**, 1781–1784 (1997)

E.R. Oxburgh, E.M. Parmentier, J. Geol. Soc. Lond. **133**, 343–355 (1977)

E.M. Parmentier, P.C. Hess, Geophys. Res. Lett. **19**(20), 2015–2018 (1992)

S.M. Peacock, Science **248**, 329–337 (1990)

R.J. Phillips, V.L. Hansen, Science **279**, 1492–1497 (1998)

R.J. Phillips, M.A. Bullock, S.A. Hauck II, Geophys. Res. Lett. **28**, 1779–1782 (2001a)

R.J. Phillips, M.T. Zuber, S.C. Solomon, M.P. Golombek, B.M. Jakosky, W.B. Benerdt, D.E. Smith, R.M.E. Williams, B.M. Hynek, O. Aharonson, S. Hauck II, Science **291**, 2587–2591 (2001b)

C. Pinet, C. Jaupart, J.C. Mareschal, C. Gariepy, G. Bienfait, R. Lapointe, J. Geophys. Res. **96**, 19941–19963 (1991)

T. Plank, C.H. Langmuir, Chem. Geol. **145**, 325–394 (1998)

F. Raulin, T. Owen, Space Sci. Rev. **104**(1–4), 377–394 (2002). doi:10.1023/A:1023636623006

C.C. Reese, V.S. Solomatov, L.N. Moresi, J. Geophys. Res. **103**(E6), 13643–13657 (1998)

C.C. Reese, V.S. Solomatov, L.N. Moresi, Icarus **139**, 67–80 (1999)

K. Regenauer-Lieb, D.A. Yuen, J. Branlund, Science **294**, 578–580 (2001)

P.R. Renne, Z. Zichao, M.A. Richards, M.T. Black, A.R. Basu, Science **269**, 1413–1416 (1995)

J.A. Resing, J.E. Lupton, R.A. Feely, M.D. Lilley, Earth Planet. Sci. Lett. **226**, 449–464 (2004)

L.E. Ricou, Tectonophysics **384**(1–4), 285–300 (2004)

R. Riding, Sedimentology **47**, 179–214 (2000)

J.R. Rogers, P.C. Bennett, Chem. Geol. **203**, 91–108 (2004)

D.B. Rowley, Geol. Soc. Am. Bull. **114**(8), 927–933 (2002)

L.H. Rüpke, J.P. Morgan, M. Hort, J.A.D. Connolly, Earth Planet. Sci. Lett. **223**, 17–34 (2004)

A.E. Saal, E.H. Hauri, C.H. Langmuir, M.R. Perfit, Nature **419**, 451–455 (2002)

Y. Sano, S.N. Williams, Geophys. Res. Lett. **23**, 2749–2752 (1996)

Y. Sano, Y. Nishio, B. Marty, Geophys. Res. Lett. **25**, 2289–2292 (2001a)

Y. Sano, N. Takahata, Y. Nishio, T. Fischer, S. Williams, Chem. Geol. **171**, 263–271 (2001b)

C. Santelli, S. Welch, H. Westrich, J. Banfield, Chem. Geol. **180**, 99–115 (2001)

S. Saxena, H.P. Liermann, G. Shen, Phys. Earth Planet. Inter. **146**, 313–317 (2004)

G.G. Schaber, R.G. Strom, H.J. Moore, L.A. Soderblom, R.L. Kirk, D.J. Chadwick, D.D. Dawson, L.R. Gaddis, J.M. Boyce, J. Russel, J. Geophys. Res. **97**(E8), 13257–13301 (1992)

A. Scherstén, T. Elliot, C. Hawkesworth, M. Norman, Nature **427**, 234–237 (2004)

B. Schott, A.P. Van den Berg, D.A. Yuen, Slow secular cooling and long lived volcanism on Mars explained. Lunar and Planetary Science Conference 33, 2002. http:/www.lpi.usra.edu/meetings/lpsc2002

G. Schubert, Annu. Rev. Earth Planet. Sci. **7**, 289–342 (1979)

G. Schubert, T. Spohn, J. Geophys. Res. **95**(B9), 14095–14104 (1990)

G. Schubert, D. Stevenson, P. Cassen, J. Geophys. Res. **85**(B5), 2531–2538 (1980)

G. Schubert, V.S. Solomatov, P.J. Tackley, D.L. Turcotte, in *Venus II: Geology, Geophysics, Atmosphere, and Solar Wind Environment*, ed. by S.W. Bougher, D.M. Hunten, R.J. Philips (University of Arizona Press, Tucson, 1997) p. 1245

S. Schumacher, D. Breuer, J. Geophys. Res. **111**(E02006) (2006). doi:10.1029/2005JE002429

D.W. Schwartzman, T. Volk, Glob. Planet. Chang. **90**, 357–371 (1991)

J.G. Sclater, C. Jaupart, D. Galson, Rev. Geophys. Space Phys. **18**, 269–312 (1980)

N.H. Sleep, J. Geophys. Res. **99**(E3), 5639–5655 (1994)

N.H. Sleep, J. Geophys. Res. **105**(E7), 17563–17578 (2000)

N.H. Sleep, B.F. Windley, J. Geol. **90**, 363–379 (1982)

D.E. Smith, M.T. Zuber, S.C. Solomon, R.J. Phillips, J.W. Head, J.B. Garvin, W.B. Banerdt, D.O. Muhleman, G.H. Pettengill, G.A. Neumann, F.G. Lemoine, J.B. Abshire, O. Aharonson, C.D. Brown, S.A. Hauck, A.B. Ivanov, P.J. McGovern, H.J. Zwally, T.C. Duxbury, Science **284**(5419), 1495–1503 (1999)

S.E. Smrekar, E.R. Stofan, Icarus **139**(1), 100–115 (1999)

V.S. Solomatov, Phys. Fluids **7**(2), 266–274 (1995)

V.S. Solomatov, J. Geophys. Res. **109**(B01412) (2004). doi:10.1029/2003JB002628

S.C. Solomon, O. Aharonson, J.M. Aurnou, W.B. Banerdt, M.H. Carr, A.J. Dombard, H.V. Frey, M.P. Golombek, S.A. Hauck, B.M. Head, J.W. Jakosky, C.L. Johnson, P.J. McGovern, G.A. Neumann, R.J. Phillips, D.E. Smith, M.T. Zuber, Science **307**, 1214–1220 (2005). doi:10.1126/science.1101812

T. Spohn, Icarus **90**, 222–236 (1990)

T. Spohn, F. Sohl, D. Breuer, Phys. Astron. **8**(3), 181–235 (1998)

T. Spohn, M.H. Acuña, D. Breuer, M. Golombek, R. Greeley, A. Halliday, E. Hauber, R. Jaumann, F. Sohl, Space Sci. Rev. **96**, 231–262 (2001)

S.W. Squyres, R.E. Arvidson, J.F. Bell III, J. Brockner, N.A. Cabrol, W. Calvin, M.H. Carr, P.R. Christensen, B.C. Clark, L. Crumpler, D.J.D. Marais, C. d'Uston, T. Economou, J. Farmer, W. Farrand, W. Folkner, M. Golombek, S. Gorevan, J.A. Grant, R. Greeley, J. Grotzinger, L. Haskin, K.E. Herkenhoff, S. Hviid, J. Johnson, G. Klingelhöfer, A.H. Knoll, G. Landis, M. Lemmon, R. Li, M.B. Madsen, M.C. Malin, S.M. McLennan, H.Y. McSween, D.W. Ming, J. Moersch, R.V. Morris, T.J.W. Parker, J. Rice, L. Richter, R. Rieder, M. Sims, M. Smith, P. Smith, L.A. Soderblom, R. Sullivan, H. Wänke, T. Wdowiak, M. Wolff, A. Yen, Science **306**(5702), 1698–1703 (2004)

T. Staudacher, C.J. Allègre, Earth Planet. Sci. Lett. **89**, 173–183 (1988)

C. Stein, J. Schmalzl, U. Hansen, Phys. Earth Planet. Inter. **142**, 225–255 (2004)

R.J. Stern, Earth Planet. Sci. Lett. **226**, 275–292 (2004)

R.J. Stern, Geology **33**(7), 557–560 (2005)

D. Stevenson, T. Spohn, G. Schubert, Icarus **54**, 466–489 (1983)

D.J. Stevenson, Earth Planet. Sci. Lett. **208**, 1–11 (2003a)

D.J. Stevenson, Comptes Rendus Geosci. **335**, 99–111 (2003b)

H. Svensen, S. Planke, A. Malthe-Sorenssen, B. Jamtveit, R. Myklebust, T. Rasmussen Eldem, S.R. Rey, Nature **429**, 542–545 (2004)

T.N. Tingle, Chem. Geol. **147**, 3–10 (1998)

D.L. Turcotte, J. Geophys. Res. **98**(E9), 17061–17068 (1993)

D.L. Turcotte, J. Geophys. Res. **100**(E8), 16931–16940 (1995)

D.L. Turcotte, E.R. Oxburgh, J. Fluid. Mech. **28**, 29–42 (1967)

D.L. Turcotte, G. Schubert, *Geodynamics, Applications of Continuum Physics to Geological Problems*, 2nd edn. (Cambridge University Press, Cambridge, 2002)

D.L. Turcotte, G. Morein, D. Roberts, B.D. Malamud, Icarus **139**, 49–54 (1999)

W.J. Ullman, D.L. Kirchman, S.A. Welch, P. Vandevivere, Chem. Geol. **132**, 11–17 (1996)

P. Van Keken, B. Kiefer, S. Peacock, Geochem. Geophys. Geosyst. **3**(10) (2002). doi:10.1029/2001GC000256

P. van Thienen, A.P. van den Berg, N.J. Vlaar, Tectonophysics **394**(1–2), 111–124 (2004a)

P. van Thienen, A.P. van den Berg, N.J. Vlaar, Tectonophysics **386**(1–2), 41–65 (2004b)

P. van Thienen, N.J. Vlaar, A.P. van den Berg, Phys. Earth Planet. Inter. **142**(1–2), 61–74 (2004c)

P. van Thienen, N. Vlaar, A. van den Berg, Phys. Earth Planet. Inter. **150**, 287–315 (2005)

P. van Thienen, A. Rivoldini, T. Van Hoolst, P. Lognonné, Icarus **185**, 197–210 (2006)

J.C. Varekamp, R. Kreulen, R.P.E. Poorter, M.J. van Bergen, Terra Nova **4**, 363–373 (1992)

C. Vasconcelos, J.A. McKenzie, S. Bernasconi, D. Grujic, A.J. Tien, Nature **377**, 220–222 (1995)

N.J. Vlaar, A.P. Van den Berg, in *Glacial Isostasy, Sea-Level and Mantle Rheology*, ed. by R. Sabadini, K. Lambeck, E. Boschi (Kluwer, Dordrecht, 1991)

W. Von Bloh, S. Franck, C. Bounama, H.J. Schellnhuber, Geomicrobiol. J. **20**, 501–511 (2003)

J. Wade, B.J. Wood, Earth Planet. Sci. Lett. **236**, 78–95 (2005)

K. Wallmann, Geochimica et Cosmochimica Acta **65**(15), 2469–2485 (2001)

H. Wänke, G. Dreibus, Philos. Trans. Roy. Soc. Lond. **A349**, 2134–2137 (1994)

A.F. White, S.L. Brantley, Chem. Geol. **202**, 479–506 (2003)

A.F. White, A.E. Blum, T.D. Bullen, D.V. Vivit, M. Schulz, J. Fitzpatrick, Geochimica et Cosmochimica Acta **63**, 3277–3291 (1999)

R.V. White, Philos. Trans. Roy. Soc. Lond. **360**, 2963–2985 (2002)

R.V. White, A.D. Saunders, Lithos **79**, 299–316 (2005)

P.B. Wignall, Earth – Sci. Rev. **53**, 1–33 (2001)

D.R. Williams, Mars fact sheet. Tech. rep., NASA, 2004. http://nssdc.gsfc.nasa.gov/planetary/factsheet/marsfact.html

D.R. Williams, Venus fact sheet. Tech. rep., NASA, 2005. http://nssdc.gsfc.nasa.gov/planetary/factsheet/venusfact.html

H.M. Williams, C.A. McCammon, A.H. Peslier, A.N. Halliday, N. Teutsch, S. Levasseur, J.-P. Burg, Science **304**, 1656–1659 (2004)

H.M. Williams, A.H. Peslier, C.A. McCammon, A.N. Halliday, S. Levasseur, N. Teutsch, J.-P. Burg, Earth Planet. Sci. Lett. **235**, 435–452 (2005)

Q. Williams, R.J. Hemley, Annu. Rev. Earth Planet. Sci. **29**, 365–418 (2001)

J.T. Wilson, Nature **211**, 676–681 (1966)

M. Wilson, J. Geol. Soc. **150**, 923–926 (1993)

A.C. Withers, B.J. Wood, M.R. Carroll, Chem. Geol. **147**, 161–171 (1998)

B.J. Wood, Earth Planet. Sci. Lett. **117**, 593–607 (1993)

D.T. Wright, A. Oren, Geomicrobiol. J. **22**, 27–53 (2005)

S.E. Zaranek, E.M. Parmentier, J. Geophys. Res. **109** (2004). doi:10.1029/2003JB002462

Y. Zhang, A. Zindler, Earth Planet. Sci. Lett. **117**, 331–345 (1993)

V.N. Zharkov, T.V. Gudkova, Phys. Earth Planet. Inter. **117**, 407–420 (2000)

M.T. Zuber, Nature **412**, 220–227 (2001)

M.T. Zuber, D.E. Smith, S.C. Solomon, J.B. Abshire, R.S. Afzal, O. Aharonson, K. Fishbaugh, P.G. Ford, H.V. Frey, J.B. Garvin, J.W. Head, A.B. Ivanov, C.L. Johnson, D.O. Muhleman, G.A. Neumann, G.H. Pettengill, R.J. Phillips, X. Sun, H.J. Zwally, W.B. Banerdt, T.C. Duxbury, Science **282**(5396), 2053–2060 (1998)

Space Sci Rev (2007) 129: 205–206
DOI 10.1007/s11214-007-9190-6

Introduction to Chapter 6: Planetary/Sun Interactions

Helmut Lammer · Véronique Dehant · Oleg Korablev ·
Rickard Lundin

Received: 23 November 2006 / Accepted: 9 April 2007 /
Published online: 15 May 2007
© Springer Science+Business Media B.V. 2007

Keywords Atmospheric loss · Young Sun · Magnetospheric protection

To understand what has generated Earth's long-time habitable environment, compared with other terrestrial planets like Venus and Mars, one cannot neglect the complex interplay between geology, water inventories, the generation of magnetic dynamos, the climate, the atmosphere, and their connection to the evolutionary influence of the Sun. Only stable and dense enough atmospheres that allow water to be liquid over geological time periods, and magnetospheres that are able to protect the surface from hostile radiation and the upper atmosphere against non-thermal loss processes will allow the evolution of complex biospheres like that on the Earth. The evolving solar energy and particle fluxes play a major role not only in climate change but also in photochemistry, escape of atoms and molecules from upper atmospheres, and hence in the evolution of planetary atmospheres. The long-term modification of planetary atmospheres and the importance of magnetospheric protection of the upper atmosphere against the solar wind are discussed in detail within the following three articles, so that a broader interdisciplinary scientific community will understand the complex connections between astrophysics, solar physics, space plasma physics, planetary atmospheres, planetology, and habitability as a whole. For instance, it is known from enriched heavy isotopes in the present atmospheres of Solar System planets that the early atmospheres should

H. Lammer (✉)
Space Research Institute, Austrian Academy of Sciences, Schmiedlstr. 6, 8042 Graz, Austria
e-mail: helmut.lammer@oeaw.ac.at

V. Dehant
Royal Observatory of Belgium, Av. Circulaire, 3, 1180 Brussels, Belgium

O. Korablev
Space Research Institute (IKI), 117997, 84/32 Profsoyuznaya Str, Moscow, Russian Federation

R. Lundin
Swedish Institute of Space Physics (IRF), Box 812, 98128 Kiruna, Sweden

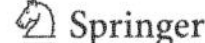 Springer

have experienced high escape rates to space, which could only be triggered by a much more active young Sun (*Kallenbach et al.* 2003, and articles and references therein).

The nuclear evolution of the Sun is generally well understood from stellar evolutionary theory, and solar structure models are in good agreement with modern helioseismology observations. Because of ever accelerating nuclear reactions in the core, the Sun is a slowly evolving variable star that has undergone an increase of about 30% in luminosity over the past 4.5 Gyr. Although the total luminosity is important for the climate on a planet, the relevant wavelengths for the heating, expansion, and escape of the upper atmospheres and of the planetary water inventory are the ionizing wavelengths less than 100 nm, which contain only a small fraction of the total spectral power. Astrophysical observations of solar analogue stars with various ages indicate that the young Sun may have had a short wavelength (X-ray and EUV) phase, with ionizing fluxes up to 100 times more intense and stellar winds which could have been between 100–1000 times stronger than that of the present Sun after the Sun arrived at the Zero Age Main Sequence.

The first article in Chap. 6 by *Lundin et al.* focuses on the short and long-time variability of the solar radiation and plasma environment, discusses related spacecraft observations, and explains the relevance of planetary magnetic fields to solar forcing of terrestrial planets and to atmospheric evolution. This article also presents a general overview of the evolving Sun and the relevant processes related to solar interaction with magnetospheric-atmospheric planetary environments for an interdisciplinary scientific community. A second article by *Kulikov et al.* presents in more detail a comparative study of the influence of the active young Sun on the early upper atmospheres of Venus, Earth, and Mars. The article investigates from this point of view why the Earth may have evolved differently from Venus and Mars. Finally the third article of this chapter by *Dehant et al.* focuses in detail on the role of magnetic dynamos, their generation including the connection to planetary rotation, and the protection of atmospheres from the solar wind. A main focus in this work is dedicated to the role of the early Martian dynamo and to related open questions which address the riddle of the putative lost, denser, early atmosphere.

The investigations presented in the three articles emphasize the fact that it is absolutely necessary to consider the effects caused by the radiation and associated particle environment of the young Sun on the atmosphere in order to understand the evolution of the atmosphere and, ultimately, the effect of atmospheric evolution on any possible biospheres.

Reference

R. Kallenbach, T. Encrenaz, J. Geiss, K. Mauersberger, T. Owen, F. Robert (eds.). Solar System History from Isotopic Signatures of Volatile Elements. Space Science Series of ISSI, Kluwer Academic Publishers, and Space Sci. Rev., 106, No. 1–4, 2003

Space Sci Rev (2007) 129: 207–243
DOI 10.1007/s11214-007-9192-4

A Comparative Study of the Influence of the Active Young Sun on the Early Atmospheres of Earth, Venus, and Mars

Yuri N. Kulikov · Helmut Lammer · Herbert I.M. Lichtenegger · Thomas Penz ·
Doris Breuer · Tilman Spohn · Rickard Lundin · Helfried K. Biernat

Received: 1 February 2006 / Accepted: 10 April 2007 /
Published online: 8 June 2007
© Springer Science+Business Media B.V. 2007

Abstract Because the solar radiation and particle environment plays a major role in all atmospheric processes such as ionization, dissociation, heating of the upper atmospheres, and thermal and non-thermal atmospheric loss processes, the long-time evolution of planetary atmospheres and their water inventories can only be understood within the context of the evolving Sun. We compare the effect of solar induced X-ray and EUV (XUV) heating on the upper atmospheres of Earth, Venus and Mars since the time when the Sun arrived at the Zero-Age-Main-Sequence (ZAMS) about 4.6 Gyr ago. We apply a diffusive-gravitational equilibrium and thermal balance model for studying heating of the early thermospheres by photodissociation and ionization processes, due to exothermic chemical reactions and cooling by IR-radiating molecules like CO_2, NO, OH, etc. Our model simulations result in extended thermospheres for early Earth, Venus and Mars. The exospheric temperatures obtained for all the three planets during this time period lead to diffusion-limited hydrodynamic escape of atomic hydrogen and high Jeans' escape rates for heavier species like H_2, He, C, N, O, etc. The duration of this blow-off phase for atomic hydrogen depends essentially on the mixing ratios of CO_2, N_2 and H_2O in the atmospheres and could last from ~100 to several hundred million years. Furthermore, we study the efficiency of various non-thermal atmospheric loss processes on Venus and Mars and investigate the possible protecting effect of the early martian magnetosphere against solar wind induced ion pick up

Yu.N. Kulikov (✉)
Polar Geophysical Institute (PGI), Russian Academy of Sciences, Khalturina Str. 15, 183010
Murmansk, Russian Federation
e-mail: kulikov@pgi.ru

H. Lammer · H.I.M. Lichtenegger · T. Penz · H.K. Biernat
Space Research Institute, Austrian Academy of Sciences, Schmiedlstr. 6, 8042 Graz, Austria

D. Breuer · T. Spohn
Institute of Planetary Research, German Aerospace Center, Rutherfordstr. 2, 12489 Berlin,
Germany

R. Lundin
Swedish Institute of Space Physics (IRF), P.O. Box 812, 98128 Kiruna, Sweden

erosion. We find that the early martian magnetic field could decrease the ion-related non-thermal escape rates by a great amount. It is possible that non-magnetized early Mars could have lost its whole atmosphere due to the combined effect of its extended upper atmosphere and a dense solar wind plasma flow of the young Sun during about 200 Myr after the Sun arrived at the ZAMS. Depending on the solar wind parameters, our model simulations for early Venus show that ion pick up by strong solar wind from a non-magnetized planet could erode up to an equivalent amount of $\sim$250 bar of O^+ ions during the first several hundred million years. This accumulated loss corresponds to an equivalent mass of $\sim$1 terrestrial ocean (TO (1 TO $\sim$1.39 $\times$ 10^{24} g or expressed as partial pressure, about 265 bar, which corresponds to $\sim$2900 m average depth)). Finally, we discuss and compare our findings with the results of preceding studies.

Keywords Early atmospheres · Atmospheric evolution · Thermospheric heating · Solar induced atmospheric loss

1 Introduction

As discussed by Lundin et al. (2007, this issue) solar radiation and particle fluxes play a major role in all atmospheric processes. So, the evolution of planetary atmospheres can only be understood within the context of the evolving Sun. The escape of atmospheric constituents from the upper planetary atmospheres depends on the evolution of the solar X-ray and EUV (XUV) radiation ($\lambda \leq 102.7$ nm), which affects dissociation and ionization processes; the thermosphere and exosphere temperatures; and thermal and non-thermal escape rates. Observations and studies of isotope anomalies in planetary atmospheres (e.g., Lammer et al. 2000a; Lammer and Bauer 2003; Becker et al. 2003; and references therein), radiative fluxes, stellar magnetic fields, stellar winds of solar-type stars with different ages (Zahnle and Walker 1982; Sonnett et al. 1991; Ayres 1997; Guinan and Ribas 2002; Wood et al. 2002, 2005; Ribas et al. 2005; Lundin et al. 2007, this issue), and lunar and meteorite fossil records (Newkirk 1980) indicate that the young Sun underwent a highly active phase after the formation of the Solar System, which lasted about 0.5–1.0 Gyr and included frequent flare events where the particle flux and radiation intensity were several hundred times more intense than today.

Recent astrophysical multiwavelength (X-ray, EUV, FUV, UV, optical) observations of solar-type stars (solar proxies) with ages which cover most of the Sun's main sequence lifetime from 0.13–8.5 Gyr (carried out within the *Sun in Time* program (Dorren and Guinan 1994; Guinan and Ribas 2002; Ribas et al. 2005) and discussed in Lundin et al. (2007, this issue)) indicate that the coronal XUV emissions of the young main-sequence Sun were about 100 to 1000 times stronger than those of the Sun today. Thus the age-radiation relationship of the solar proxies indicates that the XUV flux of the young Sun at about 2.5 Gyr, 3.5 Gyr and 4.5 Gyr ago was about 3, 6 and 100 times higher, respectively, than today (Ribas et al. 2005), which is in general agreement with the previous studies by Iben (1965), Cohen and Kuhi (1979), Gough (1977), Zahnle and Walker (1982), Sonnett et al. (1991), and Ayres (1997), although these studies focused on stars that are not "real" solar proxies.

Besides solar radiation, the solar wind plasma flux is another important factor related to the evolution of planetary atmospheres because it induces non-thermal ion loss from weakly or non-magnetized planetary atmospheres. Recent Hubble Space Telescope high-resolution spectroscopic observations of the H Lyman-α feature of several nearby main-sequence stars carried out by Wood et al. (2002, 2005) have revealed neutral hydrogen absorption associated with the interaction between the stars' fully ionized coronal wind with the partially ionized local interstellar medium. Wood et al. (2002, 2005) modelled the absorption features formed in the astrospheres of these stars and provided the first empirically-estimated coronal mass loss rates for G-K main sequence stars (see also Lundin et al. 2007, this issue). It was found that the correlation between mass loss and the X-ray surface flux of young solar-like stars obeys a power law relationship, which indicates an average solar wind density up to 1000 times higher than today during the first 100 Myr after the Sun reached the Zero-Age-Main-Sequence (ZAMS).

However, recent observations by Wood et al. (2005) of the absorption signature of the astrosphere of the $\sim$0.55 Gyr old solar-like star, ξ Boo, indicate that there may possibly exist a high-activity cutoff regarding the stellar mass loss in a radiation-activity-relation derived in the form of a power law by Wood et al. (2002). They also find that the mass loss of that particular star is about 20 times less than the average stellar mass loss value inferred at about 4 Gyr. As pointed out by Ribas et al. (2005), Wood et al. (2005) and Lundin et al. (2007, this issue), more measurements would be needed to define better the mass loss and activity relation for "cool" main sequence stars, especially at high activity levels.

The aim of the present work is to apply thermospheric balance and diffusive equilibrium models to study the effect of the evolving solar XUV radiation and solar wind plasma on the early upper atmospheres of Earth, Venus and Mars in a comparative way. In Sect. 2, we discuss the heating of the Earth's thermosphere by ionizing radiation. We apply a thermospheric model to the present Earth atmosphere and vary the mixing ratio of CO_2, NO molecules, etc., which act as thermospheric coolers due to their IR emission. In Sect. 3, we apply our models to early Earth, Venus and Mars and study the thermospheric heating and exospheric temperatures as a function of neutral gas composition and solar XUV radiation. In Sect. 4, we discuss the impact of non-thermal atmospheric loss processes over the history of Venus and Mars and the atmospheric protection effect for the early martian atmosphere by the martian magnetic dynamo. Finally, we discuss in Sect. 5 the results of our study and implications for the early atmospheric evolution of Venus, Earth and Mars and their water inventories.

2 Thermospheric Heat Balance and Composition Modelling

The main radiation responsible for heating of upper atmospheres and the formation of planetary ionospheres is the solar XUV radiation. The part of the atmosphere where the XUV radiation is absorbed and a substantial fraction of its energy is transformed into heat, leading to a positive temperature gradient $dT/dz > 0$, is the thermosphere, which extends from about 90 to 210 km on Venus and Mars and from about 90 to 500 km on Earth.

In the lower thermosphere convection can play an important role in the transport of heat, while in the upper thermosphere heat is transported by molecular conduction, leading to an isothermal region ($T = $ const). In the part of the atmosphere which is called the exosphere, where the mean free path of the atmospheric species is large and collisions become negligible, lighter atmospheric constituents whose thermal velocity exceeds the gravitational escape velocity, can escape from the planet. The base of the exosphere is defined as an altitude level where the mean free path is about equal to the local scale height $H = kT_{\mathrm{exo}}/mg$ of the gas, with k the Boltzmann constant, T_{exo} the temperature at the exobase, g the gravitational acceleration, and m the mass of the main atmospheric species. The exobase level on Venus and Mars is located at an altitude of about 210 km and on Earth at about 500 km.

The most important heating and cooling processes in the upper atmosphere of Earth can be summarized as follows (e.g., Izakov 1971; Dickinson 1972; Chandra and Sinha 1974; Gridchin et al. 1975; Gordiets et al. 1978, 1979; Gordiets and Kulikov 1981, 1982; Gordiets et al. 1982, 1987; Gordiets 1991; Dickinson et al. 1987):

1) heating due to CO_2, N_2, O_2, and O photoionization by the solar XUV radiation ($\lambda \leq$ 102.7 nm),
2) heating due to O_2 and O_3 photodissociation by the solar UV radiation,
3) chemical heating in exothermic binary and 3-body reactions,
4) neutral gas molecular heat conduction,
5) turbulent energy dissipation and heat conduction,
6) heating and cooling due to contraction and expansion of the thermosphere,
7) IR-cooling in the vibrational-rotational bands of CO_2, NO, O_3, OH, NO^+, $^{14}N^{15}N$, CO, etc.

Gordiets et al. (1979, 1981, 1982), Gordiets and Kulikov (1981) applied a numerical model to calculate the thermal budget of the Earth's thermosphere in the altitude range of 90–500 km. Their model includes the main energy sources and sinks, such as IR-radiative cooling in the vibrational-rotational bands of optically active molecules, as well as heating and cooling arising from dissipation of turbulent energy and eddy heat transport. The heating by the solar XUV radiation of the present Earth's thermosphere yields an average exospheric temperature of $\sim$1000 K (e.g., Jacchia 1977; Crowley 1991). The results of their simulations which are in agreement with observations, revealed that the most efficient heat source in the Earth's thermosphere is due to photoionization by the XUV radiation and heating which arises from photodissociation of O_2 (e.g., Gordiets et al. 1982; Hunten 1993). These thermospheric heating processes are balanced by the main cooling processes which include IR radiative cooling in the 1.27–63 μm wavelength range and cooling due to molecular and eddy conduction (e.g., Gordiets et al. 1982).

In the present study we apply a thermospheric model based on the models of Gordiets et al. (1982) and Gordiets and Kulikov (1985) to the terrestrial, venusian, and martian atmospheres by considering the expected evolution of the solar XUV flux from 100 XUV about 4.5 Gyr ago to 1 XUV (present time normalized solar value). The thermospheric models solve in the vertical direction the 1-D time-dependent equations of continuity and diffusion for nine atmospheric constituents (CO_2, O, CO, N_2, O_2, Ar, He, NO, and H_2O), hydrostatic and heat balance equations, and the equations of vibrational kinetics for radiating molecules from below the base of the thermosphere (mesopause), up to the exobase. The model is self-consistent with respect to the neutral gas temperature and vibrational temperatures of the minor species radiating in the IR. It takes into account heating due to the CO_2, N_2, O_2, CO, and O photoionization by the XUV-radiation ($\lambda \leq$ 102.7 nm), heating due to O_2 and O_3 photodissociation by solar UV-radiation, chemical heating in exothermic 3-body reactions, neutral gas molecular heat conduction, IR-cooling in the vibrational-rotational bands of CO_2 (15 μm), CO, O_3, and in the 63 μm O line, and turbulent energy dissipation and heat conduction. We use the volume heating and cooling rates for the processes included in our simulations and the heating rates owing to photodissociation which are discussed in detail by Gordiets et al. (1978, 1979, 1982) and Gordiets and Kulikov (1981, 1982, 1985). The lower boundary conditions for the system of 1-D time-dependent hydrodynamic equations modelling the termosphere are

$$\rho = \rho_0, \qquad T = T_0, \tag{1}$$

where T is the temperature, ρ the gas density and T_0 and ρ_0 are taken from available observational data. The upper boundary of the model is at the exobase level where one can assume

$$\frac{\partial T}{\partial z} = 0, \qquad \frac{\partial^2 v_z}{\partial z^2} = 0, \tag{2}$$

here v_z is the vertical bulk gas velocity of the atmosphere.

The thermospheres of the terrestrial planets are heated mainly, as was noted above, by the absorption of the solar X-ray and extreme ultraviolet (EUV) radiation. In photoionization most of the excess solar photon energy is carried away by the electrons produced. These photoelectrons may cause secondary and further ionization, dissociation and excitation of electronic states of molecules and atoms. In photodissociation the excess energy can go into internal energy of the products, or it may be released as kinetic energy. To calculate the thermospheric heating rate due to the solar XUV radiation, it is conventional to introduce into heat balance models the "solar heating efficiency", which is the fraction of the solar energy absorbed that appears locally as heat. Among many other authors, notably Torr et al. (1979, 1980), Fox and Dalgarno (1981) and Fox (1988) have contributed to the present understanding of this key parameter for the terrestrial planets.

For the Earth's thermosphere it is now generally agreed that the heating efficiency has its maximum of 50–55% of the absorbed radiative energy at the heights of about 150–180 km, while below this altitude, where the solar heating is dominated by the Schumann–Runge continuum photodissociation, its value decreases to about 30%. Above $\sim$300 km where heating in chemical reactions becomes less important due to decreasing atmospheric collision rates, the heating efficiency also decreases down to values of about 10% (see, for example, Torr et al. 1980). Since most of the solar energy is deposited at heights below 200 km, it seems reasonable, in view of many uncertainties in the model parameters, to use for our heat balance calculations a height-averaged value of $\sim$50% for the heating efficiency in the Earth's thermosphere.

Fox and Dalgarno (1981) and later Fox (1988) calculated the EUV heating efficiencies in the thermosphere of Venus for different assumptions about the fraction of excess energy converted to vibrational excitation in elementary molecular processes. In their model Fox and Dalgarno (1981) assumed that half of the excess energy released in photodissociation of CO_2 and in chemical reactions produces vibrational excitation of the molecular products. Calculated altitude-dependent heating efficiencies were around 18% below 130 km and 22% above 135 km. The lower and upper limits according to their estimates were 10% and 30%, approximately. A later more strict analysis by Fox (1988) constrained the heating efficiency between about 22% in the lower region (below $\sim$130 km) and 25% in the upper region of the day time thermosphere of Venus. However, our estimated heating efficiencies for the present time thermospheres of Venus and Mars of about 16% and 8%, correspondingly, appear to be considerably smaller than the values obtained by Fox (1988). This problem with the heating efficiency in a CO_2 atmosphere is not new, however, and has been also encountered in the modelling works of other authors (Dickinson and Bougher 1986; Hollenbach et al. 1985). More dedicated studies are needed in the future to resolve this discrepancy.

IR emission of CO_2 in the 15 μm band is the major cooling process in the lower thermospheres of Venus, Earth and Mars (e.g., Gordiets et al. 1982; Gordiets and Kulikov 1985; Bougher et al. 1999; Bougher et al. 2000). For the calculation of the heat loss rate q_{CO_2} due to IR emission of CO_2 in the 15 μm fundamental band excitation of the $CO_2(01^\circ 0)$ bending mode vibration-rotation states in collisions with heavy particles like atomic oxygen and CO_2, N_2, etc., collisional radiative de-excitation processes and absorption of radiation

are taken into account. For the 15 μm CO_2-band we use the "cool-to-space" approximation (e.g., Dickinson 1972; Gordiets et al. 1982) which can be expressed as

$$q_{CO_2} = 1.33 \times 10^{-13} g_w e^{-960/T} n_{CO_2} \left(\sum_M k_M n_M \right) F(\tau, \xi), \qquad (3)$$

where $g_w = 2$ is a statistical weight factor for the $01^\circ 0$ CO_2 molecule states, k_M is the relaxation rate constant for collisions with molecules and atoms having density n_M (CO_2, O, O_2, N_2, etc.), n_{CO_2} is the CO_2 number density and $F(\tau, \xi)$ accounts for the absorption of radiation in the band (Gordiets et al. 1982). Here τ is the reduced optical depths of the atmosphere for the 15 μm radiation at the height in question, ξ is the ratio of the radiative to the net relaxation rate of the $01^\circ 0$ CO_2 states at the same height.

The eddy conduction heating rate q_{ec} can be calculated by

$$q_{ec} = \frac{\partial}{\partial z} \left[\rho c_p K_{eh} \left(\frac{\partial T}{\partial z} + \frac{g}{c_p} \right) \right], \qquad (4)$$

where K_{eh} is the eddy heat conductivity assumed to be equal to the eddy diffusion coefficient and c_p is the specific heat at constant pressure. Because of a stable thermospheric stratification

$$\frac{\partial T}{\partial z} + \frac{g}{c_p} > 0, \qquad (5)$$

the vertical heat flux owing to eddy conduction is always negative and directed downward to the mesosphere. Therefore, the net effect of eddy conduction is thermospheric cooling. One should note that locally at some level eddy conduction may either heat or cool the thermosphere, depending on the value of $\partial(\rho K_{eh})/\partial z$ (Gordiets et al. 1982). Furthermore, we also include in our model simulations heating due to dissipation of turbulent energy

$$q_{eh} = q_d + q_g, \qquad (6)$$

where q_d and q_g are the volume rates of turbulent energy dissipation owing to the action of viscous and buoyancy forces

$$q_d = K_{eh} \frac{\rho g}{T} \left(\frac{\partial T}{\partial z} + \frac{g}{c_p} \right) \frac{1 - R_{dyn}}{R_{dyn}}, \qquad (7)$$

and

$$q_g = K_{eh} \frac{\rho g}{T} \left(\frac{\partial T}{\partial z} + \frac{g}{c_p} \right). \qquad (8)$$

Here R_{dyn} is the dynamic Richardson number for the statistically steady turbulent motion which can be expressed as

$$R_{dyn} = \frac{\varepsilon_g}{\varepsilon_s} = \frac{\frac{K_{eh} g}{K_{em} T} \left(\frac{\partial T}{\partial z} + \frac{g}{c_p} \right)}{\left(\frac{\partial u}{\partial z} \right)^2} = \alpha_t Ri, \qquad (9)$$

where ε_s is the transfer rate of kinetic energy from the mean motion to fluctuation motion and can be written as

$$\varepsilon_s = K_{em} \left(\frac{\partial u}{\partial z} \right)^2, \qquad (10)$$

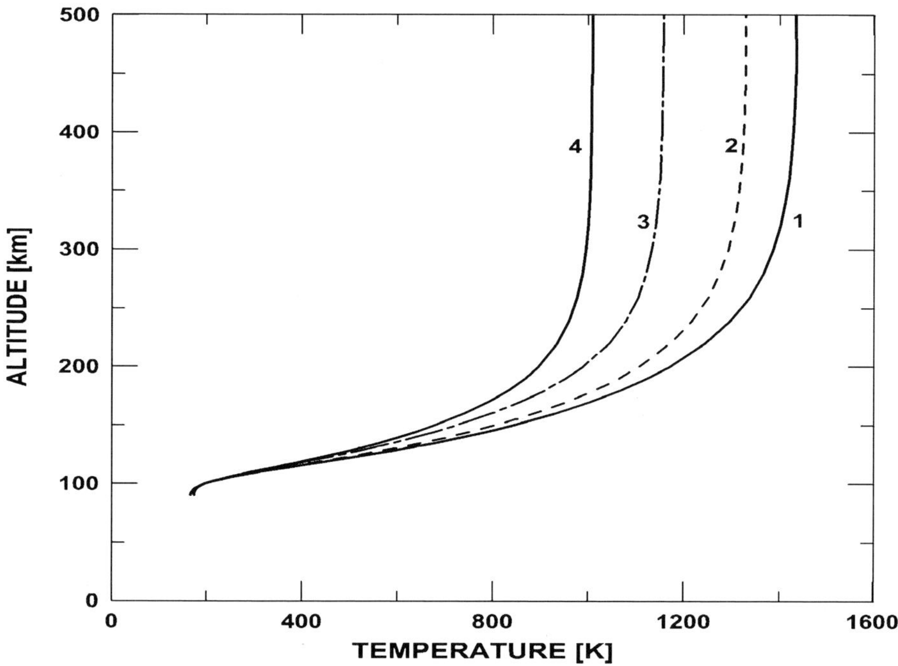

Fig. 1 Temperature of the present time Earth thermosphere showing the effect of various IR-radiating molecules on the temperature profiles. (*1*) Only 15 μm CO_2 fundamental band is cooling; (*2*) Cooling by O 63 μm IR-line plus CO_2 cooling; (*3*) NO cooling in the 5.3 μm fundamental band plus O and CO_2 cooling; (*4*) All the previously mentioned coolers including chemically excited IR bands of minor molecular constituents and cooling in the OH 2.8 μm, O_3 in 9.6 μm, and $O_2(^1\Delta_g)$ 1.27 μm bands are taken into account

with K_{em} the eddy momentum diffusion coefficient, Ri the usual Richardson number and u the horizontal component of the mean velocity of motion. The value of ε_g is the net amount of the potential gas energy associated with buoyancy forces and α_t is the reciprocal of the turbulent Prandtl number (Gordiets et al. 1982).

From the above brief description of our thermospheric models used in this study their major intrinsic limitations can be summarized as follows: 1) the present stage models do not incorporate the upper atmospheric photochemistry of CO_2, H_2O, O_2, etc.; 2) the models do not include hydrogen loss, neither Jeans nor hydrodynamic, thus they should be applied only to the upper atmospheres which are not hydrogen-rich or humid. Therefore, the major uncertainties of our simulations which are closely related to the above model limitations, can result in considerable underestimation of the thermospheric cooling due to intensive atomic hydrogen hydrodynamic loss from a "wet" terrestrial thermosphere at high exobase temperatures, especially at temperatures above the critical value for blow off. The obtained exospheric temperatures should be considered as an approximate upper limit to real thermospheric temperatures. The uncertainties associated with the absence of detailed photochemical calculations in the models may lead to systematically increasing errors in the assumed solar XUV heating efficiency for high XUV fluxes as the flux grows and, therefore, may result in substantial exobase temperature errors during the young Sun period. Also, there are uncertainties in the IR-cooling rates calculation from hot and expanded terrestrial thermospheres, as well as some others which seem, however, to be less important. Therefore, more work is in progress which will advance these thermospheric models to a higher

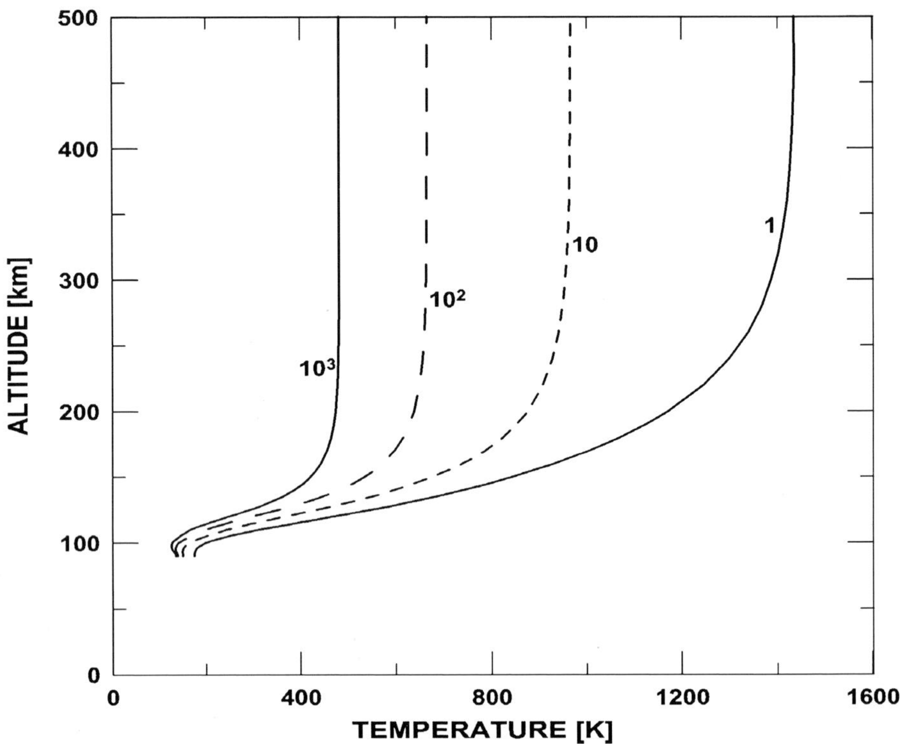

Fig. 2 Thermospheric temperature profiles for present time Earth with different CO_2 mixing ratios. The mixing ratios are expressed in units of PAL—Present Atmospheric Level: 1 PAL = 3.3×10^{-4} of the CO_2 mixing ratio

level of fidelity. However, one should note that hydrodynamic outflow conditions and effective planetary hydrogen winds depending on the available H_2O reservoirs and climate conditions may last for relatively short time periods during the planet's evolutionary stage, resulting in much "dryer" or less humid atmospheres, such as those considered in the present study.

Figure 1 shows the effect of various IR-irradiating molecules on the temperature profile of the present upper atmosphere of Earth. Figure 2 shows the effect of various CO_2 mixing ratios on the Earth's thermospheric temperature profile. One can see that a mixing ratio which is $\geq 10^3$ times higher than that of the present Earth would produce an exospheric temperature which is slightly larger than the present value for Venus (see Fig. 5 for comparison). It should be noted that the temperature profiles presented in Fig. 2 are calculated for an atmosphere containing only CO_2 as an IR-radiating molecule. All the other IR-cooling species shown in Fig. 1 are not included in the model calculations shown in Fig. 2. One can see from Figs. 1 and 2, that CO_2 can be the most efficient cooler due to IR emission in the 15 μm fundamental band for the terrestrial planetary atmospheres.

Since we are interested in the effect of higher XUV flux values produced by the young Sun (see Ribas et al. 2005; Lundin et al. 2007, this issue) on the upper atmospheres of the early terrestrial planets, we calculate in Sects. 3.1, 3.2 and 3.3 the thermospheric temperature and density profiles for early Earth, Venus and Mars.

3 Comparative Study of the Evolution of the Upper Atmospheres of Terrestrial Planets

3.1 Early Earth

For a study of the heating of the Earth thermosphere by XUV radiation from the young Sun, we apply our thermospheric model to an atmosphere with the present atmospheric composition and a related heating efficiency for the solar XUV radiation. The heating efficiency of the ionizing solar radiation in the terrestrial thermosphere can be estimated through the analysis of the photoelectron and ion energy production processes and the ways of conversion of the non-thermal energy of the photolysis products into heat. As it is known, about 50% of 34 eV, which is the average energy of a solar ionizing quantum, is spent for ion production, while the other half is carried away by primary photoelectrons which are able to produce further ionizations, gas particle excitations, etc. The photoionization and following it neutralizing ionospheric chemical reactions in the terrestrial thermosphere result in the formation of two O atoms per each ionization act which takes away from local gas heating the dissociation energy of the O_2 molecule. This energy is carried down from the thermosphere by a diffusion flux of oxygen atoms which recombine only near the base of the thermosphere or below. Also, a part of the energy released in the ionospheric chemical reactions goes into "pumping" of the internal energy levels of the atomic, ion, and molecular reaction products. A part of this internal energy can be radiated away to space in the UV, visible, and infra-red spectral ranges and thus be lost, while the other part, due to particle collisions, is thermalized and heats the ambient gas. The estimations by Gordiets et al. (1979) for the terrestrial oxygen-rich thermosphere resulted in an average value of about 50%, which is used in the present study.

All the IR-cooling mechanisms of the model are included in these calculations. Figure 3 shows the calculated exospheric temperature related to the solar XUV flux as a function of time. As shown by Ribas et al. (2005) and Lundin et al. (2007, this issue), the solar XUV flux was ∼6 times higher 3.5 Gyr ago, ∼10 times higher 3.8 Gyr ago, ∼50 times higher 4.33 Gyr ago, and reached ∼100 times the present Sun level ∼4.5 Gyr ago. One can see from Fig. 3, that the blow-off temperature for atomic hydrogen of about 5000 K would be exceeded during the first Gyr after the Sun arrived at the ZAMS, or before 3.5 Gyr ago. Here we assume that when the thermal energy of gas-kinetic motion exceeds its gravitational energy, the top of the atmosphere blows off and moves radially away from the planet (Öpik 1963). The critical temperature for the start of the blow off is then given by

$$T_c = \frac{2M_{pl}mG}{3kr}.$$ (11)

Here M_{pl} is the planetary mass, m the mass of atomic hydrogen, G Newton's gravitational constant, and r is the planetocentric distance of the exobase. For the XUV fluxes more than 10 times the present flux (>3.8 Gyr ago) on Earth one would expect very high exospheric temperatures that could result in large thermal escape rates even for heavier species like H_2, He, N, O, and C. However, we should note that the calculated, extremely high exobase temperatures ($T_{exo} \sim$ 10000–20000 K) being much above the blow off temperature for hydrogen, should be considered as rather rough estimates of the expected upper limits for T_{exo}, because of the intrinsic limitations and uncertainties of our model thermospheres, as discussed in the previous section.

Figure 4 shows the effect of the solar XUV radiation on the early Earth's exobase temperature for different levels of the CO_2 mixing ratio at the base of the thermosphere. For

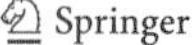

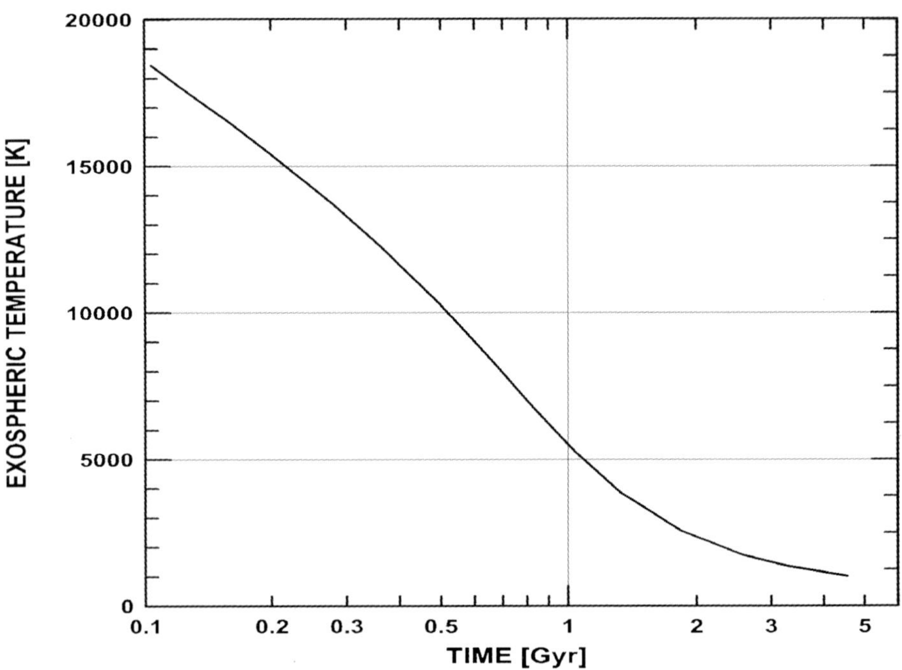

Fig. 3 Time evolution of the exospheric temperature based on Earth's present atmospheric composition over the planets' history as a function of the solar XUV flux for a strongly limited hydrogen blow-off rate

simulations with the CO_2 mixing ratios higher than its present atmospheric level (1 PAL), both the total number density at the lower boundary, which is located at the base of the thermosphere ($\sim$90 km), as well as the ratio of the N_2 to O_2 number densities have been kept constant. The ratio N_2/O_2 has been assumed to be equal to its present time value of $\sim$3.7. The atmospheric levels of Ar and He have been kept at 1 PAL. It should be noted that only the 15 μm CO_2 cooling is included in these calculations to show clearly its effect on the exobase temperature.

As can be seen from Fig. 4, a low CO_2 level implies higher exobase temperatures and an increased atmospheric loss when compared with high CO_2 abundance. Therefore, in the case of a low CO_2 level a higher loss rate of hydrogen and water vapor from the Earth's atmosphere can be expected over longer periods of time, up to about 1 Gyr, or even longer. As a result, early Earth could have lost a large amount of water if it were not protected by a strong magnetic field, due to both thermal and non-thermal escape. Atmospheric levels of CO_2 higher than in the present Earth's atmosphere, inferred from paleosols—soil samples around 2.2–2.75 Gyr ago (e.g., Raye et al. 1995; Hessler et al. 2004), would thus imply lower exobase temperatures and a reduced loss as compared to an early atmosphere with a present atmospheric level of CO_2. However, we cannot say anything definite about the CO_2 level in the ancient Earth atmosphere at this stage of research.

In their recent study Tian et al. (2005, 2006) apply a 1-D time-dependent hydrodynamic escape model to simulate thermal escape processes from a molecular hydrogen-rich early Earth's atmosphere. Because the solar XUV radiation levels were much stronger during the Archean era than today, they adopt XUV radiation levels of 1, 2.5, and 5 times the present value in their simulations. These authors also assume high CO_2 mixing ratios in the early

 Springer

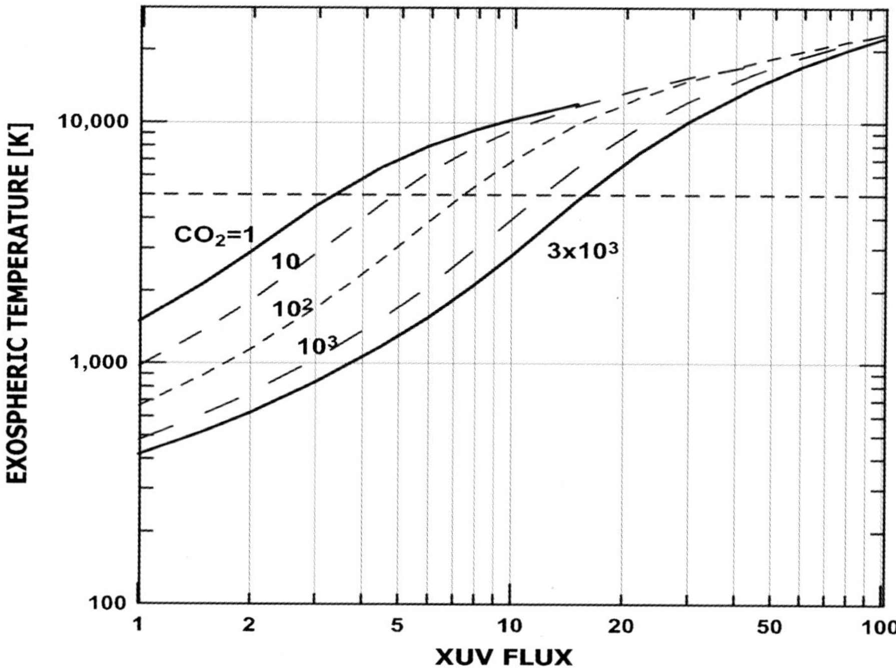

Fig. 4 Earth's exospheric temperatures for different levels of CO_2 abundance in units of PAL in the thermosphere as a function of solar XUV flux. The *numbers by the curves* correspond to CO_2 volume mixing ratios expressed in PAL (Present Atmospheric Level: 1 PAL for $CO_2 = 3.3 \times 10^{-4}$). The *horizontal dashed line* shows the blow-off temperature of atomic hydrogen

Earth's atmosphere but do not actually include IR-cooling by CO_2 in the energy balance equation of their model and argue that low oxygen and high CO_2 on early Earth yielded a cold exobase.

Applying their model, Tian et al. (2005) calculate the temperature and velocity profiles for the corresponding XUV radiation levels and obtain very low temperatures at the exobase of early Earth's atmosphere in the range of $\sim$300–600 K due to the adiabatic cooling associated with hydrodynamic escape. Although the flow velocity near the upper boundary of their model for all the three cases considered is below the escape velocity from the planet, they conclude that even in the 1 XUV level case hydrodynamic escape of molecular hydrogen would still occur. Tian et al. (2005) also calculate the hydrodynamic and Jeans escape rates for varying hydrogen homopause mixing ratios resulting from their simulations[1]. Jeans escape rates found by Tian et al. (2005) are more than one order of magnitude smaller than their simulated hydrodynamic escape rates due to the low exobase temperatures.

However, their hydrodynamic blow off solutions for an H_2-rich early Earth atmosphere at the extremely low exobase temperatures of 300–600 K are in disagreement with the predictions of the classical Jeans kinetic loss theory (Jeans 1925). According to this theory, a gravitational potential energy well is formed around a planet at the exobase level which traps the

---

[1]One should note that below the homopause level, the atmosphere is well mixed and each species adopts the same atmospheric scale height which is given by the average mass of an atmospheric particle, while above the homopause, due to molecular diffusion each atom or molecule follows its own scale height based on its mass.

 Springer

planetary atmospheric constituents whenever the exobase temperature is much lower than the critical temperature for blow off ($T_c \sim 5000$ K for H and 10,000 K for H_2 on Earth). And only particles in the high energy tail of the Maxwellian distribution function of the atmospheric gas, having kinetic energies above the escape energy at the exobase, can overcome this energy barrier ("evaporate") and leave the gravitational field of the planet (e.g., Jeans 1925; Chamberlain 1963; Öpik 1963; Gross 1972).

3.2 Present and Early Venus

On Venus neutral density measurements were carried out during 1978–1980 at the solar maximum conditions ($F_{10.7} \sim 180$–200) and again in the fall of 1992 at the solar medium conditions during the pre-entry phase of NASA's Pioneer Venus Orbiter (PVO). Von Zahn et al. (1980) derived from the He number densities measured by the PVO Bus Neutral Mass Spectrometer (BNMS) a constant neutral gas temperature on Venus' dayside between 160–500 km of about 275–290 K by taking into account the altitude variation of the gravitational acceleration. Similar exospheric temperatures of about 290–300 K were inferred from the Orbiter Neutral Mass Spectrometer (ONMS) instrument (Nieman et al. 1979a, 1979b, 1980; Hedin et al. 1983; Fox and Sung 2001).

Moderate solar activity yields on Venus an average dayside exospheric temperature of $\sim$270 K. In situ measurements in the Venus upper atmosphere during the solar minimum derived from Magellan aerobraking data yield average dayside exospheric temperatures of $\sim$240–250 K (Keating et al. 1998).

Figure 5 shows our modelled temperature profiles in the 96% CO_2 Venusian thermosphere as a function of altitude for different solar XUV flux values. One can see from Fig. 5 that our model simulations for present Venus (XUV = 1) yield the exospheric temperature for medium solar activity conditions of $\sim$270 K which is in good agreement with the global empirical model of the Venus thermosphere based on the PVO neutral mass spectrometer measurements of Hedin et al. (1983) and the neutral gas mass spectrometer of the PVO multiprobe bus (Von Zahn et al. 1980; Nieman et al. 1979a, 1979b), as well as with model simulations by Bougher et al. (1999). The average exospheric temperature rises from about 270 K at present time (1 XUV) up to $\sim$600 K 3.8 Gyr ago (10 XUV), $\sim$2300 K 4.33 Gyr ago (50 XUV), and $\sim$8200 K 4.5 Gyr ago (100 XUV). Due to the higher thermospheric temperature the thermosphere expands and the exobase level moves upward from about 200 km (present time) to about 2200 km 4.5 Gyr ago. Figure 6 shows the calculated exobase temperature on Venus for a 96%, 72%, 48% and 10% CO_2 and N_2 atmosphere and various XUV flux values as a function of time.

If the exobase temperatures are higher than $\sim$4000 K, blow-off of atomic hydrogen occurs which results in diffusion-limited hydrodynamic outflow even for a 96% CO_2 atmosphere during $\sim$130 Myr after the Sun arrived at the ZAMS.

For lower CO_2 mixing ratios the thermospheric temperatures could be above the critical temperature at the exobase for atomic hydrogen of $\sim$4000 K for much longer time and extremely high exobase temperatures in excess of 10000 K $\sim$4 Gyr ago (15 XUV) could develop.

Chassefière (1996b), by applying a hybrid hydrodynamic-kinetic model for a pure atomic hydrogen atmosphere of a Venus-like planet, found self-consistent steady state solutions with a high Jeans loss at the exobase, elevated up to one planetary radius altitude (or more). His model incorporates hydrodynamic approach below the exobase and Jeans escape at the exobase. Apart from the solar XUV energy heating, it includes heating by the solar wind (energetic neutrals, ENs) which takes place near the exobase and supplies about 2/3 of the

 Springer

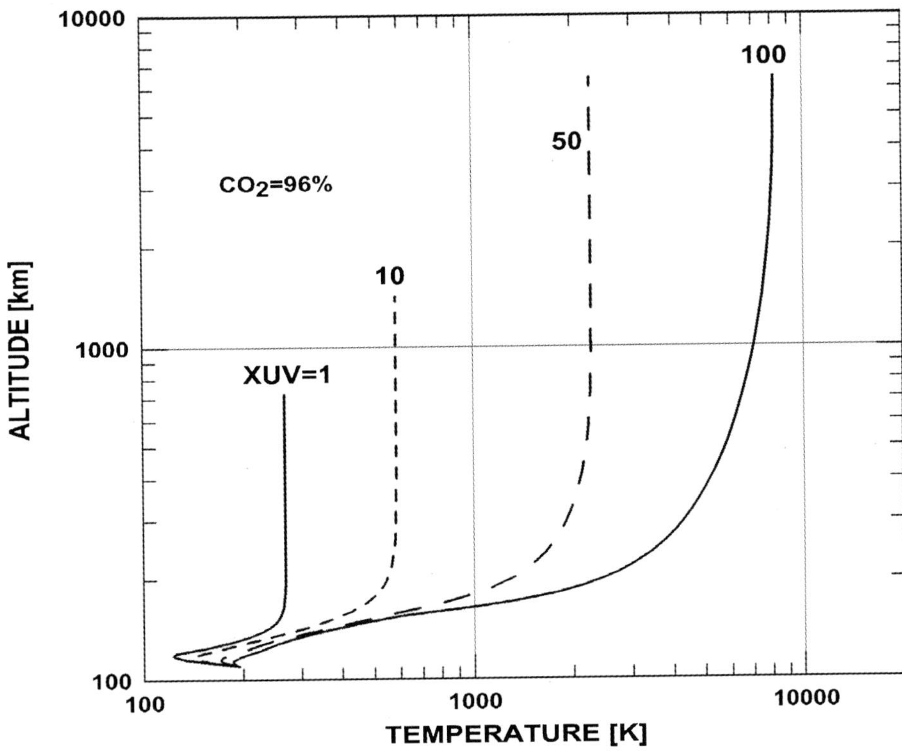

Fig. 5 Modelled temperature profiles in a 96% CO_2 thermosphere of Venus as a function of altitude for 1 XUV (present time), 10 XUV ($\sim$3.8 Gyr ago), 50 XUV ($\sim$4.33 Gyr ago), and 100 XUV ($\sim$4.5 Gyr ago) flux values

escape energy. The model also includes a thermal energy input by the hydrogen upward flux through the lower boundary. The cooling mechanisms considered in the model are adiabatic cooling due to hydrogen escape flow and downward thermal conduction loss through the lower boundary at $z_0 = 200$ km. This model does not include IR-radiation cooling which can be present in a hydrogen-rich thermosphere due to H_3^+ ion (Yelle 2004) and also in the lower thermosphere due to CO_2. In the nominal case of the model with a moderate temperature T_{exo} at the exobase of about 750 K, Chassefière (1996b) found that intense hydrodynamic upward flow of hydrogen exists below the exobase for the present solar XUV conditions relayed by high Jeans escape flux at the exobase of 2×10^{11} cm^{-2} s^{-1}. This exobase temperature is not much higher, however, than $T_{exo} \approx 500$ K obtained from our Venus thermosphere model for present time solar condition and a low CO_2 level of 10%, as it is shown in Fig. 6. Also, it is not surprising that there is no hydrogen bulk outflow from our model thermosphere, since the atmosphere is assumed to be "dry", as explained before. However, the results of Chassefière (1996b) clearly indicate that an intense Jeans escape can provide an important contribution to the energy budget of a hydrogen-rich early Venus upper atmosphere and, thus, be an important factor for finding self-consistent hydrogen outflow solutions even for a present time solar XUV flux.

The present solar flux at Venus' orbit is about 1.91 times the solar constant of today, S_{Sun} ($S_{Sun} = 1360$ W m^{-2}), while the radiation flux according to the solar standard model on Venus orbit 4.5 Gyr ago was about 1.34 S_{Sun} (e.g., Gough 1977; Kasting et al. 1984;

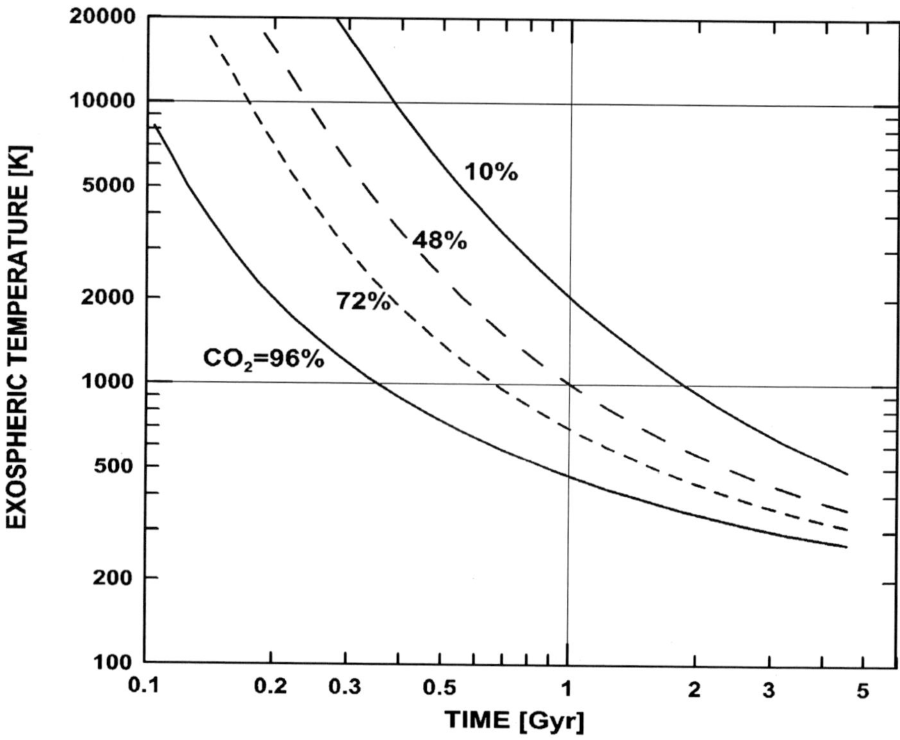

Fig. 6 Modelled venusian exobase temperatures for a 96% CO_2, for a 72% CO_2, for a 48% CO_2, and for a 10% CO_2 atmosphere as a function of the solar age and XUV flux

Lundin et al. 2007, this issue). Pollack (1971), Kasting et al. (1984) and Kasting (1988) studied runaway and moist greenhouse cases and their implications for the evolution of the atmospheres of early Earth and Venus.

Kasting (1988) found that the critical solar flux at which a runaway greenhouse effect occurs, or when a water ocean evaporates entirely, is about 1.4 S_{Sun} at Earth's orbit. This value is close to the expected solar flux incident on early Venus. Moreover, the climate models of Kasting (1988) indicate that the runaway greenhouse effect occurs under these conditions nearly independently of the amount of CO_2 in the atmosphere. For a CO_2 pressure, $P_{CO_2} \leq 10$ bar the critical solar flux value remains equal to 1.4 S_{Sun} and for a $P_{CO_2} \approx 100$ bar the S_{Sun} value increases only a slightly to 1.42 S_{Sun}.

The critical solar flux for which a runaway greenhouse effect could occur is found to be highly dependent on the presence of clouds (Kasting 1988). It was found that for clouds located at a pressure level of a few tens of a bar, the critical solar flux for a 50% and 100% cloud cover increases to about 2.2 S_{Sun} and ~4.8 S_{Sun}, respectively. From these results Kasting (1988) suggested that early Venus' surface was likely "cool" enough, so that a liquid water ocean could have been maintained before it evaporated.

Kasting and Pollack (1983) studied the hydrodynamic loss of water from a primitive H_2O-rich venusian atmosphere as a function of the H_2O mixing ratio at the mesopause level. They found that for H_2O mixing ratios of more than 7×10^{-4} atomic hydrogen becomes the major species at the exobase which moves then to greater distances, so that the exosphere becomes unstable to expansion, Jeans escape becomes inappropriate and hydrodynamic con-

 Springer

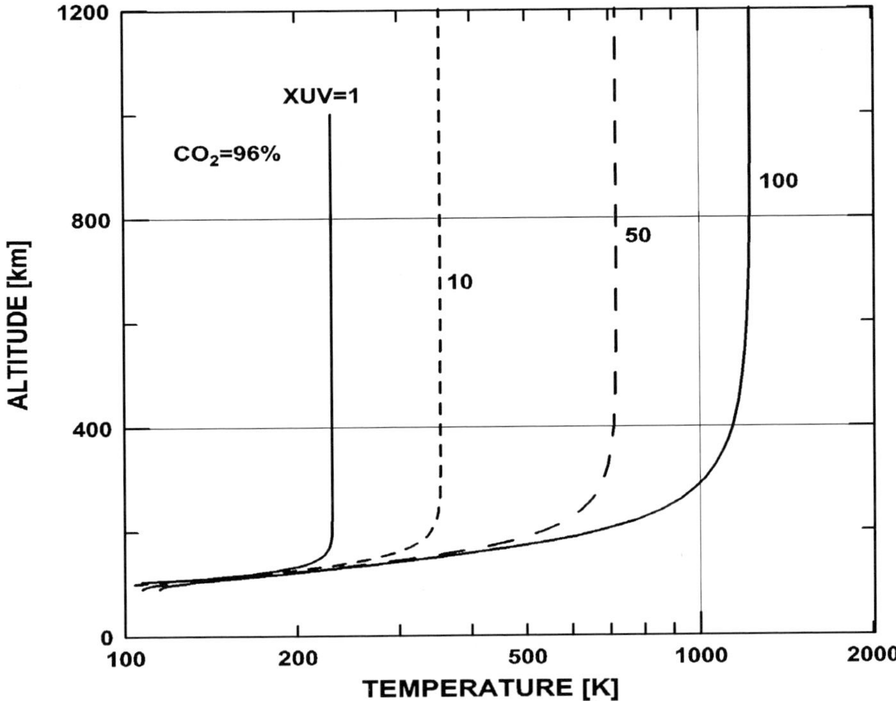

Fig. 7 Modelled martian temperature profiles in a 96% CO_2 thermosphere as a function of altitude for 1 XUV (present), 10 XUV (3.8 Gyr ago), 50 XUV (4.33 Gyr ago), and 100 XUV (4.5 Gyr ago) flux values for a low heating efficiency of $\sim$8%

ditions have to be considered. They also found that on early Venus atomic hydrogen could escape for the H_2O mixing ratio of 0.0063 at 10 XUV with a flux of $\sim$3.5 $\times$ 10^{11} cm^{-2} s^{-1}. Extrapolation of their results for the 100 XUV level yields a flux of $\sim$3.8 $\times$ 10^{12} cm^{-2} s^{-1}. If the mixing ratio of H_2O were higher (0.055–0.46) because of the runaway greenhouse, hydrogen escape flux values of the order of $\sim$1–3.5 $\times$ 10^{12} cm^{-2} s^{-1} could be reached for 8–16 times higher XUV fluxes (Kasting and Pollack 1983) and of $\sim$1.3–2.7 $\times$ 10^{13} cm^{-2} s^{-1} for 100 XUV (Kulikov et al. 2006).

By using these H escape fluxes, it can be estimated that the full amount of a terrestrial ocean could have escaped over a time period of about 50 Myr if the H_2O mixing ratio was as high as 0.46 and the solar XUV flux was about 70–100 times the present solar value. However, if large amounts of water had evaporated on early Venus, the remaining oxygen would need to be incorporated in the crust as FeO (Lewis and Prinn 1984; Rosenqvist and Chassefière 1995), or would need to be lost to space due to intense erosion by the dense solar wind of the young Sun (Kulikov et al. 2006).

3.3 Preresnt and Early Mars

Exospheric temperatures inferred from the mass spectrometer data during low solar activity by the NASA Viking Landers 1 and 2 yielded values of about 180 K and 150 K, respectively (e.g., Nier and McElroy 1977; Hanson et al. 1977; Fox and Dalgarno 1979; Barth et al. 1992; Fox and Sung 2001). A similar temperature value during low solar activity of 153

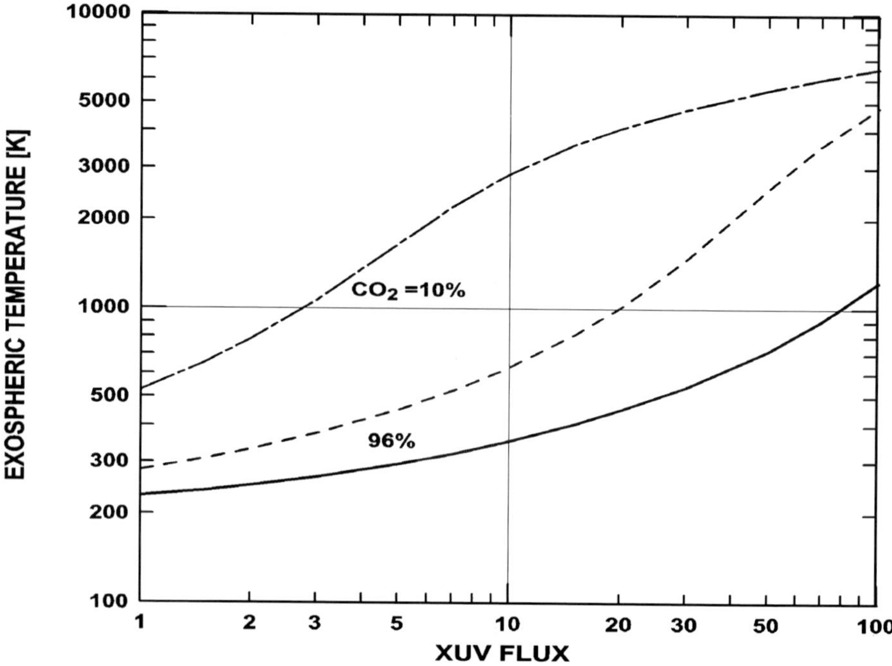

Fig. 8 Modelled martian exospheric temperatures as a function of solar XUV flux values for a 96% CO_2 atmosphere and for a heating efficiency of $\sim$8% (*solid line*) and $\sim$32% (*dashed line*). The *dashed-dotted line* corresponds to a martian atmosphere with a lower CO_2 mixing ratio of $\sim$10%. The *horizontal thin solid line* corresponds to the H blow-off temperature of $\sim$1000 K and the *vertical solid line* marks the conditions $\sim$3.8 Gyr ago

K was modelled by using thermospheric neutral gas data from the descent measurement of NASA's Mars Pathfinder in 1997 (Schofield et al. 1997; Magalhaes et al. 1999; Bougher et al. 2000). By reproducing aerobraking data of NASA's Mars Global Surveyor (MGS) with thermospheric models, one obtains exospheric temperatures of $\sim$220–230 K during moderate solar activity conditions (e.g., Keating et al. 1998; Bougher and Keating 1999) and $\sim$240 K for high solar activity.

Figure 7 shows our modelled temperature profiles in a 96% CO_2 martian thermosphere as a function of altitude for different XUV flux values. One can see from Fig. 7 that our model simulations for present Mars result in the exospheric temperature for medium solar activity conditions of $\sim$220 K which is in good agreement with the temperatures inferred from the aerobreaking data of MGS.

Figure 8 illustrates the exobase temperature on Mars for a 96%, and 10% CO_2 and N_2 atmosphere as a function of solar XUV flux for the lower and upper limits of the heating efficiency from $\sim$8% through $\sim$32% adopted in our simulations as discussed by Fox (1988). These derived estimates of the exobase temperature for early Mars show a large degree of uncertainty which results from our present poor knowledge of many essential parameters needed for modelling, including atmospheric composition and variation of minor species, heating efficiency, etc.

The blow-off temperature for atomic hydrogen on Mars is about $\sim$1000 K. Depending on an initial CO_2 mixing ratio and exobase altitude, atomic hydrogen could have been under diffusion-limited hydrodynamic blow-off conditions even for an extremely low heating effi-

 Springer

ciency of about 8% and a 96% CO_2 atmosphere during about 100 Myr after the Sun arrived at the ZAMS. For the maximum heating efficiency of $\sim$32% the blow-off period could have lasted for $\sim$400 Myr. If an early martian atmosphere had a lower CO_2 abundance with a 10% mixing ratio, for example, shortly after its volatile outgassing, then the blow-off period could have lasted much longer and would have resulted in higher exobase temperatures and larger thermal loss rates of heavier species like H_2, He, C, N and O. However, due to greater orbital distance, as compared to Venus and Earth, which results in a lower solar luminosity of $\sim$0.25 S_{Sun} on early Mars, the hydrodynamic loss of atomic hydrogen during a blow-off period could have been less efficient for Mars than for both these planets.

4 Non-Thermal Atmospheric Loss Processes From Non- or Weakly Magnetized Planets

4.1 Photochemically Produced "Hot" Neutrals and Planetary Coronae

It is known from spacecraft observations and model simulations that photochemical reactions like dissociative recombination where a positive molecular ion recombines with an electron so that the neutral molecule dissociates, and photodissociation where a photon dissociates molecular neutrals, are important sources of suprathermal or energetic H, C, N, O, and CO in the exospheres of present Venus and Mars (e.g., McElroy et al. 1982; Nagy et al. 1981; Rodriguez et al. 1984; Ip 1988; Nagy et al. 1990; Lammer and Bauer 1991; Zhang et al. 1993a; Jakosky et al. 1994; Fox and Haĉ 1997; Luhmann et al. 1997; Fox and Haĉ 1997; Kim et al. 1998; Lammer et al. 2000b; Fox and Bakalian 2001; Lammer et al. 2003a, 2006).

The released "hot" or energetic atoms may reach eventually the same temperature as the background atmosphere through a series of elastic collisions with the main background gas such as CO_2 or O. The effective cross-section area for these inelastic collisions is, however, negligibly small at these low energies. After their release the atoms may collide with the neutral background gas, may change directions, loose their energy, or may travel long distances in the upper atmosphere without collisions. Finally, the newly generated "hot" atoms which reach the exobase with energies larger than the escape energy from the planet, are lost, while particles with lower energy are responsible for the formation of extended neutral exospheres, so-called planetary coronae, which interact with the solar wind plasma.

Figure 9 displays the sum of "hot" and "cold" O number density distributions which originate from dissociative recombination of O_2^+ molecular ions for 1 XUV (present), 10 XUV (3.8 Gyr ago), 50 XUV (4.33 Gyr ago) and 100 XUV (4.5 Gyr ago) flux values on Venus (Kulikov et al. 2006). The hot oxygen component is calculated from the modelled O_2^+ ion number density by means of the two-stream Monte Carlo model of Lammer et al. (2000b). For the calculation of the O_2^+ ion profiles as a function of the XUV flux the ionospheric model of Shinagawa et al. (1987) is applied to our calculated neutral density profiles (see also Kulikov et al. 2006), with the rate coefficients for chemical reactions updated by using the data of Fox and Sung (2001). Photoionization rates of neutral gases are calculated by using the solar radiation model, photoabsorption cross sections, and photoionization cross sections of Shunk and Nagy (2000). The photoionization rates for different XUV fluxes are calculated using the photoabsorption and photoionization cross sections and multiplying the present day solar XUV spectra. The photoelectron impact ionization rates of N_2 and O are taken from Richards and Torr (1988). Higher solar XUV radiation produces denser ionospheric layers and more dissociated "hot" atoms result in a denser corona. The densities

 Springer

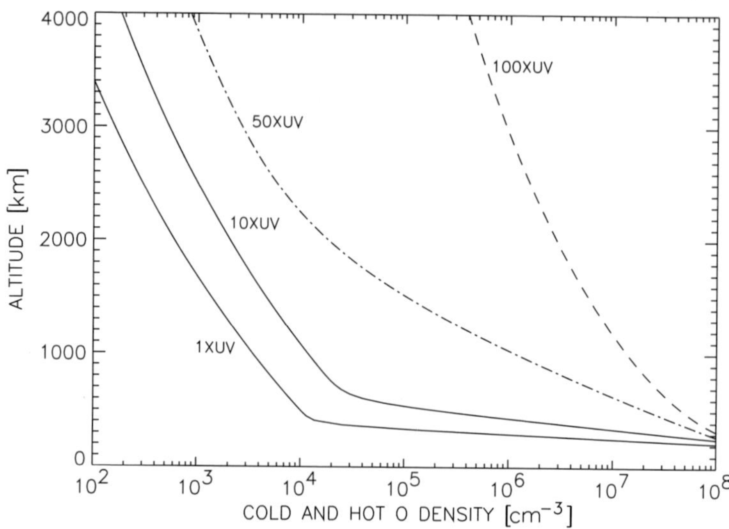

Fig. 9 Sum of "cold" and "hot" O number densities on Venus as a function of altitude for 1 XUV (present), 10 XUV (3.8 Gyr ago), 50 XUV (4.33 Gyr ago), and 100 XUV (4.5 Gyr ago) fluxes

of "cold" and "hot" oxygen populations above the exobase are obtained by extrapolations based on Chamberlain's equations (Chamberlain 1963). The density contribution of "hot" O atoms to the formation of planetary "hot" corona is mainly important for flux values ≤ 50 XUV, as can be seen from Fig. 9. The "flat" part of the profiles corresponds to the colder background atmosphere, while the "steep" part is due to the more energetic "hot" particles. For XUV fluxes > 50, the temperature of the heated background gas approaches that of the hot particles and eventually the temperatures of both populations become indistinguishable.

4.2 Solar Wind Induced Ion Pick up

The main difference between present Venus and Mars compared to Earth is the lack of a significant intrinsic magnetic field, allowing the incident energetic solar wind particles to directly interact with their upper atmospheres. The solar wind plasma flow around planetary obstacles with no or weak intrinsic magnetic field have been studied extensively by using gas-dynamic convection magnetic field models (e.g., Spreiter and Stahara 1980), semi-analytical magnetohydrodynamic (MHD) flow models (e.g., Biernat et al. 2001), or hybrid models (e.g., Terada et al. 2002; Kallio and Janhunen 2003). The model results have been compared with observational data obtained from spacecraft, especially for Mars and Venus.

Neutral atoms and molecules above the ionopause can be transformed to ions via charge exchange with solar wind particles, XUV radiation and electron impact. These newly generated planetary ions are accelerated to higher altitudes and energies by the interplanetary electric field and incorporated or picked up by the solar-wind plasma flow past the planetary obstacle to space, where they are lost from the planet (e.g., Lundin et al. 1989, 1990, 1991, 2004; Lichtenegger and Dubinin 1998; Biernat et al. 2001; Lammer et al. 2003a; Terada et al. 2002; Kulikov et al. 2006; Lammer et al. 2006). These picked up ions can also collide with the background gas and may sputter atmospheric species to energies above the escape energy from the planet.

4.3 Atmospheric Sputtering

Sputtering refers to a mechanism by which incident energetic particles (mostly ions) interact with the upper atmosphere, resulting in ejection of atmospheric species. Sputtering has been recognized as an important source of atmospheric loss in the case of Mars and of less importance for larger planets like Venus (Luhmann and Kozyra 1991). In particular during the early phase of the Solar System, atmospheric sputtering is thought to be significant for loss of martian water and other atmospheric constituents (e.g., Luhmann et al. 1992; Kass and Yung 1995, 1996; Johnson and Liu 1996; Hutchins et al. 1997; Luhmann 1997; Johnson et al. 2000; Leblanc and Johnson 2001, 2002; Lammer et al. 2003a; Chasefière and Leblanc 2004).

4.4 Viscous Processes

Apart from non-thermal loss processes like ion pick up and atmospheric sputtering, ionospheric erosion due to plasma instabilities and momentum transfer (solar wind plasma forcing) are additional important loss processes from non-magnetized planetary bodies. Measurements by the PVO spacecraft (orbiting Venus) revealed a number of characteristic ionospheric structures that may be signatures of solar wind-ionosphere interaction processes (e.g., Brace et al. 1982; Russell et al. 1982). Among them are wavelike plasma irregularities observed at the top of the dayside ionosphere and plasma clouds observed above the ionopause, primarily near the terminator and further downstream. More detailed analysis of several detached plasma clouds has shown that the ions within the clouds themselves are ionosphere-like in electron temperature and density (Brace et al. 1982). When such plasma clouds are seen far above the ionosphere, they are clearly separated by an intervening region of ionosheath plasma. Note that the ionosheath is the region between the bow-shock and the ionopause. The latter is the transition region between tangential solar wind flow around the ionosphere and the ionosphere itself. The properties of the ionosheath are strongly affected by the presence of the neutral atoms extending above the ionopause.

This large separation in a direction perpendicular to the ionosheath flow suggests that the ionospheric plasma in the clouds must have originated in the ionosphere upstream on the dayside, indicating that plasma instabilities may occur at the venusian ionopause. In the region adjacent to the sunward side of the magnetopause, where the magnetic field significantly increases and simultaneously the plasma density decreases (the so-called magnetic barrier), plasma is accelerated by a strong magnetic tension directed perpendicular to the magnetic field lines. This magnetic tension forms specific types of plasma flow stream lines near the ionopause, which are orthogonal to the magnetic field lines. This process favors the appearance of the Kelvin–Helmholtz instability[2] that can detach ionospheric plasma from a planet in the form of detached ion clouds. In studies related to terrestrial planets one can treat the solar wind flow past the planetary obstacle using a magnetohydrodynamic model which was applied successfully for the case of the solar wind flow around Venus and Mars (Lammer et al. 2003b; Penz et al. 2004; Lammer et al. 2006).

---

[2] A Kelvin–Helmholtz instability can develop in case of a velocity difference across the interface between two fluids. At Venus, this instability may occur close to the ionopause due to the interaction between the solar wind and the ionosphere.

 Springer

Table 1 Escape rates in units of $[s^{-1}]$ of H, H_2, O and O^+ for moderate solar activity conditions at present Venus (see Lammer et al. 2006, and references therein). The asterisk in the notations H* and O* indicates "hot" neutral atoms produced in ionospheric photochemical reactions

Loss process / species	Loss rate
Jeans: H	2.5×10^{19}
Photochemical reactions: H*	3.8×10^{25}
Photochemical reactions: O*	negligible
Electric field force: H^+	$\leq 7 \times 10^{25}$
Ion pick up: H^+	1×10^{25}
Ion pick up: H_2^+	$< 10^{23}$
Pick up: O^+	1.6×10^{25}
Detached plasma clouds: O^+	$5 \times 10^{24} - 1 \times 10^{25}$
Sputtering: O	6×10^{24}

4.5 Atmospheric Loss over Venus' History

4.5.1 Atmospheric Escape from Present Venus

Atmospheric escape from the upper atmosphere of Venus is mainly influenced by the loss of hydrogen and oxygen caused by the interaction of the solar XUV radiation and particle flux with the unprotected upper atmosphere. Lammer et al. (2006) estimated the total hydrogen and oxygen loss rates from present Venus shown in Table 1 and found that the ion pick up and ionospheric erosion caused by the Kelvin–Helmholtz plasma instability which is of the order of 10^{25} s^{-1} may be the most efficient escape processes for O^+ ions on Venus. Thermal atmospheric escape processes and atmospheric loss by photochemically produced O atoms yield negligible loss rates. On the other hand, photochemical production of hot H atoms is a very efficient loss mechanism for hydrogen on Venus with a global average total loss rate of about 3.8×10^{25} s^{-1}. This estimate is in agreement with a Donahue and Hartle (1992) result and of the same order, but less than estimated by Hartle and Grebowsky (1993) for an H^+ ion outflow from the Venus' nightside of about 7.0×10^{25} s^{-1} due to acceleration by an outward electric polarization force related to ionospheric holes.

Their study indicates that on Venus, due to its larger mass and size compared to Mars, the most important atmospheric escape processes of oxygen involve ions, and they are due to the interaction with the solar wind. The obtained results indicate that the ratio of H/O escape to space from the venusian upper atmosphere is about 4, and is in a much better agreement with the stoichiometrical H/O escape ratio of 2 : 1, which is not the case on Mars (Lammer et al. 2003a). However, we expect that a detailed analysis of the outflow of ions from the Venus' upper atmosphere as is expected to be measured by the Automatic Space Plasma Experiment with a Rotating Analyzer-4 (ASPERA-4) and the Venus Express magnetometer (VEX-MAG) instruments aboard ESA's Venus Express will lead to more accurate atmospheric loss estimations and better understanding of the planetary water inventory.

4.5.2 Solar Wind Induced Atmosphere Erosion and Water Loss over Venus' History

Present Venus is an extremely dry planet containing very little H_2O vapor (about 200–300 ppm Hoffman et al. 1980; Moroz et al. 1979; Johnson and Fegley 2000) in its atmosphere. The analysis of the Pioneer Venus Large-probe Neutral Mass Spectrometer (LNMS) data indicated that Venus' atmosphere is enriched in D over H relative to Earth by a factor of $\sim 120 \pm 40$, implying that Venus was once more "wet". McElroy et al. (1982) used the present hydrogen escape rates and showed that these loss rates over 4.5 Gyr would imply

a lower limit of water on Venus of $\sim0.3\%$ of a terrestrial ocean (TO). However, Donahue and Hartle (1992) and Hartle et al. (1996) deduced the amount of water lost from Venus from the measured D/H ratio based on ion mass spectrometer measurements in a range of an equivalent TO-depth of several meters to tens of meters.

For the prediction of the amount of primordial water on early Venus one has to consider two possibilities. The first hypothesis assumes that early Venus was formed from condensates in the solar nebula that contained little water (e.g., Holland 1963; Lewis 1970, 1972, 1973, 1974; Lewis and Prinn 1984), while the second hypothesis argues, in agreement with previous theories (e.g., Dayhoff et al. 1967; Walker 1975), for a water abundance more comparable to that on Earth and Mars (e.g., Wetherill 1981; Donahue and Pollack 1983; Kasting and Pollack 1983; Morbidelli et al. 2000; Raymond et al. 2004).

The supply of water to the venusian atmosphere by comets was studied by Lewis (1974), Grinspoon and Lewis (1988) and more recently by Chyba et al. (1990). Grinspoon and Lewis (1988) argued that the present water content of Venus may be in a steady state where the loss of hydrogen to space is balanced by a continuous input of water from comets or from delayed juvenile outgassing. It is important to note that in the case of an external water delivery no increase of Venus' past water inventory is required to explain the present day observed D/H isotope fractionation. The enrichment of D could conceivably have started out of more or less "dry" conditions, as originally was suggested by Lewis (1972).

However, the initial water inventory on early Venus may have been larger, because a substantial amount of water is required to explain the onset of the large greenhouse effect observed at present (e.g., Shimazu and Urabe 1968; Rasool and Bergh 1970; Donahue et al. 1982; Kasting and Pollack 1983; Chassefière 1996a; Chassefière 1996b). From these considerations one may expect that the early outgassing of water from Venus should have been much more efficient than it is at present to generate an equivalent amount of a TO (e.g., Hunten et al. 1987).

Kulikov et al. (2006) investigated for the first time the O^+ ion pick up loss rates over Venus' history and found that a dense solar wind of the young Sun could easily remove the expected amount of oxygen which could have been left from an XUV-driven hydrodynamically escaping ocean (see Table 2). Chassefière (1996a) showed that if Venus' H_2O-rich atmosphere hydrodynamically evaporated during the first 100 Myr, about 120 bar of H_2O and ~36 bar of oxygen could be lost to space along with the planetary hydrogen wind.

However, Chassefière (1996a) expected that about 85 bar of oxygen should have remained in the atmosphere and could have oxidized the surface minerals from FeO to Fe_2O_3 to a depth from a few kilometers to tens of kilometers, depending on the original water amount. On the other hand, Venus' present surface is relatively young (~500 Myr), which implies periodic efficient volcanic reforming processes. The capability to oxidize the surface by these processes is yet unknown and should be studied in the future.

As seen from Table 2, a solar wind with plasma densities exceeding that of the moderate solar wind case of Wood et al. (2002) could have removed such amounts of oxygen before 4.5 Gyr even if we neglect the pick up loss process during the expected hydrogen blow-off period in the first 100 Myr. However, the loss rate estimates by Kulikov et al. (2006) may represent a lower limit, because only the ion pick up process was studied. It can be expected that atmospheric sputtering by a dense solar wind and erosion of ionized atmospheric species by plasma instabilities might increase the total loss rate from the early Venus' atmosphere.

If one assumes that Venus was "dry" over its past, our results have several implications. If the solar wind of the young Sun was emitted within the ecliptic plane, the unmagnetized early Venus could have lost up to 200–300 bar of oxygen due to ion pick up by expected moderate to high solar wind plasma fluxes. Because present Venus has its expected original

Table 2 Pick up O^+ ion loss in units of [bar] by the solar wind integrated backwards over time for the period from 3.6 to 4.6 Gyr ago

t [Gyr] ago	3.6	3.8	4	4.2	4.4	4.5	4.6
Maximum solar wind	0.067	0.13	0.3	1.1	14	117	276
Moderate solar wind	0.036	0.066	0.17	0.43	3.5	25	59
Minimum solar wind	0.016	0.028	0.062	0.16	1	6.4	14.5

Note that the solar wind plasma interaction boundary with the extended early Venus atmosphere is assumed to be one atomic oxygen scale height below the exobase level. The shown cases are for the maximum, moderate and minimum solar wind densities of Wood et al. (2002) and Lundin et al. (2007, this issue)

CO_2 inventory of about 100 bar in the atmosphere, it is unlikely that the initial CO_2 reservoir was about 200–300 bar higher.

For overcoming this problem one can postulate that the early venusian atmosphere may have been protected during the active period of the young Sun by a strong Earth-like magnetic field, or that the early solar wind was emitted off the ecliptic plane during the first 300–500 Myr after the Sun arrived at the ZAMS (Wood et al. 2005). For the history of a possible intrinsic magnetic field on early Venus, we rely completely on theories of planetary magnetism. Stevenson et al. (1983) calculated the thermal evolution of the core by assuming that the energy available for a dynamo generation is equal to the ohmic dissipation and found that it may have been possible that the geophysical conditions on early Venus generated a magnetic moment 4.5–4.6 Gyr ago of about 0.8–1.4 $\mathcal{M}_e$, where $\mathcal{M}_e$ is the present magnetic moment of the Earth. The obtained magnetic moments decreased very fast to values of about $\leq 0.4\ \mathcal{M}_e$ after 200 Myr. Another alternative would be that Venus outgassed its main atmosphere after the first 200 Myr which would be much later than expected for early Earth.

4.6 Atmospheric Loss over the Mars' History

4.6.1 Atmospheric and Water Loss During the Past 3.5 Gyr

The present thin martian atmosphere with an average surface pressure of $\sim$7 mbar has been one of the great puzzles in our Solar System. Ancient fluvial networks on the surface of Mars suggest that it might have been warmer and more wet billions of years ago, implying a much higher atmospheric surface pressure (e.g., Kasting 1991; Forget and Pierrehumbert 1997). Surface features resembling massive outflow channels provide evidence that the martian crust contained the equivalent of a planet-wide reservoir of H_2O up to 150–200 meters deep (see, e.g., Carr 1987; Head III et al. 1999).

Since Mars did not have an appreciable intrinsic magnetic field during the past 4 Gyr (e.g., Acuña et al. 1998, 1999; Connerney et al. 1999), its atmosphere could be eroded due to the solar radiation and plasma impact.

Table 3 summarizes the modelled non-thermal atmospheric loss rates of heavy atmospheric species (O, C, CO, CO_2, O^+) from Mars over the past 3.5 Gyr. Assuming a fully active self-regulating coupling mechanism between O and H and integrating the water loss rates from Mars during the past 3.5 Gyr one obtains an amount of water equivalent to a global H_2O ocean with a depth of $\leq$12–15 m 3.5 Gyr ago (Lammer et al. 2003a, 2003b). Note that this H_2O amount could have been present also as ice if the climate conditions during the past 3.5 Gyr did not allow water to be liquid on the surface (Squyres and Kasting 1994;

Table 3 Escape rates in [s^{-1}] of O, O$^+$, CO$_2$ and CO are shown for present (1 XUV), 2.7 Gyr ago (3 XUV) and 3.5 Gyr ago (6 XUV) epochs for moderate levels of solar activity (Lammer et al. 2003a). The sputter loss rates for O, CO$_2$ and CO are calculated from the sputter yields of Table 1 of Leblanc and Johnson (2002) and incident particle pick up fluxes from Lammer et al. (2003a). The O loss rates produced by dissociative recombination for 1 XUV are based on the results of Kim et al. (1998), while the 3 XUV and 6 XUV loss rate values of hot O atoms are from Luhmann (1997). The present C loss rates which originate from dissociative recombination are taken from Fox and Bakalian (2001) and are extrapolated for the 3 and 6 XUV cases. O$^+$ ion loss rates estimated for momentum transport and detached ionospheric plasma clouds are taken from Lammer et al. (2003b) and Penz et al. (2004)

Loss process / species	Present	2.7 Gyr ago	3.5 Gyr ago
Pick up: O$^+$	3.0×10^{24}	4.0×10^{25}	8.3×10^{26}
Dissociative recombination: O	6.0×10^{24}	3.0×10^{25}	8.0×10^{25}
Dissociative recombination: C	8.0×10^{23}	4.2×10^{24}	1.3×10^{25}
Sputtering: O	3.5×10^{23}	1.3×10^{25}	1.5×10^{27}
Sputtering: CO$_2$	5.0×10^{22}	2.3×10^{24}	4.0×10^{25}
Sputtering: CO	3.7×10^{22}	2.0×10^{24}	2.5×10^{25}
Momentum transport: O$^+$	$\leq 10^{25}$	$\leq 2 \times 10^{26}$	$\leq 3 \times 10^{27}$
Plasma clouds: O$^+$	$\sim 10^{24}$	$\sim 8 \times 10^{24}$	$\sim 3 \times 10^{27}$

Wänke and Dreibus 1994). However, all atmospheric escape processes where ions are involved (ion pick up, atmospheric sputtering, momentum transport, plasma clouds) should have been reduced significantly due to the protection by the putative early martian magnetic field (Hutchins et al. 1997).

4.6.2 Magnetic Field Protection of the Early Martian Atmosphere

Planetary magnetic dynamo theory predicts that the strength of a magnetic moment depends on the planetary rotation rate, the core radius and input energy (either thermal, chemical or gravitational) to drive vigorous internal convection inside the core (e.g., Busse 1976; Stevenson et al. 1983; Schubert and Spohn 1990; Dehant et al. 2007, this issue).

The available data obtained by MGS on an early martian magnetic field are mainly restricted to measurements related to the presence of significant local, small-scale, crustal remnant magnetization (Acuña et al. 1998, 1999; Connerney et al. 1999). Investigations of martian meteoroids (Weiss et al. 2002) and observations and studies related to early martian plate tectonics and crustal evolution (Sleep 1994; Breuer and Spohn 2003) also support the idea of an early intrinsic martian magnetic field. The local magnetization appears mainly in the ancient southern highlands and is absent in the regions where large impacts occurred (like Hellas and Argyre). Since these impact basins were formed about 4 Gyr ago, it is generally argued that the martian dynamo ceased before this time (e.g., Acuña et al. 1999; Breuer and Spohn 2003).

Furthermore, magnetic studies of the martian SNC meteorite ALH 84001 revealed that about 4 Gyr old carbonates contained magnetite and pyrrhotite, which carried a stable natural remnant magnetization (Weiss et al. 2002). This result implies that Mars may have established a magnetic dynamo within 450–650 Myr after its formation with an intensity of an order of that on the present Earth. Moreover, these results support the theory that an ancient strong dynamo ceased about 4 Gyr ago. Simple model simulations based on ohmic dissipation suggest that the early martian magnetic moment could have been between the

maximum and minimum expected magnetic moment of $\sim$0.1–10 $\mathcal{M}_e$ (Stevenson et al. 1983; Schubert and Spohn 1990).

Hutchins et al. (1997) studied the effect of a paleomagnetic field, which was assumed to be cut off between 2.5 and 3.6 Gyr ago, on sputtering loss of martian Ar and Ne, as well as CO_2, from the present epoch (1 XUV) to 3.6 Gyr (6 XUV) ago. For the study of the magnetospheric compression due to the solar wind at earlier time periods they adopted the solar wind velocity model of Newkirk (1980) and the minimum expected magnetic moment of $\sim$0.1 $\mathcal{M}_e$ (Schubert and Spohn 1990). By applying the atmospheric model of Zhang et al. (1993a) and the sputter model of Luhmann et al. (1992), they found that even such a relatively weak magnetic moment, resulting in a small magnetosphere, could significantly decrease the ion production rate in the solar wind interaction region and, hence, the sputter loss.

However, it should be noted that Hutchins et al. (1997) considered only faster solar wind but they did not increase its number density to higher values as expected from the recent stellar observations by Wood et al. (2002, 2005). And a higher solar wind mass flux during the early time periods after the Sun arrived at the ZAMS could have moved the magnetopause closer to the planet, which may have resulted together with a more XUV heated and extended early martian atmosphere in higher non-thermal loss rates during the first Gyr.

To investigate the efficiency of the atmosphere protection by such an intrinsic magnetic moment as a function of the expected time-dependent stronger mass flux of the young Sun (Wood et al. 2002, 2005) in our modelled heated and extended early martian atmosphere, we calculated the subsolar magnetopause stand-off distance r_s by determining the magnetic and solar wind ram pressure balance condition at the subsolar point (e.g., Spreiter 1975; Slavin and Holzer (Slavin and Holzer 1979); Grießmeier et al. 2004, 2005)

$$r_s = \left[\frac{\mu_0^2 f_0^2 \mathcal{M}^2}{4\pi^2 (2\mu_0 n_{sw} m v_{sw}^2 + B_{IMF}^2)} \right]^{1/6}, \tag{12}$$

where $\mathcal{M}$ is the early martian magnetic moment, B_{IMF} is the interplanetary magnetic field (IMF), m is a proton mass, μ_0 and f_0 are magnetic permeability and magnetospheric form-factors, respectively. Note that the magnetic moment $\mathcal{M}$ and the IMF are assumed to be constant in our simulations, so that time variations of the stand-off distance may be only due to variations of the solar wind density n_{sw} and velocity v_{sw}.

Figure 10 shows the magnetopause stand-off distance r_s for the moderate expected solar wind density (Wood et al. 2002; Lammer et al. 2003a; Lundin et al. 2007, this issue) in the martian radii during the time period between 0.1–1 Gyr after the Sun arrived at the ZAMS for various values of magnetic moments in units of the present Earth's value ($\mathcal{M}_e$). The value of the solar wind velocity v_{sw} for the young Sun is taken from Newkirk (1980). One should note that the minimum solar wind plasma densities inferred from observations by Wood et al. (2002) will move the magnetic barrier further away from the planet, resulting in better atmospheric protection. If early Mars had a magnetic moment similar to that of present Earth, r_s 4.5 Gyr ago would have been $\sim$2.8 martian radii above the surface. An initial magnetic moment of $\sim$10 $\mathcal{M}_e$ (Schubert and Spohn 1990) results in r_s of $\sim$7 martian radii above the planetary surface. The lowest expected magnetic moment on Mars 4.5 Gyr ago of $\sim$0.1 $\mathcal{M}_e$ (Schubert and Spohn 1990) yields a magnetopause stand-off distance of $\sim$0.75 martian radii above the surface.

Only a very low magnetic moment of $\sim$0.01 $\mathcal{M}_e$ would result in a direct solar wind-atmosphere interaction like at present Mars during the first 200 Myr after the Sun arrived at the ZAMS. Due to the decreasing solar wind plasma density and velocity, r_s moves to

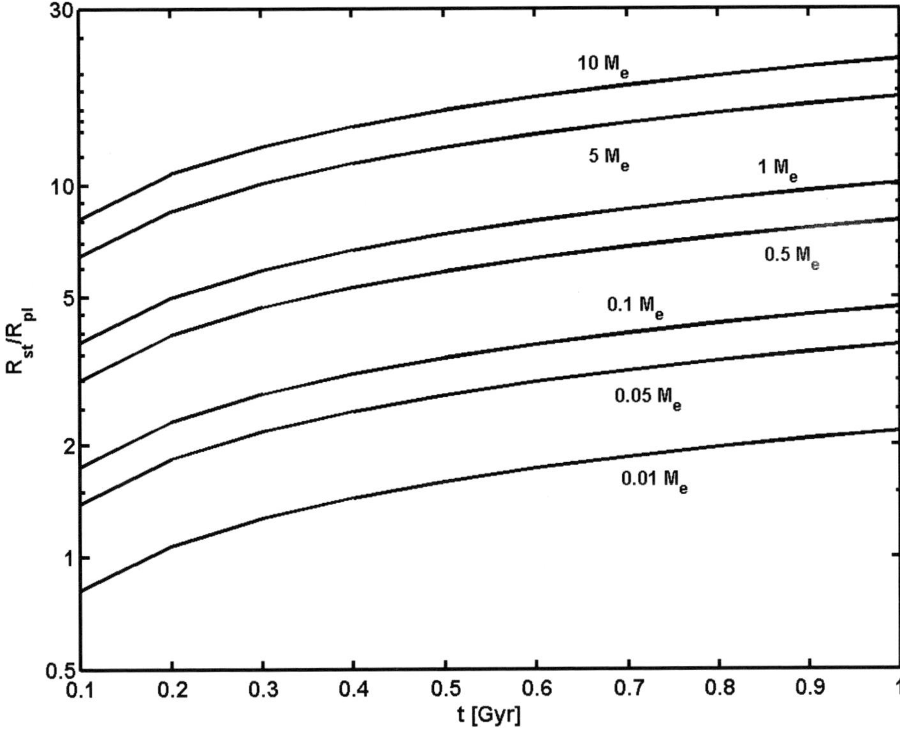

Fig. 10 Martian subsolar magnetopause stand-off distance for moderate solar wind density (Wood et al. 2002; Lammer et al. 2003a; Lundin et al. 2007, this issue) in units of the martian radii over a time period of 0.1–1 Gyr after the Sun arrived at the ZAMS for various values of expected martian magnetic moments

further distances above the martian surface, resulting in r_s for 0.01 $\mathcal{M}_e$ of $\sim$1.2 martian radii 4 Gyr ago.

Figure 11 shows the magnetopause stand-off distance r_s related to an assumed weak magnetic moment which decreases from 0.1 $\mathcal{M}_e$ to 0.01 $\mathcal{M}_e$ during the first Gyr after the Sun arrived at the ZAMS for the maximum (dotted-line), moderate (solid-line) and minimum (dashed-line) expected solar wind plasma density of Wood et al. (2002) and Lundin et al. (2007, this issue). One can see that the maximum expected solar wind mass flux 4.5 Gyr ago would move the magnetopause stand-off location to a distance of $\sim$1.4 R_{st}/R_{pl} in units of the martian radii, and to $\sim$1.85 R_{st}/R_{pl} and $\sim$2.45 R_{st}/R_{pl} for the moderate and minimum expected mass flux. One can see that r_s during the first 0.5 Gyr due to the decreasing solar wind plasma density moves to much higher altitudes above the planetary surface. Under the above assumptions, $\sim$3.5 Gyr ago due to the decreasing magnetic moment, r_s moves to altitudes which are comparable to those during the first 200 Myr after the Sun arrived at the ZAMS.

For studying the atmospheric protection effect of the early martian magnetic field on O^+ ion pick up loss we use a numerical test particle model that includes particle motion in the environmental electric and magnetic fields based on the Spreiter-Stahara gas-dynamic model (Spreiter and Stahara 1980).

 Springer

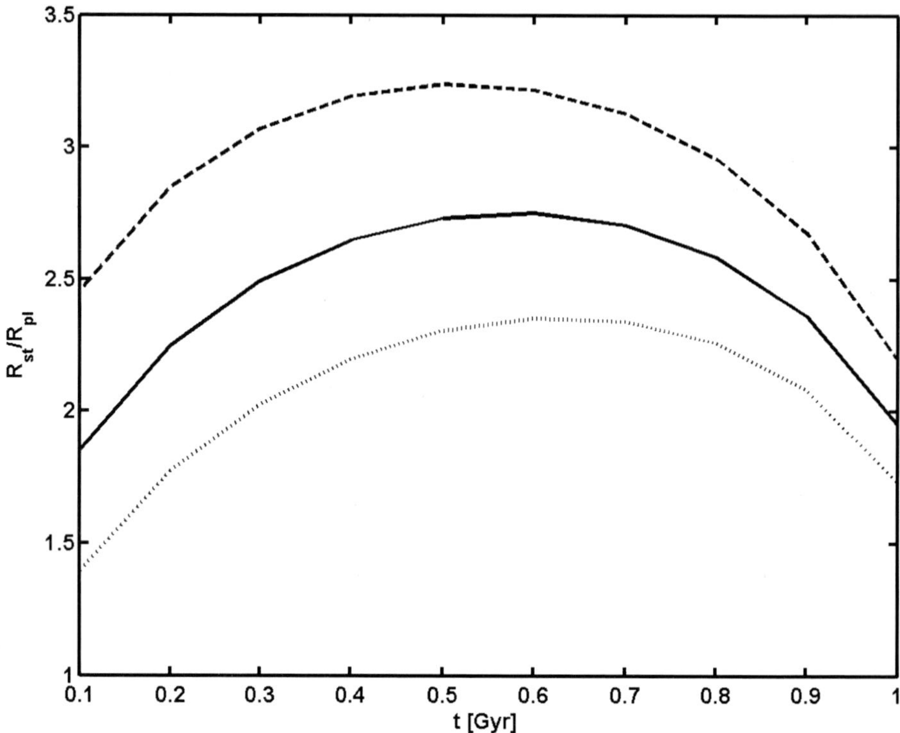

Fig. 11 Subsolar magnetopause stand-off distance for an early weak martian magnetic moment which decreases from 0.1 $\mathcal{M}_e$ to 0.01 $\mathcal{M}_e$ over the first Gyr. Maximum (*dotted-line*), moderate (*solid-line*) and minimum (*dashed-line*) expected solar wind plasma densities are taken from Wood et al. (2002), Lammer et al. (2003a), Lundin et al. (2007, this issue)

The production rates due to photoionization, electron impact and charge exchange are given by the following equations

$$Q_\gamma = \Phi_\gamma \sigma_\gamma n_O, \qquad Q_{ce} = \Phi_{sw}\sigma_{ce}n_O, \qquad Q_{ei} = \nu_e n_e n_O, \tag{13}$$

where Φ_γ and Φ_{sw} is the photon and solar wind flux, respectively, σ_γ and σ_{ce} is the ionization and charge exchange cross section of atomic oxygen, ν_e is the ionization frequency per incident electron, n_e the solar wind electron density and n_O the density of atmospheric oxygen. The energy dependent charge exchange cross sections are taken from Kallio et al. (1997) and the electron impact ionization frequencies from Cravens et al. (1987), where the electron temperature is approximated according to Zhang et al. (1993b). The total production rate Q of planetary O^+ is the sum

$$Q = Q_\gamma + Q_{ce} + Q_{ei} \tag{14}$$

and it is assumed that the ionized species are lost from the planet. This model reproduces successfully several ion distributions observed by the PVO on Venus (Luhmann 1993; Lammer et al. 2006) and the Soviet Phobos 2 spacecraft plasma measurements on Mars (Lundin et al. 1989, 1990, 1991; Lichtenegger and Dubinin 1998; Lichtenegger et al. 2002; Lammer et al. 2003a). For the calculation of the O^+ ion pick up fluxes 4.5 Gyr ago (100

 Springer

XUV), 4.33 Gyr ago (50 XUV), and 3.8 Gyr ago (10 XUV) we use the modelled "cold" and "hot" neutral O densities shown in Fig. 12, which are obtained in a similar way as in Fig. 9.

One can see from Table 4 that the O^+ ion pick up loss rates 4.33 Gyr ago (50 XUV) are of the same order of magnitude and are most likely even lower than those about 3.5 Gyr ago due to the magnetic protection of the extended upper atmosphere. For the 100 XUV case 4.5 Gyr ago the moderate solar wind mass flux obtained by Wood et al. (2002) would yield O^+ ion pick up loss rates which are comparable with the expected O atom sputter loss rates 3.5 Gyr ago. The maximum expected solar wind mass flux would move the magnetic standoff distance to about 0.4 martian radii above the surface (4.5 Gyr b.p.). The corresponding O^+ ion pick up loss rate during this early period which is shown in Table 4, was about a factor of 10 larger than the O sputter loss rate $\sim$3.5 Gyr ago (see Table 3), resulting in oxygen loss

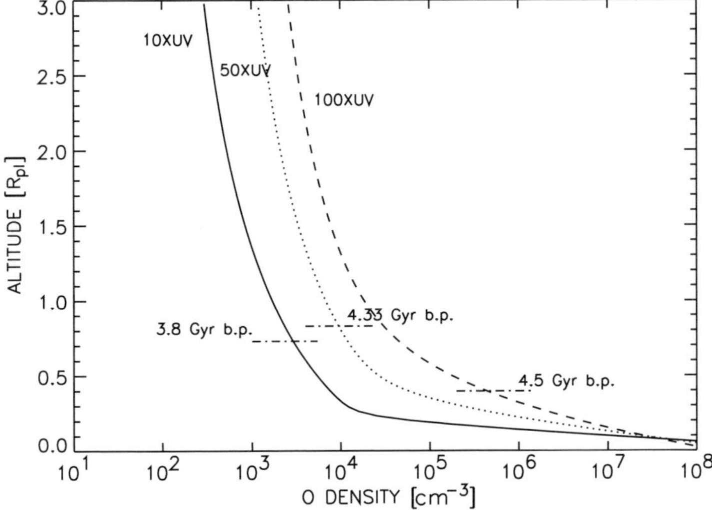

Fig. 12 Sum of "cold" and "hot" O number densities on Mars as a function of altitude in martian radii above the planetary surface for 10, 50, and 100 XUV fluxes (*solid, dotted* and *dashed lines*, respectively). The *dashed-dotted short lines* correspond to the magnetopause stand-off distances for the maximum expected solar wind mass flux (Wood et al. 2002; Lundin et al. 2007, this issue) and the expected minimum possible magnetic moment on early Mars of $\sim$0.1 $\mathcal{M}_e$ which is assumed to be decreasing to $\sim$0.01 $\mathcal{M}_e$ during the first Gyr after the Sun arrived at the ZAMS

Table 4 O^+ ion pick up escape rates in [s^{-1}] 4.5 Gyr ago (100 XUV) and 4.33 Gyr ago (50 XUV) and corresponding subsolar magnetopause stand-off distances r_s shown in Fig. 11 for maximum, moderate and minimum solar wind conditions (Wood et al. 2002; Lammer et al. 2003a; Lundin et al. 2007, this issue)

	4.5 Gyr ago [100 XUV]	4.33 Gyr ago [50 XUV]
$r_{s_{min}}$	$\approx 1.4 R_{pl}$	$\approx 2 R_{pl}$
Pick up: O^+	9.0×10^{26}	1.3×10^{26}
$r_{s_{mod}}$	$\approx 0.8 R_{pl}$	$\approx 1.4 R_{pl}$
Pick up: O^+	2.5×10^{27}	4.3×10^{26}
$r_{s_{max}}$	$\approx 0.4 R_{pl}$	$0.8 R_{pl}$
Pick up: O^+	1.5×10^{28}	1.5×10^{27}

over this time period of about 0.5 bar. Early martian magnetic moments more than 0.1 $\mathcal{M}_e$ or moderate solar wind plasma densities of the young Sun would yield negligible loss rates.

A test particle run for a 10 XUV case corresponding to a time period of $\sim$3.8 Gyr ago and a magnetic moment of the order of $\sim$0.01 $\mathcal{M}_e$ yields an O^+ ion pick up loss rate of the same order as that for an unprotected upper atmosphere at $\sim$3.5 Gyr ago (6 XUV). One should note that higher magnetic field strength of the IMF during the young Sun period (Newkirk 1980) would contribute to a higher compression of the magnetosphere. And a smaller magnetospheric stand-off distance given by (1) would result in an enhanced loss from the martian upper atmosphere.

Our model simulations for an unprotected martian upper atmosphere yield O^+ pick up loss rates, depending on the chosen solar wind mass flux values from $\sim$0.5 bar (minimum) up to $\sim$10 bar (maximum) during the first 200 Myr. However, one should note that an unprotected early martian upper atmosphere would also experience additional non-thermal loss processes like sputtering, plasma-instability-induced detached ionospheric clouds, and cool ion outflow due to the momentum transport effect (see Lundin et al. 2007, this issue). Therefore, in case of a late onset of the martian dynamo (Schubert et al. 2000) after the first 200 Myr combined with high exobase temperatures of more than 1000 K the planet could have lost up to several tens of bar.

Our results suggest that an early magnetic dynamo stronger than 0.1 $\mathcal{M}_e$ is necessary to protect the martian atmosphere during the first 200–300 Myr. Therefore, the results of our study are in general agreement with the Hutchins et al. (1997) results and indicate that an early magnetic dynamo could, probably, be strong enough to reduce the radiation and particle induced non-thermal atmospheric loss rates even for the XUV heated and expanded upper atmosphere by significant amounts.

4.6.3 Loss of CO_2 and Change in Surface Pressure

It is important to note that the total integrated mass loss of O atoms and ions shown in Table 3 may have not contributed to the total surface pressure change over the martian history, because the majority of escaping O and O^+ ions most likely had their origin from H_2O vapor but not from atmospheric CO_2 (see Lammer et al. 2003a and references therein; Amerstorfer et al. 2004). However, one should note that recent observations (Carlsson et al. 2006) and model simulations of atmospheric escape of carbon bearing species by various authors (Fox and Hać 1997; Leblanc and Johnson 2002; Lammer et al. 2003a and references therein; Chassefière and Leblanc 2004; Barabash et al. 2007) make it problematic that the surface pressure of the martian atmosphere could have reached values >0.3 bar at the end of the heavy bombardment, necessary for a greenhouse effect resulting in standing bodies of water on the martian surface.

Particle sputtering and photochemical reactions (dissociative recombination and photodissociation), however, could remove carbon bearing species (C, CO and CO_2), so that these processes would contribute to the loss of the martian atmosphere during the planets' past. Since ion pick up and sputtering depend on the solar wind mass flux, which was higher in the martian past (Lammer et al. 2003a; Lundin et al. 2007, this issue), sputtering could probably be more efficient during the early periods as long as no magnetospheric protection of the atmosphere was present.

By inspection of Table 3 it is seen that atmospheric loss due to dissociative recombination of ion species is more important for present Mars but was probably less important during earlier periods compared to sputtering and ion pick up, because the newly generated suprathermal neutrals have to move a longer distance in the XUV-heated and expanded

thermosphere before they can reach the exobase level. In such a case the "hot" particles experience more collisions which result in energy loss and lower escape rates. However, future studies should investigate in more detail the production of suprathermal neutrals by photodissociation, which could enhance the loss rates of "hot" atoms.

Impact erosion (e.g., Melosh and Vickery 1989; Brain and Jakosky 1998) and late impact accretion (e.g., Chyba et al. 1990) are additional processes that may have played an important role in the early martian atmosphere. Brain and Jakosky (1998) estimated the atmospheric loss due to impact erosion since the onset of the martian geologic record about 4 Gyr ago by using the size-frequency crater distribution N of craters having sizes more than 4 km ($N = 7 \times 10^{-4}$–3×10^{-3}) of Hartmann et al. (1981). They found an enhancement factor for the surface pressure $\sim$4 Gyr ago in the range of $\sim$2–9.

The result of Brain and Jakosky (1998) indicates that the total amount of atmosphere on Mars that was present at the onset of the geological record has been reduced by a factor of 2–9 due to large impact erosion, showing that early Mars may have lost from 50% to 90% of its original atmosphere due to impact erosion. Their result is different from the results of the previous studies by Melosh and Vickery (1989) who expected atmospheric loss rates due to impact erosion from early Mars in a range from 98% to 99.5%. However, for obtaining this result the crater density would need to be larger than $\sim$0.01 km^{-2}, which is almost one order of magnitude larger than the highest crater density values found on Mars.

It is important to note that Melosh and Vickery (1989) and Brain and Jakosky (1998) studied only the effect of impact loss and not impact delivery to the atmosphere. It may be possible that impact erosion and impact delivery are more or less in balance (private communications to H. Lammer by D. Brain and G. Neukum). If this is the case, then the major loss process should be the non-thermal escape related to a weak protecting early magnetic field or its late onset.

However, it may have been possible that Mars may have had a surface pressure at the end of the late heavy bombardment 3.8–4 Gyr ago of about $\leq$0.3–0.5 bar. The amount above the present 7 mbar surface pressure could have been lost due to sputtering and photochemical processes (e.g., Jakosky et al. 1994) and maybe due to ion pick up loss related to CO_2. Furthermore, CO_2 could be also lost due to weathering resulting in regolith deposits of the order $\sim$30–40 mbar (Zent and Quinn 1995) or of even higher levels of more than 100 mbar (Bandfield et al. 2003). Additionally, some tens of mbar could be sequestered as ice or clathrate in the polar areas (Mellon 1996). Such relatively small CO_2 surface-deposits would also be in agreement with the recent data obtained by ESA's Mars Express OMEGA-instrument that show that there may be no large amounts of carbonates on the martian surface (Biebring 2005). This result also indicates a strong early escape of most of the planets' CO_2 during the heavy bombardment period in agreement with our results.

5 Why Has the Earth Evolved Differently from Venus and Mars?

Figure 13 illustrates the results obtained from our comparative study of the influence of the XUV radiation and particle environment of the young Sun on the early upper atmospheres of Earth, Venus and Mars. From our study it seems likely that the early atmospheres ($>$4 Gyr ago) of the terrestrial planets contained high amounts of CO_2 ($\sim$90–95%) that could substantially inhibit atmospheric loss, because CO_2 appears to be the most efficient IR-radiating molecule in planetary thermospheres. A high CO_2 mixing ratio could reduce high thermospheric temperatures, that would have a strong impact on thermal and non-thermal escape rates from the upper atmospheres. Therefore, we expect that the early atmosphere of

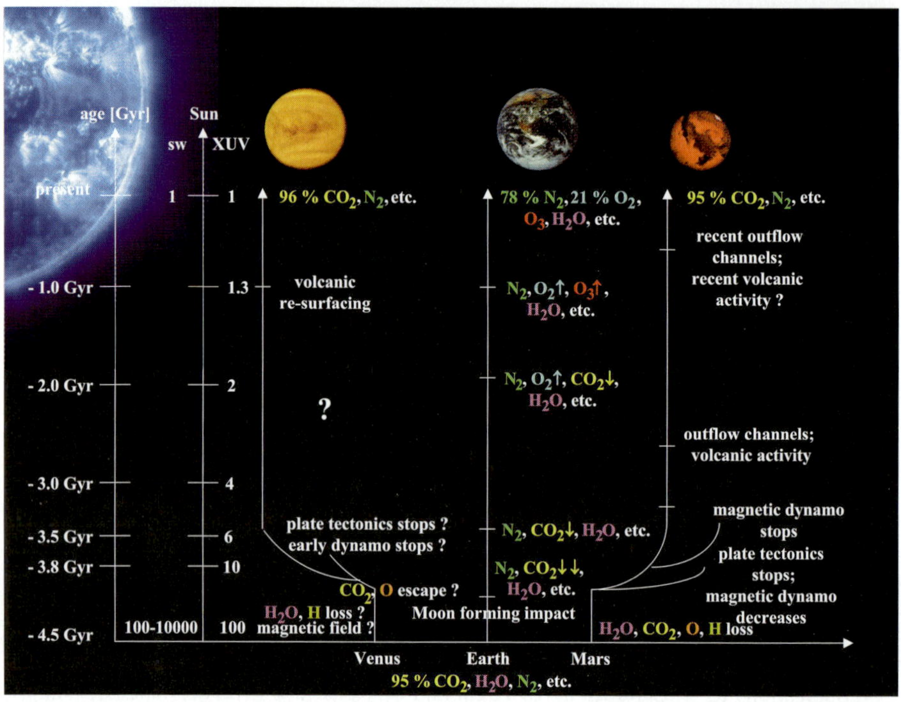

Fig. 13 Comparison of the expected evolutionary paths of the atmospheres and geophysical parameters of Venus, Earth and Mars under the influence of the solar particle and radiation environment

Earth $\sim$4.5 Gyr ago could contain the present time Venus' and Mars' CO_2 mixing ratio of $\sim$95%.

There are basically two scenarios conceivable for the generation of an early magnetic dynamo on Venus. The first possibility is that Venus had a superheated core with respect to the mantle due to the formation process. The cooling of such a superheated core would generate thermal convection in the core and consequently dynamo action. Such a thermal dynamo can be assumed to be active only during the first few hundred Myr. The second possibility could be that the young Venus experienced an early plate tectonic era which could have been driven by large H_2O inventories. Large original water inventories are essential for driving plate tectonics and keeping a magnetic moment active (Breuer and Spohn 2003). Active plate tectonics efficiently cools the mantle and core of terrestrial planets, so that an inner solid core and an outer liquid core can be formed and an early magnetic field can be easily generated. Unfortunately due to global volcanic reshaping of Venus' surface $\sim$700–800 Myr ago it is difficult to find evidence that the planet once had active plate tectonics.

An unmagnetized early Venus may have lost a large amount of its atmosphere by non-thermal escape processes like ion pick up due to the dense solar wind plasma of the young Sun, or during an efficient blow-off phase for atomic hydrogen which originated probably from an evaporating H_2O ocean (Kulikov et al. 2006). In such a case the hydrodynamically escaping hydrogen wind would form a corona around early Venus similar to that discovered around evaporating short periodic hot exoplanets like HD209458 b (Vidal-Madjar et al. 2003). Dense solar wind of the young Sun would pick up mainly the evaporating atomic hydrogen, but heavier species like CO_2 at lower atmospheric levels would be protected and

stay mainly unaffected by non-thermal loss processes until the end of the hydrogen blow-off phase. Because the activity of the young Sun decreased very fast in time, Venus was able to keep the major part of its atmosphere until today. From these results we expect that an onset of a magnetic dynamo on early Earth was essential for protecting its hot and extended early upper atmosphere against erosion by the solar wind of the young Sun. In view of our present findings a late appearance of a strong magnetic dynamo on early Earth, suggested by Ozima et al. (2005) for the explanation of implanted terrestrial atomic nitrogen and noble gases in Lunar soils, seems rather unlikely. In such a case one might expect that without a protecting magnetic field, early Earth, like early Venus, could have lost up to a terrestrial ocean of water due to solar wind erosion, as O^+ ion pick-up loss calculations for early unmagnetized Venus by Kulikov et al. (2006) indicate. However, as drawing firm conclusions from our preliminary results is still not easy, a late onset of the terrestrial magnetic dynamo cannot be completely ruled out. But it would imply then that early Earth could have lost huge amounts of its water and atmosphere due to intense solar wind erosion.

There is observational evidence that early Mars (≥ 4 Gyr ago) had a denser CO_2 atmosphere (e.g., McKay and Stocker 1989; Haberle et al. 1994; Brain and Jakosky 1998), a warmer climate (e.g., Forget and Pierrehumbert 1997; Haberle 1998; Manning et al. 2006), most likely liquid water running over its surface (e.g., Carr 1987; Head III et al. 1999; Baker 2001).

Depending on assumed solar wind parameters of the young Sun, the results of our study show that an early martian magnetic field could reduce the escape rate of atmospheric ions due to non-thermal loss processes very efficiently. On the other hand, our study indicates that Mars could have lost its whole atmosphere from several bar to several tens of bar due to a combined effect of the solar XUV heated upper atmosphere and a dense solar wind plasma flow of the young Sun if the planet were not protected by its magnetosphere during the first 200 Myr.

However, if one neglects the delivery of the atmosphere by impacts, one can expect that impact erosion may have been an efficient atmospheric loss process on early Mars during the heavy bombardment period (4–4.5 Gyr ago). This could have removed from about 50% to 90% of the early martian atmosphere (Brain and Jakosky 1998). On the other hand, if impact erosion was nearly in balance with impact atmospheric delivery, non-thermal loss processes combined with a weak early magnetic dynamo or a late onset of the dynamo (Schubert et al. 2000) (later than ~200 Myr after the planet's origin) can explain the loss of the initial martian atmosphere. Future impact effect studies should, therefore, focus on the net delivery and loss of atmosphere during the heavy bombardment period. In addition, the atmospheric erosion and production rates due to asteroid and cometary impacts need to be combined with thermal and non-thermal loss calculations for early Mars models.

The remaining atmospheric species after the end of the heavy bombardment period may have been partly lost to space from the unprotected upper atmosphere due to non-thermal loss processes and the surface weathering into the ground, ice and clathrates (e.g., Mellon 1996; Zent and Quinn 1995; Bandfield et al. 2003).

Interestingly, Jakosky et al. (1994) found that efficient sputtering till the end of the heavy bombardment period can explain the isotopic $^{36}Ar/^{38}Ar$ ratio observed in the present martian atmosphere without requiring an epoch of very intensive hydrodynamic blow-off. However, the fractionation of Xe isotopes still requires an efficient hydrodynamic blow-off period (Jakosky et al. 1994; Pepin 1994; Becker et al. 2003).

As one can see from our Figs. 7 and 8, the early martian thermosphere and exosphere were hot enough due to the high XUV radiation, so that atomic hydrogen could experience blow-off. Depending on assumed atmospheric mixing ratios, available IR-coolers, warmer

climate in the lower atmosphere, and heavy impactors which contributed to thermospheric heating, large amounts of H_2O-related hydrogen may have escaped. During an early efficient blow-off period even heavy elements like Xe isotopes could have been dragged away with the intense hydrogen wind (e.g., Zahnle et al. 1990; Chassefière 1996a, 1996b; Becker et al. 2003). These blow-off conditions and high loss rates of hydrogen which could drag along with it heavier species, might have been much more important during the martian accretion phase (Dreibus and Wänke 1987).

It should be noted that our present understanding of loss processes which are induced by photodissociation, dissociative recombination, and atmospheric sputtering is based on estimates, since the loss rates are still not measured by spacecraft. Thus, the obtained loss rates are model dependent and have to be compared with future observational data and measurements by a Mars aeronomy orbiter. Also, Mars remains an attractive target for future space missions planned by NASA, ESA, Russia, etc., but so far no successful mission, which was dedicated to a study of the martian thermosphere-exosphere environment was launched. Such a mission should study in detail temperature, isotopes, cold and hot particle populations, escape fluxes of photochemically produced "hot" neutrals, ions, including isotopes, also ionopause loss processes and how they depend on the solar XUV activity. However, the missing data which may help us to understand the evolution of the early martian magnetic dynamo, its surface pressure, atmospheric sputtering and photochemical loss processes, etc. over the planets' history can only be procured by using a comprehensive package of instruments such as the one proposed for a Mars Magnetic and Environmental Orbiter (MEMO).

By comparing Venus and Mars with early Earth (Fig. 13), we find that expected high CO_2 abundance on early Earth might have been substantially reduced due to weathering from the atmosphere onto the surface and seafloor mainly during the first 500 Myr. Fortunately, the high XUV radiation of the young Sun decreased from $\sim$100 times higher value 4.5 Gyr ago (see Ribas et al. 2005; Lundin et al. 2007, this issue) to a $\sim$10 times higher value 3.8 Gyr ago and about 6 times higher value 3.5 Gyr ago than that of today.

At these lower XUV radiation fluxes 3.5–3.8 Gyr ago much lower CO_2 mixing ratios ($\sim$1%) were able to substantially cool the upper atmosphere of Earth and to keep the exobase temperature below the critical level for hydrogen blow-off, so that the atmosphere and planetary water inventory could not be completely lost. Interestingly, these time periods correlate with the expected origin of the first life-forms which supposedly changed the Earth atmospheric composition to its present state.

6 Conclusion

We presented a comparative study of the influence of the solar radiation and particle environment of the young Sun on the early atmospheres of Earth, Venus and Mars. The main results of our study can be summarized as follows:

1) The early atmosphere of Earth about 4.0–4.5 Gyr ago may have had a Venus- or martian-like high content of CO_2, which is the most efficient IR-radiating cooler in planetary thermospheres, important to protect the upper atmosphere from high exobase temperatures and high loss rates.

2) An intrinsic planetary magnetic field like on present Earth, or early Mars is essential for the protection of an extended upper atmosphere from the XUV and solar wind induced erosion. An early martian dynamo larger than $\sim$0.1 $\mathcal{M}_e$ could reduce the ion erosion from the early martian atmosphere in an efficient way.

 Springer

3) The modelled exobase temperatures obtained for early Earth, Venus and Mars after the young Sun arrived at the ZAMS indicate that periods of blow-off and diffusion-limited hydrodynamic escape of atomic hydrogen and high Jeans escape rates for heavier species like H_2, He, C, N, O, etc., should have occurred. The duration of this blow-off period for atomic hydrogen on each planet might essentially depend on the mixing ratios of CO_2, N_2, and H_2O in the atmosphere and could last from ~ 100 to several hundred Myr.

4) Our model simulations show that atmospheric erosion by the solar wind ion pick up from a non-magnetized Venus could effectively erode O^+ ions during the first Gyr after the Sun arrived at the ZAMS. Depending on the solar wind parameters expected for the young Sun, an equivalent amount of up to one terrestrial ocean could be lost (see also Kulikov et al. 2006).

The comparative study of the early upper atmospheres of Earth, Venus, and Mars suggests that we will not be able to understand the evolution of Earth-type biospheres and habitability of terrestrial planets in general if we neglect atmospheric effects caused by the radiation and particle environment of the young Sun/stars, which change dramatically over the history of the solar/stellar system. Therefore, a better knowledge of early solar wind parameters including estimations of magnetic fields of young stellar systems and plasma outflow from young Sun-like stars is urgently needed. Thus, more projects/missions related to observations of solar proxies should be undertaken in the near future.

In conclusion we must also note that our findings presented in this study should be considered as preliminary and awaiting further confirmation of the hypotheses on which our reasoning is mainly based, and of the progress in relevant stellar and planetary observations and development of more advanced atmospheric loss models for analyzing observational data and making realistic predictions.

Acknowledgements   The authors would like to thank Eric Chassefière and another anonymous reviewer for their constructive criticism of the manuscript and helpful suggestions. We also wish to thank the Editor, Kathryn Fishbaugh, for her valuable comments and recommendations. Yu.N. Kulikov, H. Lammer, H.I.M. Lichtenegger, T. Penz, and H.K. Biernat thank the "Österreichischer Austauschdienst" (ÖAD), which supported this work by the project I.12/04. The authors acknowledge also the support by the Austrian Academy of Sciences, "Verwaltungsstelle für Auslandsbeziehungen", by the Russian Academy of Sciences (RAS), for supporting working visits to the PGI/RAS in Murmansk, Russian Federation. Yu.N. Kulikov, H. Lammer and H.I.M. Lichtenegger also thank the Russian Foundation for the Basic Research, which supported this study as a joint Russian-Austrian project No. 03-05-20003 "Solar-planetary relations and space weather". H.K. Biernat and T. Penz also thanks the Austrian "Fonds zur Förderung der wissenschaftlichen Forschung" which supported this study under project P17099-N08. Yu.N. Kulikov, H. Lammer, and H.K. Biernat acknowledge also the International Space Science Institute (ISSI; Bern, Switzerland), as this study was partly carried out within the framework of the ISSI Team "Evolution of Habitable Planets".

References

M.H. Acuña et al., Science **279**, 1676–1680 (1998)

M.H. Acuña et al., Science **284**, 790–793 (1999)

U.V. Amerstorfer, H. Lammer, T. Tokano, F. Selsis, C. Kolb, A. Bérces, G. Kovács, M.R. Patel, C.S. Cockell, Gy. Rontó, T. Penz, N.V. Erkaev, H.K. Biernat, in *Proc. 3rd European Workshop on Exo/Astrobiology*, ed. by R.A. Harris, L. Ouwehand. ESA SP, vol. 545 (2004), pp. 165–166

T.R. Ayres, J. Geophys. Res. **102**, 1641–1651 (1997)

V.R. Baker, Nature **412**, 228–236 (2001)

J.L. Bandfield, T.D. Gloch, P.R. Christensen, Science **301**, 1084–1086 (2003)

S. Barabash, A. Fedorov, R. Lundin, J.-A. Sauvaud, Science **315**, 501–514 (2007)

C.A. Barth, A.I.F. Stewart, S.W. Bougher, D.M. Hunten, S.J. Bauer, A.F. Nagy, in *Mars* (Univ. Arizona Press, 1992), pp. 1054–1089

R.H. Becker, R.N. Clayton, E.M. Galimov, H. Lammer, B. Marty, R.O. Pepin, R. Weiler, Space Sci. Rev. **106**, 377–410 (2003)

J.-P. Biebring the OMEGA team, Science **307**, 1576–1581 (2005)

H.K. Biernat, N.V. Erkaev, C.J. Farrugia, Adv. Space Res. **28**, 833–839 (2001)

S.W. Bougher, G.M. Keating, Structure of the Mars Upper Atmosphere: MGS Aerobraking Data and Model Interpretation. The Fifth International Conference on Mars, July 19–24, 1999, Pasadena, California, abstract no. 6010, 1999

S.W. Bougher, S. Engel, R.G. Roble, B. Foster, J. Geophys. Res. **104**, 16,591–16,611 (1999)

S.W. Bougher, S. Engel, R.G. Roble, B. Foster, J. Geophys. Res. **105**, 17669–17692 (2000)

L.H. Brace, R.F. Theis, W.R. Hoegy, Planet. Space Sci. **30**, 29–37 (1982)

D.A. Brain, B.M. Jakosky, J. Geophys. Res. **103**, 22689–22694 (1998)

D. Breuer, T. Spohn, J. Geophys. Rev. **108**, 5072 (2003). doi: 10.1029/2002JE001999

F.H. Busse, Phys. Earth Planet. **12**, 350–358 (1976)

E. Carlsson, F. Fedorovc, S. Barabash, E. Budnik, A. Grigorieva, H. Gunell, H. Nilssona, J.-A. Sauvaud, R. Lundin, Y. Futaanaa, M. Holmström, H. Anderssona, M. Yamauchi, J.-D. Winningham, R.A. Frahmd, J.R. Sharber, J. Scherrer, A.J. Coates, D.R. Linder, D.O. Kataria, E. Kallio, H. Koskinen, T. Säles, P. Riihela, W. Schmidt, J. Kozyra, J. Luhmann, E. Roelof, D. Williams, S. Livi, C.C. Curtis, K.C. Hsiehj, B.R. Sandel, M. Grande, M. Carter, J.-J. Thocaven, S. McKenna-Lawlor, S. Orsini, R. Cerulli-Irelli, M. Maggi, P. Wurz, P. Bochsler, N. Krupp, J. Wocho, M. Fraenz, K. Asamura, C. Dierker, Icarus **182**, 320–328 (2006)

M.H. Carr, Nature **326**, 30–34 (1987)

J.W. Chamberlain, Planet. Space Sci. **11**, 901–996 (1963)

S. Chandra, A.K. Sinha, J. Geophys. Res. **79**, 1916–1921 (1974)

E. Chassefière, Icarus **124**, 537–552 (1996a)

E. Chassefière, J. Geophys. Res. **101**, 26039–26056 (1996b)

E. Chassefière, F. Leblanc, Planet. Space Sci. **52**, 1039–1058 (2004)

C.F. Chyba, P.J. Thomas, L. Brookshaw, C. Sagan, Science **249**, 366–373 (1990)

M. Cohen, L.V. Kuhi, Astrophys. J. Suppl. **41**, 743–843 (1979)

J.E.P. Connerney, M.H. Acuña, P. Wasilewski, N.F. Ness, H. Rème, C. Mazelle, D. Vignes, R.P. Lin, D. Mitchell, P. Cloutier, Science **284**, 794–798 (1999)

T.E. Cravens, J.U. Kozyra, A.F. Nagy, T.I. Gombosi, M. Kurtz, J. Geophys. Res. **92**, 7341–7353 (1987)

G. Crowley, Rev. Geophys. Suppl. **29**, 1143–1165 (1991)

M.O. Dayhoff, R. Eck, E.R. Lippincott, C. Sagan, Science **155**, 556–557 (1967)

V. Dehant, H. Lammer, Yu.N. Kulikov, J.-M. Grießmeier, D. Breuer, O. Verhoeven, Ö. Karatekin, T. van Hoolst, O. Korablev, P. Lognonné, Space Sci. Rev. (2007, this issue). doi: 10.1007/s11214-007-9163-9

R.E. Dickinson, J. Atmos. Sci. **29**, 1531–1556 (1972)

R.E. Dickinson, S.W. Bougher, J. Geophys. Res. **91**, 70–80 (1986)

R.E. Dickinson, R.G. Roble, S.W. Bougher, Adv. Space Res. **7**, (10)5–(10)15 (1987)

T. Donahue, J.H. Hoffman, A.J. Watson, Science **216**, 630–633 (1982)

T.M. Donahue, J.B. Pollack, in *Venus*, ed. by D.M. Hunten, L. Colin, T.M. Donahue, V.I. Moroz (University of Arizona Press, Tucson, 1983), pp. 1003–1036

T.M. Donahue, E. Hartle, Geophys. Res. Lett. **12**, 2449–2452 (1992)

J.D. Dorren, E.F. Guinan, in *The Sun as a Variable Star*, ed. by J.M. Pap, C. Frölich, H.S. Hudson, S.K. Solanki (Cambridge University Press, Cambridge, 1994), pp. 206–216

G. Dreibus, H. Wänke, Icarus **71**, 225–240 (1987)

F. Forget, R.T. Pierrehumbert, Science **278**, 1273–1276 (1997)

J.L. Fox, A. Dalgarno, J. Geophys. Res. **84**, 7315–7333 (1979)

J.L. Fox, A. Dalgarno, J. Geophys. Res. **86**, 629–639 (1981)

J.L. Fox, Planet. Space Sci. **36**, 37–46 (1988)

J.L. Fox, A. Hać, J. Geophys. Res. **102**, 24005–24011 (1997)

J.L. Fox, K.Y. Sung, J. Geophys. Res. **106**, 21305–21335 (2001)

J.L. Fox, F.M. Bakalian, J. Geophys. Res. **106**, 28785–28795 (2001)

J.-M. Grießmeier, U. Motschmann, A. Stadelmann, T. Penz, H. Lammer, F. Selsis, I. Ribas, E.F. Guinan, H.K. Biernat, W.W. Weiss, Astron. Astrophys. **425**, 753–762 (2004)

J.-M. Grießmeier, A. Stadelmann, U. Motschmann, N.K. Belisheva, H. Lammer, H.K. Biernat, Astrobiology **5**, 587–603 (2005)

G.V. Gridchin, E.A. Zhadin, A.I. Ivanovsky, V.A. Marchevsky, Geomagn. Aeron. **15**, 93–100 (1975)

D.H. Grinspoon, J.S. Lewis, Icarus **74**, 21–35 (1988)

S.H. Gross, J. Atmos. Sci. **29**, 214–218 (1972)

E.F. Guinan, I. Ribas, in *The Evolving Sun and Its Influence on Planetary Environments*, ed. by B. Montesinos, A. Giménz, E.F. Guinan. ASP, vol. 269 (San Francisco, 2002), pp. 85–107

B.F. Gordiets, M.N. Markov, L.A. Shelepin, Planet. Space Sci. **26**, 933–948 (1978)

B.F. Gordiets, Yu.N. Kulikov, M.N. Markov, M.Ya. Marov, Preprint FIAN, N 112 (1979)

B.F. Gordiets, Yu.N. Kulikov, Kosm. Issled. **19**, 249–260 (1981)

B.F. Gordiets, Yu.N. Kulikov, Kosm. Issled. **19**, 539–550 (1981)

B.F. Gordiets, Yu.N. Kulikov, Trudy FIAN **130**, 29–47 (1982)

B.F. Gordiets, Yu.N. Kulikov, M.N. Markov, M.Ya. Marov, J. Geophys. Res. **87**, 4504–4514 (1982)

B.F. Gordiets, A.I. Osipov, L.A. Shelepin, *Kinetic Processes in Gases and Molecular Lasers* (Nauka, Moscow, 1980), 512 pp. (English translation: Gordon and Breach, New York, 1987, 688 pp.)

B.F. Gordiets, Yu.N. Kulikov, Adv. Space Res. **5**, 113–117 (1985)

B.F. Gordiets, Trudy FIAN **212**, 109–122 (1991)

D.O. Gough, in *The Solar Output and Its Variations*, ed. by O.R. White (University of Colorado Press, Boulder, 1977), pp. 451–473

R.M. Haberle, D. Tyler, C.P. McKay, W.L. Davis, Icarus **109**, 102–120 (1994)

R.M. Haberle, J. Geophys. Res. **103**, 28467–28479 (1998)

W.B. Hanson, S. Santani, D.R. Zuccaro, J. Geophys. Res. **82**, 4351–4363 (1977)

R.E. Hartle, J.M. Grebowsky, J. Geophys. Res. **98**, 7437–7445 (1993)

R.E. Hartle, T.M. Donahue, J.M. Grebowsky, H.G. Mayr, J. Geophys. Res. **101**, 4525–4538 (1996)

W.K. Hartmann et al., in *Basaltic Volcanism on the Terrestrial Planets* (Pergamon, Tarrytown, 1981)

J.W. Head III, H. Hiesinger, M.A. Ivanov, M.A. Kreslavsky, S. Pratt, B.J. Thomson, Science **286**, 2134–2137 (1999)

A.E. Hedin, H.B. Nieman, W.T. Kasprzak, A. Seiff, J. Geophys. Res. **88**, 73–83 (1983)

A.M. Hessler, D.R. Lowe, R.L. Jones, D.K. Bird, Nature **428**, 736–738 (2004)

J.H. Hoffman, R.R. Hodges Jr., T.M. Donahue, M.B. McElroy, J. Geophys. Res. **85**, 7882–7890 (1980)

H.D. Holland, in *The Origin and Evolution of Atmospheres and Oceans*, ed. by P.J. Brancasio, A.G.W. Cameron (Wiley, New York, 1963), pp. 86–101

D.J. Hollenbach, S.S. Prasad, R.C. Witten, Icarus **64**, 205–220 (1985)

D.M. Hunten, T.M. Donahue, J.C.G. Walker, J.F. Kasting, in *Origin and Evolution of Planetary and Satellite Atmospheres*, ed. by S.K. Atreya, J.B. Pollack, M.S. Matthews (University of Arizona Press, Tucson, 1987), pp. 386–423

D.M. Hunten, Science **259**, 915–920 (1993)

K.S. Hutchins, B.M. Jakosky, J.G. Luhmann, J. Geophys. Res. **102**, 9183–9189 (1997)

W.-H. Ip, Icarus **76**, 135–145 (1988)

I. Iben Jr., Astrophys. J. **141**, 993–1018 (1965)

M.N. Izakov, Space Sci. Rev. **12**, 261–298 (1971)

L.G. Jacchia, Thermospheric temperature, density and composition: New models. Spec. Rep., 375, Smithson. Inst. Astrophys. Obs., Cambridge, 1977

B.M. Jakosky, R.O. Pepin, R.E. Johnson, J.L. Fox, Icarus **111**, 271–288 (1994)

J.H. Jeans, *The Dynamical Theory of Gases*, 4th edn. (Cambridge University Press, Cambridge, 1925)

R.E. Johnson, M. Liu, J. Geophys. Res. **101**, 3649–3647 (1996)

N.M. Johnson, B. Fegley, Icarus **146**, 301–306 (2000)

R.E. Johnson, D. Schnellenberger, M.C. Wong, J. Geophys. Res. **105**, 1659–1670 (2000)

E. Kallio, J.G. Luhmann, S. Barabash, J. Geophys. Res. **102**, 22183–22197 (1997)

E. Kallio, P. Janhunen, Annales Geophysicae **21**, 2133–2145 (2003)

D.M. Kass, Y.L. Yung, Science **268**, 697–699 (1995)

D.M. Kass, Y.L. Yung, Science **274**, 1932–1933 (1996)

J.F. Kasting, J.B. Pollack, Icarus **53**, 479–508 (1983)

J.F. Kasting, J.B. Pollack, T.P. Ackerman, Icarus **57**, 335–355 (1984)

J.F. Kasting, Icarus **74**, 472–494 (1988)

J.F. Kasting, Icarus **94**, 1–13 (1991)

G.M. Keating, R.H. Tolson, T.J. Schellenberg, N.C. Hsu, S.W. Bougher, Study of Venus upper atmosphere using Magellan drag measurements. Second Ann. Progress Rep., NAG5-6081, NASA, Washington DC, 1998

J. Kim, A.F. Nagy, J.L. Fox, T. Craven, J. Geophys. Res. **103**, 29,339–29,342 (1998)

Yu.N. Kulikov, H. Lammer, H.I.M. Lichtenegger, N. Terada, I. Ribas, C. Kolb, D. Langmayr, R. Lundin, E.F. Guinan, S. Barabash, H.K. Biernat, Planet. Space Sci. **54**, 1425–1444 (2006)

H. Lammer, S.J. Bauer, J. Geophys. Res. **96**, 1819–1825 (1991)

H. Lammer, W. Stumptner, G.-J. Molina-Cuberos, S.J. Bauer, T. Owen, Planet. Space Sci. **48**, 529–543 (2000a)

H. Lammer, W. Stumptner, S.J. Bauer, Planet. Space Sci. **48**, 1473–1478 (2000b)

H. Lammer, S.J. Bauer, Space Sci. Rev. **106**, 281–292 (2003)

H. Lammer, H.I.M. Lichtenegger, C. Kolb, I. Ribas, E.F. Guinan, R. Abart, S.J. Bauer, Icarus **106**, 9–25 (2003a)

H. Lammer, C. Kolb, T. Penz, U.V. Amerstorfer, H.K. Biernat, B. Bodiselitsch, Int. J. Astrobiol. **2**, 1–8 (2003b)

H. Lammer, H.I.M. Lichtenegger, H.K. Biernat, N.V. Erkaev, I.L. Arshukova, C. Kolb, H. Gunell, A. Lukyanov, M. Holmstrom, S. Barabash, T.L. Zhang, W. Baumjohann, Planet. Space Sci. **54**, 1445–1456 (2006)

F. Leblanc, R.E. Johnson, Planet. Space Sci. **49**, 645–656 (2001)

F. Leblanc, R.E. Johnson, J. Geophys. Res. **107**, 1–6 (2002)

J.S. Lewis, Earth Planet. Sci. Lett. **10**, 73–80 (1970)

J.S. Lewis, Icarus **16**, 241–252 (1972)

J.S. Lewis, Space Sci. Rev. **14**, 401–410 (1973)

J.S. Lewis, Science **186**, 440–443 (1974)

J.S. Lewis, R.G. Prinn, *Planets and Their Atmospheres: Origin and Evolution* (Academic, New York, 1984)

H.I.M. Lichtenegger, E.M. Dubinin, Earth Planets Space **50**, 445–452 (1998)

H.I.M. Lichtenegger, H. Lammer, W. Stumptner, J. Geophys. Res. **107**, 1279 (2002). doi: 10.1029/2001JA000322

J.G. Luhmann, J.U. Kozyra, J. Geophys. Res. **96**, 5457–5467 (1991)

J.G. Luhmann, R.E. Johnson, M.G.H. Zhang, Geophys. Res. Lett. **19**, 2151–2154 (1992)

J.G. Luhmann, J. Geophys. Res. **98**, 17615–17621 (1993)

J.G. Luhmann, J. Geophys. Res. **102**, 1637 (1997)

R. Lundin, A. Zakharov, R. Pellinen, B. Hultquist, H. Borg, E.M. Dubinin, S. Barabash, N. Pissarenko, H. Koskinnen, I. Liede, Nature **341**, 609–612 (1989)

R. Lundin, A. Zakharov, R. Pellinen, S.W. Barabash, H. Borg, E.M. Dubinin, B. Hultquist, H. Koskinen, I. Liede, N. Pissarenko, Geophys. Res. Lett. **17**, 873–876 (1990)

R. Lundin, E.M. Dubinin, H. Koskinen, O. Norberg, N. Pissarenko, S.W. Barabash, Geophys. Res. Lett. **18**, 1059–1062 (1991)

R. Lundin et al., Science **305**, 1933–1936 (2004)

R. Lundin, H. Lammer, I. Ribas, Space Sci. Rev. (2007, this issue). doi: 10.1007/s11214-007-9176-4

J.A. Magalhaes, J.T. Schofield, A. Seiff, J. Geophys. Res. **104**, 8943–8955 (1999)

C.V. Manning, C.P. McKay, K.J. Zahnle, Icarus **180**, 38–59 (2006)

P. McKay, C.R. Stocker, Rev. Geophys. **27**, 189–214 (1989)

M.T. Mellon, Icarus **124**, 268–279 (1996)

H.J. Melosh, A. Vickery, Nature **338**, 487–489 (1989)

M.B. McElroy, M.J. Prather, J.M. Rodriguez, Science **215**, 1614–1615 (1982)

A. Morbidelli, J. Chambers, J.I. Lunine, J.M. Petit, F. Robert, G.B. Valsecchi, K. Cyr, Meteorit. Planet. Sci. **35**, 1309–1320 (2000)

V.I. Moroz, N.A. Parfentev, N.F. Sanko, Kosm. Issled. **17**, 727–742 (1979)

A.F. Nagy, T.E. Cravens, J.H. Yee, A.I.F. Stewart, Geophys. Res. Lett. **8**, 629–632 (1981)

A.F. Nagy, J. Kim, T.E. Cravens, Ann. Geophys. **8**, 251–256 (1990)

H.B. Nieman, R.E. Hartle, W.T. Kasprzak, N.W. Spencer, D.M. Hunten, G.R. Carignan, Science **203**, 770–772 (1979a)

H.B. Nieman, R.E. Hartle, A.E. Hedin, W.T. Kasprzak, N.W. Spencer, D.M. Hunten, G.R. Carignan, Science **205**, 54–56 (1979b)

H.B. Niemann, W.T. Kasprzak, A.E. Hedin, D.M. Hunten, W. Spencer, J. Geophys. Res. **85**, 7817–7827 (1980)

A.O. Nier, M.B. McElroy, J. Geophys. Res. **82**, 4341–4349 (1977)

G. Newkirk Jr., Geochim. Cosmochim. Acta Suppl. **13**, 293–301 (1980)

M. Ozima, K. Seki, N. Terada, Y.N. Miura, F.A. Podosek, H. Shinagawa, Nature **436**, 655–659 (2005)

E.J. Öpik, Geophys. J. Roy. Astron. Soc. **7**, 490–526 (1963)

T. Penz, N.V. Erkaev, H.K. Biernat, H. Lammer, U.V. Amerstorfer, H. Gunell, E. Kallio, S. Barabash, S. Orsini, A. Milillo, W. Baumjohann, Planet. Space Sci. **52**, 1157–1167 (2004)

R.O. Pepin, Icarus **111**, 289–304 (1994)

J.B. Pollack, Icarus **14**, 295–306 (1971)

S.I. Rasool, C. De Bergh, Nature **226**, 1037–1039 (1970)

R. Raye, O. Kuo, H.D. Holland, Nature **378**, 603–605 (1995)

S.N. Raymond, T. Quinn, J.I. Lunine, Icarus **168**, 1–17 (2004)

I. Ribas, E.F. Guinan, M. Güdel, M. Audard, Astrophys. J. **622**, 680–694 (2005)

P.G. Richards, D.G. Torr, J. Geophys. Res. **93**, 4060–4066 (1988)

J.M. Rodriguez, M.J. Prather, M.B. McElroy, Planet. Space Sci. **32**, 1235–1255 (1984)

J. Rosenqvist, E. Chassefière, Planet. Space Sci. **43**, 3–10 (1995)

C.T. Russell, J.G. Luhmann, R.C. Elphic, F.L. Scarf, L.H. Brace, Geophys. Res. Lett. **9**, 45–48 (1982)

J.T. Schofield et al., Science **278**, 1752–1758 (1997)

G. Schubert, T. Spohn, J. Geophys. Res. **95**, 14095–14104 (1990)

G. Schubert, C.T. Russell, W.B. Moore, Nature **408**, 666–667 (2000)

Y. Shimazu, T. Urabe, Icarus **9**, 498–506 (1968)

H. Shinagawa, T.E. Cravens, A.F. Nagy, J. Geophys. Res. **92**, 7317–7330 (1987)

R.W. Shunk, A.F. Nagy, *Ionospheres – Physics, Plasma Physics, and Chemistry* (Cambridge University Press, Cambridge, 2000)

J.A. Slavin, R.E. Holzer, J. Geophys. Res. **84**, 2076–2082 (1979)

N.H. Sleep, J. Geophys. Res. **99**, 5639–5655 (1994)

J.R. Spreiter, NASA Spec. Publ. **SP-397**, 135–149 (1975)

J.R. Spreiter, S.S. Stahara, J. Geophys. Res. **98**, 17,251–17,262 (1980)

S.W. Squyres, J.F. Kasting, Science **265**, 744–748 (1994)

D.J. Stevenson, T. Spohn, G. Schubert, Icarus **54**, 466–489 (1983)

C.P. Sonnett, M.S. Giampapa, M.S. Matthews, *The Sun in Time* (University of Arizona Press, Tucson, 1991)

N. Terada, S. Machida, H. Shinagawa, J. Geophys. Res. **107**, 1471–1490 (2002). doi: 10.1029/2001JA009224

F. Tian, O.B. Toon, A.A. Pavlov, H. De Sterck, Science **308**, 1014–1017 (2005)

F. Tian, O.B. Toon, A.A. Pavlov, Science **311**, 38b (2006)

D.G. Torr, M.R. Torr, J. Atmos. Terr. Phys. **41**, 799–839 (1979)

M.R. Torr, P.G. Richards, D.G. Torr, J. Geophys. Res. **85**, 6819–6826 (1980)

A. Vidal-Madjar, A. Lecavelier des Etangs, J.-M. Désert, G.E. Ballester, R. Ferlet, G. Hébrard, M. Mayor, Nature **422**, 143–146 (2003)

U. Von Zahn, K.H. Fricke, D.M. Hunten, D. Krankowsky, K. Mauersberger, A.O. Nier, J. Geophys. Res. **85**, 7829–7840 (1980)

J.C.G. Walker, J. Atmos. Sci. **32**, 1248–1256 (1975)

H. Wänke, G. Dreibus, Philos. Trans. Roy. Soc. London Ser. A. **349**, 285–293 (1994)

B.P. Weiss, H. Vali, F.J. Baudenbacher, J.L. Kirschvink, S.T. Stewart, D.L. Shuster, Earth Planet. Sci. Lett. **201**, 449–463 (2002)

G.W. Wetherill, Icarus **46**, 70–80 (1981)

B.E. Wood, H.-R. Müller, G. Zank, J.L. Linsky, Astrophys. J. **574**, 412–425 (2002)

B.E. Wood, H.-R. Müller, G.P. Zank, J.L. Linsky, S. Redfield, Astrophys. J. **628**, L143–L146 (2005)

R. Yelle, Icarus **170**, 167–179 (2004)

K.J. Zahnle, J.C.G. Walker, Rev. Geophys. **20**, 280–292 (1982)

K. Zahnle, J.B. Pollack, J.F. Kasting, Icarus **84**, 503–527 (1990)

A.P. Zent, R.C. Quinn, J. Geophys. Res. **100**, 5341–5349 (1995)

M.H.G. Zhang, J.G. Luhmann, S.W. Bougher, A.F. Nagy, J. Geophys. Res. **98**, 10,915–10,923 (1993a)

M.H.G. Zhang, J.G. Luhmann, A.F. Nagy, J.S. Spreiter, S.S. Stahara, J. Geophys. Res. **98**, 3311–3318 (1993b)

Space Sci Rev (2007) 129: 245–278
DOI 10.1007/s11214-007-9176-4

Planetary Magnetic Fields and Solar Forcing: Implications for Atmospheric Evolution

Rickard Lundin · Helmut Lammer · Ignasi Ribas

Received: 4 April 2006 / Accepted: 22 March 2007 /
Published online: 3 July 2007
© Springer Science+Business Media, Inc. 2007

Abstract The solar wind and the solar XUV/EUV radiation constitute a permanent forcing of the upper atmosphere of the planets in our solar system, thereby affecting the habitability and chances for life to emerge on a planet. The forcing is essentially inversely proportional to the square of the distance to the Sun and, therefore, is most important for the innermost planets in our solar system—the Earth-like planets. The effect of these two forcing terms is to ionize, heat, chemically modify, and slowly erode the upper atmosphere throughout the lifetime of a planet. The closer to the Sun, the more efficient are these process. Atmospheric erosion is due to thermal and non-thermal escape. Gravity constitutes the major protection mechanism for thermal escape, while the non-thermal escape caused by the ionizing X-rays and EUV radiation and the solar wind require other means of protection. Ionospheric plasma energization and ion pickup represent two categories of non-thermal escape processes that may bring matter up to high velocities, well beyond escape velocity. These energization processes have now been studied by a number of plasma instruments orbiting Earth, Mars, and Venus for decades. Plasma measurement results therefore constitute the most useful empirical data basis for the subject under discussion. This does not imply that ionospheric plasma energization and ion pickup are the main processes for the atmospheric escape, but they remain processes that can be most easily tested against empirical data.

Shielding the upper atmosphere of a planet against solar XUV, EUV, and solar wind forcing requires strong gravity and a strong intrinsic dipole magnetic field. For instance, the strong dipole magnetic field of the Earth provides a "magnetic umbrella", fending of the solar wind at a distance of 10 Earth radii. Conversely, the lack of a strong intrinsic magnetic field at Mars and Venus means that the solar wind has more direct access to their topside

R. Lundin (✉)
Swedish Institute of Space Physics (IRF), Box 812, 98128 Kiruna, Sweden
e-mail: rickard@irf.se

H. Lammer
Space Research Institute, Austrian Academy of Sciences, Schmiedlstr. 6, 8042 Graz, Austria

I. Ribas
Facultat de Ciències, Campus UAB, Institut de Ciències des l'Espai (CSIC-IEEC), Torre
C5-parellE, 08193 Bellaterra, Spain

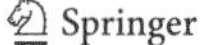 Springer

atmosphere, the reason that Mars and Venus, planets lacking strong intrinsic magnetic fields, have so much less water than the Earth?

Climatologic and atmospheric loss process over evolutionary timescales of planetary atmospheres can only be understood if one considers the fact that the radiation and plasma environment of the Sun has changed substantially with time. Standard stellar evolutionary models indicate that the Sun after its arrival at the Zero-Age Main Sequence (ZAMS) 4.5 Gyr ago had a total luminosity of $\approx$70% of the present Sun. This should have led to a much cooler Earth in the past, while geological and fossil evidence indicate otherwise. In addition, observations by various satellites and studies of solar proxies (Sun-like stars with different age) indicate that the young Sun was rotating more than 10 times its present rate and had correspondingly strong dynamo-driven high-energy emissions which resulted in strong X-ray and extreme ultraviolet (XUV) emissions, up to several 100 times stronger than the present Sun. Further, evidence of a much denser early solar wind and the mass loss rate of the young Sun can be determined from collision of ionized stellar winds of the solar proxies, with the partially ionized gas in the interstellar medium. Empirical correlations of stellar mass loss rates with X-ray surface flux values allows one to estimate the solar wind mass flux at earlier times, when the solar wind may have been more than 1000 times more massive.

The main conclusions drawn on basis of the Sun-in-time-, and a time-dependent model of plasma energization/escape is that:

1. Solar forcing is effective in removing volatiles, primarily water, from planets,
2. planets orbiting close to the early Sun were subject to a heavy loss of water, the effect being most profound for Venus and Mars, and
3. a persistent planetary magnetic field, like the Earth's dipole field, provides a shield against solar wind scavenging.

Keywords Planetary magnetospheres · Solar forcing · Young Sun · Ionospheric plasma escape · Loss of planetary water

1 Introduction

Solar forcing affects the planets in the inner solar system in different ways, the most obvious being the solar gravitation force. But there are also other forcing terms affecting the planets: solar irradiation and the solar plasma outflow. The solar irradiation, with a spectrum from X-ray to infrared, provides the highest input power to the planetary environment, corresponding to a power between 490–720 W/m^2 for Mars, with its elliptic orbit, for the Earth 1370 W/m^2, and 2620 W/m^2 for Venus. The power input from solar plasma outflow/the solar wind, is quite variable, but the average power is six orders of magnitude smaller (0.001–0.003 W/m^2). Yet, one may argue that the solar plasma forcing has a more significant effect on a planetary atmosphere than solar irradiation alone. For instance, thermal escape (due to solar irradiation) primarily favours light atoms and molecules (e.g. hydrogen) while non-thermal escape processes (due to solar plasma forcing) are much less mass sensitive. Nonthermal escape in general results in an order of magnitude higher mass loss for the Earth-like planets. This apparent anomaly, with orders of magnitude difference in input power, demonstrates the non-linear behaviour of nature, i.e. sheer power is not sufficient to explain physical phenomena. It is the process that matters, such as in this case for the escape of planetary volatiles.

 Springer

Thermal escape from an atmosphere is determined from a Maxwellian (thermalized) particle distribution, the escape rate given by the temperature of the distribution at the exobase and the escape velocity of the object at the exobase. Theoretically all particles within the Maxwellian distribution of particles having velocities above the escape velocity will be lost. *Nonthermal escape* may be defined as all other processes where the energization and escape of particles is related with (microscopic) nonthermal processes. Excluded are all (macroscopic) processes of catastrophic nature such as e.g. impact erosion by in falling objects from space. Non-thermal escape is not unrelated to thermal escape, because most non-thermal escape processes are based on the existence of photo ionization processes. The electromagnetic radiation from the Sun also determines the scale-height of the atmosphere, and correspondingly also the ionosphere and cross-sectional area for e.g. solar wind forcing. However, we separate the two mechanisms mainly because of their differences with respect to solar forcing. Thermal escape is (mainly) due to solar radiation, whereas non-thermal escape is related with a broader aspect of solar wind forcing, such as sputtering, ion pickup, ionospheric plasma energization etc. The basic argument is that solar wind energy and momentum, electromagnetic or corpuscular, defines the forcing regardless of individual processes inferred. Following the definitions by Chassefière et al. (2007), there are two classes of thermal escape:

1. Jeans escape, driven by EUV and XUV heating of the upper atmosphere. Atmospheric atoms having velocities above escape velocity at the exobase level are free to escape into space.
2. Hydrodynamic escape, consisting of a bulk expansion of the upper atmosphere due to intense solar EUV/XUV fluxes, allowing atoms to overcome the gravitational binding force. Hydrodynamic escape plays an important role in low gravity environments (e.g. comets), but is also considered to have played a major role in outflow during early Noachian on Mars (e.g. Chassefière and Leblanc 2004).

With regard to non-thermal escape we present a slightly modified definition as compared to that proposed by Chassefière et al. (2007). The following four processes here identify non-thermal escape:

1. Photochemical escape, associated with dissociative recombination. Ions produced by photo ionization may reach higher temperatures than the neutral atmosphere in the ionosphere. Recombination/charge-exchange produce energetic neutrals, some of them have sufficient velocity to escape the planet (e.g. Luhmann et al. 1992; Lammer et al. 1996; Fox and Hac 1997; Kim et al. 1998; Chassefière and Leblanc 2004).
2. Ion sputtering produced by ions impacting the upper atmosphere/corona leads to the ejection of neutral particles (e.g. Luhmann et al. 1992; Jakosky et al. 1994; Johnson et al. 2000; Leblanc and Johnson 2002).
3. Ionospheric plasma energization and escape driven by direct solar wind forcing (e.g. Pérez-de Tejada 1987, 1998; Lundin and Dubinin 1992). The process is more complex for a planet with an intrinsic magnetic dynamo such as the Earth (see e.g. Moore et al. 1999, for a review). Plasma waves are important for the transfer of energy and momentum from the solar wind to planetary magnetospheres (see e.g. Chaston et al. 2005). In a similar manner, waves observed in the shocked solar wind plasma are likely to take part in the energization of ionospheric ions near Mars (e.g. Espley et al. 2004; Winningham et al. 2006; Lundin et al. 2006b).
4. Ionospheric ion pickup, a process caused by the protrusion of the solar wind motional electric field into a planetary ionosphere. The combined solar wind electric and magnetic

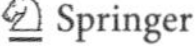

field ($\underline{E} \times \underline{B}$) results in a cycloid motion of energized ionospheric ions (Luhmann and Kozyra 1991; Dubinin et al. 1993; Kallio et al. 1998; Kallio and Janhunen 2002; Ma et al. 2004; Nagy et al. 2004; Kallio et al. 2006; Dubinin et al. 2006; Luhmann et al. 2006). Mass loading leads to a local weakening of the motional electric field, and a correspondingly lower energization (Lundin et al. 1991; Kallio et al. 1998).

Notice that the first two non-thermal escape processes are associated with the escape of neutral atoms, while the last two processes are associated with the energization and escape of ionospheric plasma. In what follows we will focus on the energization and escape of ionospheric plasma. One important reason for this, obvious from the title of this report, is that a planetary magnetic field has implications for the ionospheric plasma escape and the corresponding atmospheric evolution. Another important reason is that ionospheric plasma escape is a topic where theory may be compared with numerous direct in situ observations. Photochemical escape and sputtering are processes that by and large lack adequate in situ measurements. Studies of the latter two processes are therefore based on models and simulations. In a similar way, quantitative results of thermal escape (Jeans escape and hydrodynamic escape) are based on models and simulations.

The magnetic field plays an important role in controlling ionized gases—plasmas. The solar wind, a wind of plasma escaping from the Sun, is in fact embedded in (frozen-in) the solar magnetic field (Alfvén 1950; Parker 1958). In the same manner a plasma flow, like the solar wind, cannot easily protrude into a strong planetary magnetic field. For instance, the Earth's dipole magnetic field acts as a "magnetic umbrella" fending off the solar wind (Fig. 1). The standoff distance from the Earth is typically some 70 000 km away in the sub solar region. Conversely, planets lacking strong intrinsic magnetic fields such as Mars and Venus have no "magnetic umbrella", and the solar wind can directly access the upper atmosphere. Recent measurements from Mars (Lundin et al. 2004b) show that the solar wind may impact as low as 270 km above the dayside surface of Mars. This illustrates the problem for planetary atmospheres without magnetic shielding. The relative erosion rate of ionospheric plasma is consequently lower for the Earth than for Mars, for example. The total outflow rate for the Earth, the mass flux dominated by O^+ ions, is 1–3 kg/s (Chappell et al. 1987). However, recent data and arguments suggest that most of the outflowing ions

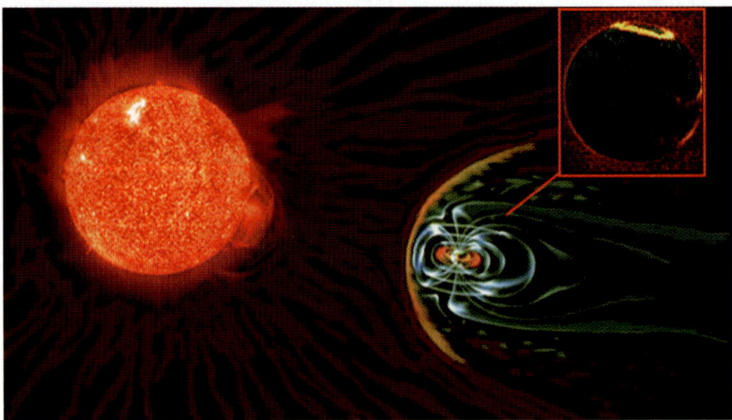

Fig. 1 The magnetic field of the Earth acting as a shield against direct solar wind forcing of the Earths atmosphere and ionosphere. The aurora in the close-up view of the Earth (*upper right*) illustrates that a limited amount of solar wind forcing occurs in the polar region

Fig. 2 The comet tail, an example of direct forcing by the Sun, the solar EUV/UV and the solar wind

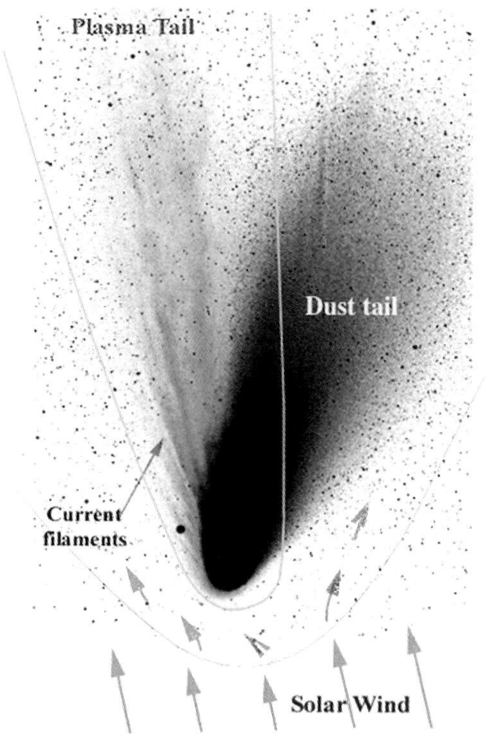

are recaptured by the Earth (Seki et al. 2001), i.e. less than 1 kg/s are lost to space. For Mars, the mass loss of O^+ and O_2^+ during solar maximum has been estimated to be ≈ 1 kg/s (Lundin et al. 1989). Recent data from Mars Express imply a much lower escape during solar minimum (Barabash et al. 2007), with orders of magnitude variability connected with solar wind dynamic pressure changes (Lundin et al. 2006b). Mars has a very tenuous atmosphere, the average atmospheric pressure being two orders of magnitude lower than on the Earth. Moreover, the area of the solar wind obstacle (the solid planet) is about four times larger for the Earth compared to Mars. Therefore, considering the volatile inventory on both planets, Mars is losing atmospheric O^+ and O_2^+ much faster compared to the Earth.

Weakly magnetized planets like Mars and Venus, behave like comets, the nightside cavities forming elongated tails of escaping planetary plasma originating from the upper atmosphere/ionosphere. The planetary wind is stretching out in the antisolar direction in the same way as for comets (Fig. 2) yet at a rate lower than for a typical comet during solar approach. The main difference between a planet and a comet in terms of volatile escape is mass/gravity. Venus and Mars have much stronger gravity, which retains volatile substances for billions of years before they are significantly eroded away by the solar wind. The low gravity of comets means that their atmosphere builds up and expands while approaching the Sun, leading to a gradually expanding obstacle for the erosive solar wind. The loss from a km-size comet (e.g. Halley) is $1-10 \times 10^6$ kg/day at 1 AU, while the loss from the Venus and Mars is typically 100 times lower. The heavy loss of matter is a reason why the tails of comets are visible and the plasma tails of planets are not.

The present rate of escape observed from a weakly magnetized planet such as Mars corresponds to significant losses of volatiles throughout the planetary lifetime. For instance,

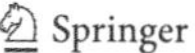 Springer

Lammer et al. (1996, 2003) and Lundin and Barabash (2004a) estimated the loss of volatiles during the last 3.5 billion years; the water loss corresponds to a global equivalent layer GEL of 10–30 meters. These figures were obtained by considering the evolution of the Sun (Wood et al. 2002) and of the planetary atmosphere. On the other hand, morphological surface features suggest that more surface and near-surface water was present during Noachian and Hesperian times (GEL 100–500 m) (McKay and Stoker 1989; Baker 2001; Lunine et al. 2003; Bibring et al. 2004; Neukum 2005; Bertaux et al., this volume; Nisbet et al., this volume). In a similar manner Venus may have been subject to a heavy loss of water (Kulikov et al. 2006, 2007). The interaction of the solar wind with Venus (Russell et al. 2006) and the corresponding rate of outflow by ion pickup processes (Luhmann and Bauer 1992; Luhmann et al. 2006) based on Pioneer Venus Orbiter PVO measurements suggest a modest solar wind interaction with Venus. On the other hand, the comet-like ionospheric features found by Brace et al. (1987) imply a rather strong ionospheric response to solar forcing.

Under the assumption that all Earth-like planets accreted from matter of essentially the same chemical origin, the differences we observe today may be related with evolutionary processes. The evolution of volatiles is one particular aspect of that. The Earth is the only planet where a significant hydrosphere remains, while Mars and particularly Venus are strongly dehydrated in comparison.

In this report we discuss the evolution of planetary volatiles, with a focus on water, under varying solar forcing conditions with time. Our focus is on the acceleration and escape of ionospheric plasma, for reasons already mentioned, but also because we believe that plasma escape to a large extent couples to the other escape processes. We start by describing the variability of the solar radiation and the solar plasma environment. We continue by defining and describing the internal forcing that leads to the solar output and the external forcing that the Sun imposes on the planets. We then discuss why magnetic shielding plays such an important role in protecting a planetary atmosphere. Finally, we present a model and a scenario of the loss of water and CO_2 from Mars, Venus and the Earth. We conclude on basis of this that a strong intrinsic magnetic field is important for maintaining water on a planet. A wealth of water is central for the evolution of life, for making a planet habitable for more advanced life-forms.

2 Variability of the Solar Radiation and Plasma Environment

The Sun is the main source of surface and atmosphere energy for the Earth-like planets, interior energy/heat flow playing a negligible role. Without a dependable (stable) star like the Sun, the Earth would not have developed a rich and diverse biosphere, the home to millions of living species. This raises two questions: Why only on the Earth, and not also on Mars and Venus? Has the Sun always been "dependable"? We focus in this section on the second question, addressing in the following continuing sections an issue related with a biosphere on the Earth-like planets—the water inventory.

2.1 Short-Time Solar Variability

High precision radiometric observations of the Sun carried out by several satellites since the late Seventies have shown that the Sun undergoes small cyclic variations in brightness. These brightness changes are closely related with the $\approx$11-year sunspot cycle. Lean (1997) found that the observed bolometric luminosity of the Sun varied over the recent solar cycles 21, 22, and 23 (from about 1978) by about 0.12%.

The Sun is brightest during the times of maximum sunspot numbers and faintest during the sunspot minima. This can be explained by the larger changes in the area coverage and intensity of magnetic white light facular regions that peak near sunspot maximum. The observed light variations of the Sun arise from the blocking effect of sunspots and increased facular contribution to brightness, which slightly offsets the former. Even though the observed light bolometric variations of the Sun are very small over its activity cycle, variations over the sunspot cycle are much larger at shorter wavelengths (e.g., Lean 1997; Guinan and Ribas 2002).

For example, typical variations of solar coronal X-ray emissions from the minimum to the maximum of the $\approx$11-year activity cycle are $\approx$500% (Guinan and Ribas 2002). The cyclic changes arising from variations in the chromospheres and transition region emission range from 10–200% at NUV, FUV and EUV wavelengths. Also the frequencies and intensities of flaring events and coronal mass ejections (CME) are strongly correlated with the Sun's activity cycle. For example, the rate of CME occurrences is larger during the sunspot maxima than during sunspot minima (Webb and Howard 1994).

These cyclic short-time variations play a role in Earth's global climate, and numerous studies have been carried out on the possible influence of the 11-year solar cycle on climate, frequency of storms and cyclones, rainfall, droughts, vegetation, insect populations, etc. (see, e.g., Hoyt and Schatten 1997).

However, Earth's expected cyclic temperature variation of about 0.1°C at the surface, computed from simple black body considerations and the small change of irradiance of 0.12%, does not seem sufficient in itself to produce observable climate changes (e.g. Foukal et al. 2006). On the other hand, the changes in the XUV flux of the Sun over a typical activity cycle are significantly larger that the total irradiance changes, and these high energy solar emissions are absorbed, heating the Earth's stratosphere and thermosphere. Although the deposited energy is small, non-linear feedback mechanisms could amplify the affects on climate by altering, for example, the tropospheric heat exchange between the equator and the polar regions.

An interesting and surprisingly close connection between the solar cycle effects and the climate presented by Friis-Christensen and Lassen (1991) has been a matter of controversy for some time. The finding was followed by an equally surprisingly good connection between cloud cover and cosmic radiation (Svensmark and Friis-Christensen 1997; Svensmark 2000; Kanipe 2006). The authors argue that certain clouds are formed as a result of the precipitation of galactic cosmic ray particles. Since the solar magnetic field and the solar wind acts as a shield and damper for galactic cosmic ray particles, the variability of the solar magnetic field will modulate the cosmic ray flux in the inner part of the solar system. In this way the solar activity, and the corresponding generation/escape of solar magnetic flux may represent a climate controller for the Earth-like planets. Very recently Svensmark et al. (2006) showed results from a laboratory experiment that may very well be the final proof for the hypothesis of cloud formation by cosmic ray fluxes.

2.2 Temporal Variation of the Sun's Total Luminosity

Because of ever accelerating nuclear reactions in its core, the Sun is a slowly evolving variable star that has undergone a $\approx$30% increase in luminosity over the past 4.5 Gyr. Solar evolution models show that in $\approx$1 Gyr from now, when the Sun is about 10% brighter, the Earth will be heated enough so that its oceans start to evaporate. About 6 Gyr from now the Sun expands beyond the Earth orbit to become a red giant.

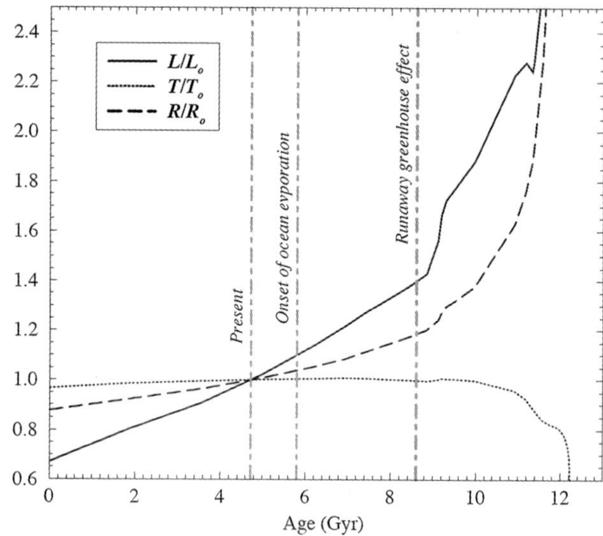

Fig. 3 The evolution of the effective temperature, radius, and luminosity of the Sun from the zero-age main sequence (ZAMS) to the start of its red giant phase. The *vertical lines* mark the approximate expected occurrences of Earth-related phenomena such as the onset of ocean evaporation and the start of the runaway greenhouse effect. Based on the evolution models of Bressan et al. (1993) and the predictions by Kasting (1988)

The nuclear evolution of the Sun is generally well understood from stellar evolutionary theory. We also have a good understanding of the internal solar structure thanks to helioseismology, although the former excellent agreement between observation and theory has been disturbed by recent updates in the abundances of some key elements (Guzik et al. 2006). For example, models for the ZAMS Sun given by Bressan et al. (1993) indicate values of $L = 0.67 L_{Sun}$, $R = 0.89 R_{Sun}$, and $T_{eff} = 0.96 T_{eff\,Sun}$ (where $T_{eff\,Sun} = 5777$ K). A plot showing the evolution of the luminosity, radius, and temperature of the Sun according to Bressan et al. (1993) is shown in Fig. 3.

One finds from these standard stellar evolutionary models that the young Sun, about 4.6 Gyr ago, was ≈ 200 K cooler and $\approx 10\%$ smaller than today and had an initial luminosity of $\approx 70\%$ of the present Sun so that in the early stages of the Solar System, the young Sun's irradiance (Solar constant) is expected to have been significantly lower. The lower luminosity of the young Sun should have lead to a much cooler Earth in the past. However, paleoclimate studies show that the young Earth always had liquid water and was, therefore, typically warmer (possibly heated by greenhouse gases such as CO_2 and CH_4), than during recent times. The increase in solar luminosity over the history of geologic time periods and its effect on the Earth's climate have been discussed by various authors in the past (e.g., Sagan and Mullen 1972; Newman and Rood 1977; Owen 1979). On the other hand, a young Sun with a slightly higher initial solar mass than previously assumed would produce a slightly higher total luminosity (Kasting et al. 1993; Whitmire et al. 1995; Sackmann and Boothroyd 2003) and could also be a viable explanation for warm temperatures on early Earth and Mars, otherwise are difficult to account for, in particularly for Mars. Sackmann and Boothroyd (2003) pointed out that such a higher initial solar mass leaves a fingerprint on the Sun's present internal structure that is large enough to be detectable, in principle, via helioseismic observations. Their computations demonstrated that several mass-losing solar models are more consistent with the helioseismic observations than is the standard solar model (Sackmann and Boothroyd 2003). However, there are several uncertainties in the observed solar composition and in the input physics on which solar models are based and these uncertainties have a slightly larger effect on the Sun's present internal structure than a possible fingerprint left from an early efficient solar mass loss. Future improvements by a

 Springer

factor of 2 in the accuracy of input parameters for solar mass loss models could reduce the size of the uncertainties below the level of the fingerprints left by a more massive, brighter young Sun. This would allow us to determine whether early solar mass loss took place or not. More observations of mass loss rates from other young solar-like stars similar to the young Sun are needed.

The impact of the solar radiation and particle fluxes on the evolution of planetary atmospheres and their water inventories has received much less attention than studies of the climate change and greenhouse effect. Previous observation-based studies have indicated that the young Sun was a far stronger source of energetic particles and electromagnetic radiation (e.g., Newkirk 1980; Zahnle and Walker 1982; Ayres 1997; Güdel et al. 1997) than today's Sun. These early studies are now supported by a large number of multiwavelength (X-ray, EUV, FUV, UV, optical) observations of solar proxies providing solid evidence that the early Sun was a much more active star than it is at present (e.g., Ribas et al. 2005).

2.3 Evolution of Solar X-Rays and EUV Radiation

Because ionization, thermospheric heating, production of extended neutral gas corona, and thermal and non-thermal escape of atmospheric constituents all depend on the evolution of the solar X-ray and EUV radiation (XUV) and the solar wind, one cannot neglect the changing solar radiation and plasma environment in the study of the evolution of planetary atmospheres. The relevant wavelengths for the heating of upper atmospheres of planets are the ionizing ones $\lambda \leq 102.7$ nm (e.g., Gordiets et al. 1982; Hunten 1993; Bauer and Lammer 2004), which contain only a small fraction of the present solar spectral power. The high radiation levels of the young Sun were triggered by strong magnetic activity. The magnetic activity of the Sun is expected to have greatly decreased with time (Skumanich 1972; Simon et al. 1985; Güdel et al. 1997; Guinan and Ribas 2002; Ribas et al. 2005) as the solar rotation slowed down through angular momentum loss. Observational evidence and theoretical models suggest that the young Sun rotated about 10 times faster than today and had significantly enhanced magnetically generated coronal and chromospheric activity (Keppens et al. 1995; Guinan and Ribas 2002; Ribas et al. 2005).

The NASA sponsored "Sun in Time" program (Dorren and Guinan 1994) was established to study the magnetic evolution of the Sun using a homogeneous sample of single nearby G0-V main sequence stars which have known rotation periods and well-determined physical properties, including temperatures, metal abundances and ages that cover most of the Sun's main sequence lifetime from 130 Myr to 8.5 Gyr. The observations, obtained with various satellites like ASCA, ROSAT, EUVE, FUSE and IUE satellites, cover a range between 0.1- and 330 nm, except for a gap at between 36–92 nm, which is a region of very strong interstellar medium absorption. Details of the datasets and the flux calibration procedure employed are provided in Ribas et al. (2005). The results indicate that the young main sequence Sun was about 100–1000 times stronger in XUV emissions than at present. Similarly, the transition region and chromospheric FUV–UV emissions of the young Sun are expected to be 10–100 and 5–10 times stronger, respectively, than at present, and the flux variation over age is therefore a steep wavelength function. Figure 4 shows the time evolution of the spectral range with 0.1 nm $< l <$ 100 nm, which includes X-rays and EUV and the *Lyman-α* line at 121.56 nm (Lammer et al. 2003; Ribas et al. 2005) at a distance of 1 AU.

In the 100–0.1 nm interval, the fluxes follow a power-law relationship $I_{\mathrm{XUV}}(t)/I_{\mathrm{XUV}} = 6.16 \cdot (t[\mathrm{Gyr}]^{-1.19})$. At longer wavelengths, the *Lyman-α* emission feature can contribute to a significant fraction of the XUV flux. High-resolution *Hubble Space Telescope* (HST)

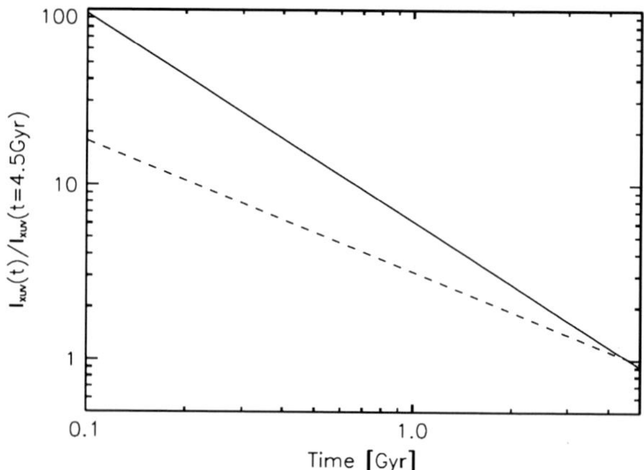

Fig. 4 Time evolution of the I_{XUV} energy flux for solar-like G stars (*solid line*: $\lambda = 100$–0.1 nm; *dashed line*: *Lyman-α* $= 121.56$ nm) (Ribas et al. 2005)

spectroscopic observations were used to estimate the net stellar flux. These measurements, together with the observed solar *Lyman-α* define the following power-law relationship with high correlation $I_{L\alpha}(t)I_{L\alpha} = 3.17 \cdot (t[\text{Gyr}])^{-0.75}$.

In both power-laws, the XUV and *Lyman-α* expressions are valid for ages between 0.1–7 Gyr, I_{XUV} and $I_{L\alpha}$ are the present integrated fluxes at 1 AU and $I_{XUV}(t)$ and $I_{L\alpha}(t)$ are the integrated fluxes as a function of time. One finds fluxes of $\approx 6 \cdot I_{XUV}$ and $\approx 3 \cdot I_{L\alpha}$ about 3.5 Gyr ago, and $\approx 100 \cdot I_{XUV}$ and $\approx 20 \cdot I_{L\alpha}$ about 100 Myr after the Sun arrived on the ZAMS.

2.4 The Solar Wind of the Young Sun

HST high-resolution spectroscopic observations of the H *Lyman-α* feature of several nearby main-sequence G and K stars carried out by Wood et al. (2002) have revealed neutral hydrogen absorption associated with the interaction between the stars' fully ionized coronal winds with the partially ionized local interstellar medium. Wood et al. (2002) modelled the absorption features formed in the astrospheres of these stars and provided the first empirically estimated coronal mass loss rates for solar-like G and K main sequence stars. They estimated the mass loss rates from the system geometry and hydrodynamics and found from their small sample of G and K-type stars, where astrospheres which can be observed, that mass loss rates increase with stellar activity. This study suggests that the young Sun had a much more dense solar wind than today. The correlation between mass loss and X-ray surface flux follows a power law relationship, which indicates an average solar wind density up to 1000 times higher than today during the first 100 Myr after the Sun reached the ZAMS.

Mass loss rates of cool, main sequence stars depend on their rotation periods, which are in turn correlated with the stars' ages. To obtain the evolution of the stellar wind velocity v_{sw} and stellar wind density n_{sw} of solar-like stars, one can use the X-ray activity-stellar mass loss and solar wind velocity power-law relations of Wood et al. (2002) and Newkirk (1980). One obtains the time-behaviour of the solar wind parameters (Griemeier et al. 2004) $v_{sw} = v^*[1 + (t/\tau)]^{-0.4}$ and particle density $n_{sw} = n^*[1 + (t/\tau)]^{-1.5}$.

The proportionality constants are determined by average present-day solar wind conditions (Griemeier et al. 2004). With $v_{sw} = 400$ km s^{-1} and $n_{sw} = 10^7$ m^{-3} for $t = 4.6$ Gyr and at $d = 1$ AU (Schwenn 1990) one obtains $v^* = 3200$ km s^{-1}, $n^* = 2.4 \cdot 10^{10}$ m^{-3} (density at 1 AU). The time constant is $\tau = 2.56 \cdot 10^7$ yr (Newkirk 1980). For distances other than 1 AU, the solar wind parameters can be scaled with an r^{-2} dependency. The time variation of $n(t)$ at Earth orbit of 1 AU is shown in Fig. 5. One can see from Fig. 5 that the observational

 Springer

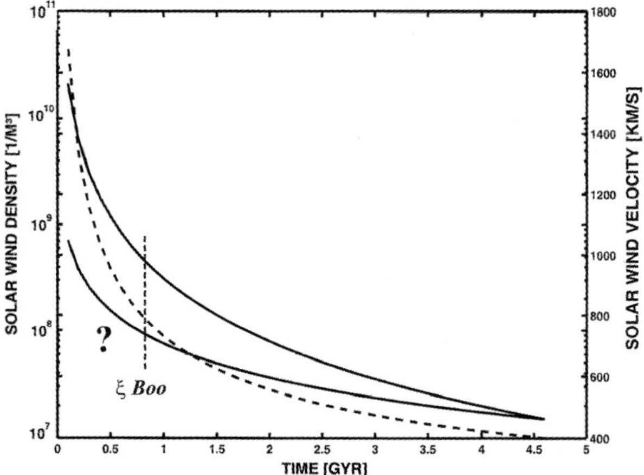

Fig. 5 Evolution of the observational based minimum and maximum stellar wind densities scaled to 1 AU (*left scale: solid lines*) obtained from several nearby solar-like stars. On the *right scale* one can see the evolution of the stellar wind velocity (*dashed curve*). More observations or early active stars with ages less than 700 Myr are needed to obtain a better picture of the mass-loss/activity relation shortly after the Sun arrived at the ZAMS

data for solar-like G and K stars suggest that more active stars have higher mass loss rates and solar wind number density. However, observations of the active M dwarf star *Proxima Cen* and the active *RS CVn* system λ *And* (G8 IV + M V) are inconsistent with this relation and show lower mass loss rates. A recent observation of the solar-like young star ξ *Boo* also shows a lower mass loss rate, consistent with the mass loss rates previously found for *Proxima Cen* and λ *And* (Wood et al. 2005).

The common feature of these three stars is that they are all very active in X-ray surface fluxes at about 10^6 erg cm^{-2} s^{-1}, about a factor of 30 larger than the Sun and corresponding to a time of about 700 Myr after solar-like stars arrived at the ZAMS. The results of these observations show the uncertainty of early mass loss and solar wind estimations, because ξ *Boo* is a usual but very active young solar-like star, while *Proxima Cen* and λ *And* are different types than normal G and K stars. One can speculate that there may be a high-activity cut-off to the mass-loss/activity relation obtained by Wood et al. (2002), although more active young solar-like G and K stars with X-ray surface fluxes larger than 10^6 erg cm^{-2} s^{-1} should be observed and studied in the future.

3 External and Internal Forcing

3.1 The Solar Output—Internal Forcing

Internal forcing is here defined as forcing emanating from the body itself—from internal energy sources. The solar output (the solar irradiance and the solar corpuscular radiation) is an example of internal forcing. Also the variability of the internal forcing as discussed in Sect. 2 is driven by the interior of the Sun, or at least dominated by internal processes. However, one cannot completely rule out the possibility of external influences on the solar output. An example of this is Jupiter, where internal forcing matches, or even exceeds external (solar wind) forcing. Numerous examples may also be found in astrophysics, such as binary stars.

As mentioned in Sect. 2.2, the solar luminosity is expected to vary little with time, on a long—as well as short term base. However, solar X-rays and EUV, changing orders of magnitude on longer-terms, are also subject to a high short-term variability. Radiation in this frequency range represents the most variable part of the solar irradiance. Equally, or even

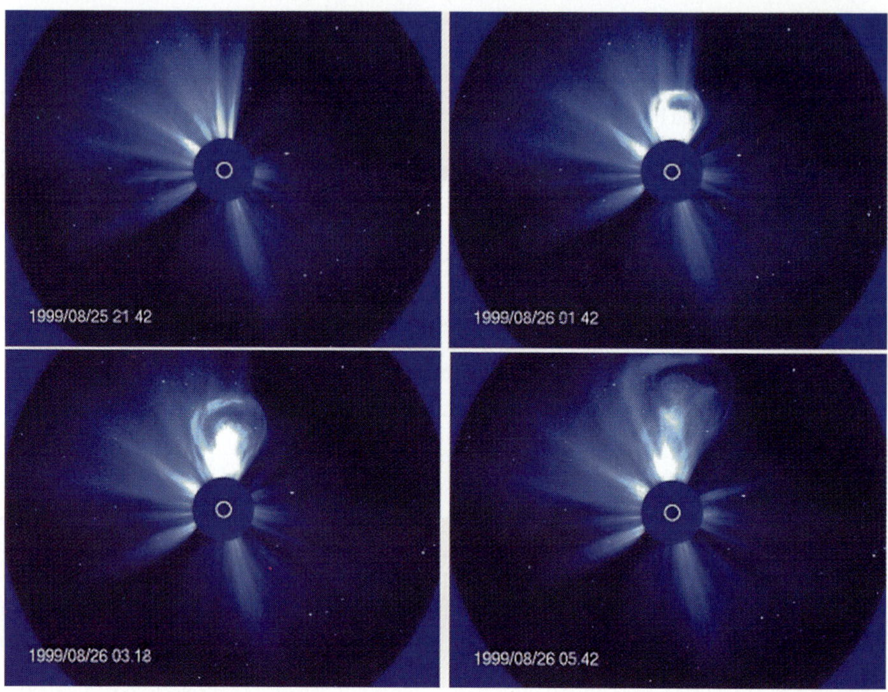

Fig. 6 SOHO/LASCO image of a time sequence showing the variable outflow of solar coronal plasma, characterized by streamers and a CME

more variable, is the coronal plasma escape, the solar wind and solar energetic particles. This is demonstrated by the SOHO image of the solar corona (Fig. 6). The coronal outflow displays a strong and dynamic variation in space and time. The structured coronal plasma escape may vary by several orders of magnitude, especially during solar maximum—the most active period of the solar cycle. This is also the time of maximum X-ray and EUV flux. The difference in coronal plasma escape and X-ray, EUV fluxes, is substantial between solar maximum and solar minimum. For instance, the solar wind dynamic pressure reaches about an order of magnitude higher during solar maximum compared to solar minimum, and the occurrence frequency of CMEs increases by a factor of five or more (e.g. Bird and Edenhofer 1990, and references therein). Similarly, solar EUV fluxes are a factor 2–4 higher during solar maximum compared to solar minimum.

The connection between internal processes, the solar irradiance, the coronal mass escape, and their relation to the solar variability is beyond the scope of this report. Various aspects connected with internal forcing are discussed in Sect. 2. We simply conclude here that solar internal forcing results in variable and structured phenomena undergoing short-term ($\approx$11, 22, 80, 200, 1000, ... year, Lundstedt et al. 2006) cyclic variations as well as long-term ($\approx$100–4600 million years) changes (e.g. Wood et al. 2002, 2005, and Ribas et al. 2005)

3.2 External Forcing—Direct and Indirect

External forcing of a celestial object exposed to the solar wind may be categorized into two groups, *direct forcing* and *indirect forcing*. By direct forcing we mean that the solar wind has direct access to the obstacle, whereby energy and momentum can be transferred locally.

 Springer

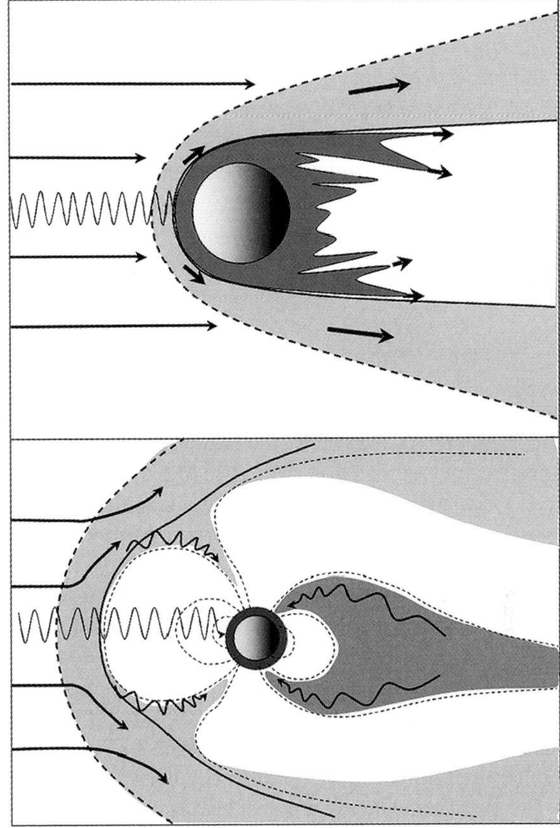

Fig. 7 Diagram illustrating the two types of external forcing, direct forcing (*upper*) and indirect forcing (*bottom*). In the latter case the atmosphere/ionosphere is protected by a strong intrinsic magnetic field

Conversely, indirect forcing implies that there is no direct contact. Energy and momentum may yet be transferred/mediated to the obstacle by other means such as waves, electric fields and electric currents. Figure 7 illustrates these two categories of solar wind forcing. In the bottom figure a magnetic field acts as a shield against direct plasma forcing (indirect forcing dominates). Direct forcing, (direct contact between the planetary ionosphere/atmosphere and solar wind plasma) may also occur for a strongly magnetized planet like the Earth, but then only in narrow dayside "cusp" regions, as indicated in Fig. 7.

A comet approaching the Sun illustrates direct forcing. A low gravity and the lack of a protective "shield" means that a comet is subject to massive forcing by solar radiation and the solar wind. The solar radiation heats the surface, leading to a release of particles/gas constituting the coma. The solar EUV may lead to such a high degree of ionization, the $N_e/N_{neutral}$ ratio exceeding 10^{-6}, that electric and magnetic fields become the dominating forcing terms in the local gas dynamics. The picture of comet Hale-Bop (Fig. 2) illustrates this quite well. Notice that a comet of the Hale-Bop composition and size has two tails— a high-speed plasma tail induced by the solar wind interaction pointing away from the Sun (less the aberration) and—a dust tail comprising debris along the comet trajectory. The plasma tail is a cometary wind consisting of ionized matter/gas from the comet—plasma and energized neutrals. The cometary wind gradually reaches solar wind velocities in the deep tail ($\approx$300–900 km/s). Such high velocities as compared to the comet orbital velocity (tens of km/s) represent a complete loss, i.e. the particles have no chance to return to the

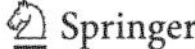 Springer

comet. The "rays" observed in the dust tail, suggests that the solar wind may remove also large parts of the dust tail.

Indirect forcing may occur also for shielded/protected objects. In this case the forcing may be considered a "leakage" through the shield, but it may also be considered a global process that couples to external conditions in the solar wind. Magnetic reconnection (Dungey 1961) is an example of a global coupling between an external magnetic field and the internal magnetic field of a celestial object. For suitable boundary conditions this may lead to indirect external forcing. A dense, re-radiating, atmosphere may also serve as a shield for solar irradiation, while only a strong dipole magnetic field is capable of protecting the atmosphere/ ionosphere from fast charged particles.

3.3 Magnetic Shielding

Global magnetic shielding is well known from space plasma- and cosmic ray physics. Magnetic shielding is governed by the Lorenz-force, prohibiting incident charged particles from protruding deeper than two Larmor radii perpendicular to the local magnetic field ($\mathbf{B}$). The exception is when an electric field ($\mathbf{E}$) is applied perpendicular to the magnetic field, i.e. $\mathbf{F} = e\mathbf{E} + e[\mathbf{v} \times \mathbf{B}]$, and charged particles may protrude/convect across the magnetic field. In reality the physics is more complex, with second order forcing terms contributing, and particles may traverse deeper than two Larmor radii. Nevertheless, a sufficiently strong magnetic field acts as a stopper, a magnetic shield, fending off charged particles and prohibiting them from accessing deep regions of a magnetic field domain. Figures 1 and 7 (lower) illustrate how magnetic shielding deflects the solar wind. Energy and momentum may yet cross the stopping boundary by various processes (magnetic reconnection, diffusion, viscous interaction, Kelvin–Helmholtz instabilities, impulsive penetration, see e.g. (Keyser et al. 2005) for a review) and propagate along magnetic field lines to the ionosphere/atmosphere.

The magnetized celestial object best known to us is the Earth. The strong terrestrial magnetic dipole serves as an excellent shield, fending off the solar wind at about 10 Earth radii (Re) in the solar direction. A "bubble", *a magnetosphere*, is created, separating the solar wind plasma from the near Earth environment (Fig. 1). The terrestrial magnetosphere exemplifies the indirect forcing. The transfer of solar wind energy and momentum across the dayside-stopping boundary, the magnetopause, provide essentially all plasma forcing of the Earth's ionosphere and atmosphere by indirect means. Penetrating plasma energy and momentum are converted in a dynamo to electric fields, currents and waves that subsequently propagate down, leading to forcing in the ionosphere and upper atmosphere.

Global magnetic shielding is therefore not sufficient to prohibit external plasma forcing. However, a strong magnetic dipole field markedly reduces the efficiency of energy and momentum transfer; the stronger the intrinsic dipole field, the lower the efficiency of external forcing.

Without magnetic shielding by an intrinsic magnetic dipole (e.g. Venus and Mars) the atmosphere and ionosphere are directly exposed to the solar wind, leading to a more direct solar wind forcing. The combined EUV and solar wind forcing lead to ionization and efficient removal of atmospheric atoms and molecules from the topside atmosphere. On the other hand the solar wind ram pressure transmitted to the ionosphere of Mars and Venus induces a magnetic barrier (e.g. Zhang et al. 1991, and Crider et al. 2004), sometimes referred to as a solar wind magnetic field pile-up. This induced magnetic barrier also serves as a magnetic shield such that the solar wind is swept sideways along the flanks as illustrated in Fig. 7. However, the magnetic barrier is not a solid obstacle. The upper ionosphere is in direct contact with shocked solar wind plasma; the motional solar wind electric field,

waves and protruding solar wind ions allow energy and momentum transfer to the topside ionosphere. In the next section we describe how solar wind energy and momentum is transferred to planetary ions as a result of direct forcing.

3.4 Energy and Momentum Transfer

The solar wind interaction with the unprotected upper atmosphere and ionosphere of a planet results in a transfer of energy and momentum to planetary particles. A number of processes may be conceived, but the removal of matter boils down to kinetics. The available energy and momentum of the solar wind limit the total outflow by all non-thermal escape processes. Here we do not discuss irradiative solar forcing (ponderomotive forcing) as an energy and momentum transfer process. Radiative forcing is associated with heating and ionization of a planetary atmosphere, indirectly responsible for thermal (Jeans, hydrodynamic) escape and dissociated recombination. In what follows we focus on the two ionospheric plasma escape processes related with non-thermal escape.

The efficiency of non-thermal escape is related to the energy and momentum transfer efficiency and the cross-sectional area exposed to the solar wind. A low gravity and lack of a magnetic shield imply an extended atmosphere and a larger cross-sectional area for direct solar wind scavenging. Besides solar X-rays and EUV ionization other ionization processes, such as electron impact ionization (Zhang et al. 1993), take place in the exosphere of non-magnetized planets. These ionization processes lead to further heating of the upper atmosphere, enhanced ionization and increased transfer of energy and momentum to the ionosphere and exosphere.

The transfer of energy and momentum by the solar wind takes place in the topside atmosphere and ionosphere, where the solar wind may interact with ionospheric plasma and neutral gas. Interaction with ionospheric plasma leads to energization and escape, as described in the Introduction. Solar wind interaction with neutral gas may lead to further ionization from electron impact and to ion sputtering and the ejection of neutral particles (see Introduction). The interaction region lies within the altitude range above the planet where the transfer of solar wind energy and momentum takes place. A simple fluid dynamic description of the interaction region above a non-magnetized planet is described in Fig. 8. Two cross section areas, A_M and A_{max}, are marked out within the interaction volume that stretches out along the tail. Interaction (i.e., transfer of solar wind energy and momentum to the planetary plasma embedded in the flow) is expected to continue until balance in momentum flux is achieved between the solar- and planetary tail plasma. The cross section areas in Fig. 8 are selected because they have been experimentally determined, For instance, Lundin and Dubinin (1992), used A_M ($A_M = \pi(R_o^2 - R_i^2)$) as the minimum cross-sectional area to estimate the theoretical escape from Mars. They used R_o as the altitude of experimentally determined "mass loading boundary" and R_i as the unperturbed/cold Martian ionosphere, as a minimum cross section of the interaction volume along the flanks. The interaction volume expands towards the tail, the tailward flaring leading to the other cross-section area extreme $A_{max} = \pi R_{Tail}^2$. Assuming a symmetric distribution of the solar wind and planetary wind through the flow channels determined by A_M and A_{max}, we may compare model outflow with experimental data.

The solar wind energy and momentum transfer responsible for the energization and loss of matter from, for example the upper Martian atmosphere and ionosphere has been estimated based on fairly straightforward considerations (Pérez-de Tejada 1987, 1998; Lundin

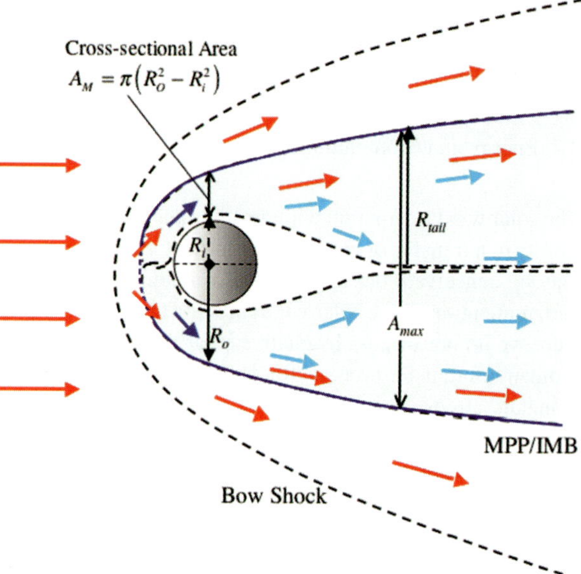

Fig. 8 Diagram of solar wind plasma forcing of a non-magnetized planetary atmosphere/ionosphere. The cross-sectional area and depth constitute the region of solar wind energy and momentum transfer. *Blue arrows* illustrate planetary ion escape. Notice that energy and momentum transfer (by solar wind motional electric field, waves, etc.) is expected to occur within the flow volume extending from the subsolar region to the deep tail

and Dubinin 1992). From the conservation of energy and momentum in the transfer of energy and momentum flux for the solar wind (Φ_{SW}) and Martian ions (Φ_M) one may obtain:

$$\Phi_M = \frac{v_{SW} \cdot m_{SW}}{v_M \cdot m_M}\left(\Phi_{SW} - \frac{v_{i,SW}}{v_{SW}}\Phi_{i,SW}\right) \cdot \frac{\delta_{SW}}{\delta_M}, \tag{1}$$

where $\Phi_{i,SW}$ and, $v_{i,SW}$ is the local/decelerated solar wind flux and velocity respectively. The ratio δ_{SW}/δ_M defines the relative momentum exchange thickness, here assumed to be ≈ 1. The above equation illustrates that the Martian ion flux is strongly coupled to speed (v_M) and mass (m_M) of the outflowing ions. The ratio between the solar wind velocity and the planetary ions v_{SW}/v_M, provides a flux amplification for ions of equal masses. This means that slowly escaping ions lead to higher mass losses. As illustrated in Fig. 9, the maximum escape flux for a solar wind velocity of 400 km/s may be amplified by up to 80 times the solar wind flux in the Martian environment ($v_{esc} = 5$ km/s), and by a factor of 5 if the escaping ions have 16 times higher mass than the solar wind protons. Thus, due to the high solar wind velocity and the relatively low escape velocity, a very high escape rate is theoretically feasible for Mars.

The net escaping mass flux from a planet subject to direct solar wind forcing, S_M, may now be obtained from:

$$S_M = A \cdot m_M \cdot \Phi_M \ (\text{kg/s}), \tag{2}$$

where Φ_M is the ion outflow of mass m_M, and A is the cross-sectional area ($A_M \rightarrow A_{max}$).

Lundin and Dubinin (1992) made a simple estimate of the input versus output mass flux on Mars by comparing measurement data with average solar wind input. The estimated range of cross-sectional areas $A_M \approx 1.6 \times 10^{14}$ and $A_{max} = 6.5 \times 10^{14}$ m^2 were based on measurement data of the planetary heavy ion escape (e.g. O$^+$). Using a solar wind proton density and velocity of 5×10^6 m^{-3} and 400 km/s respectively, the total solar wind flux through the cross-sectional is $\Phi_{SW} = 2 \times 10^{12}$ (m^{-2} s^{-1}). The corresponding range of solar wind mass flux input then corresponds to 0.53 kg/s and 2.2 kg/s respectively. This is in the

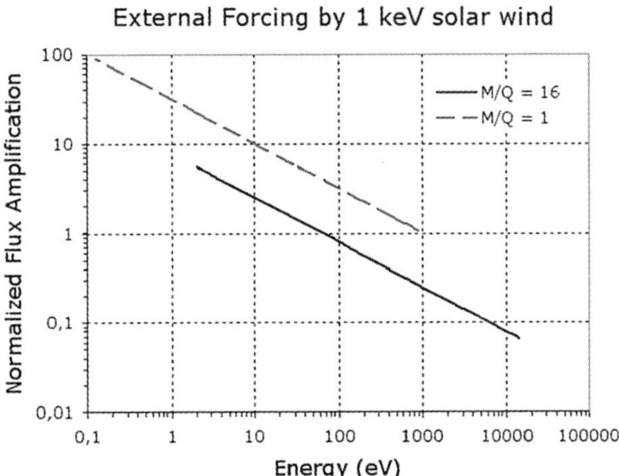

Fig. 9 Normalized momentum exchange by external solar wind forcing, illustrating the amplification of escape flux with decreasing outflow velocity (1)

range of Martian ion escape observed (Lundin et al. 1989), i.e. ≈ 1 kg/s. However, considering the amplification factor (Fig. 9) the escape mass flux may be much higher for low outflow velocities. If for instance O^+ ions are escaping at a velocity of 10 km/s (≈ 8.4 eV), about twice the escape velocity at Mars, a typical solar wind energy and momentum transfer would lead to a theoretical O^+ outflow in the range 6–25 kg/s.

4 Solar Radiative and Particle Forcing of the Earth-Like Planets

The distance to the Sun matters for both the irradiance and the solar wind forcing. For the two extremes, Venus and Mars, the unit radiative and particle forcing differs by a factor of ≈ 4. Despite this, the differences in the present solar forcing are relatively small. Moreover, the difference in atmospheric composition and density is less obvious when studying the ionospheres of the Earth-like planets. The ionization by solar EUV leads to classical Chapman layers, largely similar in electron density and ion composition, the highest ion abundance being O_2^+ in the lower ionosphere and O^+ in the upper ionosphere. The corresponding mass outflow/escape from the Earth-like planets are therefore likely to be dominated by the above constituents. This is certainly the case for the Earth (e.g. Chappell et al. 1987) and Mars (Lundin et al. 1989; Carlsson et al. 2006), and apparently also for Venus (Lundin et al. 2007). It raises the question about the molecular source of oxygen, water or CO_2? No doubt the Earth's outflow of H^+ and O^+ originates from water, but what about Venus and Mars, both planets having atmospheres dominated by CO_2? We will return to this in a forthcoming section.

Table 1 illustrates the power input by solar radiation and particles.

The solar wind input power for the Earth is obtained by assuming a limited (1–2%) transfer of energy through the magnetopause. An interesting and important aspect of the input power is the six order of magnitude difference between solar power of the solar irradiation and the solar wind power. The difference intuitively suggests that solar irradiation is the main driver for ionospheric and atmospheric processes. This is certainly the case for heating/expansion and dayside ionization of the atmosphere. However, this difference is not the case for the outflow/escape of matter. As already noted in Sect. 1, accelerated ionospheric O^+ dominates the escaping mass flux from the Earth. As indicated by the title of this report, our focus is on the implications of a planetary magnetic field for solar forcing. Our

Table 1 Solar input power to the Earth-like planets

	Solar irradiation power (W)	Solar wind input on the "obstacle" (W)
Venus	3.1×10^{17}	1.6×10^{11}
Earth	1.8×10^{17}	2.6×10^{11}
Mars	2.2×10^{16}	1.1×10^{11}

escape model is based on empirical data of ionospheric plasma escape, demonstrating that plasma escape is expected to be more severe for non-magnetized planets. This does not rule out other processes such as e.g. hydrodynamic escape (Chassefière and Leblanc 2004) and dissociated recombination (Luhmann et al. 1992). On the contrary, adding all processes together suggests an even more dramatic past for non-magnetized planets in the inner solar system.

5 The Present Loss of Volatiles from the Earth-Like Planets

The volatile content, abundances and states of the Earth-like planets, remains an intriguing issue in planetology. Consider the following alternative theories:

(1) the present differences in volatile content of the planets is directly related to the accretion process, or
(2) the present differences in volatile content is a consequence of the long-term evolution of the planets (gain/loss).

The first theory assumes an initial accretion differentiation of the planets. The second theory suggests long-term evolutionary differentiation of the planetary volatile content.

Considering the overall similarities such as proximity to the Sun and average density, it seems reasonable to assume that the Earth-like planets aggregated from the same matter in the early solar nebula under rather similar conditions. Earth and Mars apparently acquired significant hydrospheres, indicating that the same may have occurred on Venus. The issue therefore appears to be—why did Mars, Venus and Earth evolve so differently? The distance to the Sun is an important factor, but it can hardly explain the present vast differences in atmospheric and hydrospheric conditions.

If evolutionary differentiation is the main cause, the following questions may be raised: by what mechanisms, when and how quickly? Catastrophic impact by large celestial objects in the early phase of the solar system is an obvious, yet circumstantial mechanism. The Earth was subject to such a bombardment, but it was also able to retain a vast hydrosphere and a dense atmosphere. Dynamical escape by solar forcing appears to be the most likely process responsible for a gradual depletion of the atmosphere and hydrosphere of the Earth-like planets. As discussed in Sects. 1–3, solar forcing leads to two types of escape processes— thermal escape and non-thermal escape.

Thermal escape is a process that favours light atoms. Thermal expansion and heating of a multispecies planetary atmosphere implies highest scale-height for light atoms, and therefore preference to escape (e.g. McElroy et al. 1977; Lewis and Prinn 1984; Pepin 1994). Hydrodynamic escape (Chassefière 1996) should be effective at Mars during an early/wet period (first $\approx$500 My) of intensified solar UV/EUV/XUV radiation. As will be demonstrated in this section such an expansion will also increase the cross-sectional area for the solar EUV/XUV, and solar wind forcing, the subsequent ionization leading to strong

non-thermal escape (dissociated recombination, sputtering, ionospheric plasma acceleration, and ion pickup) as described in Sect. 1. In this way thermal expansion and non-thermal escape are strongly coupled. Furthermore, as already demonstrated, ionospheric plasma energization and escape processes (Lundin and Dubinin 1992; Luhmann and Bauer 1992; Jakosky et al. 1994; Lammer et al. 1996, 2003) are more powerful in bringing large mass fluxes up to well beyond escape velocities, thus contributing to dehydrating unprotected bodies in the inner solar system. Before substantiating this further we consider the present volatile inventory on Earth, Venus and Mars (see e.g. Chassefière et al. 2007), where our focus is on the H_2O and CO_2 inventory.

Earth

The present volatile inventory according to Lewis and Prinn (1984) is 1.67×10^{21} kg H_2O, 9.2×10^{19} kg CO_2, and 4.3×10^{18} kg N_2 The Earth's total volatile mass is therefore 1.76×10^{21} kg, corresponding to ≈ 345 Bar. Note that there is an uncertainty related with the volatile content of CO_2 in rocks. The Earth is geologically active and plate tectonics makes estimates based on rock inventories difficult.

Mars

The present volatile inventory on Mars can be estimated by taking an average atmospheric pressure of 0.007 bars, mainly CO_2, with an admixture of N_2 (2.7%), and Ar (1.6%), for example. The atmospheric water content is highly variable but is generally less than 0.1%. Considering a surface area of 1.5×10^{14} m^2 we obtain a total atmospheric mass of $\approx 1.0 \times 10^{16}$ kg. The remaining volatile content, such as water, remains a matter of discussion (e.g. Carr and Head 2003; Bibring et al. 2004). Using a global equivalent layer (GEL) of water equal to 20 m, based estimates of the amount of water in the polar cap (Zuber et al. 1988) we obtain $\approx 3 \times 10^{18}$ kg H_2O. Assuming the same ratio between water and CO_2 on Mars as for the Earth ($\approx 6\%$) we obtain $\approx 2 \times 10^{17}$ kg CO_2, the latter assumed to be in the form of CO_2—ice in the south polar region and/or potentially in subsurface minerals, gas, and ice.

Venus

The volatile inventory in the mantle is unknown on Venus, so we can only rely on the volatile inventory in the atmosphere. The present volatile inventory in the Venus 90 bar atmosphere, essentially all CO_2, is according to Lewis and Prinn (1984): 4.1×10^{20} kg CO_2, with N_2 coming second in terms of abundance (1.6×10^{19} kg). The H_2O content amounts to $\approx 10^{16}$ kg. The total current water inventory on Venus is therefore some five orders of magnitude lower than that of the Earth.

Comparing Earth and Venus we note that Venus has more CO_2 and N_2 than the Earth, while the Earth has a higher total volatile inventory than Venus. Therefore, lacking a plausible chemical separation mechanism during the formation of the planetary atmospheres—favouring water on the Mars and CO_2 on Venus, one may hypothesize that the early Venus had a similar water inventory to the Earth. The question is therefore why the present Martian and Venusian volatile inventories are so different from the Earth.

On the other hand, if Venus had the same initial global water inventory as the Earth, with all water subsequently removed at a constant rate, the average loss rate would be about 10 000 kg/s. Recently Kulikov et al. (2006, 2007) studied how much of the H_2O-related oxygen could have been lost to space by the ion pick up process due to the stronger solar

wind and higher XUV fluxes of the young Sun and found, depending on the used solar wind parameters, that ion pick up by a strong early solar wind on a non-magnetized Venus could erode during 4.6 Gyr more than about 250 bar of atomic oxygen ions, corresponding to an equivalent amount of one terrestrial ocean. However, until we have better empiric data available on the ionospheric and atmospheric outflow from Venus we are left with just speculations.

5.1 Atmospheric Loss from Earth

The present outflow of atmospheric oxygen/hydrogen/nitrogen by nonthermal escape from the Earth is 1–3 kg/s (e.g. Chappell et al. 1987). The outflow is highly dependent on solar activity, varying on longer time scales between solar maximum and solar minimum, but also strongly on shorter times scales (CMEs, flare, etc., e.g. Moore et al. 1999). The outflow is induced by indirect forcing as previously discussed, the forcing mapping down to the auroral ionosphere guided by the Earth's dipole magnetic field (Fig. 10). If the outflow were considered as a loss at 3 kg/s and if that constituted N_2 and O_2, it would take about 50 billion years to evacuate the atmosphere. The corresponding time to deplete the ocean would be about 15 000 billion years. The margin for the present Earth is therefore comfortingly high. Furthermore, because of the strong Terrestrial magnetic dipole field, a recycling of outflowing planetary ions takes place (e.g. Seki et al. 2001). A conservative estimate of the total atmospheric loss is about 1 kg/s or less for the Earth.

Figure 11 demonstrates the variability of the average ionospheric plasma escape (for solar minimum and solar maximum), the right-hand side showing that the O^+/H^+ ratio corresponds to an erosion of plasma from above $\approx$1000 km during solar minimum. Solar cycle dependence is also evident since O^+ ions dominate during solar maximum. The latter implies a higher mass flux as well; more mass is "lost" during solar maximum. There are two main reasons for the enhanced O^+ outflow: (1) increased scale height from enhanced X-ray and EUV fluxes, and (2) enhanced solar wind plasma forcing. As will be discussed later, these are also the two prime reasons for the long-term erosion of planetary volatiles. Notice that the escape of nitrogen is omitted. Nitrogen escape as N^+ and N_2^+ is generally low compared to O^+ escape. The N^+/O^+ escape ratio reported (e.g. Yau and Whalen 1991; Yau et al. 1993) is of the order $\approx$0.1, implying that the loss of volatiles from the Earth by nonthermal escape primarily originates from water.

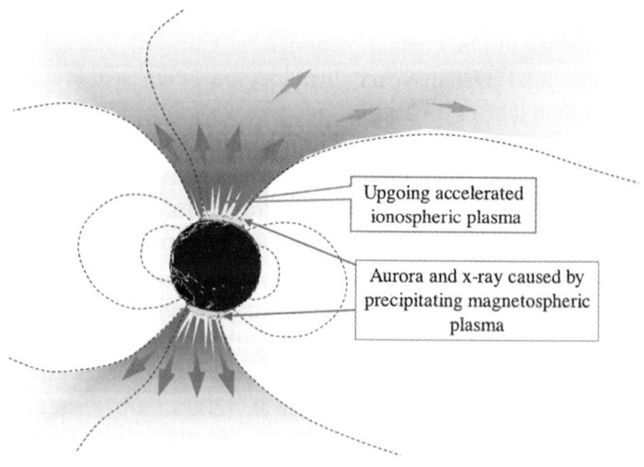

Fig. 10 Effects of external forcing on a magnetized object like the Earth, causing polar aurora and polar region plasma outflow

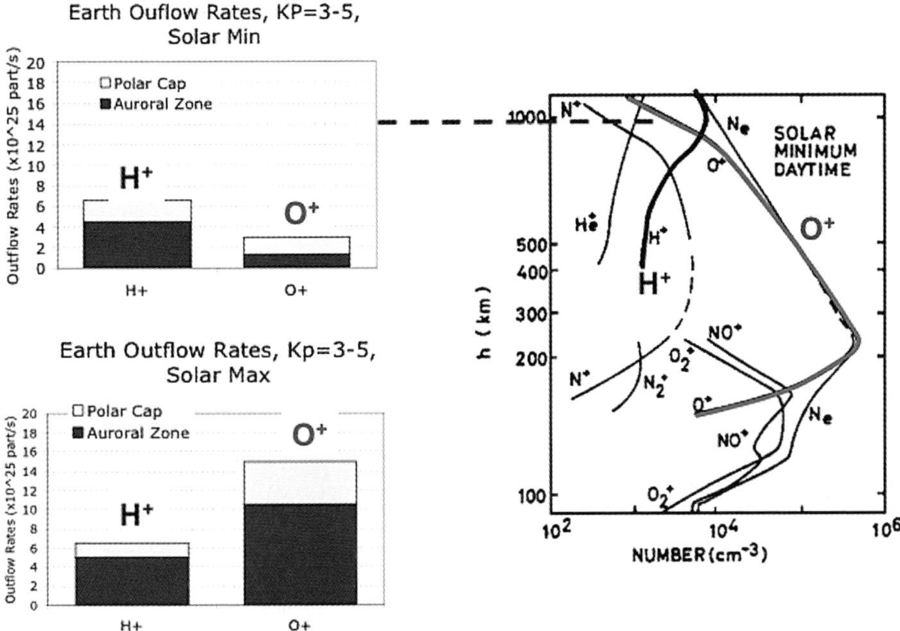

Fig. 11 *Left*: Solar minimum/maximum and ionospheric plasma outflow rate (Yau and Whalen 1991; Yau et al. 1993). *Right*: Solar minimum ionospheric ion profile illustrating that the outflow during solar minimum corresponds to an equivalent altitude of 1,000 km

5.2 Atmospheric Loss from Present MARS—Phobos-2, Mars Express

Phobos-2, launched in July 1988, was the first spacecraft enabling a quantitative estimate of the volatile escape form Mars. Based on the energized outflow of H^+, O^+, O_2^+ and CO_2^+ detected by the ASPERA experiment on Phobos-2 (Lundin et al. 1989) it was possible to estimate the volatile loss from Mars at ≈ 1 kg/s. More recent data taken during solar minimum by Mars Express, although with less adequate low-energy (<100 eV) coverage, implies a much lower outflow (<30 g, Barabash et al. 2007). The outflow rate also depends strongly on solar disturbances (solar EUV and solar wind dynamic pressure, Lundin et al. 2006b). The dominating escape flux based on Phobos-2 results comprise H^+, O^+, and $O_2^+ + CO_2^+$ (e.g. Lundin et al. 1989; Norberg et al. 1993). Recent Mars Express results (e.g. Carlsson et al. 2006) demonstrated that the escape flux constituted O^+ and O_2^+ in almost equal concentration, with a $\approx 10\%$ admixture of CO_2^+. The predominant oxygen heavy ion escape is similar to that from the Earth (less the O_2^+ escape). This indicates a preference for the loss of water. CO_2 molecules disappear at a much slower rate despite the fact that the CO_2/H_2O content ratio in the atmosphere is about 1000. Dehydration therefore appears to be a most efficient process in accelerating and removing ionospheric ions, to the extent that it may even dominate the escape of water from arid planets like Mars. However, this does not preclude other processes from making substantial contribution to the escape such as e.g. photochemical escape and ion sputtering, for example (see Introduction and references therein).

The solar wind interaction with Mars and the corresponding energization and escape of ionospheric plasma, has been subject of theory and simulations for many years. The review

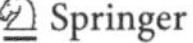 Springer

by Nagy et al. (2004) provides a good summary of the progress made in understanding the plasma environment of Mars based the Phobos-2 and Mars Global Surveyor (MGS) results. Figure 12a, based on Fig. 1.1 from Nagy et al. (2004), summarizes the insight gained in the analysis of data from Phobos-2 and MGS. Based on magnetic field data (Phobos-2, MGS), electron data (MGS) and 3 months of ion data (Phobos-2) the following features were identified:

1. The bow shock, which marks an obstacle to the solar wind expected from gas dynamics and/or magneto hydrodynamics. Inside the bow shock one finds shocked solar wind plasma, the magnetosheath.
2. The magnetic pileup Region, MPR, (Vignes et al. 2000), a region dominated by planetary ions. A Magnetic Pileup Boundary (MPB) separates MPR from the magnetosheath. MPB is a boundary that has had many names (see Nagy et al. 2004).
3. The ionosphere, separated to MPR by a boundary that resembles the Ionopause on Venus.

MPB, the boundary between the magnetosheath plasma and the plasma contained in the induced magnetosphere, has also been termed Induced Magnetosphere Boundary (IMB) (Lundin et al. 2004b). Moreover, the notion of the Photoelectron Boundary (PEB) has been introduced on the basis of Mars Express electron data (Lundin et al. 2004b) to mark the boundary separating the cold ionospheric plasma produced in the dayside and the hotter, induced magnetospheric plasma. PEB does not mark a strict boundary, though, because photoelectrons may reach high altitudes in the Martian tail (Frahm et al. 2006).

A review of modelling and simulations of the plasma environment of Mars is given by Nagy et al. (2004). Although in many aspects idealized, simulations provide an instantaneous global perspective of the plasma escape. Important progress has been made in understanding the ion-pickup process using, for instance, modern 3D hybrid models (e.g. Kallio and Janhunen 2002; Modolo et al. 2005) and 3D MHD fluid models (e.g. Ma et al. 2004; Harnett and Winglee 2005, and Luhmann et al. 2006).

Mars Express, launched in 2 June 2003, has provided a follow-up of the non-thermal escape from Mars. Results from the ion and electron analyzers of the ASPERA-3 experiment have considerably improved our understanding of solar forcing effects on Mars. For instance, Lundin et al. (2004b) noted that the solar wind penetrates deep into the dayside atmosphere, and may rapidly accelerate ions to high energies there. This implies that the entire dayside atmosphere of Mars is under intense solar wind forcing. The data also suggests a very corrugated contact surface between the ionosphere and the solar wind, in part due to the existence of patchy crustal magnetizations at Mars (Acuña et al. 1998) and the associated magnetic "cusps" (Krymskii et al. 2002). These "cusp/cleft" magnetic field lines are draped tailward (e.g. Harnett and Winglee 2005), promoting energization and outflow of ionospheric ions (Lundin et al. 2006b). The crustal magnetic field therefore acts as a stopper for incident solar wind plasma (closed field lines), as well as an acceleration region (Auroral Acceleration region, AA, Lundin et al. 2006c). Aside from the magnetic "anomalies", ion data from the flank and tail of Mars suggest many similarities between Mars and comets; the bulk of the flow following the external solar wind flow in the antisunward direction (e.g. Lundin and Dubinin 1992; Dubinin et al. 2006). A diagram summarizing the morphology of the ionospheric plasma outflow from Mars is illustrated in Fig. 12. AA stands for the Auroral Acceleration region in the tail. Except for pickup ions, most of the accelerated ionospheric ion outflow is contained inside IMB.

The composition of the ion outflow reflects the depth of solar forcing into the ionosphere. This is illustrated in Fig. 13, showing ASPERA-3 ion composition data (Lundin et al. 2006a)

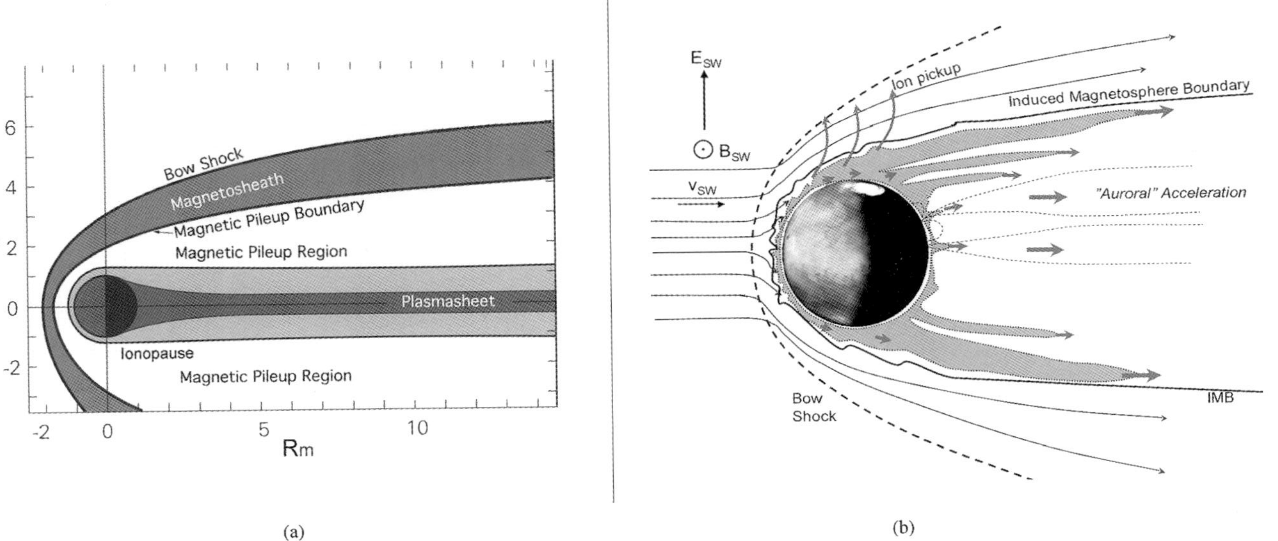

Fig. 12 Solar wind interaction with Mars. (**a**) (From Nagy et al. 2004) shows the structure of the Martian plasma environment based on primarily Phobos-2 and MGS results. (**b**) Demonstrates the energization and ionospheric plasma escape from Mars based on ion and electron measurements from Mars Express. Ion pickup (by the solar wind electric field) and ionospheric plasma energization (by waves and/or parallel electric fields) are illustrated schematically in (**b**)

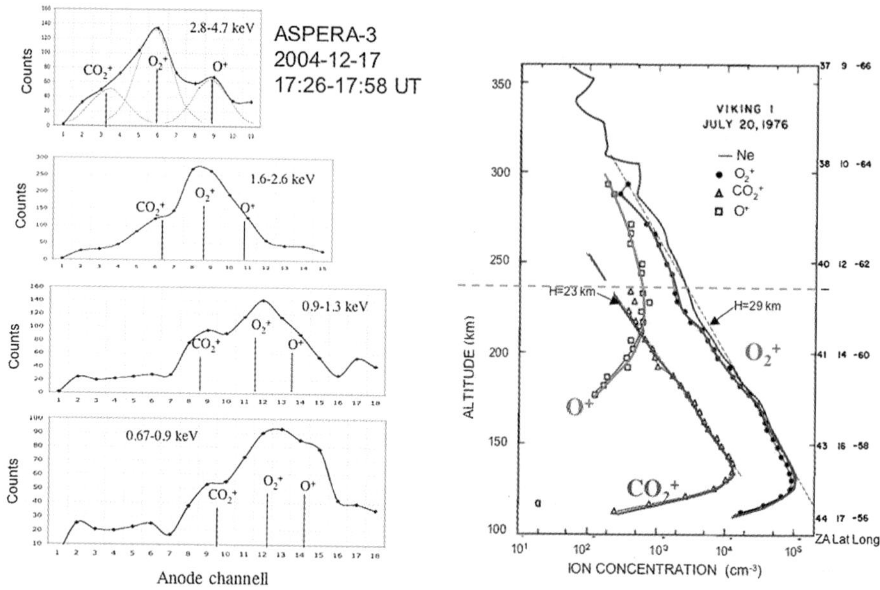

Fig. 13 *Left*: Ion mass spectra illustrating the composition of ionospheric outflow from Mars, with O_2^+ being most abundant, O^+ coming second and CO_2^+ coming third. The composition is similar to that observed by the Viking-1 spacecraft at an altitude of about 240 km (figure obtained from Hanson et al. 1977)

combined with an ionospheric density and an ion composition profile determined by the Viking-1 lander (Hanson et al. 1977). The accelerated ion composition of the outflow in 2004 is similar to the ionospheric composition in the altitude range 230–250 km as measured by Viking-1 in 1974, observations made 30 years apart but during the declining/minimum phase of solar activity. Figure 13 implies a somewhat higher CO_2^+/O^+ ratio than the average value of 8 found by Carlsson et al. (2006). However, the abundance ratios are quite variable, the ratio by Carlsson et al. (2006) marking an average. The CO_2^+/O^+ ratio may be interpreted as an altitude scalar for ion energization, a high ratio implying a low altitude for ion energization and vice versa.

Under the assumption that outflow of O^+ originates from water, the loss rate, acting over 4.5 billion years, Lundin et al. (1989) estimated the loss to a GEL of water of ≈ 1 m at Mars. This is certainly insufficient to explain a wetter early Mars (e.g. Carr and Head 2003). However, the 1 m estimate GEL of water level was based on constant solar forcing, and a constant supply. A wet early Mars also implies a dense atmosphere and grosser cross-sectional area (the outflow was higher in the past) leading to a higher total loss also for constant solar forcing. The Wood et al. (2002) findings about the evolution of sun-like stars, discussed in sections 2.3 and 2.4, have a major impact on solar forcing models. Based on these findings Lammer et al. (2003) estimated that Mars might have lost some 15–30 m GEL of water during the last 3.5 billion years. Considering also the impact of the early intense X-ray and EUV fluxes (Ribas et al. 2005) on the photochemistry and thermal properties of an early Martian atmosphere, the loss may have been even higher.

Assuming that all hydrogen lost to space from Mars originates from H_2O, theoretical studies and spacecraft observations indicate that the stoichiometrically H : O escape ratio of 2 : 1 to space cannot be maintained (Lammer et al. 2003; Patel et al. 2003). This result

implies an oxygen surface sink and a strong atmosphere-surface interaction. The oxygen surface sink, which is needed for establishing the 2 : 1 ratio between the H and O loss over the past 2 Gyr, may be responsible for enhanced soil/surface oxidation processes. Depending on different models of meteoritic gardening, the expected range for the oxidant extinction depth should be between 2–5 m. These constraints on the oxidant extinction depth are important for the search of organic material since in situ excavation of samples from the Martian subsurface.

5.3 Atmospheric Loss from Present Venus

Atmospheric escape from the upper atmosphere of Venus is mainly influenced by the loss of hydrogen and oxygen caused by the interaction of solar radiation and particle flux with the unmagnetized planetary environment. The loss of CO_2 and N_2 is estimated to be less significant, for reasons similar to those applicable for Mars (CO_2-loss) and the Earth (N_2-loss). The ionospheric ion composition altitude profile in Fig. 14 illustrates this quite well, the outflow replicating the ion abundance in the topside ionosphere.

Luhmann and Bauer (1992) presented a first estimate of the O^+ escape, $\approx 10^{25}$ s^{-1} (≈ 0.3 kg/s), based on Pioneer Venus Orbiter (PVO) ion data. Lammer et al. (2006) using a gas dynamic test particle model for average solar activity conditions deduced similar loss rates: 1×10^{25} s^{-1} and 1.6×10^{25} s^{-1} (≈ 0.4 kg/s) for H^+, and O^+ pick up ions, respectively. Estimations of ion loss rates due to detached plasma clouds that were observed by the Pioneer Venus Orbiter, yield O^+ ion loss rates in the order of about 10^{25} s^{-1} (Terada et al. 2002; Lammer et al. 2006). Due to the higher gravitational acceleration compared to Mars, thermal atmospheric escape and atmospheric loss by photo-chemically produced oxygen atoms yield negligible loss rates on Venus. Sputtering by incident pick up O^+ ions give O atom loss rates in the order of about 6×10^{24} s^{-1}. On the other hand, photo-chemically-produced, hot hydrogen atoms are a very efficient loss mechanism for hydrogen on Venus with a global average total loss rate of about 3–8×10^{25} s^{-1} (Donahue and Hartle 1992). This loss is of the same order as, but less than, the estimated H^+ ion outflow on the Venus nightside of about 7.0×10^{25} s^{-1} due to acceleration by an outward electric polarization force related to ionospheric holes (Hartle and Grebowsky 1993).

One finds that on Venus, due to its larger mass and size compared to Mars, the most relevant atmospheric escape processes of oxygen involve ions and are caused by the interaction with the solar wind and related processes. However, a detailed analysis of the outflow of ions from the Venus upper atmosphere by the ASPERA-4 and VEX-MAG instruments

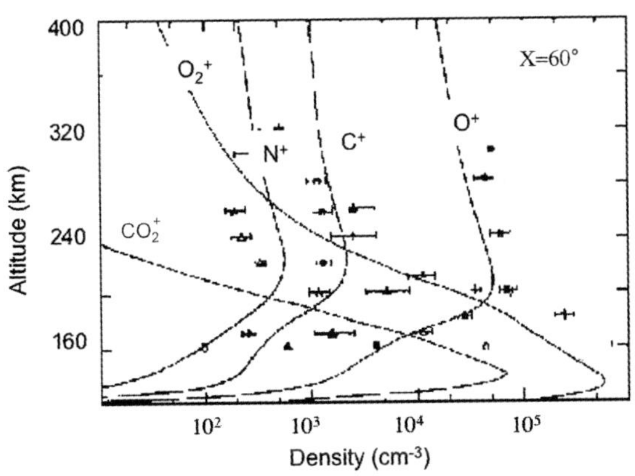

Fig. 14 Ion composition of the Venus ionosphere, illustrating the dominances of O_2^+ at low altitudes and O^+ atoms at high altitudes (after Nagy et al. 1980, and Hartle et al. 1980)

aboard ESA's Venus Express VEX will lead to more accurate atmospheric loss estimations and a better understanding of the planet's water inventory. New measurement results from 42 tail crossings with Venus Express (Lundin et al. 2007) suggest an even higher O^+ outflow (1.6×10^{26} s^{-1}), corresponding to a mass loss of 3.8 kg/s.

5.4 Amplified Particle Outflow

The evolution of the particle outflow from the Earth-like planets by solar forcing is a matter of time (solar evolution), internal properties (atmosphere, hydrosphere), and proximity to the Sun. The solar evolution as discussed in Sects. 2.2 and 2.3 implies very strong forcing in the early time period of the solar system. We now present a model of the planetary particle outflow based on solar mass losses (Wood et al. 2002, 2005) and solar EUV/XUV radiation (Ribas et al. 2005) versus time. The assumptions and boundary conditions are similar to those used in determining the theoretical outflow discussed in Sect. 3.4, incorporating the solar evolution and its effect on existing and past atmospheres of the Earth-like planets. We assume for the sake of simplicity that only the Earth had a protecting magnetic umbrella, while Venus and Mars remained essentially unmagnetized throughout. The model is fully analytical, starting with the presently measured/inferred planetary outflow and going back in time.

We start with the case of Mars, using as present outflow H^+, O^+, and O_2^+ ions (dissociated and ionized H_2O). Recent data from Mars Express indicates a highly solar EUV and solar wind dependent outflow, therefore also solar cycle dependent. The Phobos-2 data of 1 kg/s (solar maximum) is an order of magnitude higher than the MEX (solar minimum) values. We therefore assume an average escape flux of 0.5 kg/s from the present atmosphere, removed from a mean cross-sectional area $A_M \approx 1.6 \times 10^{14}$ (Sect. 3.4). The cross-sectional area is in effect an extension of the Martian atmosphere and ionosphere, specifically by the scale heights of H, O, and O_2, and the solar forcing (solar wind and EUV). Enhanced solar forcing leads to increased cross-sectional area and enhanced planetary outflow. This should apply forward as well as backward in time. From Wood et al. (2002, 2005) and Ribas et al. (2005) we obtain the following evolution of the solar mass loss/solar wind versus time.

$$\Phi_{SM}(t) \approx \Phi_0 t^{-\alpha_1}, \quad \alpha_1 \approx 1.8, \tag{3}$$

where $\Phi_{SM}(t)$ is the solar wind flux at a given time t and Φ_0 is the solar wind flux at 0.1 Gyr.

In a similar manner the following power-law relation is used for the decay of solar EUV radiation versus time:

$$\phi_{EUV}(t) \approx \phi_0 t^{-\alpha_2}, \quad \alpha_2 = 1.2. \tag{4}$$

The value for alpha is an approximation of the Ribas et al. (2005) and Lammer et al. (2003) value for the solar EVU decay ($\alpha_2 \approx 1.2$). Equation (4) implies a 98 times higher solar EUV flux at 0.1 Gyr than today (4.6 Gyr). These should be considered mean EUV flux values with time. In fact, the EUV flux values during a solar cycle (≈ 11 years) may vary by a factor 2–4. This variability demonstrates the dynamic influence of the solar EUV/UV radiation on the planetary atmosphere and ionosphere.

The EUV radiation causes, besides ionization, heating and expansion of the Martian upper atmosphere and ionosphere and, consequently, an increased cross-section area for the solar wind interaction. Moreover, a higher atmospheric pressure for early Mars (0.3–0.5 bar, Kulikov et al. (2007), this issue) would further increase the solar wind interaction region.

 Springer

Altogether, it seems reasonable to assume that the area, $A(t)$, also varies with time, i.e.

$$A(t) \approx A_\mathrm{M} t^{-\alpha_2}, \quad \alpha_2 = 1.2, \tag{5}$$

where A_M is the area discussed in Sect. 3.4. Notice that the size of A_M is expected to scale with the atmospheric and ionospheric scale height. Atomic species (H^+, O^+) have larger scale heights, the interaction region extending much further into space compared to molecular species (O_2^+ and CO_2^+). In further analysis, we focus on the water-related heavy ions, O^+ and O_2^+, and consider in a very tentative way the escape of the main atmospheric constituent (CO_2^+). For O^+ we use the empirical value of the cross-sectional area (Sect. 3.4) $A_\mathrm{M} = A_{O+}$. Considering the approximately factor of four larger scale height for CO_2^+ compared to O^+, (e.g. Hanson et al. 1977) the CO_2^+ area would scale as: $A_{CO_2^+} \approx A_{O+}/16$. Furthermore, the sensitivity and thermal expansions from increased solar UV/EUV for CO_2 is low compared to that for O_2 and O. We here assume that the effective area for CO_2^+ scales with time as $A(t) \approx A_{CO_2^+}{}^{-\alpha_2}$, where $\alpha_2 = 0.6$.

We noted from (1) the connection between the mass-loaded escape flux from Mars (Φ_M) and the input solar wind flux (Φ_SW). The mass flux of the escaping ions will be substantially higher if the flow velocities are just a few times the escape velocity (≈ 5 km/s), as compared to the incident solar wind velocity (≈ 400 km/s) (Sect. 3.4, (1)). Nevertheless, we assume for the sake of simplicity that the incident solar wind mass flux results in an equivalent mass flux for the outflowing planetary wind. We also assume a relative momentum exchange thickness of 1 in (1).

Combining expressions (2) and (5) we now get the following species-dependent (M) expression describing the mass flux versus time escaping from Mars:

$$s_\mathrm{M}(t) = m_\mathrm{M} A_\mathrm{M}(t) \Phi_\mathrm{M}(t). \tag{6}$$

where m_M is the escaping ion mass, $A(t)$ is the cross-sectional area (species dependent) and $\Phi_M(t) = \Phi_0{}^{-\alpha_1}$ is the escape flux versus time. Inserting an average total escape rate (0.5 kg/s), with composition 45% O^+, 45% O_2^+, and 10% CO_2^+ (e.g. Carlsson et al. 2006) into expression (6) gives the time-dependent escape flux as depicted in Fig. 15. The cross-sectional size for O_2^+ is here scaled as $\alpha_2 = 0.9$, i.e. an intermediate value between O^+ and

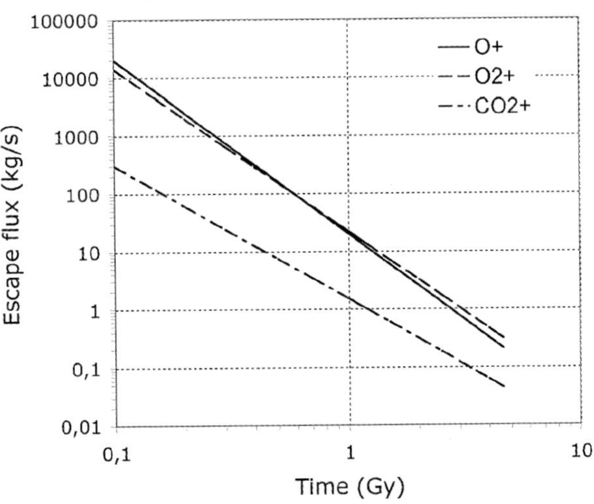

Fig. 15 Model describing the O^+, O_2^+, and CO_2^+ mass escape from Mars by solar forcing

CO_2^+. The much lower escape of CO_2^+ compared to O^+ and O_2^+, leads to atmospheric carbon enrichment. The selectivity of the process is best described by the fact that $\approx 10\%$ of today's escape from Mars is carbon-based despite that $\approx 95\%$ of the atmosphere is CO_2. 90% of the outflow originates from the mere 0.1% of water in the atmosphere. Recent VEX measurements indicate that the same applies for Venus, i.e. H^+ and O^+ dominates the ionospheric escape (Barabash et al. 2007).

A rather modest mass loss during the last ≈ 3 Gyr is expected from the model, the total inventory loss of O^+, O_2^+, and CO_2^+ corresponding to ≈ 0.1 Bar. The most critical time appears to be the first 500 million years. During that period the model predicts a loss corresponding to several tens of Bars. We emphasize that such a dramatic consequence of the model critically relies on the solar history (Wood et al. 2002, 2005 and Ribas et al. 2005). Nevertheless, if their findings are correct, it also implies that the early Noachian of Mars was wet, the early water inventory rapidly decreasing the first 1000 million years.

Notice that the above model of the volatile evolution on Mars is entirely based on solar EUV and particle forcing, addressing transport and photochemistry very qualitatively. We simply assume an upward atmospheric transport of water molecules capable of sustaining the 0.5 kg/s escape of H^+, O^+ and O_2^+. In view of the total atmospheric H_2O content ($\approx 10^{13}$ kg), this assumption is not unreasonable. Solar EUV and X-ray forcing is capable of providing sufficient thermal expansion, albeit following the solar variability. For reasons already discussed we also assume water to dominate the volatile loss process. Moreover, the model assumes that Mars lacked a sufficiently strong magnetic field to fend off the solar wind throughout its history. There is evidence for a magnetic dynamo acting on early Mars, but this dynamo may have ceased as early as 3.9–4.2 Gyr ago (e.g. Schubert et al. 2000). However, the solar forcing during this early period may have been so powerful that even a strong magnetic protection, such as that governed by the Earth's magnetic dynamo, may have been inadequate, leading to strong losses also from the Earth.

Based on the model one may also estimate the volatile loss from Earth and Venus. Venus resembles Mars in that it lacks a magnetic dynamo, let alone strong crustal magnetizations. The gravity is stronger; the CO_2 upper atmosphere is cooler, with a water-mixing ratio of $< 10^{-5}$. Altogether, this leads to a less extended upper atmosphere/ionosphere and a reduced cross-sectional area for energy and momentum transfer by the solar wind. The first estimates of the ionospheric O^+ escape by Luhmann and Bauer (1992) indicated a tail loss of $\approx 10^{25}$ ions/s, corresponding to a mass escape rate of ≈ 0.3 kg/s. This should be regarded a lower limit since PVO lacked a mass resolving ion spectrometer, and O^+ ions were inferred from double peaks in ion energy spectra. Assuming ion pickup, the O^+ ions constitutes the high-energy peak (> 1 keV) and H^+ ions the low-energy peak ($\approx 1/16$ of the O^+ peak). Recent measurements from VEX (Barabash et al. 2007) shows that large fluxes of O^+ ions escape at very low energies, the bulk outflow frequently peaking in the range 10–100 eV along the flanks of the Venus inner tail. Using ASPERA-4 ion data from 42 VEX tail traversals in August 2006 (over 1000 mass resolved ion spectra), and taking the average flux within a circular cross-sectional area of diameter $3R_V$, we obtain an average O^+ mass flux of 3.8 kg/s (Lundin et al. 2007). We use in the model as a conservative estimate 2 kg/s as the present average O^+ loss rate, assuming no magnetic shielding throughout the history of Venus, plus an expanding cross-sectional area (5) permanently exposed to the solar wind (3). Whereas we can use preliminary results for the O^+ loss rate in our model, we are still lacking an analysis of the CO_2^+ escape for Venus. We therefore use the same CO_2^+ percentage of the ion escape rate as for Mars (10% of total ion escape). This results in an accumulated CO_2-loss of $\approx 21\%$ from 0.1 Byr up to now, barely visible on a logarithmic scale as in Fig. 16. This

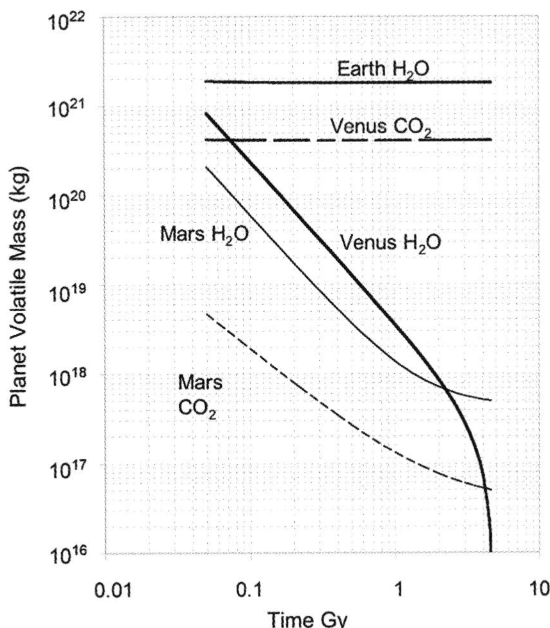

Fig. 16 Model describing the mass loss of H_2O and CO_2 on Venus, Earth and Mars by solar forcing. Strong magnetic shielding assumed for the Earth, no magnetic shielding for Venus, and options with and without magnetic shielding (up to 0.5 Gyr) for Mars. The H_2O inventory on the Earth remains essentially unaltered, while Mars and Venus have been subject to major losses

implies that if the CO_2^+ percentage ion escape is the same as for Mars the CO_2 inventory has changed only marginally by solar forcing throughout the lifetime of Venus.

Finally, the Earth's O^+ loss may be estimated based on the presently measured average loss rate of ≈ 1 kg/s. The total value may be even lower, considering the "recycling" back to the atmosphere (Seki et al. 2001). The Earth's atmosphere is also subject to an enhanced solar wind forcing (3) like for Venus and Mars. We assume for the sake of simplicity the same EUV forcing ($\alpha = 1.0$ in (5)) like that for Mars and Venus.

The total mass escaping from each of the Earth-like planets between a time t_0 (0.1 Gyr) and t_1 ($> t_0$) is obtained from the expression:

$$S_{\sum M t_1} = \sum_M \int_{t_0}^{t_1} s_M(t)\, dt \ (\text{kg}). \qquad (7)$$

The mass loss may be described as a decrease of the initial volatile inventory mass versus time. We focus here on water and CO_2, assuming that the escaping O^+ and O_2^+ ions originate from dissociated water. Adding the estimated water and CO_2 inventory described in Sect. 5 to the mass given by expression (7), we obtain the total inventory the history of water and CO_2 on the Earth-like planets as shown in Fig. 16. Notice that Earth and Venus have remained volatile-rich, albeit with predominantly water on the Earth and CO_2 on Venus. Mars has apparently lost most of its water and CO_2, but some still remains frozen in the polar cap deposits, for example. For Mars two curves are introduced regarding loss of water, with or without an Earth-like dipole magnetic field during the first 500 Myr. The magnetic protection is assumed to be as effective as the Earth's dipole field, illustrating that the water loss rate was substantial for Earth as well. Despite magnetic shielding on Mars the water loss is only reduced by about a factor of two during the first ≈ 500 Myr. The model implies that the young (0.1 Gyr) Earth-like planets had rather similar water and CO_2 inventories. However, Mars with its lower gravity and a smaller surface than

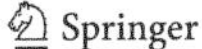

Earth and Venus had a correspondingly smaller inventory. Moreover, if the escape from Venus in the past was primarily H^+, O^+, and O_2^+, as for Mars (e.g. Norberg et al. 1993; Carlsson et al. 2006), Venus could have started with a GEL of water similar to that of the Earth (Kulikov et al. 2006). The lack of protection, and its proximity to the Sun, made most of the water disappear during the first billion years. The subsequent runaway greenhouse effect, governing a powerful hydrodynamic escape, would effectively remove any remaining hydrosphere on the surface of Venus.

6 Conclusions

The present difference in water and CO_2 inventories for Venus, Earth and Mars are substantial, water being the most abundant volatile on Earth, Venus dominated by a dense CO_2 atmosphere, and while Mars having a mix of water and CO_2, most of it supposedly as ice. "Why?", is a question still requiring a good answer. Most theories assume that a fierce early Sun (T-Tauri phase) removed all volatiles from the inner planets in the solar system, the volatiles subsequently restored by cometary impact. However, this requires a differentiated volatile insertion, with preference of water for the Earth. Another possibility is that a sufficiently high water and CO_2 inventory accumulated during the accretion phase, and a sufficiently strong gravity and magnetic shielding, may have been enough to retain an atmosphere and hydrosphere even after the early fierce phase of the Sun. If the latter is the case a secondary restoration of water and CO_2 is not necessary, and the present inventory of the Earth-like planets is the result of a long-term evolution. The water and CO_2 erosion model dating back in time to 0.1 Gyr after the planet formation, suggests that Earth, Venus and Mars may have had similar relative water and CO_2 inventories. Lacking magnetic shielding, Venus and Mars would have been more vulnerable than Earth, rapidly loosing the most abundant volatile molecule, water, by thermal and non-thermal escape. The model based on ionospheric plasma escape shows that non-thermal escape by solar forcing is sufficiently effective to remove some 40 Bar of water from Mars and at least 50 Bar of water from Venus. Similarly, the loss for Mars is consistent with estimates mentioned by Chassefière et al. (2007) and others. On the other hand, the model predicts that the loss of water from the Earth, a planet that retained its magnetic shielding, has been insignificant after 0.1 Gyr of the planet formation.

Notice that the loss of water and CO_2 may have been higher than presented here. A number of processes driven by solar forcing contribute to the escape of volatiles from the atmosphere of the Earth-like planets. Solar EUV/XUV forcing causes heating, expansion, and ionization of the atmosphere. Hydrodynamic escape (e.g. Chassefière 1996) is a process that may have been effective in the early removal of hydrogen. Photochemical escape (Nagy et al. 1980; Luhmann et al. 1992) due to dissociative recombination of heated ionospheric ions may have been effective in an early removal of oxygen. Solar wind forcing leads to further energization and escape of ionospheric plasma, but also to sputtering (Luhmann and Kozyra 1991). The latter results in the escape of neutrals. Altogether, there are good reasons to believe that solar forcing in the early solar system was capable of removing even more volatiles, specifically water, from Mars and Venus, than that implied from our model.

The ultimate fate for free/unbound atoms and molecules immersed in the inner solar system is to become picked up by the solar wind and brought to the solar system periphery. The solar wind interaction with a comet (Fig. 2) illustrates this quite well. The ionospheric

 Springer

escape model presented here predicts a higher outflow for lower outflow velocity (expression (1) and Fig. 9). Compare for instance with the mass-loaded plasma outflow in the near tail of a comet, reaching velocities of 10–50 km/s. This velocity is an order of magnitude lower than the average solar wind velocity. With an early average solar wind velocity of $\approx$3000 km/s (Wood et al. 2005), and a correspondingly higher dynamic pressure (goes as square of the velocity) the amplification factor becomes substantially higher for the Earth-like planets. One may easily conceive a ten times higher mass loss than that perceived from the model.

In summary, we have presented important aspects of the long-term and short-term variability of solar forcing—solar X-ray EUV radiation and the solar wind. Solar forcing is expected to have played an important role in the atmospheric evolution of planets orbiting close to the Sun. We have here focussed on how solar forcing affects the ionosphere of the Earth-like planets; on a long-term perspective; with or without magnetic shielding. There should be no principle differences in physics governing the present short-term variability of ionospheric plasma escape from the Earth (e.g. Chappell et al. 1987; Yau and André 1997) and the long-term solar forcing effects on the atmosphere and ionosphere. The terrestrial escape flux may vary by an order of magnitude during a short duration (hours) solar disturbance (e.g. a CME). A similar trend, albeit much slower, is observed in the course of a solar cycle ($\approx$11 years). If the findings by Wood et al. (2002, 2005) and Ribas et al. (2005) on the long-term variability of the Sun are correct it would imply a major differentiation of the volatile inventory of the Earth-like planets. Non-thermal escape acting in the early period of the solar system is capable of removing vast amounts of water on Mars and Venus. The first 500–1000 million years were probably the most critical period. Without sufficient magnetic shielding a planetary atmosphere would be subject to fierce solar forcing, effectively removing major fractions of the most volatile element—water. Unless being the only recipient of a vast additional water supply following the early solar forcing period the Earth must have endured the most dangerous period, thanks to a persistent intrinsic magnetic shield.

We conclude that a strong planetary dipole magnetic field has important implications for the evolution of life and for the habitability of a planet. A strong dipole magnetic field helps setting up a magnetic "umbrella", a stand off distance between the atmosphere and the eroding plasma wind from a star. Magnetic protection is particularly critical for planets near a young solar-like star during the first few hundred to one billion years. This is a period when solar-like stars irradiate planets with intense XUV/EUV and a fierce plasma wind. Without sufficient shielding this may lead to a climate crisis, with loss of water as major implication. Mars and Venus represent two extremes of the consequence of un-shielding, Mars a cold dehydrated planet with tenuous atmosphere, Venus a dehydrated planet with a hot dense atmosphere. Mars may have had a sufficiently dense atmosphere and water inventory the first billion years to qualify as "habitable" for primitive life forms. Also Venus is expected to have had a wealth of water during the first few hundred million years, potentially sufficient to qualify as "habitable" for at least primitive life forms. However, the first hundred million years of the Sun must have been an extremely harsh time for life. The solar variability a billion year later was more modest, with less implication on climate. With most of the hydrospheres on Venus and Mars gone, with a tenuous atmosphere on Mars and potentially already a runaway greenhouse on Venus, the habitability would anyway be at best marginal. Episodic habitability governed by solar and planetary conditions that changes climate in a favourable direction is a possibility. For instance, an atmospheric density and temperature above the triple point of water on Mars, combined with stable solar conditions, may

evolve into episodic habitability. Conversely, critical episodes with a combination of vanishing planetary magnetic protection and dramatically enhanced solar activity, may have had negative effects on a "favoured planet"—the Earth.

Acknowledgements R. Lundin acknowledges support from the Swedish National Space Board. H. Lammer and I. Ribas acknowledges support from the "Büro für Akademische Kooperation und Mobilität" of the Austrian Academic Exchange Service under the AD-Acciones Integradas project No. 12/2005. I. Ribas acknowledges also support from the Spanish Ministerio de Ciencia y Tecnologa through a Ramn y Cajal fellowship.

References

M.H. Acuña et al., Science **279**, 1676 (1998)

H. Alfvén, *Cosmical Electrodynamics* (Oxford University Press, Oxford, 1950)

T.R. Ayres, J. Geophys. Res. **102**, 1641–1651 (1997)

V.R. Baker, Nature **412**, 228–236 (2001)

S.J. Bauer, H. Lammer, *Planetary Aeronomy: Atmosphere Environments in Planetary Atmospheres* (Springer, Berlin, 2004)

S. Barabash, A. Fedorov, R. Lundin, J.-A. Sauvaud, Science (2007)

J.-P. Bibring, Y. Langevin, F. Poulet, A. Gendrin, B. Gondet, M. Berthé, A. Souflot, P. Drossart, M. Combes, G. Bellucci, V. Moroz, N. Mangold, B. Schmidt, OMEGA team, Nature **428**(6983), 627–630 (2004)

M.K. Bird, P. Edenhofer, in *Physics of the Inner Heliosphere I*, ed. by R. Schwenn, E. Marsch, vol. XI (Springer, Berlin, 1990), pp. 13–97

A. Bressan, F. Fagotto, G. Bertelli, C. Chiosi, Astron. Astrophys. Suppl. Ser. **100**, 647–664 (1993)

L.H. Brace, W.T. Kasprzak, H.A. Taylor, R.F. Theis, C.T. Russell, A. Barnes, J.D. Mihalov, D.M. Hunten, J. Geophys. Res. **92**, 15 (1987)

E. Carlsson et al., Icarus **182**(2), 320 (2006)

M.H. Carr, J.W. Head, J. Geophys. Res. **108**, 5042 (2003). doi:10.1029/2002JE001963

C.R. Chappell, T.E. Moore, J.H. Waite Jr, J. Geophys. Res. **92**, 5896 (1987)

E. Chassefière, Icarus **124**, 537–552 (1996)

E. Chassefière, F. Leblanc, Planet. Space Sci. **52**, 1039–1058 (2004)

E. Chassefière, F. Leblanc, B. Langlais, Planet. Space Sci. (2007). doi:10.1016/j.pss.2006.02.003

C.C. Chaston, L.M. Peticolas, C.W. Carlson, and 16 coauthors, J. Geophys. Res. **110**, A02211 (2005). doi:10.1029/2004JA010483

D.H. Crider, D.A. Brain, M.H. Acuña, D. Vignes, C. Mazelle, C. Bertucci, Space Sci. Rev. **111**, 203–221 (2004)

T.M. Donahue, R.E. Hartle, Geophys. Res. Lett. **19**, 2449–2452 (1992)

J.D. Dorren, E.F. Guinan, in *The Sun as a Variable Star*, ed. by J.M. Pap, C. Frolich, H.S. Hudson, S. Solanki (Cambridge Univ. Press, Cambridge, 1994), p. 206

E. Dubinin, R. Lundin, H. Koskinen, N. Pissarenko, J. Geophys. Res. **98**, 3991 (1993)

E. Dubinin, D. Winningham, M. Fränz, the ASSPERA-3 team, Icarus **182**(2), 343 (2006)

J.W. Dungey, Phys. Rev. Lett. **6**, 47–48 (1961)

J.R. Espley, P.A. Cloutier, D.H. Crider, D.A. Brain, M.H. Acuña, J. Geophys. Res. (2004). 2004AGUFMSA13A1120E

P. Foukal, C. Fröhlich, H. Spruit, T.M.L. Wigley, Nature **443**, 161–166 (2006). doi:10.1038

J.L. Fox, A. Hac, J. Geophys. Res. **102**, 24005–24011 (1997)

R.A. Frahm, J.R. Sharber, J.D. Winningham, the ASSPERA-3 team, Space Sci. Rev. **126**, 389–402 (2006)

E. Friis-Christensen, K. Lassen, Science **254**, 698 (1991)

B.F. Gordiets, Yu.N. Kulikov, M.N. Markov, M.Ya. Marov, J. Geophys. Res. **87**, 4504–4514 (1982)

J.-M. Griemeier, A. Stadelmann, T. Penz, H. Lammer, F. Selsis, I. Ribas, E.F. Guinan, U. Motschmann, H.-K. Biernat, W.W. Weiss, Astron. Astrophys. **425**, 753–762 (2004)

E.F. Guinan, I. Ribas, in *The Evolving Sun and its Influence on Planetary Environments*, ed. by B. Montesinos, A. Giménez, E.F. Guinan. ASP, vol. 269 (San Francisco, 2002), pp. 85–107

J.A. Guzik, L.S. Watson, A.N. Cox, Memorie della Societa Astronomica Italiana **77**, 389 (2006)

M. Güdel, E.F. Guinan, S.L. Skinner, Astrophys. J. **483**, 947–960 (1997)

E.M. Harnett, R.M. Winglee, J. Geophys. Res. **110** (2005). doi:10.1029/2003JA010315

W.B. Hanson, S. Sanatani, D.R. Zuccaro, J. Geophys. Res. **82**, 4351–4363 (1977)

R.E. Hartle, H.A. Taylor, S.J. Bauer, L.H. Brace, C.T. Russell, R.E. Daniell, J. Geophys. Res. **85**, 7739–7746 (1980)

R.E. Hartle, J.M. Grebowsky, J. Geophys. Res. **98**, 7437–7445 (1993)

D.V. Hoyt, K.H. Schatten, *The Role of the Sun in Climate Change* (Oxford University Press, Oxford, 1997)

D.E. Hunten, Science **259**, 915–920 (1993)

B.M. Jakosky, R.O. Pepin, R.E. Johnson, J.L. Fox, Icarus **111**, 271 (1994)

R.E. Johnson, D. Schnellenberger, M.C. Wong, J. Geophys. Res. **105**, 1659–1670 (2000)

E. Kallio, J.G. Luhmann, J.G. Lyon, J. Geophys. Res. **103**, 4753–4754 (1998)

E. Kallio, P. Janhunen, J. Geophys. Res. **107**(A3) (2002). doi:10.1029/2001JA000090

E. Kallio, A. Fedorovm, E. Budnik, the ASPERA-3 team, Icarus **182**(2), 448 (2006)

J. Kanipe, Nature **443**, 141–143 (2006)

J.F. Kasting, Icarus **74**, 472–494 (1988)

J.F. Kasting, D.P. Whitmire, R.T. Reynolds, Icarus **101**, 108–128 (1993)

R. Keppens, K.B. MacGregor, P. Charbonneau, Astron. Astrophys. **294**, 469–487 (1995)

J. Keyser, M.W. Dunlop, C.J. Owen, B.U.Ö. Sonnerup, S.E. Haaland, A. Vaivads, G. Paschmann, R. Lundin, L. Rezeau, Space Sci. Rev. **118**, 231–320 (2005)

J. Kim, A.F. Nagy, J.L. Fox, T. Craven, J. Geophys. Res. **103**(29), 29,339–29,342 (1998)

A.M. Krymskii, T.K. Breus, N.F. Ness, M.H. Acuña, J.E.P. Connerney, D.H. Crider, D.L. Mitchell, S.J. Bauer, J. Geophys. Res. **107**(A9), 1245 (2002). doi:10.1029/2001JA000239

Yu.N. Kulikov, H. Lammer, H.I.M. Lichtenegger, N. Terada, I. Ribas, C. Kolb, D. Langmayr, R. Lundin, E.F. Guinan, S. Barabash, H.K. Biernat, Planet. Space Sci. **54**, 1425–1444 (2006)

Yu.N. Kulikov, H. Lammer, H.I.M. Lichtenegger, T. Penz, D. Breuer, T. Spohn, R. Lundin, H.K. Biernat, Space Sci. Rev. (2007, this issue). doi: 10.1007/s11214-007-9192-4

H. Lammer, W. Stumptner, S.J. Bauer, Geophys. Res. Lett. **23**, 3353–3356 (1996)

H. Lammer, H.I.M. Lichtenegger, C. Kolb, I. Ribas, E.F. Guinan, R. Abart, S.J. Bauer, Icarus **106**, 9–25 (2003)

H. Lammer, H.I.M. Lichtenegger, H.K. Biernat, N.V. Erkaev, I.L. Arshukova, C. Kolb, H. Gunell, A. Lukyanov, M. Holmstrom, S. Barabash, T.L. Zhang, W. Baumjohann, Planet. Space Sci. (2006, in press)

J. Lean, Ann. Rev. Astron. Astrophys. **35**, 33–67 (1997)

F. Leblanc, R.E. Johnson, J. Geophys. Res. (2002). doi:10.1029/2000JE001473

J.S. Lewis, R.G. Prinn, *Planets and Their Atmospheres: Origin and Evolution* (Academic, Orlando, 1984)

J.G. Luhmann, J.U. Kozyra, J. Geophys. Res. **96**, 5457 (1991)

J.G. Luhmann, R.E. Johnson, M.H.G. Zhang, Geophys. Res. Lett. **19**, 2151 (1992)

J.G. Luhmann, S.J. Bauer, in *Venus and Mars: Atmospheres, Ionospheres, and Solar Wind Interactions*. AGU Monograph, vol. 66 (1992), pp. 417–430

J.G. Luhmann, S.A. Ledvina, J.G. Lyon, C.T. Russell, Planet. Space Sci. **54**, 1457–1471 (2006)

R. Lundin, A. Zakharov, R. Pellinen, B. Hultqvist, H. Borg, E.M. Dubinin, S. Barabasj, N. Pissarenko, H. Koskinen, I. Liede, Nature **341**, 609 (1989)

R. Lundin, E.M. Dubinin, S.V. Barabash, H. Koskinen, O. Norberg, N. Pissarenko, A.V. Zakharov, Geophys. Res. Lett. **18**, 1059 (1991)

R. Lundin, E.M. Dubinin, Adv. Space Res. **12**(9), 255 (1992)

R. Lundin, S. Barabash, Planet. Space Sci. **52**, 1059–1071 (2004a)

R. Lundin, S. Barabash, H. Andersson, M. Holmström, the ASPERA3 team, Science **305**, 1933 (2004b)

R. Lundin, D. Winningham, S. Barabash, R. Frahm, the ASPERA-3 team, Icarus **182**(2), 308 (2006a)

R. Lundin, S. Barabash, the ASPERA 3 team (2006b, manuscript under preparation)

R. Lundin, D. Winningham, S. Barabash, the ASPERA 3 team, Science **311**, 980–983 (2006c)

R. Lundin, S. Barabash, J.-A. Sauvaud, the ASPERA-4 team, Science (2007, submitted)

J.I. Lunine, J. Chambers, A. Morbidelli, L.A. Leshin, Icarus **165**, 1–8 (2003)

H. Lundstedt, L. Liszka, R. Lundin, R. Muschler, Annales Geophysicae **24**, 1–10 (2006)

M.B. McElroy, T.Y. Kong, Y.L. Yung, J. Geophys. Res. **82**, 4379–4388 (1977)

Y. Ma, A.F. Nagy, I.V. Sokolov, K.C. Hansen, J. Geophys. Res. **109**, A07211 (2004). doi:10.1029/2003JA010367

C.P. McKay, C.R. Stoker, Rev. Geophys. **27**, 189–214 (1989)

R. Modolo, G.M. Chanteur, E. Dubinin, A.P. Matthews, Ann. Geophys. **23**, 433–444 (2005)

T.E. Moore, R. Lundin, D. Alcayde, M. Andre, S.B. Ganguli, M. Temerin, A. Yau, Space Sci. Rev. **88** (1999)

A.F. Nagy, T.E. Cravens, S.G. Smith, H.A. Taylor, H.C. Brinton, J. Geophys. Res. **85**, 7795–7801 (1980)

A.F. Nagy, D. Winterhalter, K. Sauer et al., Space Sci. Rev. **111**(1), 33–114 (2004)

G. Neukum, New view of Mars after two years of Mars Express high resolution stereo camera data acquisition and analysis. American Geophysical Union, Fall Meeting 2005, abstract #P13C-03

G. Newkirk Jr., Geochimica Cosmochimica Acta Suppl. **13**, 293–301 (1980)

M.J. Newman, R.T. Rood, Science **198**, 1035–1037 (1977)

O. Norberg, R. Lundin, S. Barabash, in *COSPAR Colloquium 4, Plasma Environments of Non-magnetic Planets*, ed. by T.I. Gombosi (1993), pp. 299–304

T. Owen, in *Evolution of Planetary Atmospheres and Climatology of the Earth*. International Conference, Nice, France, A79-33839 13-42, Toulouse, CNRS, 1979, pp. 1–10

E.N. Parker, Astrophys. J. **128**, 664 (1958)

M.R. Patel, A. Berc¶es, C. Kolb, H. Lammer, P. Rettberg, J.C. Zarnecki, F. Selsis, Int. J. Astrobiol. **2**, 21–34 (2003)

R.O. Pepin, Icarus **111**, 289–304 (1994)

H. Pérez-de Tejada, J. Geophys. Res. **92**, 4713 (1987)

H. Pérez-de Tejada, J. Geophys. Res. **103**, 31499–31508 (1998)

I. Ribas, E.F. Guinan, M. Güdel, M. Audard, Astrophys. J. **622**, 680–694 (2005)

C.T. Russell, J.G. Luhmann, R.J. Strangeway, Planet. Space Sci. **54**, 1482–1495 (2006)

I.-J. Sackmann, A.I. Boothroyd, Astrophys. J. **583**, 1024–1039 (2003)

C. Sagan, G. Mullen, Science **177**, 52–56 (1972)

K. Seki, R.C. Elphic, M. Hirahara, T. Terasawa, T. Mukai, Science **291**, 1939–1941 (2001)

G. Schubert, C.T. Russell, W.B. Moore, Nature **408**, 666 (2000)

R. Schwenn, in *Large-Scale Structure of the Interplanetary Medium, Physics of the Inner Heliosphere I*, vol. XI, ed. by R. Schwenn, E. Marsch (Springer, Berlin, 1990), p. 99

T. Simon, E.F. Boesgaard, G. Herbig, Astrophys. J. **293**, 551–570 (1985)

A. Skumanich, Astrophys. J. **171**, 565–567 (1972)

H. Svensmark, E. Friis-Christensen, J. Atmosph. Sol.-Terr. Phys. **59**, 1225–1232 (1997)

H. Svensmark, Space Sci. Rev. **93**, 175–185 (2000)

H. Svensmark, J.O.P. Pedersen, N. Marsh, M. Enghoff, U. Uggerhøj, *Proceedings of the Royal Society A*, October 3rd, 2006

N. Terada, S. Machida, H. Shinagawa, J. Geophys. Res. **107**, 1471–1490 (2002)

D. Vignes, C. Mazelle, H. Rème, M.H. Acuña, J.E.P. Connerney, R.P. Lin, D.L. Mitchell, P. Cloutier, D.H. Crider, N.F. Ness, Geophys. Res. Lett. **27**(1), 49 (2000)

D.F. Webb, R.A. Howard, J. Geophys. Res. **99**, 4201–4220 (1994)

D.P. Whitmire, L.R. Doyle, R.T. Reynolds, J. Matese, J. Geophys. Res. **100**, 5457–5464 (1995)

J.D. Winningham, R.A. Frahm, J.R. Sharber, the ASPERA-3 team Icarus **182**(2), 360 (2006)

B.E. Wood, H.-R. Müller, G. Zank, J.L. Linsky, Astrophys. J. **574**, 412–425 (2002)

B.E. Wood, H.-R. Müller, G.P. Zank, J.L. Linsky, S. Redfield, Astrophys. J. **628**, L143–L146 (2005)

A.W. Yau, B.A. Whalen, Geophys. Res. Lett. **18**, 345–348 (1991)

A.W. Yau, B.A. Whalen, C. Goodenaough, E. Sagawa, T. Mukai, J. Geophys. Res. **98**, 11205–11224 (1993)

A.W. Yau, M. André, Space Sci. Rev. **37**, 1 (1997)

K.J. Zahnle, J.C.G. Walker, Rev. Geophys. **20**, 280–292 (1982)

T.L. Zhang, J.G. Luhmann, C.T. Russell, J. Geophys. Res. **96**, 11145 (1991)

M.H. Zhang, J. Luhmann, A.F. Nagy et al., J. Geophys. Res. **98**, 3311 (1993)

M.T. Zuber, D.E. Smith, S.C. Solomon, J.B. Abshire, R.S. Afzal et al., Science **282**, 2053–2060 (1988)

Space Sci Rev (2007) 129: 279–300
DOI 10.1007/s11214-007-9163-9

Planetary Magnetic Dynamo Effect on Atmospheric Protection of Early Earth and Mars

V. Dehant · H. Lammer · Y.N. Kulikov · J.-M. Grießmeier · D. Breuer · O. Verhoeven ·
Ö. Karatekin · T. Van Hoolst · O. Korablev · P. Lognonné

Received: 1 August 2006 / Accepted: 20 February 2007 /
Published online: 21 June 2007
© Springer Science+Business Media, Inc. 2007

Abstract In light of assessing the habitability of Mars, we examine the impact of the magnetic field on the atmosphere. When there is a magnetic field, the atmosphere is protected from erosion by solar wind. The magnetic field ensures the maintenance of a dense atmosphere, necessary for liquid water to exist on the surface of Mars. We also examine the impact of the rotation of Mars on the magnetic field. When the magnetic field of Mars ceased to exist (about 4 Gyr ago), atmospheric escape induced by solar wind began. We consider scenarios which could ultimately lead to a decrease of atmospheric pressure to the presently observed value of 7 mbar: a much weaker early martian magnetic field, a late onset of the dynamo, and high erosion rates of a denser early atmosphere.

Keywords Habitability · Rotation · Magnetic field · Atmospheric escape

V. Dehant (✉) · O. Verhoeven · Ö. Karatekin · T. Van Hoolst
Royal Observatory of Belgium, Brussel, Belgium
e-mail: v.dehant@oma.be

H. Lammer
Space Research Institute, Austrian Academy of Sciences, Graz, Austria

Y.N. Kulikov
Polar Geophysical Institute (PGI), Russian Academy of Sciences, Apatity, Russia

J.-M. Grießmeier
Observatoire de Paris, Paris, France

D. Breuer
DLR, Berlin, Germany

O. Korablev
Space Research Institute, Moscow, Russia

P. Lognonné
Institut de Physique du Globe de Paris, Paris, France

1 Introduction

Habitability is usually believed to be a function of the size of a planet, the distance of the planet to its star (or the Sun for the Solar system), the brightness of its star, the absorption and reflection energy balance of the planet, the composition of its atmosphere, plate tectonics, presence of a magnetic field, gravitational interactions, protection from asteroids and comets, orbital and rotational dynamics, and more, as discussed in this book. Here we address rotation dynamics and their relationship with the magnetic field, as well as their relationship with atmospheric escape. In particular, we need to consider the influence of the solar wind on the atmosphere, as was discussed in the first part of this chapter. Solar wind is made up of high-energy protons and electrons that are emitted from the upper regions of the Sun and stream toward the Earth. Since protons and electrons are charged particles, they are deflected by a magnetic field. Living organisms on the Earth are protected from the effects of the solar wind because Earth's magnetic field deflects the charged particles away from the Earth or toward the poles. The rotation of a planet is an important mechanism often believed to be responsible for the enhancement of the magnetic field. Protection from the solar wind may therefore depend on the planetary rotation. Both magnetic fields and atmospheres protect living organisms from cosmic rays. A sufficiently dense atmosphere may also provide conditions favorable for the presence of liquid water on a planet, an essential prerequisite for the beginnings of life and its further evolution.

The interaction of the Sun's energy with the atmospheres of Earth and Mars has differences and similarities, not only related to the existence of a magnetic field, but also to their radii, rotation rates, gravitational accelerations, topography, surface pressure, and surface and internal compositions. The absence of a large intrinsic magnetic field on Mars is a current difference, but there was a magnetic field in the early history of Mars, as suggested by data of the MAG/ER (Magnetometer/Electron Reflectometer) instrument of Mars Global Surveyor (MGS) spacecraft (Connerney et al. 1999; Acuña et al. 2001). After the MGS orbit injection in 1997 and during the aerobraking phase and science-phasing orbit phase, it was possible to get magnetic measurements at periapses ranging from 85 to 170 km above Mars' surface. Connerney et al. (1999) (see also Purucker et al. 2000; Acuña et al. 2001) have discovered strongly magnetized regions in its crust, closely associated with the ancient, cratered terrain of the highlands in the southern hemisphere. There has been a great interest in the literature in the meaning of the spatial pattern of magnetization (see e.g. Jurdy and Stefanick 2004; Arkani-Hamed 2004), including possible lineations (these crustal magnetizations even suggest reversals) that were believed to suggest an analogy to plate tectonic (like the lineations derived from magnetization on Earth's ocean floor). Recent models favor thermal remanance acquired during the time these ancient rocks cooled, in conjunction with dike intrusions (Nimmo 2000) as the best way to explain the lineations. The MGS remnant magnetic field measurements show that the magnitude of the magnetic moment observed at satellite altitudes implies a source in the martian crust with a magnetic moment of $\sim 1.6 \times 10^{16}$ A m^2 $= 1.6 \times 10^{19}$ G cm^3 (Acuña et al. 1998): MGS has detected surface magnetic anomalies of up to 1500 nT. The magnetic sources are about an order of magnitude stronger than those of the Earth's continents (e.g., Toft and Arkani-Hamed 1992; Arkani-Hamed and Dyment 1996) and comparable in magnitude with the remanent magnetization of fresh extrusive basalt near the oceanic ridge axes of Earth (e.g., Bleil and Petersen 1983). These magnetic anomalies indicate the existence of a strong, ancient, intrinsic martian magnetic moment with a field strength 0.1 to 10 times of the present Earth field (Ness et al. 1999; Mitchell et al. 1999, 2001), first suspected from Viking probes data (Hargraves

 Springer

et al. 1977, 1979). The strength of the martian magnetic anomalies could in addition reflect a strong magnetic susceptibility due to the composition of the rocks.

Little is currently known about the interior of Mars. The only in situ observations important for the interior are those of the martian gravity field, the tidal effect on an orbiter, and the polar moment of inertia, which are derived from radio tracking of orbiting and landed spacecraft (e.g., Smith et al. 1998; Folkner et al. 1997; Konopliv et al. 2006). These observations are the main constraints for interior models additionally based on analysis of SNC meteorites or chondrite data and extrapolation of the Earth's internal structure to the lower pressures of Mars' interior (Sohl and Spohn 1997; Sanloup et al. 1999; Bertka and Fei 1998; Sohl et al. 2005; Verhoeven et al. 2005). The question of whether the core is presently liquid or solid is still an open question although the measured tidal effect on the orbits of the MGS, Mars Odyssey, and MEX suggests an at least partially liquid core (Yoder et al. 2003; Balmino et al. 2006; Konopliv et al. 2006; Duron et al. 2007).

The rotation and orientation of Mars at present are very similar to those of the Earth. But they may have been more different in their past, which can have other implications for planetary and atmosphere evolution. We know that the Earth has kept very similar orientation with a few degree differences, but as explained by Laskar et al. (2004b, see also 1994 1994, 1996, 1997; Laskar et al. 2002, 2004a), the obliquity of Mars may have changed considerably during its history. This issue will be addressed in Sect. 3, while Sect. 2 addresses the rotation of the planets and Sects. 4 and 5, the implications for the magnetic field and the associated effect on the magnetopause distances (atmosphere protection). The last section, Sect. 6 addresses the resulting atmosphere escape (see also the other two parts of the chapter).

2 Rotation of Early Earth–Moon System (Faster than Today) and Mars

Due to tidal dissipation and angular momentum conservation, the Earth–Moon distance is increasing presently (the Moon moves away from the Earth at a rate of approximately 4 cm/year, see Dickey et al. 1994), and the rotation of the Earth is decreasing (the length-of-day is longer at present than in the past; it increases by about 20 microseconds per year due to the Moon, and by another 25 microseconds per year due to the Sun). The reason for these changes is the so-called tidal friction effect. Tidal friction is the time-averaged, global dissipation mechanism of rotational energy and angular momentum by tides. It is caused by the non-instantaneous effects of tidal forces on mass redistribution and on ocean currents and heights (if there is an ocean). Today, the precise mechanism of tidal energy dissipation is still an open question and, despite a better knowledge of time-scales and sophisticated mathematical modelling techniques, ocean tides are still not known with the desired accuracy to compute this effect, but are believed to be the major actors in this process. If m_p and m_s are the masses of the primary and secondary bodies, Ω is the rotation rate of the primary planet, D is their semi-major axis in the mean ellipse of their relative distances, Q_p is the dissipation factor inside the planet ($Q_{\mathrm{Earth}} \approx 12$), k_2 is the tidal Love number, a is the radius of the primary planet, and n the angular rotation rate of the secondary body (the Moon, in the Earth's case) around the planet, one has (Murray and Dermott 2000):

$$\dot{D} = sign(\Omega - n)3k_2\frac{m_s}{m_p}a^5\frac{\sqrt{G(m_s + m_p)}}{Q_p D^{11/2}}.$$

Using ancient observations of solar eclipses, Stephenson and Morrison (1984) have shown that the Earth length of the day is increasing by 2 milliseconds per century, on average.

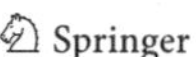

The day was shorter by about 2 hours and the year "longer" by about 35 days during the Devonian period (400 million years ago, i.e. 0.4 Gyr ago). There was thus a more rapid rotation of the Earth in the geologic past. A linear extrapolation shows that the length of day would be 19 h, 1 Gyr ago, and that the Moon was very close to the Earth 1.2 Gyr ago (15,000–20,000 km).

A full computation, taking into account that Earth is not a perfect sphere (Kopal 1972), that its rotation axis is not perpendicular to the orbit of the Moon, and the contribution of the Sun, leads to a predicted decrease of the rotation of the Earth of approximately 2.3 milliseconds per century. Extrapolating all this to the past indicates that the Earth's rotation period was at a few hours after its birth. However, the linear extrapolation with the present dissipation rate is not very realistic (Williams 2000; Varga et al. 2006). The dissipation rate in the oceans depends on the rotation rate of the planet as well as on the distribution of the continents. The motion of the continents (Kvale et al. 1999), intense global glaciations (Evans et al. 1997; Kirschvink et al. 2000), or higher mantle activities (Melezhik et al. 2003; Greff-Lefftz and Legros 1999), are among the geophysical processes that could have further influenced the rotation rate of Earth. It is believed that the initial rotation period of the Earth could have been as low as 13.1 h (MacDonald 1964), and an initial rotation rate of 10 hours is consistent with the relation between the planetary angular momentum, density, and planetary mass observed in the solar system (Hubbard 1984, Chap. 4). These numbers provide a good starting point for our simulations.

The present rotation of Mars may be, in contrast, much more similar to the planet's primitive rotation. Phobos and Deimos are indeed much smaller than the Moon compared to Earth–Moon system due to the smaller masses of the martian satellites ($m_{Phobos} = 1.47 \times 10^{-7}\ m_{Moon}$ and $m_{Deimos} = 1.47 \times 10^{-8}\ m_{Moon}$). Hence, the rotation period of Mars is essentially unaffected by Phobos and Deimos. The solar tides have changed the rotation rate of Mars by only a couple of minutes over the age of solar system. Tidal friction could have reduced the rotation significantly only if Mars had a larger moon in the past. Large impacts near the end of the accretion phase are likely to determine the initial spin position and rotation rate. These initial conditions remain unknown (Lissauer et al. 2001), and, therefore, one cannot exclude the possibility that Mars has kept the present rotation rate throughout its past. In the Noachian (3.5–4.5 Gyr ago), the planet probably had a magnetic field, a more massive atmosphere and a wetter climate. During the same period, Mars suffered heavy impact bombardment, and creation of a large impact basin could have important consequences on the climate as well as on the rotation. Geophysical processes such as volcanic events and uplifts, construction of the large Tharsis province or of giant impact basins, or formation of mass anomalies associated with mantle convection could also have affected the rotation rate of Mars. Recently discovered geological evidences point towards important climatic changes as well (Head et al. 2003, 2005, 2006; Neukum et al. 2004). Mars present obliquity is similar to that of the Earth, but it has been probably much larger in the past (Laskar et al. 2004b; see also Laskar 1994, 1996, 1997; Laskar et al. 2002, 2004a). Climate models have predicted that at high obliquities ($>45°$, polar ice may sublimate rapidly from the poles and be re-deposited in the tropics, resulting in periodic large-scale surface mass redistributions (Levrard et al. 2004; Forget et al. 2006). However, the effects of climate change-induced surface ice redistribution on the rotation are likely to be small. Zuber and Smith (1999) showed that if the whole north polar deposits (with an estimated volume of $1.2 \times 10^6\ km^3$) (Zuber et al. 1998) were removed, the change in Mars' mean moment of inertia would be $\Delta I/I = 3.88 \times 10^{-6}$. Consequently, it is difficult to explain any diminution in the rotation rate of Mars throughout its past unless a large impact scenario (such as the one formed Hellas basin 4 Gyr ago) is considered.

 Springer

3 Obliquity, Insolation and Atmosphere

Insolation (the flux emitted by the Sun and received at the surface of the planet) and obliquity (the angle between the equator and the orbit plane of a planet) are essential for the understanding of the past climate of the Earth and Mars. The insolations and obliquities of planets are changing with time as gravitational interaction between solar system bodies causes their orbits and spin axes to evolve. The long-term orbital and spin dynamics of the planets depends also on internal planetary processes, such as internal dissipation, interactions of the solid planet with its fluid layers (atmosphere, ocean or core) and deformation of the planet as a response to time-variable external forcing and surface loads. Besides their apparent effect on the climate, variations of orbital and spin parameters (more precisely the precession, see below) could possibly also affect the planetary dynamo, although there is no consensus as explained in the next section.

The present-day obliquity and rotation period of Mars and the Earth are similar, and daily and seasonal insolation variations are therefore comparable. The seasonal insolation variations on Mars are larger due to the about five times larger eccentricity of Mars compared to the Earth, which implies a 40% difference between the solar flux received at perihelion and aphelion. Moreover, the seasonal and diurnal temperature variations compared to the insolation variations are relatively larger for Mars than for the Earth due to the presence of oceans and a larger atmosphere on Earth. Although the present-day obliquity and rotation period of Mars and the Earth are similar, the long-term spin variations of the Earth and Mars differ substantially. Due to its closer distance to the Sun and the presence of a large moon, the rotation rate of the Earth has changed much more than that of Mars as a result of tidal dissipation (see Laskar and Joutel 1993; Laskar and Robutel 1993, and the next section). Mars' rotation rate can be considered as close to primordial. Due to the presence of a large moon, Earth's obliquity has remained in the range of 22.1–24.5° over the last 18 Myr (Laskar et al. 1993a, 1993b). Mars' obliquity shows much larger variations: during the last 20 Myr, the obliquity of Mars is statically likely to have varied between about 10° and 45° (Laskar and Robutel 1993; Laskar et al. 2004a). Figure 1 presents the variations of the eccentricities, obliquities, and insolations of Mars and the Earth as a function of time. The present-day values of these parameters are summarized in Table 1

The Earth owes its spin axis stability to the lunar torque, which decreases its precession period from 8.1×10^4 to 2.6×10^4 years (Ward 1973). As a result, the motion of the spin axis is much faster than the motion of the orbit normal, and the spin axis follows the instantaneous orbit pole, keeping the obliquity nearly constant. For Mars, the precession period of the spin axis is close to periods of slow secular changes in its orbit, and large chaotic obliquity variations can occur as a result of this secular resonance overlap. A Moon-like contribution to Mars' precession would not be stabilizing, but would instead force it deeper into the spin–orbit resonance (Laskar et al. 2004a). Over the entire history of the Solar System, only statistical analyses can be made because of the mathematically chaotic nature of the system. Evaluated over 4 Gyr, the averaged value of the obliquity of Mars is estimated as 37.62°, suggesting that the present obliquity is uncharacteristically low and that the present climate is not representative for the martian climate. In the distant future (some billions of years), as the Moon continues to evolve outward from the Earth due to the tides, a resonance is likely to occur. This will drag the Earth into the chaotic zone, and its obliquity will no longer be stable (Ward et al. 1979; Ward 1982, 1992; Ward and Rudy 1991; Tomasella et al. 1996; Néron de Surgy and Laskar 1997).

Table 1 The current orbital parameters of Mars and Earth

	Mars	Earth
Inclination, I	1°66	1°58
Eccentricity	0.093	0.017
Obliquity, θ	25.2°	23.5°
Precession of spin axes	173 kyr	26 kyr

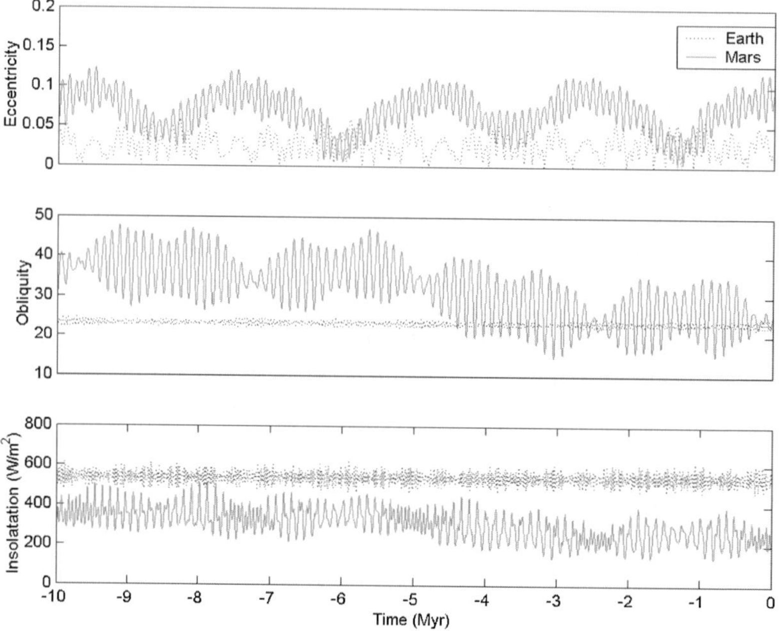

Fig. 1 Variations of the Martian orbital parameters and of the Martian North Pole summer insolation over the last 10 Myr (from the calculations of Laskar et al. 2002, 2004a)

4 Rotational Effect on Magnetic Dynamo—Expected Stronger Magnetic Field

In this section, we examine the effect of rotation on dynamo strength and in particular the effect of rapid rotation.

4.1 Magnetic Field Conditions

The magnetic field of the Earth is driven by dynamo action in a fluid, conducting core. In a hydrodynamo, an electrically conducting material moves through a magnetic field, which generates a current, creating an additional magnetic field. After an initial magnetic field has started the dynamo, the induced magnetic field becomes self-sustaining, and requires no further outside input. Dynamo action is believed to be the cause of global scale planetary magnetic fields. What a planet needs in order to produce a magnetic field is a liquid, conducting interior in motion. This would be the case in particular when rotation or large scale motion occurs in a large region (where the motion is produced) (Stevenson 1983, 2003; Stevenson et al. 1983).

While there is no clear evolution scenario that can be applied to all the planets, the following sequence of events for Mars (Stevenson 2001) is the most probable. An iron-rich liquid core formed during or rapidly after the hot accretion of Mars at about 4.5 billion years (4.5 Gyr) ago by gravitational separation of mostly liquid, immiscible iron from the partly solid silicates and subsequently cooled at a rate dictated by the overlying mantle. To form a liquid core, the energy of gravitational formation of the planet is converted into heat that is sufficient to heat Mars' formative material to several thousand degrees above the melting point. Even with the loss of heat by radiation, a magma ocean is likely (Stevenson 2001). The liquid iron then accumulates at the bottom of the magma ocean and descends to the planet's center as large blobs, by Stokes flow (Navier–Stokes equations for a highly viscous fluid with the inertial force equals to zero), through the possibly viscous and high-pressure deep mantle. After core formation, dynamo action is started due to large-scale motion of a liquid iron alloy inside the core. These large-scale motions can be produced by two different mechanisms that are believed to appear subsequently:

(1) *Thermal convection*: When conduction inside the core is not sufficient to evacuate the heat out of the core, motions in the iron alloy appear; this case corresponds to a cooling core that is superheated with respect to the mantle. Such a superheated core is the consequence of a rapid and vigorous core formation process. As long as the core can cool sufficiently (i.e., the heat flow out of the core is larger than the heat flow along the core adiabat-critical heat flow) electric currents caused by the thermal convection in the core can generate a magnetic field. Such a thermal dynamo is in general only active during the first few hundred million years of planetary evolution because the core heat flow decreases rapidly to the critical heat flow below which heat is transported simply by conduction.

(2) *Compositional convection*: As the consequence of the planet's cooling, the core temperature can reach the liquidus of the iron alloy. At that temperature the iron starts to precipitate and forms an inner core. This precipitation of iron results in motion inside the liquid core and the dynamo is now maintained by compositionally driven convection currents. The dynamo is sustained as long as the core is cooling and the inner core continues to grow sufficiently fast. Eventually, the dynamo system will collapse if the core alloy composition reaches the eutectic composition[1], or if the core has completely solidified. In both cases, compositional convection is not possible anymore. Dynamo action could also fail if the compositionally driven convection is weak.

These scenarios allow a planet to possess a magnetic field early in its history first by thermal convection and later by compositional convection due to the formation of an inner core. It is further possible but not necessary that there is a time gap between the two mechanisms of dynamo generation depending on the cooling efficiency and the liquidus temperature of the core. The core-freezing hypothesis states that by default there must have been a period when the planet's core was partially molten, and that the continuing solidification of the inner core (and the associated compositional convection) would allow the existence of a magnetic field in that period (see Stevenson 2001 for further details on these hypotheses).

4.2 Existence of a Magnetic Field for Mars

For Mars, the observed magnetic field in its early history could arise from both kinds of dynamo generation, thermal or compositional. An early thermal dynamo, however, is favored

---

[1]The eutectic temperature is the lowest temperature at which a mix of two materials will melt or solidify directly. A eutectic mixture of two or more constituents is a mixture that solidifies simultaneously out of the liquid—or melts simultaneously out of the solid—, at a minimum freezing/melting point.

 Springer

by thermal evolution models (e.g. Hauck and Phillips 2002; Breuer and Spohn 2003, 2006; Williams and Nimmo 2004).

4.3 Strength of a Magnetic Field Related to Rotation

An important issue is the strength of the magnetic field generated either by thermal or compositional convection. One concept is based on a force balance between Coriolis and Lorenz forces (Elsasser number, i.e., the ratio of Lorenz forces to Coriolis forces, is equal to unity). As a consequence of this force balance, the magnetic field strength is assumed to be proportional to the rotation rate; the faster a planet rotates, the stronger the magnetic field is. Considering further that the convection velocity is proportional to the size of the region where the motion is taking place, the magnetic field strength increases with the core radius. In addition, dynamo simulations show that the dipole field strength can decrease even further at a critical outer liquid shell thickness (if the solid inner core becomes very large, e.g. Stanley et al. 2005). We, therefore, can suggest that the fast rotation of Earth or of any other planet and the possibility to have large motions in the fluid core at the beginning of its life was a very favorable ingredient for generation of its magnetic field. In other words, one can roughly assume that the magnetic field is proportional to the rotation rate, and if the rotation rate of early Earth was twice as large as today, the magnetic field would be two times larger than now (see below). As summarized by Grießmeier et al. (2005), there are several analytical models (deriving the planetary magnetic field from a dipole moment) relating the planetary magnetic dipole moment M and the rotation ω. The surface magnetic field is proportional to M and inversely proportional to r^3. This relation allows us to express the magnetic field strength either in terms of M or in terms of dipole strength. The models provided in Grießmeier et al. (2005) yield the following scaling laws for the magnetic moment:

$$M \propto \rho_c^{1/2}\omega r_c^4, \tag{1}$$

$$M \propto \rho_c^{1/2}\omega^{1/2} r_c^3 \sigma^{-1/2}, \tag{2}$$

$$M \propto \rho_c^{1/2}\omega^{3/4} r_c^{7/2} \sigma^{-1/4}, \tag{3}$$

$$M \propto \rho_c^{1/2}\omega r_c^{7/2}, \tag{4}$$

where (1) comes from Busse (1976), (2) from Stevenson (1983, or Stevenson et al. 1983) and Mizutani et al. (1992), (3) from Mizutani et al. (1992), and (4) from Sano (1993). The parameters ρ_c stands for the core density, r_c, for the core radius, and σ, for the conductivity.

4.4 Theoretical Computation of the Strength of the Magnetic Field for Mars and the Earth

These models yield proportionality either directly to the rotation as, for instance, for the models of Busse (1976) and Sano (1993), or to the square root of the rotation (model of Stevenson 1983 and Mizutani et al. 1992), or even to the rotation to the power 3/4 (see Mizutani et al. 1992). Thus, the shorter the planetary rotation period is (i.e. the faster its rotation), the larger the resulting magnetic moment is. It must be noted that the models also introduce a dependence on the dynamo region radius (to the power 3, 3.5, or 4) and on the conductivity of the region.

Figures 2 and 3 show the planetary magnetic moment of early Earth and early Mars, as a function of possible early rotation periods. The values used for Figs. 2 and 3 have been deduced from the above scaling laws. We assume the following initial conditions (taken

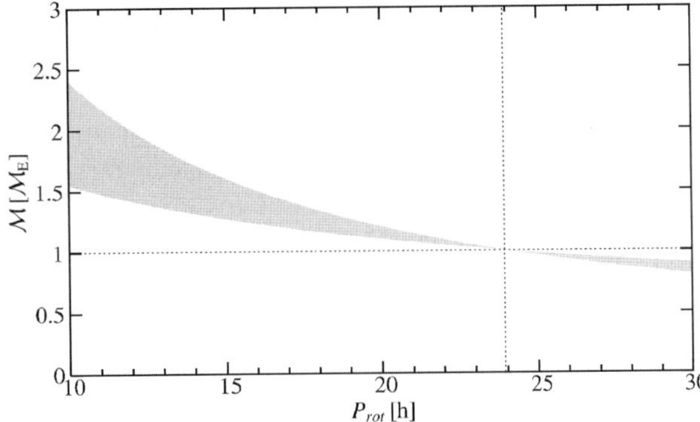

Fig. 2 Estimated magnetic moment of early Earth, relative to its present-day value, as a function of possible early planetary rotation periods in hours. The *dotted vertical line* denotes the present-day sidereal rotation rate of the Earth

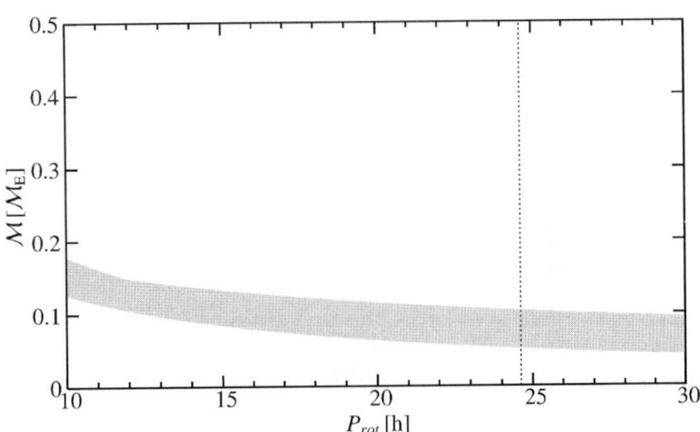

Fig. 3 Estimated magnetic moment of early Mars during the phase when it sustained a dynamo, as a function of possible early planetary rotation periods in hours. The magnetic moment is given in units of the present-day magnetic moment of Earth. The *dotted vertical line* denotes the present-day sidereal rotation rate of Mars

from Cain et al. 1995): (i) the planetary radius for Mars r_{Mars} is 0.53 r_{Earth}, (ii) the planetary density is 7.57 10^3 kg/m^3, (iii) the size of the planetary core is $r_c = 0.52 r_{\mathrm{Mars}}$, and (iv) the conductivity σ is roughly the same as on Earth. A typical estimation of the Earth iron core conductivity is 5×10^5 S m^{-1}, corresponding to a magnetic diffusivity of 1.6 m^2 s^{-1} (Braginsky and Roberts 1995). The resulting magnetic moment is given in units of the Earth's current magnetic moment $M_E = 8 \times 10^{22}$ A m^2. The areas shaded in grey represent the ranges of results obtained by the different magnetic moment scaling laws. For an initial rotation period of 10 hours or more, one finds that the Earth's magnetic moment was at most 2.5 M_E (the Earth's current magnetic moment). The results demonstrate that the magnetic moment was considerably larger than it is today, as a result of the shorter rotation periods (higher rotation rates).

The magnetic moment of ancient Mars is estimated to reach up to 0.2 M_E. It can be seen that the smaller size of the planet and core when compared to the Earth results in a magnetic moment that is a factor of about 10 smaller. Consequently, as expected, for the range of rotation periods used, we always find a magnetic moment of Mars smaller than that of Earth. Also note that these values are consistent with the estimated lower bound of the magnetic field deduced from MGS measurements, i.e. 0.1 to 10 times that of the Earth (see Sect. 1). As will be discussed later on the removal of the martian atmosphere, a magnetic moment equal to or smaller than about 0.1–0.2 M_{Earth} during the first 250 Myr is preferred to explain the evolution of the martian atmosphere, as it would imply strongly increased loss rates. Higher magnetic field values would be a problem for the loss of the martian atmosphere and would consequently provide a too thick present day atmosphere. It

is remarkable that the two absolutely different approaches originating from the two different branches of planetary science (planetary magnetic dynamo theory and atmospheric science) when applied to the same subject—the early martian magnetic moment evaluation, have lead us to almost coinciding conclusions that early Mars may have had during about the 1st 200 Myr an intrinsic magnetic moment not stronger then $\sim 0.2 M_{\text{Earth}}$.

4.5 Strength of a Magnetic Field from Other Phenomena

Although the analytical models considered here predict an increase of magnetic moment with increasing rotation rate, recent numerical experiments indicate that the magnetic moment may be independent of the angular frequency and the electrical conductivity. Christensen and Aubert (2006) and Olson and Christensen (2007) have studied numerically an extensive set of dynamo models in rotating spherical shells, varying all relevant control parameters by at least two orders of magnitude. These simulations have shown that the magnetic field is basically controlled by the buoyancy flux (mass transported per time unit by buoyancy). This result contradicts the previously discussed concept (Figs. 2 and 3). The numerical dynamo simulations done by Christensen and Aubert (2006) fit quite well the magnetic field observations for Earth and Jupiter but cannot explain the magnetic field of Mercury. Future studies are needed to answer the question of the relation between magnetic field strength and its dependence on the rotation rate, electrical conductivity, and buoyancy flux.

A last remark concerns the precession and precessionally-driven dynamo models. The feasibility of a precessionally-driven dynamo has been investigated by Malkus (1968), Stacey (1973), and Loper (1975) for the Earth (see also Rochester 1974; Vanyo 1991). The relative orientation of the angular-velocity vectors of the mantle and core are determined from angular momentum conservation or from interaction torques resulting from pressure, viscous, and magnetic stresses at the core–mantle interface. Loper (1975) finds that the dissipative torques are weaker by a factor of 10^{-4} than those estimated by Malkus (1968) and Stacey (1973), resulting in only a small amount of energy that is, in addition, dissipated in the boundary layers at the core–mantle interface. Thus no energy is avfailable to drive the geodynamo. Other, more recent studies (e.g. Kerswell 1996), however, find no reason to dismiss a precessionally-driven dynamo on energetic grounds. The main problem is that it is unclear how the rapidly varying precessional forcing can generate slow motion in the outer core that would drive the dynamo.

4.6 Observed Strength of the Past Martian Magnetic Field

The detection of Mars surface anomalies by the MAG/ER experiment on MGS (see Sect. 1) indicates the existence of a strong ancient intrinsic martian magnetic moment. There are several ways in which the martian crust could have been magnetized. One of the most probable ways is Thermo-Remnant Magnetization (TRM). With TRM, the magnetization is produced when rock cools below a critical temperature (Curie temperature) in the presence of a magnetizing field. This is an effective mode for producing an intense remnant field. The magnitude of magnetization produced in the rock depends on the strength of the internal field, the mineralogy and the magnetic microstructure. There is a trade-off between the concentration of magnetic carriers and the strength of the magnetic field: The weaker the magnetic field the more magnetic carriers are required to explain an observed magnetization. Assuming an early martian magnetic field similar to the present day Earth

field, the concentration of magnetic carriers might be comparable to the amount in extrusive basalt. Bearing in mind the possible reduction of the remanent magnetization through viscous decay and chemical alterations during this very long geologic time (the magnetization of the oceanic basalt decreases by a factor of 4–5 during the first 20 Myr, largely by chemical alterations, e.g., Bleil and Petersen 1983), the initial magnetization of the martian source bodies must have been even stronger. However, there is ample evidence that FeO content of the martian mantle is about twice that of the Earth's (e.g., Sohl and Spohn 1997; Sanloup et al. 1999). Whether this high concentration of FeO translates to a high concentration of magnetic minerals depends on the oxidation state of the martian mantle and lower parts of the crust. However, for a high concentration of magnetic minerals, the early magnetic field could have been smaller than that of the present Earth field. Due to all these uncertainties, the early martian core field intensity can be assumed to be in a range of a magnetic moment of 0.1–10 times that of present Earth. Note that a factor of 10 is consistent with the scaling laws of Fig. 3.

The majority of the crustal magnetic sources detected by MGS of up to 1500 nT are located in an ancient, densely cratered terrain of the martian highlands, south of the crustal dichotomy boundary (Connerney et al. 1999; Acuña et al. 2001). This dichotomy boundary is the geologic division between the heavily cratered highlands to the south and the relatively young, smooth plains to the north where the martian crust is thinner. No magnetic anomalies were detected in the major martian volcanic areas. The large impacts that formed the Hellas and Argyre basins and that are believed to have formed about 4 Gyr ago are also not associated with magnetic crustal anomalies. The absence of crustal magnetic fields in these areas implies that the martian dynamo had already ceased to operate when these impact basins were formed. This evidence supports most thermal evolution models that consider the magnetic field history.

4.7 Thermal Evolution, Mars Interior, and Mars Magnetic Field

The thermal evolution models assume a hot early Mars immediately after accretion followed by rapid cooling and crust formation (e.g., Schubert and Spohn 1990; Spohn et al. 1998; Breuer and Spohn 2003, 2006; Williams and Nimmo 2004). A thermally driven dynamo during the first few hundred million years is then the consequence of the early rapid cooling if the core is superheated with respect to the mantle. As an alternative, the core cooled rapidly below its melting temperature to start an inner core growth associated with a chemical dynamo. In the case of inner core growth, the concentration of a light constituent like sulfur in a mainly iron core is essential for the temporal evolution of the dynamo. A high core sulfur abundance of 14.2% suggested for Mars from geochemical analyzes of the SNC (Shergottites, Nakhlites, Chassigny) meteorites (meteorites found on Earth and originating from Mars) (e.g., Dreibus and Wänke 1985; McSween 1985) lowers its melting point considerably. Model calculations show that no inner core freeze-out occurs during the first 4.5 Gyr if a sulfur content $>15\%$ is assumed (see Stevenson et al. 1983). If the sulfur content, however, is much lower than 15%, a solid inner core would have formed and grows on a geologic time scale once it begins to freeze out. The lack of a present-day magnetic field can be explained in several ways: (1) the core may be entirely solid, (2) the core may be liquid but is stably stratified and not convecting, and (3) the outer liquid core may be too thin or the compositional convection too weak for an operative dynamo.

Geodetic measurements of the mass, moment of inertia, and Love numbers are hardly compatible with a pure iron core, therefore confirming the presence of a light element. This is also consistent with the inhomogeneous accretion hypothesis in which the volatile sulfur becomes more abundant in the solar nebula as distance increases from the Sun. Since Mars is located in a more distant orbit than any other terrestrial world, it is likely to contain more sulfur. If this is true, then heat-flow convection should have ceased about 4 billion years ago, with no subsequent inner-core solidification. A liquid core is consistent with the large value of the k_2 Love number computed very recently by Konopliv et al. (2006, see also Yoder et al. 2003; Balmino et al. 2006) using MGS and Odyssey tracking data. Indeed, this value (0.152 ± 0.009) is large enough to rule out a completely solid core and shows that at least the outer part of the core must be liquid. Although the mineralogy of Mars is presently poorly constrained, models of the internal structure of Mars built from very different compositions cannot fit this latest estimate of the k_2 Love number without a hot temperature profile in the mantle (Rivoldini et al. 2005, see also Verhoeven et al. 2005). More precisely, the mantle must be associated with a mean temperature greater than at least 1800 K. This estimate does not depend much on the composition of the mantle.

Core solidification, however, cannot be completely ruled out. Indeed the martian mantle, richer in volatiles compared to Earth, may possess a lower viscosity. The effect leads to a more efficient heat transfer from the core through mantle convection, suggesting a rapid cooling of the planet and probably an inner core growth. Nevertheless, the core solidification scenario is much less probable than a totally molten core for various reasons. First, a solidified inner-core has to be large enough so that the remaining fluid region is too thin to sustain a present dynamo (or the compositional convection has become too weak), but also small enough to be compatible with the observed large value of the k_2 Love number. Second, thermal evolution models show that such a compositional convection would most likely be active over a much longer period of time than observed for Mars (Breuer and Spohn 2003; Williams and Nimmo 2004). Third, the quality coefficient of the tide, as deduced from observations for deriving Phobos ephemerides, is about 85 (Bills et al. 2005; Lainey et al. 2007); however, this value cannot be explained easily unless a very hot and highly attenuating mantle is considered (Lognonné and Mosser 1993; Zharkov and Gudkova 1997). Fourth, the absence of plate tectonics on Mars at present suggests that the heat transfer from the interior of the planet is presently less efficient than on Earth. An alternative evolution scenario consistent with a present day inner core but not with a dynamo activity, is that Mars had a change in its heat transport mechanism. Plate tectonics during the first few hundred million years might have cooled the core efficiently to form an inner core. This early period of plate tectonics is then followed by a period without plate tectonic (mono-plate situation), which heats up the interior again as a consequence of inefficient heat transfer. In case the core temperature does not decrease again below the minimum core temperature, a present inner core exists but dynamo activity is unlikely as this inner core cannot further grow. There are no strong geological indications for an early plate tectonics regime (e.g., Solomon 1994; Sleep and Tanaka 1995; Pruis and Tanaka 1995) and furthermore, such a model has some problems explaining the crustal evolution of Mars (Breuer and Spohn 2003).

5 Early Solar Wind (min–max)—Effect on Magnetopause Distances (Atmosphere Protection)

In this section, we consider that the magnetic field of Mars has existed up to 4 Gyr ago. Atmospheric escape processes are considered to begin at that time. The efficiency of thermal and non-thermal atmospheric escape processes depends mainly on the history of the intensity of the solar X-ray and EUV radiation and of the solar wind mass flux. As discussed by

Ribas et al. (2005) and Lundin et al. (2007) observations with the ASCA, ROSAT, EUVE, FUSE and IUE satellites of a homogeneous sample of single nearby Sun-like main sequence stars have revealed age dependent energy emissions (Guinan et al. 2005). The considered stars have known rotation periods and well-determined physical properties, including temperatures, luminosities, metal abundances and ages from 130 Myr to 8.5 Gyr. Their emission intensities ($\lambda = 0.1$–120 nm) were 10 times the present solar value about 4 Gyr ago and about 100 times that of today during the first 100 Myr after the young Sun arrived at the Zero-Age-Main-Sequence (ZAMS[2]). As shown by Kulikov et al. (2006, 2007) the strong XUV emissions of the young Sun had a major effect on the heating and expansion of the upper atmospheres, the ionospheres and, hence, on the evolution of planetary atmospheres.

It is important to note that not only the radiation intensity of the Sun has been changing during its lifetime (Lundin et al. 2007), but also the solar wind mass flux was higher during the active period of the young Sun (Newkirk Jr. 1980; Wood et al. 2002, 2005). For example, 3.9 Gyr ago, the solar wind density was higher by a factor of 16, and the solar wind velocity was approximately twice its current value (see Grießmeier et al. 2004). Hubble Space Telescope high-resolution spectroscopic observations of the H Lyman-α feature of several nearby main-sequence K- and G-type stars carried out by Wood et al. (2002, 2005) have revealed neutral hydrogen absorption associated with the interaction between the stars' fully ionized coronal winds and the partially ionized local interstellar medium. They modeled the absorption features observed in the astrospheres of these stars and provided the first empirically-estimated coronal mass loss rates of main sequence stars with ages younger than that of the Sun. The studies of Wood et al. (2002, 2005) indicate that the young Sun had a much denser solar wind mass flux than today. The correlation between mass loss and X-ray surface flux follows a power law relationship, which indicates an average solar wind density up to 100–1000 times higher than today during the first 100 Myr after the Sun reached the ZAMS. However, recent observations of the absorption signature of the astrosphere of the $\sim$500 Myr old solar-type G star 'ξ Boo' imply that there exists, possibly, a high-activity cutoff time regarding the mass loss in the radiation activity relation (Wood et al. 2005; Lundin et al. 2007). The observations of 'ξ Boo' indicate that the mass loss of that star is about 20 times less than the average stellar mass loss value estimated for 4 Gyr old G-type stars. From this uncertainty for young stars ($<$0.5 Gyr) one has to conclude that the stellar sample of young solar-like stars is at present not large enough to give accurate solar wind mass flux estimations during the first 500 Myr after the Sun arrived at the ZAMS.

The active young Sun might have had a strong influence on the early atmospheres of the Earth and Mars. Taking into account the evolution of the star, Kulikov et al. (2007) showed that the presence of an ancient intrinsic magnetic field on Mars could have had significant consequences for the evolution of the atmosphere, especially by reducing the amount of certain atmospheric constituents lost to space. To investigate the efficacy of the atmospheric protection by the intrinsic magnetic field against the solar wind erosion one has to calculate the subsolar magnetopause stand-off distance R_s by determining the magnetic and solar wind ram pressure balance condition at the subsolar point (e.g., Grießmeier et al. 2004, 2005; Kulikov et al. 2007[3]). It is evident that if the magnetopause stand-off distance moves closer to a planet due to a decrease of its dynamo strength, the solar wind will be able

---

[2] One defines the Zero-Age Main Sequence (ZAMS) as the curve in the Hertzsprung–Russell diagram where stars are at the beginning of hydrogen fusion.

[3] Details of the calculation of the pressure balance between the solar wind plasma and the magnetic pressure of the planetary magnetic field can be found in these references.

Table 2 Magnetopause stand-off distances in martian radii for the minimum ($R_S^{\text{min sw}}$), moderate ($R_S^{\text{mod sw}}$) and maximum ($R_S^{\text{max sw}}$) solar wind mass flux conditions (Wood et al. 2002) and different times after the planet's origin. The initial magnetic moment is assumed to be similar to that of present Earth after the planets origin, but declines quickly, as modelled by Schubert and Spohn (1990)

Stand-off distance	100 Myr (100 XUV)	220 Myr (70 XUV)	280 Myr (50 XUV)
$R_S^{\text{min sw}}$	14.0 r_{Mars}	1.9 r_{Mars}	0.7 r_{Mars}
$R_S^{\text{mod sw}}$	10.8 r_{Mars}	1.3 r_{Mars}	0.4 r_{Mars}
$R_S^{\text{max sw}}$	8.0 r_{Mars}	0.8 r_{Mars}	0.1 r_{Mars}

Here R_S^{min} (R_S^{max}) is the magnetopause stand-off distance corresponding to the minimum (maximum, resp.) solar wind mass flux

Table 3 Magnetopause stand-off distances in Earth radii at 1 AU for the minimum ($R_S^{\text{min sw}}$), moderate ($R_S^{\text{mod sw}}$), and maximum ($R_S^{\text{max sw}}$) solar wind mass flux conditions and a time period corresponding to the faster rotation and for different magnetic moments proportional to the present Earth value

Stand-off distance	1 M_{Earth}	1.5 M_{Earth}	2.5 M_{Earth}
$R_S^{\text{min sw}}$	1.8 r_{Earth}	2.2 r_{Earth}	2.7 r_{Earth}
$R_S^{\text{mod sw}}$	1.1 r_{Earth}	1.4 r_{Earth}	1.8 r_{Earth}
$R_S^{\text{max sw}}$	0.6 r_{Earth}	0.8 r_{Earth}	1.1 r_{Earth}

to penetrate deeper into the upper atmosphere where neutral and ion densities are higher. As a result, atmospheric neutral atoms and molecules can be more rapidly transformed into ions through charge-exchange reactions with solar wind particles and, hence, more planetary ions can be accelerated to higher energies by the solar wind electromagnetic fields, picked-up by the solar wind plasma flow around the planet, and lost into space at a higher rate.

Table 2 shows the expected magnetopause stand-off distances from the planetary surface R_s, estimated for early Mars. These estimates use a strong initial magnetic moment (10 times the moment of the present Earth), declining in accordance with the model of Schubert and Spohn (1990), and the minimum, moderate and maximum expected solar wind mass flux of the young Sun from Wood et al. (2002).

As one can see from Table 2, under these assumptions the sub-solar magnetopause stand-off distance 4.5 Gyr ago (i.e., at 100 Myr) for early Mars would be comparable with that of present Earth ($\sim$10 r_{Earth}). Kulikov et al. (2007) showed that in such a case early Mars would lose a negligible amount of the atmosphere. Note that the magnetic moment decreases during the first 280 Myr, but the solar wind mass flux is higher than at present, therefore, the magnetopause stand-off distance is closer to the planet at 280 Myr than at 100 Myr. By assuming a weaker magnetic moment of about 10% that of the present Earth, one obtains sub-solar magnetopause stand-off distances of about 1.9, 1.3 and 0.8 r_{Mars} at 100 Myr after the Sun arrived at the ZAMS, for minimum, moderate, and maximum solar wind mass flux of Wood et al. (2002), respectively. The ion pick-up loss calculations by Kulikov et al. (2007) show that in the case of a weak early magnetic field (about 10% of M_{Earth}) or a late onset of the martian dynamo (after the first 200 Myr) a denser early atmosphere of Mars could have been lost, including from several bar to a few tens of bar of oxygen produced by photolysis of atmospheric water vapor.

We calculated (Table 3) the magnetopause standoff distance for Earth at 1 AU with the magnetic moments related to a faster rotation period of early Earth (Fig. 2).

In this computation we considered only the extremely early period during the first 100 Myr after the Earth's origin or the Sun's arrival at the ZAMS. The sub-solar magnetopause distance for present Earth is at about 10 Earth-radii and loss due to solar wind erosion is negligible. The table shows that the sub-solar magnetopause stand-off distance for early Earth was smaller and that rotation tends to increase it to larger distances.

Models of core evolution are consistent with both an early magnetic field occurrence and a late onset of the magnetic field (Connerney et al. 2004). Before the dynamo was on or once the intrinsic field had decayed, Mars could lose atmosphere due to non-thermal atmospheric loss processes (e.g., Lammer et al. 2003; Chassefière and Leblanc 2004; Kulikov et al. 2007). The onset of the dynamo is known from the post-dates of the youngest observed impact basins on Mars or of post-emplacement of crustal modification on any acquired magnetic remanance (Acuña et al. 2001; Connerney et al. 2004).

6 Effect of the Early Martian Magnetic Field on the Atmospheric Loss

The present thin martian atmosphere with a surface pressure of about 7–10 mbar is one of the great puzzles in our Solar System. Ancient fluvial networks (for example) on the surface of Mars suggest that the planet was warmer and wetter at least 4 billion years ago (see Bertaux et al. 2007, this issue), although this water could have been only present at intermittent periods (Baker 1997, 1999, 2000). Surface features resembling massive outflow channels provide evidence that the martian crust contained a planet wide reservoir of H_2O equivalent to a several hundred meters deep global ocean (Carr 1987, 1996). There are several possibilities for the fate of early H_2O and CO_2: they exist as ice and/or liquid somewhere on or within the planet or within hydrated minerals (in case of water), they have been lost to space, or a combination of the above.

The evolution of the martian atmosphere, with regard to water, is influenced by non-thermal atmospheric loss processes of heavy atmospheric constituents. Since Mars does not have an appreciable intrinsic magnetic field at present and has a comparatively small gravitational acceleration, all known atmospheric loss processes are at work, and several important atmospheric constituents, namely H, H_2, N, O, C, CO, O_2 and CO_2 have been continuously lost from the atmosphere (e.g. Lammer et al. 2003 and references therein, Lundin et al. 2007; Kulikov et al. 2007).

After the young Sun arrived at the ZAMS, heavy noble gasses may have been hydrodynamically fractionated during the accretion phase of the planet to their present composition, with corresponding depletions and fractionations of lighter primordial atmospheric species (Zahnle and Walker 1982, 1990; Pepin 1994; Donahue 1995; Chassefière 1996; Donahue 2004). Subsequently the CO_2 pressure history and the isotopic evolution of atmospheric species during this early period were determined by an interplay between impact erosion and impact delivery, carbonate precipitation, outgassing and carbonate recycling, and perhaps also by feedback stabilization under greenhouse conditions. This period was also influenced by thermal and non-thermal atmospheric loss processes which depended partly on the time of the onset of the martian magnetic dynamo, its field strength, the decrease-time of the magnetic dynamo and the radiation and particle environment of the young Sun (Kulikov et al. 2007).

Thermal atmospheric escape occurs when atoms move upward with velocities greater than the escape velocity to an atmospheric level, where the collision probability is low (i.e. the critical level or exobase).

Nearly all of the non-thermal atmospheric escape mechanisms involve ions. Many of these mechanisms, such as charge exchange, dissociative recombination and sputtering processes release energy, which appears in the form of excitation and kinetic energy of neutral products.

The kinetic energy of these newly born hot atoms in some cases is of the order of several electron volts. If the starting altitude for escape is close to the exobase or above it, then the escaping fraction of hot particles is about 1, corresponding to atmospheric levels where the collision probability is low. Thus, nearly all atoms with energies greater than the escape energy and having thermal velocity directed above the local horizon can escape. A strong intrinsic magnetic field similar to that of Earth would protect the planet from most of the eroding effect of the solar-wind on the upper atmosphere. Without such magnetic shielding, the solar-wind transfers momentum to atoms and ions on high ballistic trajectories, and they can be swept away with the solar-wind. Hutchins and Jakosky (1997) have found that even a weak magnetosphere can significantly decrease the ion production and the resulting non-thermal atmospheric loss rates by about orders of magnitude.

Another important effect of the ancient magnetic field and a much denser atmosphere is the shielding of the martian surface from cosmic rays and UV radiation (Molina-Cuberos et al. 2001; Rontó et al. 2003). Cosmic rays, which consist of charged particles, are deflected by the magnetic field depending on their energy, and only high energy particles are able to penetrate and reach the surface. UV radiation can also be effectively absorbed in a dense planetary atmosphere. By investigating surface protection during the history of the martian atmosphere one can see that the atmospheric conditions on Mars may have been comparable to those on Earth about 4.5 Gyr ago. Given that life on Earth may have appeared as early as 3.5 billion years ago, life forms may also have developed on early Mars, under those favorable atmospheric conditions.

Recent studies of the H_2O loss from early Mars (Lunine et al. 2003) show that tens of bar of hydrogen could have been lost by hydrodynamic escape and from several bar to tens of bar of oxygen could remain in the early martian atmosphere. This remaining oxygen should have been lost from Mars due to surface weathering (Lammer et al. 2003), oxidation of carbon compounds (Rosenqvist and Chassefière 1995), and atmospheric escape processes during the planet's history. One should note that this oxygen could have contributed to the total surface pressure on early Mars as long as it remained in the atmosphere. Once the oxygen was lost to space or the surface, outgassed atmospheric species like CO_2 and N_2 determined the surface pressure value.

Given calculated oxygen solar wind-induced loss rates several studies find that an amount of water equivalent to a global ocean of 50 meters deep has been lost over the last 3.5 billion years (Luhmann et al. 1992; Kass and Yung 1996). More recently Johnson and Leblanc (2001) used a 3-D Monte-Carlo model to calculate the sputter oxygen loss between 4 Gyr ago and present time and obtained lower escape rates. Lammer et al. (1996) used the hydrogen loss rate as the limiting factor for the loss of water from the early martian atmosphere with its higher exospheric temperatures. According to this model, maximum loss of H_2O to space over the past 3.5 billion years is equivalent to a global ocean with a depth of less than 10 meters. Lammer et al. (2003) studied solar wind pick-up loss rates of oxygen ions with a numerical test-particle model during the past 3.5 Gyr on Mars. The authors included the stellar XUV flux data from observations of solar proxies (Ribas et al. 2005; Lundin et al. 2007) and the moderate solar wind mass flux obtained from absorption measurements of hot hydrogen gas surrounding stars in Hubble Space Telescope spectra by Wood et al. (2002). By adding the loss of atomic oxygen, which can directly escape form

the martian upper atmosphere via dissociative recombination (Luhmann 1997) and sputtering of atomic oxygen, Lammer et al. (2003) estimate an average total loss of H_2O over the past 3.5 Gyr equivalent to a global ocean with a depth of only $\sim$12 m. This result is in good agreement with recent 2-D and 3-D global hybrid non-thermal ion loss simulations by Terada et al. (2007), as well as with those of Kulikov et al. (2007).

The original amount of H_2O is estimated from geological evidence (e.g., size and number of ancient fluvial beds, see Carr 1987, 1996; Bertaux et al. 2007, this issue). An approximate amount of a 0.5–1 km deep global ocean of ancient liquid water is generally suggested. Observational evidence for the subsequent escape comes from the analysis of the D/H ratio in the martian atmosphere and in the Zagami SNC meteorite (see, e.g., Donahue 1995). Isotope studies suggest that about 4–10 meters of H_2O were lost to space over the past 3.5 Gyr, leaving today a reservoir of crustal water in the form of ice and permafrost equivalent to a global ocean of several tens of meters. A large uncertainty still remains about the present inventory, because considerable extrapolation is involved and because a massive escape (hundreds meters of water) due to impact erosion and/or hydrodynamic escape is hypothesized in the first 0.5–1 Gyr to reconcile observations (Carr 1987).

The recent 2-D and 3-D model simulations and more realistic input parameters discussed above have resulted in significantly lower water loss rates than the previous, simpler calculations. The more recently estimated loss equivalent to a 10–12 m global ocean is in good agreement with the 4–10 m loss from the isotopic ratios analysis. It is important to note that the total, integrated oxygen mass loss calculated using the above models should not affect the surface pressure during the martian history, because the majority of these oxygen atoms and molecules have most likely originated from photolysis of atmospheric H_2O vapor, not from atmospheric CO_2.

Impact erosion (Walker 1986; Brain and Jakosky 1998; Melosh and Vickery 1989) and late impact accretion (Chyba 1990) are other processes which could have played an important role in the evolution of the early martian atmosphere. However, large impacts not only erode the atmosphere but also stimulate outgassing and might also deliver volatiles to a planet. As a result it is difficult to model the net effect on atmosphere production and loss. Future studies should focus on the net impact production and loss of atmosphere during the heavy bombardment period. The calculations of thermal and non-thermal loss for early Mars should take into account the atmospheric erosion and production rates due to asteroid and cometary impacts.

We have shown that loss of atmospheric species to space has played an important role in the evolution of the martian atmosphere. The detection of surface magnetic anomalies by MGS, which implies the existence of an ancient intrinsic magnetic field, may have important consequences for various non-thermal escape mechanisms in the past. Extrapolation of the results of the MGS mission also indicates that the martian atmosphere has been unprotected by its magnetosphere since 4 Gyr ago. At the same time a strong ionizing flux of the early Sun might have produced very important atmospheric sputtering, as suggested by calculations using early solar ionizing flux and solar-wind models.

An ancient intrinsic magnetic field could protect the atmosphere from sputtering and solar wind-induced loss in several ways. First, the magnetic field deflects the solar-wind around the planet and limits the ion production rate in the upper atmosphere by eliminating solar-wind induced ionization processes. Therefore, loss of atmospheric ions and atoms by pick-up is minimized during Earth-like magnetic field periods. Secondly, a strong intrinsic magnetic field would be a barrier for atmospheric collision sputtering processes too. Only loss of neutral O and N atoms originating from dissociative recombination and upward flowing ionospheric CO_2^+ and O_2^+ ions over the martian polar caps of the magnetosphere would be important (Kar 1990; Lammer and Bauer 1992;

Lammer et al. 1996). From our simulations using a rotation rate similar to that of Earth, we found that if early Mars had a time period when its magnetic moment was larger than 0.1–$0.2\ M_{Earth}$, the magnetosphere would have efficiently protected the atmosphere against sputtering and solar wind interaction processes. A strong planetary field would have shielded ions produced in the upper atmosphere from capture by the solar-wind magnetic field. As a result, loss of atmospheric ions and neutral atoms by direct sweeping and/or collisional sputtering would be strongly slowed down.

7 Conclusions

We have considered a strong magnetic field of early Earth and Mars in the scenarios described above. A faster rotation of a planet during its early history might provide a means of creating a strong field. A strong magnetic field is believed to be efficient in substantially reducing planetary atmospheric loss during the most critical early period of solar evolution. For Mars, the shut down of the dynamo has been estimated from spacecraft observations to be at about 4 Gyr ago.

According to a first scenario, the initial martian CO_2 surface pressure could have been several bars, which is sufficient for having liquid water stable on the surface over long time periods in spite of ever present atmospheric loss processes. However, model simulations of atmospheric escape for Mars from the present time till the end of the heavy bombardment about 3.8 Gyr ago show that it was very difficult to remove a dense, initial CO_2 atmosphere during that time period. The large amount of CO_2 in the early atmosphere would have produced a higher initial surface pressure than today.

For a scenario wherein the upper atmosphere of Mars has been protected during the first 200–250 Myr after the planet's origin by a strong early magnetic field comparable in magnitude to the present time Earth's field, our study suggests a more dense atmosphere on present Mars. In this case, non-thermal loss processes would not be efficient enough to decrease the early surface pressure to the presently observed value of about 7 mbar. On the other hand, a much weaker early martian magnetic field (<10–20% that of the present Earth), or a late onset of the dynamo after the first 200 Myr, in combination with high exobase temperatures of more than 1000 K, could have been responsible for erosion of a much denser early atmosphere. Our estimates of the strength of the early martian magnetic field obtained by using analytical scaling relations between the planetary magnetic dipole moment, planet's rotation rate, and other parameters show that its ancient magnetic field could indeed have been that weak. However, as a consequence of the weak internal field, a strong concentration of magnetic carriers is required to explain the observed strong remnant magnetization.

A denser ancient atmosphere during the first 300 Myr after the planet's origin would be very effective in shielding the martian surface from a harmful impact of the solar UV radiation due to photoabsorption processes. Generation of an early magnetosphere of Mars could provide direct shielding of the atmosphere by deflecting solar wind away from the planet, and therefore, could play an important role in the formation of a habitable environment on the planet during an early phase of its evolution.

The existence of a wet period in the early history of Mars and the protection of the atmosphere by the magnetic field (which maintains a pressure high enough for liquid water to exist at the surface of Mars) are fully supported by observations of the recent spacecrafts. The present thin atmosphere of Mars makes it necessary to invoke a process for the escape of the atmosphere. The amplitude of the atmospheric pressure in the early history of Mars

or equivalently the quantity of atmosphere that has escaped are still unknown. Volcanism and meteorite impacts may have play an important role in the escape process and need to be further quantified.

Acknowledgements   We would like to thank Kathryn Fishbaugh as well as an anonymous reviewer for carefully reading the manuscript and for their helpful comments. The work of some of the authors (VD, OK, OV, TV) was financially supported by the Belgian PRODEX program managed by the European Space Agency in collaboration with the Belgian Federal Science Policy Office, as well as (for OV) by an Action-1 project financed by the Belgian Federal Science Policy Office.

References

M.H. Acuña, J.E.P. Connerney, P. Wasilewski, R.P. Lin, K.A. Anderson, C.W. Carlson, J. McFadden, D.W. Curtis, D. Mitchell, H. Reme, C. Mazelle, J.A. Sauvaud, C. d'Uston, A. Cros, J.L. Medale, S.J. Bauer, P. Cloutier, M. Mayhew, D. Winterhalter, N.F. Ness, Science **279**(5357), 1676–1680 (1998)

M.H. Acuña, J.E.P. Connerney, P. Wasilewski, R.P. Lin, D. Mitchell, K.A. Anderson, C.W. Carlson, J. Mc-Fadden, H. Reme, C. Mazelle, D. Vignes, S.J. Bauer, P. Cloutier, N.F. Ness, J. Geophys. Res. **106**(E10), 23,403–23,418 (2001). doi:10.1029/2000JE001404

J. Arkani-Hamed, J. Geophys. Res. **109**(E9) (2004). CiteID:E09005, doi:10.1029/2004JE002265

J. Arkani-Hamed, J. Dyment, J. Geophys. Res. **101**, 11,401–11,425 (1996)

V. Baker, The early martian climate was episodically warm and wet, LPI meeting on Early Mars, extended abstract 3042 (1997)

V. Baker, The MEGAOUTFLO hypothesis for long-term environmental change on Mars, invited talk. 31st meeting DPS/AAS, extended abstract 3901 (1999)

V.R. Baker, R.G. Strom, J.M. Dohm, V.C. Gulick, J.S. Kargel, G. Komatsu, G.G. Ori, J.W. Rice Jr., Mars= oceanus borealis, ancient glaciers, and the MEGAOUTFLO hypothesis. LPSS 31st meeting, extended abstract 1863 (2000)

G. Balmino, J. Duron, J.C. Marty, Ö. Karatekin, in: *Proc. IAG-IAPSO-IABO General Assembly on 'Dynamic planet'* (Cairns, 2006), pp. 895–902

Bertaux et al., Space Sci. Rev., this issue (2007). doi:10.1007/s11214-007-9193-3

C.M. Bertka, Y. Fei, Earth Planet. Sci. Lett. **157**(1–2), 79–88 (1998). doi:10.1016/S0012-821X(98)00030-2

B.G. Bills, G.A. Neumann, D.E. Smith, M.T. Zuber, J. Geophys. Res. **110**(E7) (2005) CiteID:E07004, doi:10.1029/2004JE002376

U. Bleil, N. Petersen, Nature **301**, 384–388 (1983)

S.I. Braginsky, P.H. Roberts, Geophys. Astrophys. Fluid Dyn. **79**, 1–97 (1995)

D.A. Brain, B.M. Jakosky, J. Geophys. Res. **103**, 22,689–22,694 (1998)

D. Breuer, T. Spohn, J. Geophys. Res. **108**(E7), 8–11 (2003), CiteID:5072, doi:10.1029/2002JE001999

D. Breuer, T. Spohn, Planet. Space Sci. **54**(2), 153–169 (2006)

F.H. Busse, Phys. Earth Planet. Interiors **12**, 350–358 (1976)

J.C. Cain, P. Beaumont, W. Holter, Z. Wang, H. Nevanlinna, J. Geophys. Res. **100**(E5), 9439–9454 (1995)

M.H. Carr, Nature **326**, 30–35 (1987)

M.H. Carr, *Water on Mars* (Oxford University Press, Oxford, 1996) 229 pp.

E. Chassefière, Icarus **124**, 537–552 (1996)

E. Chassefière, F. Leblanc, Planet. Space Sci. **52**, 1039–1058 (2004)

U.R. Christensen, J. Aubert, Geophys. J. Int. **166**(1) 97–114 (2006)

C.F. Chyba, Nature **343**, 129–133 (1990)

J.E.P. Connerney, M.H. Acuña, P.J. Wasilewski, N.F. Ness, H. Rème, C. Mazelle, D. Vignes, R.P. Lin, D.L. Mitchell, P.A. Cloutier, Science **284**(5415), 794–798 (1999). doi:10.1126/science.284.5415.794

J.E.P. Connerney, M.H. Acunã, N.F. Ness, T. Spoh, G. Schubert: Space Sci. Rev. **111**(1), 1–32 (2004). doi:10.1023/B:SPAC.0000032719.40094.1d

J.O. Dickey, P.L. Bender, J.E. Faller, X.X. Newhall, R.L. Ricklefs, J.G. Ries, P.J. Shelus, C. Veillet, A.L. Whipple, J.R. Wiant, J.G. Williams, C.F. Yoder, Science **265**(5171) 482–490 (1994)

T.M. Donahue, Nature **374**, 432–434 (1995)

T.M. Donahue, Icarus **167**, 225–227 (2004)

G. Dreibus, H. Wänke, Meteoritics **20**, 367–381 (1985)

J. Duron, G. Balmino, J.C. Marty, P. Rosenblatt, S. Le Maistre, V. Dehant, Icarus (2007, in preparation)

D.A. Evans, N.J. Beukes, J.L. Kirschvink, Nature **386**, 262–266 (1997)

W.M. Folkner, C.F. Yoder, D.N. Yuan, E.M. Standish, R.A. Preston, Science **278**(5344), 1749 (1997)

F. Forget, R.M. Haberle, F. Montmessin, B. Levrard, J.W. Head, Science **311**, 368–371 (2006)

M. Greff-Lefftz, H. Legros, Science **286**(5445), 1707–1709 (1999)

J.M. Grießmeier, A. Stadelmann, U. Motschmann, N.K. Belisheva, H. Lammer, H.K. Biernat, Astrobiology **5**(5), 587–603 (2005)

J.M. Grießmeier, A. Stadelmann, T. Penz, H. Lammer, F. Selsis, I. Ribas, E.F. Guinan, U. Motschmann, H.K. Biernat, W.W. Weiss, Astron. Astrophys. **425**, 753–762 (2004)

E.F. Guinan, M. Güdel, M. Audard, Astrophys. J. **622**, 680–694 (2005)

R.B. Hargraves, D.W. Collinson, R.E. Arvidson, P.M. Cates, J. Geophys. Res. **84**, 8379–8384 (1979)

R.B. Hargraves, D.W. Collinson, R.E. Arvidson, C.R. Spitzer, J. Geophys. Res. **82**, 4547–4558 (1977)

S.A. Hauck, R.J. Phillips, J. Geophys. Res. **107**(E7) (2002). doi:10.1029/2001JE001801

J.W. Head, D.R. Marchant, M.C. Agnew, C.I. Fassett, M.A. Kreslavsky, Earth Planet. Sci. Lett. **241**, 663–671 (2006)

J.W. Head, J.F. Mustard, M.A. Kreslavsky, R.E. Milliken, D.R. Marchant, Nature **426**, 797–802 (2003)

J.W. Head, G. Neukum, R. Jaumann, H. Hiesinger, E. Hauber, M. Carr, P. Masson, B. Foing, H. Hoffmann, M. Kreslavsky, S. Werner, S. Milkovich, S. van Gasselt, Nature **434**, 346–351 (2005)

W.B. Hubbard, *Planetary Interiors* (Van Nostrand Reinhold Co., New York, 1984)

K.S. Hutchins, B.M. Jakosky, J. Geophys. Res. **102**, 9183–9189 (1997)

R.E. Johnson, F. Leblanc, Planet. Space Sci. **49**, 645–656 (2001)

D.M. Jurdy, M. Stefanick, J. Geophys. Res. **109**(E10) (2004). CiteID:E10005, doi:10.1029/2004JE002277

J. Kar, Geophys. Res. Lett. **17**(2), 113–115 (1990)

D.M. Kass, Y.L. Yung, Science **274**, 1932–1933 (1996)

R.R. Kerswell, J. Fluid Mech. **321**, 335–370 (1996)

J.L. Kirschvink, E.J. Gaidos, L.E. Bertani, N.J. Beukes, J. Gutzmer, L.N. Maepa, R.E. Steinberger, Proc. Natl. Acad. Sci. USA **97**, 1400–1405 (2000)

A.S. Konopliv, C.F. Yoder, E.M. Standish, Y. Dah-Ning, W.L. Sjogren, Icarus **182**, 23–50 (2006)

Z. Kopal, Astrophys. Space Sci. **16**, 3–51 (1972)

Yu.N. Kulikov, H. Lammer, H.I.M. Lichtenegger, T. Penz, D. Breuer, T. Spohn, R. Lundin, H.K. Biernat, Space Sci. Rev., this issue (2007). doi:10.1007/s11214-007-9192-4

Yu.N. Kulikov, H. Lammer, H.I.M. Lichtenegger, N. Terada, I. Ribas, C. Kolb, D. Langmayr, R. Lundin, E.F. Guinan, S. Barabash, H.K. Biernat, Planet. Space Sci. **54**(13–14), 1425–1444 (2006)

E.P. Kvale, H.W. Johnson, C.P. Sonett, A.W. Archer, A. Zawistoski, J. Sediment. Res. **69**, 1154–1168 (1999)

V. Lainey, V. Dehant, M. Paetzold, Astron. Astrophys. **465**(3), 1075–1084 (2007). doi:10.1051/0004-6361:20065466

H. Lammer, S.J. Bauer, J. Geophys. Res. **97**(E12), 20,925–20,928 (1992)

H. Lammer, H.I.M. Lichtenegger, C. Kolb, I. Ribas, E.F. Guinan, R. Abart, S.J. Bauer, Icarus **106**, 9–25 (2003)

H. Lammer, W. Stumptner, S.J. Bauer, Geophys. Res. Lett. **23**, 3353–3356 (1996)

J. Laskar, Astron. Astrophys. **287**(1), L9–L12 (1994)

J. Laskar, Celest. Mech. Dyn. Astron **64**(1/2), 115–162 (1996)

J. Laskar, Astron. Astrophys. **317**, L75–L78 (1997)

J. Laskar, A.C.M. Correia, M. Gastineau, F. Joutel, B. Levrard, P. Robutel, Icarus **170**, 343–364 (2004a)

J. Laskar, F. Joutel, Celest. Mech. **57**, 293–294 (1993)

J. Laskar, F. Joutel, F. Boudin, Astron. Astrophys. **270**, 522–533 (1993b)

J. Laskar, F. Joutel, P. Robutel, Nature **361**, 615–617 (1993a)

J. Laskar, B. Levrard, J.F. Mustard, Nature **419**(6905), 375–377 (2002)

J. Laskar, P. Robutel, Nature **361**, 608–612 (1993)

J. Laskar, P. Robutel, F. Joutel, M. Gastineau, A.C.M. Correia, B. Levrard, Astron. Astrophys. **428**, 261–285 (2004b)

B. Levrard, F. Forget, F. Montmessin, J. Laskar, Nature **431**(7012), 1072–1075 (2004). doi:10.1038/nature03055

J.J. Lissauer, E.V. Quintana, E.J. Rivera, M.J. Duncan, Icarus **154**(2), 449–458 (2001)

P. Lognonné, B. Mosser, Surv. Geophys. **14**, 239–302 (1993)

D.E. Loper, Phys. Earth Planet. Interiors **11**(1), 43–60 (1975)

J.G. Luhmann, J. Geophys. Res. **102**, 1637 (1997)

J.G. Luhmann, R.E. Johnson, M.H.G. Zhang, Geophys. Res. Lett. **19**(21), 2151–2154 (1992)

R. Lundin, H. Lammer, I. Ribas, Space Sci. Rev. this issue (2007). doi:10.1007/s11214-007-9176-4

J.I. Lunine, J. Chambers, A. Morbidelli, L.A. Leshin, Icarus **165**, 1–8 (2003)

G.J.F. MacDonald, Rev. Geophys. **2**, 467–541 (1964)

W.V.R. Malkus, Science **160**(3825), 259–264 (1968)

H.Y. McSween Jr., Rev. Geophys. Space Phys. **23**, 391–416 (1985)

V.A. Melezhik, M.M. Filippov, A.E. Romashkin, Geochimica Cosmochimica Acta **67S**, 286 (2003)

H.J. Melosh, A.M. Vickery, Nature **338**, 487–489 (1989)

D.L. Mitchell, R.P. Lin, C. Mazelle, H. Rème, P.A. Cloutier, J.E.P. Connerney, M.H. Acuña, N.F. Ness, J. Geophys. Res. **106**(E10), 23,419–23,428 (2001)

D.L. Mitchell, R.P. Lin, H. Rème, M.H. Acuña, P.A. Cloutier, N.F. Ness, Bull. Am. Astron. Soc. **31**, 1584 (1999)

H. Mizutani, T. Yamamoto, A. Fujimura, Adv. Space Res. **12**, 265–279 (1992)

G.J. Molina-Cuberos, W. Stumptner, H. Lammer, N.I. Koemle, K. O'Brien, Icarus **154**, 216–222 (2001)

C.D. Murray, S.F. Dermott, *Solar System Dynamics* (Cambridge Univ. Press, New York, 2000)

O. Néron de Surgy, J. Laskar, Astron. Astrophys. **318**, 975–989 (1997)

N.F. Ness, M.H. Acuña, J. Connerney, P. Wasilewski, C. Mazelle, J. Sauvaud, D. Vignes, C. D'Uston, H. Reme, R. Lin, D.L. Mitchell, J. McFadden, D. Curtis, P. Cloutier, S.J. Bauer, Adv. Space Res. **23**(11), 1879–1886 (1999)

G. Neukum, R. Jaumann, H. Hoffmann, E. Hauber, J.W. Head, A.T. Basilevsky, B.A. Ivanov, S.C. Werner, S. van Gasselt, J.B. Murray, T. McCord, Nature **432**, 971–979 (2004)

G. Newkirk Jr., Geochimica Cosmochimica Acta Suppl. **13**, 293–301 (1980)

F. Nimmo, Geology **28**, 391–394 (2000)

P. Olson, U.R. Christensen, Earth Planet. Sci. Lett. **250**(3–4), 561–571 (2007). doi:10.1016/j.epsl.2006.08.008

R.O. Pepin, Icarus **111**, 289–304 (1994)

M.J. Pruis, K.L. Tanaka, Abstr. Lunar Planet. Sci. **26**, 1147 (1995)

M. Purucker, D. Ravat, H. Frey, C. Voorhies, T. Sabaka, M. Acuña, Geophys. Res. Lett. **27**, 2449–2452 (2000)

I. Ribas, E.F. Guinan, M. Güdel, M. Audard, Astrophys. J. **622**, 680–694 (2005)

A. Rivoldini, O. Verhoeven, T. Van Hoolst, A. Mocquet, V. Dehant, Mars interior structure models from tidal measurements. American Geophysical Union, Fall Meeting 2005, abstract #G51A-0801

M.G. Rochester, in: *Proc. Second Int. Symp. 'Geodesy and physics of the Earth'*, Potsdam, East Germany, May 7–11, 1973. Part 1 (A75-18106 06-46) (Deutsche Akademie der Wissenschaften, Zentralinstitut fuer Physik der Erde, 1974) pp. 77–89

G. Rontó, A. Bercés, H. Lammer, C.S. Cockell, G.J. Molina-Cuberos, M.R. Patel, F. Selsis, Photochem. Photobiol. **77**(1), 34–40 (2003)

J. Rosenqvist, E. Chassefière, Planet. Space Sci. **43**, 3–10 (1995)

C. Sanloup, A. Jambon, P. Gillet, Phys. Earth Planet. Interiors **112**, 43–54 (1999)

Y. Sano, J. Geomagn. Geoelectr. **45**, 65–77 (1993)

G. Schubert, T. Spohn, J. Geophys. Res. **95**, 14,095–14,104 (1990)

N.H. Sleep, K.L. Tanaka, Astron. Soc. Pac. Mercury **25**(5), 10 (1995)

D.E. Smith, M.T. Zuber, H.V. Frey, J.B. Garvin, J.W. Head, D.O. Muhlerman, G.H. Pettengill, R.J. Phillips, S.C. Solomon, H.J. Zwally, W.B. Banerdt, T.C. Duxbury, Science **279**(5357), 1686 (1998)

F. Sohl, G. Schubert, T. Spohn, J. Geophys. Res. **110**, E12008 (2005). doi:10.1029/2005JE002520

F. Sohl, T. Spohn, J. Geophys. Res. **102**, 1613–1635 (1997)

S.C. Solomon, Nature **369**, 606 (1994)

T. Spohn, F. Sohl, D. Breuer, Astron. Astrophys. Rev. **8**, 181–235 (1998)

F.D. Stacey, Geophys. J. Roy. Astron. Soc. **33**, 47–55 (1973)

S. Stanley, J. Bloxham, W.E. Hutchison, M.T. Zuber, Earth Planet. Sci. Lett. **234**, 27–38 (2005). doi:10.1016/j.epsl.2005.02.040

F.R. Stephenson, L.V. Morrison, Phil. Trans. Roy. Soc. London, Ser. A **313**, 47–70 (1984)

D.J. Stevenson, Rep. Prog. Phys. **46**, 555–620 (1983)

D.J. Stevenson, Nature **412**, 214–219 (2001)

D.J. Stevenson, Earth Planet. Sci. Lett. **208**(1–2), 1–11 (2003)

D.J. Stevenson, T. Spohn, G. Schubert, Icarus **54**, 466–489 (1983)

N. Terada, H. Lammer, T. Penz, H. Shinagawa, S. Machida, C. Kolb, H.I.M. Lichtenegger, H.K. Biernat, I. Ribas, Icarus (2007, submitted)

P.B. Toft, J. Arkani-Hamed, J. Geophys. Res. **97**, 4387–4406 (1992)

L. Tomasella, F. Marzari, V. Vanzani, Planet. Space Sci. **44**(5), 427–430 (1996)

J.P. Vanyo, Geophys. Astrophys. Fluid Dyn. **59**, 209–234 (1991)

P. Varga, K.R. Rybicki, C. Denis, Icarus **180**(1), 274–276 (2006). doi:10.1016/j.icarus.2005.04.022

O. Verhoeven, A. Rivoldini, P. Vacher, A. Mocquet, G. Choblet, M. Menvielle, V. Dehant, T. Van Hoolst, J. Sleewaegen, J.-P. Barriot, P. Lognonné, J. Geophys. Res. (Planets) **110**(E4), E04009 (2005). doi:10.1029/2004JE002271

J.C.G. Walker, Icarus **68**, 87–98 (1986)

W.R. Ward, Science **181**(4096), 260–262 (1973)

W.R. Ward, Icarus **50**, 444–448 (1982)

W.R. Ward, in *Mars*, ed. by H. Kieffer, B. Jakosky, C. Snyder (Univ. of Arizona, Tucson, 1991), pp. 298–320
W.R. Ward, J.A. Burns, O.B. Toon, J. Geophys. Res. **84**, 243–259 (1979)
W.R. Ward, D.J. Rudy, Icarus **94**, 160–164 (1991)
G.E. Williams, Rev. Geophys. **38**, 37–59 (2000)
J.P. Williams, F. Nimmo, Geology **32**(2), 97–100 (2004)
B.E. Wood, H.-R. Müller, G. Zank, J.L. Linsky, Astrophys. J. **574**, 412–425 (2002)
B.E. Wood, H.-R. Müller, G.P. Zank, J.L. Linsky, S. Redfield, Astrophys. J. **628**, L143–L146 (2005)
C.F. Yoder, A.S. Konopliv, D.N. Yuan, E.M. Standish, Folkner, W. M (2003). Science **300**(5617), 299–303 (2003)
K. Zahnle, J.B. Pollack, J.F. Kasting, Icarus **84**, 503–527 (1990)
K.J. Zahnle, J.C.G. Walker, Rev. Geophys. Space Phys. **20**, 280–292 (1982)
V.N. Zharkov, T.V. Gudkova, Planet. Space Sci. **45**, 401–407 (1997)
M.T. Zuber, M.D. Smith, Effect of Mars' present-day north polar cap on the moment of inertia and planetary dynamics. 30th Annual Lunar and Planetary Science Conference, Houston, TX, 1999, abstract no. 1589
M.T. Zuber, D.E. Smith, S.C. Solomon, J.B. Abshire, R.S. Afzal, O. Aharonson, K. Fishbaugh, P.G. Ford, H.V. Frey, J.B. Garvin, J.W. Head, A.B. Ivanov, C.L. Johnson, D.O. Muhleman, G.A. Neumann, G.H. Pettengill, R.J. Phillips, X. Sun, H.J. Zwally, W.B. Banerdt, T.C. Duxbury, Science **282**, 2053–2060 (1998)

Space Sci Rev (2007) 129: 301–304
DOI 10.1007/s11214-007-9239-6

Epilogue: The Origins of Life in the Solar System and Future Exploration

**Philippe Lognonne · David Des Marais ·
François Raulin · Kathryn Fishbaugh**

Received: 17 May 2007 / Accepted: 15 June 2007 / Published online: 14 July 2007
© Springer Science+Business Media B.V. 2007

The broad investigation of life in the universe must address its long-term survival (the primary focus of this book) and also its origins. Studies of survival address the evolution of the planetary environment, its habitability, the adaptability of living systems to changing conditions and possibly co-evolution of life with its host planet. Studies of origins explore the conditions necessary for the appearance of the first living organisms and the processes of prebiotic chemical evolution.

Life emerged on Earth after a period of chemical evolution that involved liquid water, inorganic nutrients, organic matter and energy. The availability of these ingredients seems a *sine qua non* for the evolution of inanimate matter to living systems capable of replication, adaptation and Darwinian evolution. Environmental conditions determine the availability and utility of these ingredients, as well as the kinetics of the associated chemical reactions. For example, the potential role of hydrothermal systems in prebiotic chemistry implies that liquid water be in contact with rocky materials, especially when environmental conditions in these systems promote redox reactions that provide chemical energy to sustain prebiotic processes.

A major gap in our understanding of the origin(s) of life is our ignorance of the environment(s) where and when it occurred and of the time scale of the processes leading to the appearance of the first organisms. Plate tectonics, crustal erosion, surface recycling and atmospheric escape have indeed erased almost all of the geological and environmental information of the Earth's first billion years. The characteristics of the primitive crust; the chemical composition and physical conditions of the atmosphere; the pH, temperature and rate of formation of the oceans; and the rate of continental emergence were recorded in the now-obliterated ancient crust and therefore remain poorly constrained. Even more uncertain is how common are primitive, Earth-like habitable conditions during the formation of telluric planets. Filling this knowledge gap requires data that can be obtained only by exploring planets with at least part of their ancient crust preserved and by a comparative approach, first

P. Lognonne · D. Des Marais · F. Raulin (✉) · K. Fishbaugh
Creteil Cedex 94010, France
e-mail: raulin@lisa.univ-paris12.fr

between the Earth and the other telluric planets of the Solar System, and second between the Solar System and other planetary systems.

In the quest for evidence of ancient habitable environments elsewhere in our solar system, Mars will be the key target. Mars' ancient environment was Earth-like, possibly with liquid water present on or near the surface, and its ancient crust is better preserved than that of Earth (Des Marais et al. 2007). Mercury, the Moon and asteroids, which are relatively geologically inactive, have not obliterated their ancient crusts, even if they have never possessed liquid water. Thus the surface and subsurface of these bodies might provide information about early evolution of the Sun (imprinted in isotope ratios in regolith grains), which clearly affected the early environmental conditions on Earth and Mars.

One major goal of Martian missions is therefore to determine the primitive Martian environment and to understand any evidence of prebiotic chemical evolution and ancient life in the oldest Martian terrains. However, several steps will be necessary to improve our understanding of the primitive Martian environment. Past missions have already provided key insights, such as the discovery of phyllosilicates by the Mars Express Visible and Infrared Mineralogical Mapping Spectrometer (OMEGA) (Bibring et al. 2005) and the in-situ analysis of sulphates by the Mars Exploration Rovers (MER) (Squyres et al. 2004), which provide a sketch of past environmental conditions on Mars. New steps will be taken by performing in situ analyses that are able to extract crucial information about physical, chemical and even geophysical (e.g., the intensity of the early magnetic field) conditions from rocks, soil or the atmosphere. Other steps can be achieved by characterization of early geological evolution. Detailed geologic maps and an improved knowledge of the structure and radiogenic contents of the crust, mantle and core will better constrain our geophysical models of early volcanism and crustal formation, volatile production and subsurface heat fluxes. Models of our early Sun and of early Martian atmospheric erosion will be crucial for constraining the losses of volatiles. Such a comprehensive approach will lead to a better understanding of the primitive climatic and environmental conditions of Mars. The search for organic matter and other clues about chemical evolution and ancient life on Mars will be difficult. This is due to the complexity of the necessary analyses and to the effects of impacts, eolian erosion and UV and radiation exposure on the old surface materials. Organic analyses must be performed in situ in the most primitive Martian terrains, and they will require exploration of the subsurface to depths sufficient to gain access to materials that have experienced habitable environments during more recent periods. Surface exploration should target exhumed materials that can reveal subsurface conditions and potential inhabitants.

Either from the point of view of the origin of life or primitive habitability or from the point of view of the evolution of habitability on terrestrial planets, Mars is also probably the only water-rich planet (in addition to Earth) that might be extensively explored with the technology and financial resources available during the next few decades. ESA's ExoMars rover will explore the context of life and habitability of the Red Planet by analyzing samples from up to two meters below the surface and by deploying a geophysical and environmental observatory. NASA's MSL rover will conduct definitive measurements of minerals, volatile species and organic matter in samples from surface materials and rock cores. Future Mars missions should continue along this path by visiting new landing sites and by deploying a more global network of observatories in order to better constrain the early habitability of the planet, its evolution and the present environment, either at the surface or in the subsurface. Clearly such a systematic search of the Martian surface for preserved evidence of habitable environments and life is a centrally important exploration strategy for both NASA and ESA. This approach will help us identify the most interesting sites for future sample return or deep drill missions.

In this book, we have concentrated on the habitability of terrestrial planets, but life might have developed on other planetary bodies in the Solar System. Europa is likely to have an internal, liquid water ocean that may be relatively close to the surface. Models of the internal structure of Europa suggest that this ocean is in contact with underlying bedrock, offering interesting possibilities for prebiotic chemistry and, if life emerged in the Europa ocean, for the sustenance of any biosphere (Reynolds et al. 1983; Raulin 2005; Greenberg 2005). Even if future missions to Europa are not able to access and explore the sub-surface ocean, they should at least be able to constrain the geometry of the sub-surface ocean, to monitor the presence of ocean-floor geological activity, to analyze samples of sub-surface material brought to the surface by geologic activity, and to search for biological signatures that may appear episodically on the surface or in the tenuous atmosphere of the Galilean satellite.

Titan is also likely to possess an internal water ocean. However the liquid water layer—at least now—is probably located between two layers of water ice and is not in contact with the bedrock. The environmental conditions are thus less favourable for the development of life than in the case of Europa, and the current presence of life, although it cannot be completely ruled out, seems less probable. Even so, the new understanding that we now have of Titan, thanks to the fantastic success of the ESA/NASA Cassini-Huygens mission, highlights the exo/astrobiological importance of the largest satellite of Saturn. Titan offers many similarities with the Earth, and in particular the primitive Earth. It is harbouring today a very well-developed organic chemistry of prebiotic interest (Fortes 2000; Raulin 2005, 2007). The planetary and exo/astrobiological communities are already envisioning a mission for systematic exploration of Titan, with horizontal mobility to study different locations of the satellite (Lorenz 2000; Sittler et al. 2006).

The Cassini Huygens mission has also identified Enceladus as a new target of crucial importance for exo/astrobiology. This small satellite of Saturn seems to contain liquid water in its internal structure that is in contact with silicate and rocks, offering conditions, like Europa, that are favourable for prebiotic chemistry and even life development. Future missions to Enceladus should also include some capabilities to explore—even indirectly from flybys—its internal ocean.

Beyond the Solar System, the extrasolar planets offer a fantastic field of investigation for searching for extraterrestrial life and for studying a potentially very wide variety of planetary objects. More than 200 exoplanets have been detected so far. Several more will probably be found in the very near future by the CNES-ESA COROT mission, launched in December 2006, which will look for star occultation generated by extra-solar planets. NASA's Kepler mission, scheduled to be launched in 2008, will also pursue this search. A new field of investigation is now fully open. These studies will provide statistics on the masses of planets (including the potential "ocean-planets") as well as their distances from their sun. Future steps should require observations of the atmospheres of these planets to infer their surface temperatures and to search for atmospheric biomarkers, which might provide clear and unambiguous evidence of a biological presence. The proposed Darwin mission would be able to systematically determine the chemical composition of the atmosphere of the extrasolar planets and to look for the possible presence of atmospheric biomarkers. Such a mission is therefore of interest to a large scientific community and will probably require a strong cooperation between ESA, NASA and other international partners.

We must recognize that our knowledge of the essential requirements for life and therefore also our concepts of habitable planets are based principally upon our understanding of the biosphere during the later stages of Earth history. Our concepts will change as we discover more about life both in ancient and in extreme environments. As we explore beyond Earth we must be prepared to recognize "life as we don't know it", namely life forms which developed

in alien environments that have imparted characteristics fundamentally different from those of life on Earth. Accordingly, to recognize alien environments that are quite different yet are habitable, we must learn how to intellectually escape the constraints of our "biosphere as we know it."

Many questions remain. How frequently and for how long do habitable environments appear on a planet after its formation? What conditions in term of planetary mass, accretion dynamics, impact history and interactions between a star and any giant planets are necessary for the development of even primitive habitable conditions in a planet's early history? How persistent and extensive must habitable environments be in order for a planet to sustain life over the long term? What is the minimum geological evolution necessary for a sustained biosphere? How common are planets with such geological evolution? Must the surface environment be habitable at least somewhere in order for life to exist anywhere on a planet? These are some of the key questions that should drive future missions of planetary exploration concerned with assessing and characterizing habitability.

References

J.P. Bibring et al., Science **307**, 1576–1581 (2005)

D.J. Des Marais, B.M. Jakosky, B.M. Hynek, in *The Martian Surface: Composition, Mineralogy and Physical Properties*, ed. by J. Bell (Cambridge University Press, Cambridge, 2007, in press)

A.D. Fortes, Icarus **146**, 444–452 (2000)

R. Greenberg, *Europa The Ocean Moon – Search for an Alien Biosphere* (Springer-Praxis, New York, 2005)

R. Lorenz, J. Br. Interplanet. Soc. **53**, 218–234 (2000)

F. Raulin, Space Sci. Rev. **116**, 471–487 (2005)

F. Raulin, *Space Sci. Rev.* (2007, in press, online)

R. Reynolds, S. Squyres, D. Colburn, C.P. McKay, Icarus **56**, 246–254 (1983)

E. Sittler et al., *Fourth International Planetary Probe Workshop* Pasadena, 2006

S.W. Squyres et al., Science **306**, 1698–1703 (2004)

Space Science Series of ISSI

1. R. von Steiger, R. Lallement and M.A. Lee (eds.): *The Heliosphere in the Local Interstellar Medium.* 1996 — ISBN 0-7923-4320-4

2. B. Hultqvist and M. Øieroset (eds.): *Transport Across the Boundaries of the Magnetosphere.* 1997 — ISBN 0-7923-4788-9

3. L.A. Fisk, J.R. Jokipii, G.M. Simnett, R. von Steiger and K.-P. Wenzel (eds.): *Cosmic Rays in the Heliosphere.* 1998 — ISBN 0-7923-5069-3

4. N. Prantzos, M. Tosi and R. von Steiger (eds.): *Primordial Nuclei and Their Galactic Evolution.* 1998 — ISBN 0-7923-5114-2

5. C. Fröhlich, M.C.E. Huber, S.K. Solanki and R. von Steiger (eds.): *Solar Composition and its Evolution – From Core to Corona.* 1998 — ISBN 0-7923-5496-6

6. B. Hultqvist, M. Øieroset, Goetz Paschmann and R. Treumann (eds.): *Magnetospheric Plasma Sources and Losses.* 1999 — ISBN 0-7923-5846-5

7. A. Balogh, J.T. Gosling, J.R. Jokipii, R. Kallenbach and H. Kunow (eds.): *Co-rotating Interaction Regions.* 1999 — ISBN 0-7923-6080-X

8. K. Altwegg, P. Ehrenfreund, J. Geiss and W. Huebner (eds.): *Composition and Origin of Cometary Materials.* 1999 — ISBN 0-7923-6154-7

9. W. Benz, R. Kallenbach and G.W. Lugmair (eds.): *From Dust to Terrestrial Planets.* 2000 — ISBN 0-7923-6467-8

10. J.W. Bieber, E. Eroshenko, P. Evenson, E.O. Flückiger and R. Kallenbach (eds.): *Cosmic Rays and Earth.* 2000 — ISBN 0-7923-6712-X

11. E. Friis-Christensen, C. Fröhlich, J.D. Haigh, M. Schüssler and R. von Steiger (eds.): *Solar Variability and Climate.* 2000 — ISBN 0-7923-6741-3

12. R. Kallenbach, J. Geiss and W.K. Hartmann (eds.): *Chronology and Evolution of Mars.* 2001 — ISBN 0-7923-7051-1

13. R. Diehl, E. Parizot, R. Kallenbach and R. von Steiger (eds.): *The Astrophysics of Galactic Cosmic Rays.* 2001 — ISBN 0-7923-7051-1

14. Ph. Jetzer, K. Pretzl and R. von Steiger (eds.): *Matter in the Universe.* 2001 — ISBN 1-4020-0666-7

15. G. Paschmann, S. Haaland and R. Treumann (eds.): *Auroral Plasma Physics.* 2002 — ISBN 1-4020-0963-1

16. R. Kallenbach, T. Encrenaz, J. Geiss, K. Mauersberger, T.C. Owen and F. Robert (eds.): *Solar System History from Isotopic Signatures of Volatile Elements.* 2003 — ISBN 1-4020-1177-6

17. G. Beutler, M.R. Drinkwater, R. Rummel and R. Von Steiger (eds.): *Earth Gravity Field from Space – from Sensors to Earth Sciences.* 2003 — ISBN 1-4020-1408-2

18. D. Winterhalter, M. Acuña and A. Zakharov (eds.): *"Mars" Magnetism and its Interaction with the Solar Wind.* 2004 — ISBN 1-4020-2048-1

19. T. Encrenaz, R. Kallenbach, T.C. Owen and C. Sotin: *The Outer Planets and their Moons* — ISBN 1-4020-3362-1

20. G. Paschmann, S.J. Schwartz, C.P. Escoubet and S. Haaland (eds.): *Outer Magnetospheric Boundaries: Cluster Results* — ISBN 1-4020-3488-1

21. H. Kunow, N.U. Crooker, J.A. Linker, R. Schwenn and R. von Steiger (eds.): *Coronal Mass Ejections* — ISBN 978-0-387-45086-5